OSTEOPONTIN: ROLE IN CELL SIGNALLING AND ADHESION

ANNALS OF THE NEW YORK ACADEMY OF SCIENCES
Volume 760

OSTEOPONTIN: ROLE IN CELL SIGNALLING AND ADHESION

Edited by David T. Denhardt, William T. Butler, Ann F. Chambers, and Donald R. Senger

The New York Academy of Sciences
New York, New York
1995

∞ *The paper used in this publication meets the minimum requirements of American National Standard for Information Sciences—Permanence of Paper for Printed Library Materials, ANSI Z39.48-1984.*

Cover: *Paraffin-embedded mouse cochlea stained with a monoclonal antibody (1 :400) against osteopontin (anti-uropontin, ZH; a kind gift from Dr. John R. Hoyer) using the Vector ABC method. Osteopontin epitopes have been detected in a variety of cell and tissue types. This section of the mouse inner ear stained with a monoclonal OPN antibody demonstrates the wide distribution within a single structure. Most of the dark brown staining patterns along the outer edges of the tissue demarcate the bony labyrinth of the cochlea. The triangular-shaped stained region near the center of the figure highlights the inner bony core of the cochlea called the modiolus which contains blue-stained nuclei of the primary-auditory neurons. Although these neurons express* opn *mRNA at high levels, protein epitopes are often undetectable with standard immunohistochemistry. The crescent-shaped stained structure that edges an open fluid space (upper left quadrant) highlights epithelial cells that compose the stria vascularis, a trilaminate tissue that regulates cochlear fluid homeostasis. Fibrocytes within the spiral ligament also contain OPN epitopes; light staining of this structure can be detected between the stria vascularis and the outer bony labyrinth. Additional information can be found in the paper by Davis, Lopez and Mou. Photograph provided by Dr. Robin L. Davis. (Adapted from Lopez et al., Hearing Res., in press.)*

Library of Congress Cataloging-in-Publication Data

Osteopontin : role in cell signalling and adhesion / editors, David T. Denhardt . . . [et al.].
p. cm. — (Annals of the New York Academy of Sciences ; v. 760)
Includes bibliographical references and index.
ISBN 0-89766-939-8 (cloth : alk. paper). — ISBN 0-89766-940-1 (paper : alk. paper)
1. Osteopontin—Congresses. I. Denhardt, David T. II. Series.
Q11.N5 vol. 760
[QP552.077]
500 s—dc20
[574.19′2454] 95-11516
CIP

Bi-Comp/PCP
Printed in the United States of America
ISBN 0-89766-939-8 (cloth)
ISBN 0-89766-940-1 (paper)
ISSN 0077-8923

ANNALS OF THE NEW YORK ACADEMY OF SCIENCES

Volume 760
April 21, 1995

OSTEOPONTIN: ROLE IN CELL SIGNALLING AND ADHESION[a]

Editors and Conference Chairs
DAVID T. DENHARDT, WILLIAM T. BUTLER, ANN F. CHAMBERS, AND DONALD R. SENGER

CONTENTS

[a] This volume contains the papers from a conference entitled *Osteopontin: Role in Cell Signalling and Adhesion,* which was held by the New York Academy of Sciences in New Brunswick, New Jersey on October 21–23, 1994.

Part VI. Role of Osteopontin in Bone Metabolism

Part VII. Function of Osteopontin in the Kidney and Inner Ear

Poster Papers

Financial assistance was received from:

Supporter

- NATIONAL INSTITUTE OF ARTHRITIS AND MUSCULOSKELETAL AND SKIN DISEASES/NIH

Contributors

- AMGEN INC.
- BIOGEN INC.
- MERCK RESEARCH LABORATORIES
- NATIONAL INSTITUTE ON AGING/NIH
- NATIONAL INSTITUTE ON DEAFNESS AND OTHER COMMUNICATION DISORDERS/NIH
- PFIZER CENTRAL RESEARCH
- RUTGERS, THE STATE UNIVERSITY OF NEW JERSEY, BUREAU OF BIOLOGICAL RESEARCH
- SMITHKLINE BEECHAM PHARMACEUTICALS
- UNIVERSITY OF WESTERN ONTARIO, DEPARTMENT OF ONCOLOGY
- WYETH-AYERST RESEARCH

Preface

This New York Academy of Sciences *Annals* is the permanent record of an exciting and timely conference devoted solely to osteopontin (OPN) and held October 21–23, 1994 in New Brunswick, New Jersey. The goal of the conference was to bring together scientists from various specialties who were actively involved in studies on OPN, a protein which increasingly has been recognized as potentially important for a variety of physiological processes as well as in the pathogenesis of several common diseases. In this multidisciplinary assembly almost one hundred scientists representing a variety of scientific disciplines (including formation and homeostasis of mineralized tissue, kidney function, immune processes, cardiovascular physiology and pathology, infectious disease, and cancer) were able to share observations and interpretations in a beneficial manner, undoubtedly resulting in a number of long-lasting collaborations.

The idea for this conference originated at a special workshop on OPN, part of the 1989 meeting of the Texas Mineralized Tissue Society. One of the outcomes of this initial meeting was an agreement to use the name osteopontin and the abbreviation OPN, although it was clear that this protein is not restricted to bone and that its tissue distribution is both broad and complex. At this meeting there was a general sentiment that we should plan a second conference within a few years. Although it took longer than first anticipated for this meeting to materialize, the passage of time allowed for the accumulation of important new data on the biology of OPN as well as increased interest in this protein. Because of the broad spread of research interests, the meeting brought about discussions representing a variety of unique perspectives. The intensity of interactions, the quality of the presentations, and the insightful and probing discussion periods elicited new ideas that will carry over from one field to another.

The organizers are particularly grateful to the New York Academy of Sciences for having the foresight to support an international meeting whose focus was "Osteopontin: Role in Cell Signalling and Adhesion." Staff members Geraldine Busacco and Sherryl Greenberg guided and assisted us in a very professional manner. We thank William Sharrock of the Bone Biology Program and the National Institutes of Health (NIA, NIDCD) for their interest and support. We are grateful to the following corporate sponsors: Amgen Inc., Biogen Inc., Merck Research Laboratories, Pfizer Central Research, SmithKline Beecham Pharmaceuticals and Wyeth-Ayerst Research. We also acknowledge contributions from the Bureau of Biological Research of Rutgers University and the Department of Oncology, University of Western Ontario. We thank the following for helping to make this meeting happen: Cecilia Giachelli, Marc McKee, Barid Mukherjee, Charles Prince, Gideon Rodan, Jaro Sodek, Martha Somerman, Gary Stein, and Arthur Veis. Finally, the staff of the Hyatt Regency Hotel in New Brunswick were also instrumental in making the meeting the success that it turned out to be.

David T. Denhardt
William T. Butler
Ann F. Chambers
Donald R. Senger

Osteopontin Overview

GIDEON A. RODAN

Department of Bone Biology and
Osteoporosis Research
Merck Research Laboratories
West Point, Pennsylvania 19486

This overview will explore if there is a common thread linking the various putative functions of osteopontin (OPN). OPN is a negatively charged, glycosylated phosphoprotein with a molecular weight of its polypeptide backbone of around 32,000. It has been independently discovered several times and as described in this volume is expressed in many tissues. In bone, it is produced by both matrix-producing osteoblasts, at the mineralization front, and by bone resorbing osteoclasts.[1] In bone sections, it is found at high concentrations on the cement lines where bone formation follows resorption.[2] In the kidney, it is found in the loop of Henle and the distal convoluted tubules[3] and increases during nephritis.[4] In blood vessels, it is found in arterial smooth muscle cells and its level increases in atherosclerotic plaques[5–8] and during restenosis, following balloon angioplasty.[9] It is also found in the inner ear[10,11] and in the epithelial cells of the gallbladder, urinary and reproductive tracts, gastrointestinal tracts, mammary gland, bronchi, salivary glands and sweat ducts.[12] OPN mRNA was visualized in the mucosa of the stomach, the colon, the liver, the gallbladder, the pancreas, the secretory endometrium and the fallopian tubes, all of which are subject to relatively continuous renewal.[12]

OPN is secreted, binds calcium and was found to bind covalently to the ubiquitous extracellular protein fibronectin,[13] via transglutaminase-catalyzed cross-linking.[14] *In vivo,* it is spontaneously elevated under conditions of sepsis and metastatic cancer and is induced by phorbol esters. In cultured cells, its production is modulated by cytokines, such as TGFβ, EGF and PDGF[15] as well as IL-2.[16] OPN is chemotactic for smooth muscle cells[17] as well as macrophages.[18] Cells transfected with antisense OPN RNA had reduced tumorigenicity.[19,20]

OPN contains the GRGDS sequence recognized by several integrins and was shown *in vitro* to serve as an attachment substrate to several cell types,[21] primarily via the $\alpha_v\beta_3$ integrin. This integrin is abundant in osteoclasts and has been implicated in osteoclast activation, since $\alpha_v\beta_3$ antibodies and ligands inhibit osteoclast activity.[22,23] Osteopontin, in its phosphorylated form,[24] was proposed to be the natural ligand for the osteoclast integrin. $\alpha_v\beta_3$ is also found in endothelial cells and is important for neovascularization[25] and cell migration,[26] osteopontin being the possible substrate. Thrombin cleaves OPN close to the RGD sequence[27] and the thrombin-cleaved OPN not only binds to $\alpha_v\beta_3$ as well as to the native OPN but also promotes better cell attachment and spreading than native OPN.[28] Integrins serve not only for cell attachment but are also signal-transducing receptors,[29] and OPN was shown to modulate intracellular calcium in $\alpha_v\beta_3$-containing cells.[30–32]

What do these various putative OPN functions and features have in common? They may all be related, either directly or evolutionarily to inflammation and tissue repair.

Cell survival depends on interaction with the environment for recognition and uptake of nutrients for avoidance of and protection from injury. The tools developed for these functions by unicellular organisms include receptors and uptake mechanisms for nutrients, and glycocalix and locomotion for protection. These elements evolved as the basis for cell-cell communication and tissue formation in multicellular organisms. The matrix plays a central role in this scheme, both as a physical scaffold within which cells which perform related functions are organized in a tissue, as well as the repository of instructive molecules which through signalling initiate or facilitate the differentiation which gives rise to the specialized tissue. This fundamental mechanism of tissue formation, which involves cell-cell communication via cytokines and matrix molecules, has evolved in multicellar organisms as the basis for responding to injury by the processes of inflammation and tissue repair. The sequence of events following injury of any kind at any site in the body includes: 1) Release of chemotactic signals by the injured cells; 2) Recruitment of monocytes/macrophages, lymphocytes and, depending on the stimulus, other inflammation-related cells, such as granulocytes (including eosinophils), mast cells, etc.; 3) Further production of cytokines by these cells, which support feed-forward reactions and coordinate the appropriate cellular responses: some of the molecules generated by these cells are matrix molecules, notably osteopontin produced by macrophages and other cells; and 4) Attachment of fibroblast lineage cells, which elaborate more matrix in a specially defined arrangement to form a scar for tissue repair or to encapsulate an infection or a tumor. Blood clotting appears as a special case of this process where the interactive cells are primarily endothelial cells and thrombocytes. The cytokines include, as in inflammation, a broad array of growth factors and arachidonic acid metabolites, and the matrix, for the sake of rapid action, is already present as fibrinogen, which needs only to be cross-linked and organized.

The bone remodelling sequence bears a striking similarity and has many features in common with the inflammation/tissue repair sequence. Bone remodelling is supposed to be initiated by a microfracture[33] or possibly by a mechanical deformation sensed by the lining cells and osteocytes.[34] This is followed by recruitment of osteoclasts, cells of the monocyte-macrophage family. Like the macrophages which engulf and digest tissue debris, osteoclasts digest the bone matrix. Other monocytic cells have been described on the resorption surface, towards the end of the resorption phase. This step is followed by recruitment of osteoblasts, the cousins of fibroblasts, which by analogy to scar formation deposit a well-organized and especially appropriately oriented matrix. As in inflammation and clotting, endothelial cells probably play an important role in bone formation since there is no osteogenesis without angiogenesis. The molecular events in these processes, many of which have been described in detail, have both common and distinct features. For example, the cytokines involved in inflammation, such as the interleukins, TGFβ, FGF, TNF, arachidonic acid metabolites, and endothelin, many of which are macrophage products, modulate cellular function in both systems. In the context of osteopontin, it is noteworthy that β_3 integrins seem to be intimately involved in inflammation, clotting and bone remodelling. In clotting, platelets are strongly activated by $\alpha_{IIb}\beta_3$ interaction with fibrinogen via its RGD-containing domain. In bone, $\alpha_v\beta_3$ is important for osteoclastic resorption, since blocking antibodies and RGD-based ligands which bind to $\alpha_v\beta_3$ inhibit resorption.[23] Similar antibodies and ligands were also reported to inhibit neovascularization,[25] which plays an important role in inflammation and tissue repair, by interacting with $\alpha_v\beta_3$ receptor on endothelial cells.

Osteopontin is one of the major ligands for the vitronectin receptor $\alpha_v\beta_3$. OPN contains the RGD sequence, which is recognized by several integrins, and cells which possess $\alpha_v\beta_3$ attach to osteopontin in culture.[21] OPN also acts as a signalling ligand for cells which possess $\alpha_v\beta_3$ receptors where it was reported to produce a transient increase in intracellular calcium[31] and to activate the calcium ATPase pump.[30] OPN thus seems to fulfill many of the functions described above in the inflammation/tissue repair cascade. It is produced and secreted in response to injury in smooth muscle cells and many other cells which are subject to wear and tear, such as epithelial cells and the lining of the gastrointestinal system and is induced in mouse epidermal cells by phorbol esters.[35] It is also produced by macrophages[27,36] and activated T cells,[37] and its production is upregulated by IL-2 in natural killer cells related to the granulated metrial gland cells which produce it.[16] It is also produced by osteoclasts,[1] the cells which correspond to macrophages in the bone remodelling sequence. OPN is chemotactic, the second step in the inflammation/tissue repair cascade. As such, it probably stimulates cell migration most likely via integrin signal transduction. Moving to the next stage, repair, we find osteopontin in the matrix, bound to fibronectin, as a suitable substrate for the attachment of $\alpha_v\beta_3$-containing cells, such as endothelial cells and fibroblasts during tissue repair. In bone, we find OPN on the cement line at the reversal phase, where bone formation follows resorption and we find it to be produced by matrix elaborating osteoblasts.[38]

OPN thus seems to have evolved as a major player in the injury/repair cascade, fulfilling distinct functions at its various stages, many mediated by its receptor $\alpha_v\beta_3$. Viewing this molecule in that context could explain its widespread localization, its diverse proposed functions and its response to cytokines and other stimuli.

REFERENCES

1. Tezuka, K., T. Sato, H. Kamioka, P. J. Nijweide, K. Tanaka, T. Matsuo, M. Ohta, N. Kurihara, Y. Hakeda & M. Kumegawa. 1992. Identification of osteopontin in isolated rabbit osteoclasts. Biochem. Biophys. Res. Commun. **186:** 911–917.
2. McKee, M. D., M. C. Farach-Carson, W. T. Butler, P. V. Hauschka & A. Nanci. 1993. Ultrastructural immunolocalization of noncollageneous (osteopontin and osteocalcin) and plasma (albumin and α_2HS-glycoprotein) proteins in rat bone. J. Bone Miner. Res. **8:** 485–496.
3. Lopez, C. A., J. R. Hoyer, P. D. Wilson, P. Waterhouse & D. T. Denhardt. 1993. Heterogeneity of osteopontin expression among nephrons in mouse kidneys and enhanced expression in sclerotic glomeruli. Lab. Invest. **69:** 355–363.
4. Giachelli, C. M., R. Pichler, D. Lombardi, D. T. Denhardt, C. E. Alpers, S. M. Schwartz & R. J. Johnson. 1994. Osteopontin expression in angiotensin II-induced tubulointerstitial nephritis. Kidney Int. **45:** 515–524.
5. Giachelli, C. M., N. Bae, M. Almeida, D. T. Denhardt, C. E. Alpers & S. M. Schwartz. 1993. Osteopontin is elevated during neointima formation in rat arteries and is a novel component of human atherosclerotic plaques. J. Clin. Invest. **92:** 1686–1696.
6. Hirota, S., M. Imakita, K. Kohri, A. Ito, E. Morii, S. Adachi, H.-M. Kim, Y. Kitamura, C. Yutani & S. Nomura. 1993. Expression of osteopontin messenger RNA by macrophages in atherosclerotic plaques. Am. J. Pathol. **143:** 1003–1008.
7. Ikeda, T., T. Shirasawa, Y. Esaki, S. Yoshiki & K. Hirokawa. 1993. Osteopontin mRNA is expressed by smooth muscle-derived foam cells in human atherosclerotic lesions of the aorta. J. Clin. Invest. **92:** 2814–2820.
8. Shanahan, C. M., N. R. B. Cary, J. C. Metcalfe & P. L. Weissberg. 1994. High expression of genes for calcification-regulating proteins in human atherosclerotic plaques. J. Clin. Invest. **93:** 2393–2402.
9. Giachelli, C., N. Bae, D. Lombardi, M. Majesky & S. Schwartz. 1991. Molecular cloning and characterization of 2B7, a rat mRNA which distinguishes smooth muscle

cell phenotypes *in vitro* and is identical to osteopontin (secreted phosphoprotein 1, 2ar). Biochem. Biophys. Res. Commun. **177:** 867–873.

10. MARK, M. P., C. W. PRINCE, S. GAY, R. L. AUSTIN & W. T. BUTLER. 1988. 44-kDal bone phosphoprotein (osteopontin) antigenicity at ectopic sites in newborn rats: kidney and nervous tissues. Cell Tissue Res. **251:** 23–30.
11. SWANSON, G. J., S. NOMURA & B. L. M. HOGAN. 1989. Distribution of expression of 2AR (osteopontin) in the embryonic mouse inner ear revealed by *in situ* hybridization. Hearing Res. **41:** 169–178.
12. BROWN, L. F., B. BERSE, L. VAN DE WATER, A. PAPADOPOULOS-SERGIO, C. A. PERRUZZI, E. J. MANSEAU, H. F. DVORAK & D. R. SENGER. 1992. Expression and distribution of osteopontin in human tissues: widespread association with luminal epithelial surfaces. Mol. Biol. Cell **3:** 1169–1180.
13. NEMIR, M., M. W. DEVOUGE & B. B. MUKHERJEE. 1989. Normal rat kidney cells secrete both phosphorylated and nonphosphorylated forms of osteopontin showing different physiological properties. J. Biol. Chem. **264:** 18202–18208.
14. BENINATI, S., D. R. SENGER, E. CORDELLA-MIELE, A. B. MUKHERJEE, I. CHACKALAPARAMPIL, V. SHANMUGAM, K. SINGH & B. B. MUKHERJEE. 1994. Osteopontin: its transglutaminase-catalyzed posttranslational modifications and cross-linking to fibronectin. J. Biochem. **115:** 675–682.
15. KASUGAI, S., Q. ZHANG, C. M. OVERALL, J. L. WRANA, W. T. BUTLER & J. SODEK. 1991. Differential regulation of the 55 and 44 kDa forms of secreted phosphoprotein (SPP-1, osteopontin) in normal and transformed rat bone cells by osteotropic hormones, growth factors and tumor promoter. Bone Miner. **13:** 235–250.
16. POLLACK, S. B., P. A. LINNEMEYER & S. GILL. 1994. Induction of osteopontin mRNA expression during activation of murine NK cells. J. Leukocyte Biol. **55:** 398–400.
17. LIAW, L., M. ALMEIDA, C. E. HART, S. M. SCHWARTZ & C. M. GIACHELLI. 1994. Osteopontin promotes vascular cell adhesion and spreading and is chemotactic for smooth muscle cells *in vitro*. Circ. Res. **74:** 214–224.
18. SINGH, R. P., R. PATARCA, J. SCHWARTZ, P. SINGH & H. CANTOR. 1990. Definition of a specific interaction between the early T lymphocyte activation (Eta-1) protein and murine macrophages *in vitro* and its effect upon macrophages *in vivo*. J. Exp. Med. **171:** 1931–1942.
19. BEHREND, E. I., A. M. CRAIG, S. M. WILSON, D. T. DENHARDT & A. F. CHAMBERS. 1994. Reduced malignancy of ras-transformed NIH 3T3 cells expressing antisense osteopontin mRNA. Cancer Res. **54:** 832–837.
20. GARDNER, H. A. R., B. BERSE & D. R. SENGER. 1994. Specific reduction in osteopontin synthesis by antisense RNA inhibits the tumorigenicity of transformed Rat1 fibroblasts. Oncogene **9:** 2321–2326.
21. SOMERMAN, M. J., C. W. PRINCE, W. T. BUTLER, R. A. FOSTER, J. M. MOEHRING & J. J. SAUK. 1989. Cell attachment activity of the 44 kilodalton bone phosphoprotein is not restricted to bone cells. Matrix **9:** 49–54.
22. DAVIES, J., J. WARWICK, N. TOTTY, R. PHILP, M. HELFRICH & M. HORTON. 1989. The osteoclast functional antigen implicated in the regulation of bone resorption is biochemically related to the vitronectin receptor. J. Cell Biol. **109:** 1817–1826.
23. CHAMBERS, T. J., K. FULLER, J. A. DARBY, A. S. PRINGLE & M. A. HORTON. 1986. Monoclonal antibodies against osteoclasts inhibit bone resorption *in vitro*. Bone Miner. **1:** 127–135.
24. EK-RYLANDER, B., M. FLORES, M. WENDEL, D. HEINEGÅRD & G. ANDERSSON. 1994. Dephosphorylation of osteopontin and bone sialoprotein by osteoclastic tartrate-resistant acid phosphatase. J. Biol. Chem. **269:** 14853–14856.
25. BROOKS, P. C., R. A. F. CLARK & D. A. CHERESH. 1994. Requirement of vascular integrin $\alpha_v\beta_3$ for angiogenesis. Science **264:** 569–571.
26. LEAVESLEY, D. I., M. A. SCHWARTZ, M. ROSENFELD & D. A. CHERESH. 1993. Integrin β_1 and β_3-mediated endothelial cell migration is triggered through distinct signaling mechanisms. J. Cell Biol. **121:** 163–170.
27. SENGER, D. R., C. A. PERRUZZI, C. F. GRACEY, A. PAPADOPOULOS & D. G. TENEN. 1988. Secreted phosphoproteins associated with neoplastic transformation: close

homology with plasma proteins cleaved during blood coagulation. Cancer Res. **48:** 5770–5774.

28. SENGER, D., C. A. PERRUZZI, A. PAPADOPOULOS-SERGIOU & L. VAN DE WATER. 1994. Adhesive properties of osteopontin: regulation by a naturally occurring thrombin-cleavage in close proximity to the GRGDS cell-binding domain. Mol. Biol. Cell **5:** 565–574.
29. HYNES, R. 1992. Integrins: versatility, modulation, and signaling in cell adhesion. Cell **69:** 11–25.
30. MIYAUCHI, A., J. ALVAREZ, E. M. GREENFIELD, A. TETI, M. GRANO, S. COLUCCI, A. ZAMBONIN-ZALLONE, F. P. ROSS, S. M. TEITELBAUM, D. CHERESH & K. A. HRUSKA. 1991. Recognition of osteopontin and related peptides by an $\alpha_v\beta_3$ integrin stimulates immediate cell signals in osteoclasts. J. Biol. Chem. **266:** 20369–20374.
31. ZIMOLO, Z., G. WESOLOWSKI, H. TANAKA, J. HYMAN, J. R. HOYER & G. A. RODAN. 1994. Soluble $\alpha_v\beta_3$ integrin ligands raise $[Ca^{2+}]i$ in rat osteoclasts and mouse-derived osteoclast-like cells. Am. J. Physiol. **266:** C376–C381.
32. SCHWARTZ, M. A. 1993. Spreading of human endothelial cells on fibronectin or vitronectin triggers elevation of intracellular free calcium. J. Cell Biol. **120:** 1003–1010.
33. MORI, S. & D. B. BURR. 1993. Increased intracortical remodeling following fatigue damage. Bone **14:** 103–109.
34. RODAN, G. A. 1993. Osteoblastic Cells. *In* Osteoporosis Proceedings. C. Christiansen & B. Riis, Eds. 218–220. Handelstrykkeriet Aalborg ApS. Aalborg, Denmark.
35. CRAIG, A. M., J. H. SMITH & D. T. DENHARDT. 1989. Ostopontin, a transformation-associated cell adhesion phosphoprotein, is induced by 12-O-tetradecanoylphorbol 13-acetate in mouse epidermis. J. Biol. Chem. **264:** 9682–9689.
36. MIYAZAKI, Y., M. SETOGUCHI, Y. HIGUCHI, S. AKIZUKI & S. YAMAMOTO. 1990. The mouse osteopontin gene. Expression in monocytic lineages and complete nucleotide sequence. J. Biol. Chem. **265:** 14432–14438.
37. PATARCA, R., G. J. FREEMAN, R. P. SINGH, F. Y. WEI, T. DURFEE, F. BLATTNER, D. C. REGNIER, C. A. KOZAK, B. A. MOCK, H. C. I. I. I. MORSE, T. R. JERRELLS & H. CANTOR. 1989. Structural and functional studies of the early T lymphocyte activation (Eta-1) gene. J. Exp. Med. **170:** 145–162.
38. CHEN, J., K. SINGH, B. B. MUKHERJEE & J. SODEK. 1993. Developmental expression of osteopontin (OPN) mRNA in rat tissues: evidence for a role for OPN in bone formation and resorption. Matrix **13:** 113–123.

Structural and Functional Domains of Osteopontin[a]

WILLIAM T. BUTLER

Department of Basic Sciences
Dental Branch
University of Texas-Houston Health Science Center
Houston, Texas 77030

Osteopontin (OPN) was first described as a phosphorylated protein secreted by transformed cells but not by their nontransformed counterparts.[1] Since this initial observation, OPN has been rediscovered in a number of independent studies utilizing a variety of tissues, cells and biological fluids.[2–5] During the past decade a remarkable amount of progress has been made concerning the occurrence and biological relevance of OPN. In this review I shall attempt to give a succinct overview of what is known about the structure of OPN and what certain aspects of this information can tell us about its function.

OPN can be described as a phosphorylated glycoprotein that is rich in sialic acid. It consists of a single chain of 264 to 301 amino acids (depending upon the species) and is rich in aspartic acid, glutamic acid and serine. Rat bone OPN is the best characterized to date; it contains about 30 monosaccharides, potentially present as 1 N-linked and 5–6 O-linked oligosaccharides.[6] This OPN also contains 12 phosphoserines and 1 phosphothreonine, as shown by careful quantitation of the phosphate liberated and the serine lost, after β-elimination experiments.[6] It has also been reported that rat OPN contains sulfate, but the nature of this substituent was not clarified.[7]

The complete amino acid sequences for rat,[8] mouse,[9] human,[10] pig,[11] cow,[12] and chicken[13] OPN have been deduced from cDNA sequences. About half of the rat sequence was verified by sequencing of OPN or its proteolytic products.[6,14] Likewise, a substantial portion of cow OPN has been sequenced and verifies the deduced sequence.[15]

A comparison of the sequences for the 6 species[16] reveals interesting features that no doubt reflect conservations related to function. OPN of the first 5 of these species (*i.e.*, rat, mouse, human, pig and cow) are identical in about 40% of their positions. Chicken OPN is the most divergent; whenever this sequence is added to the analysis, the degree of identity is only about 19%. The distribution of conserved sequences along the OPN chain are compared in FIGURE 1, using the numbering of rat OPN (since it is the longest). Three areas show the highest level of sequence conservation: 1) the NH_2-terminal fourth of the molecule; 2) a segment around the GRGDS integrin-binding sequence and 3) the extreme COOH-terminus. A region in the COOH-terminal half just past the thrombin cleavage site (segments 8–11) has few residues that are strictly conserved across the 6 species. This segment contains one stretch of 17 amino acids only present in rat OPN. This same area in chicken OPN is missing a 31-residue piece relative to rat OPN.

[a] The original experiments described in this manuscript were supported by NIH Grant 2RO1 AR39273 and by NRSA Fellowship 1 F32 DE05589 (to Dr. N. M. Ritter).

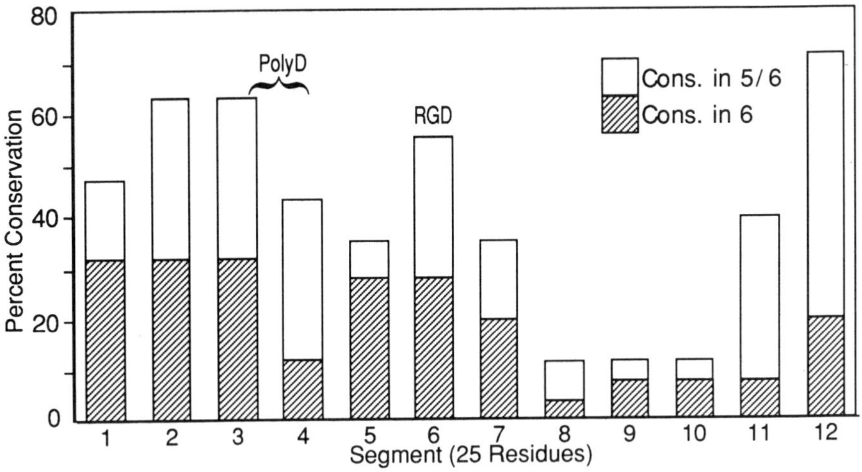

FIGURE 1. The level of conserved amino acid residues in the OPN of six species as a function of the position in the rat OPN chain (301 residues). The chain was divided into 12 segments of 25 residues each; the per cent of the conserved amino acids for each segment is given for the six species (*hatched bars*) and when one of the OPN species (usually chicken) was omitted from the calculation (*unfilled bars*). For a detailed comparison of the amino acid sequences, see Fisher.[16]

Concerning the types of conserved amino acids, serine shows the highest level of conservation; about 24% of the serines are found in all 6 species. Glutamic acid (17%) and aspartic acid (19%) have relatively high conservation rates. This pattern undoubtedly reflects the importance of 1) sequences that are substrates for kinase activities and are phosphorylated and 2) the poly-Asp stretch (see below).

A motif of LPVK or IPVK is found at the NH_2-terminus of 5 of the OPN sequences (FIG. 2), with only the PV portion preserved in chicken OPN. This motif of LPV or IPV is also found in several of the sialic acid-rich glycoproteins

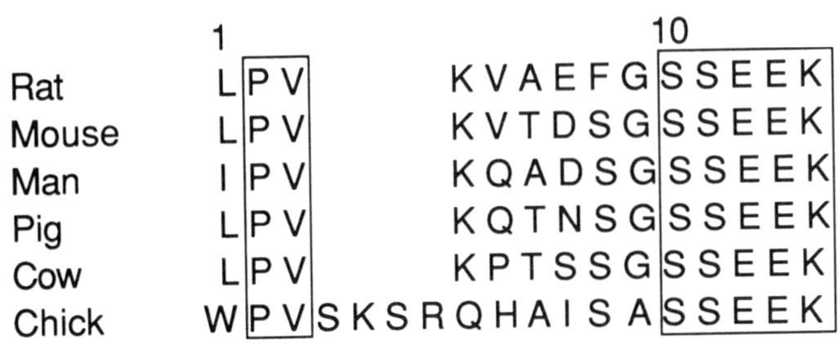

FIGURE 2. Comparison of the NH_2-terminal sequences of OPN from six species. Several sequences are conserved in this region (*boxed* amino acids), including two serines in an acidic cluster (residues 10–14) that are likely candidates for phosphorylation. Numbering for this figure and for FIGURES 3, 4 and 5, correspond to the rat sequence.

DSP (rat)	Ile -Pro-Val-Pro-Gln-Leu-Val-Pro
OPN (rat)	Leu-Pro-Val-Lys-Val-Ala-Glu-Phe
OPN (human)	Ile -Pro-Val-Lys-Gln-Ala-Asp-Ser
BAG-75 (rat)	Leu-Pro-Val-Ala-Arg-Tyr-Gln-Asn
Dmp1 (rat)	Leu-Pro-Val-Ala-Arg-Tyr-Gln-Asn
CGnA (bovine)	Leu-Pro-Val-Gln-Ser-Pro-Met-Gln

FIGURE 3. Comparison of the NH_2-terminal sequences of rat and human OPN with those of dentin sialoprotein (DSP), bone acidic glycoprotein 75 (BAG-75), dentin matrix protein 1 (Dmp1) and chromogranin A. These proteins have similar sequences at the beginning of the chains; the significance of these similarities is unknown.

of mineralized tissues (FIG. 3). Although the significance of this conserved tripeptide is not understood, it may relate to some type of cell signaling, such as that requiring the signal peptides of secreted proteins.

FIGURE 2 also shows a conserved sequence (SSEEK) at positions 10–14 that is considered to be a substrate for appropriate intracellular kinases responsible for catalyzing the phosphorylation of OPN. A number of other serines that are found in all 6 species are in similar sequences, usually involving acidic clusters. Sorensen and Petersen[15] have localized 14 phosphoserines and 1 phosphothreonine by sequencing peptides derived from bovine milk OPN. Twelve of the phosphoserines were in Ser-X-Glu/Pse sequence motifs, indicating phosphorylations catalyzed by mammary gland casein kinase. Two phosphoserines were in sequences of the type Ser-X-X-Glu suggesting recognition by casein kinase II. The conservation of serines and glutamic acids and their juxtaposition in numerous sequences indicates that phosphorylation of OPN is important to its function. This conclusion is also supported by results of studies that show differences in properties and potential functions of phosphorylated and nonphosphorylated OPN.[17–20]

We have isolated and sequenced a number of peptides from rat bone OPN after cleavage with endoprotease Lys-C.[14] One of our observations was that several peptides with the same sequence were separated into different HPLC peaks, and that they differed with respect to the presence of substituted (presumably phosphorylated) serines. These data indicate that rat bone OPN contains partially phosphorylated serine residues, similar to that seen for other posttranslational modifications. Another finding was that Lys-19 and Lys-29 are substituted; the exact nature of the lysine modifications is unclear at this time.

OPN contains a single NES sequence, a recognition site for N-glycosylation, that is conserved in 5 of the species (FIG. 4). In our sequence studies with bone OPN, we found no evidence for glycosylation of this site, since the asparagine was obtained in good yield after Edman degradation.[14] Sorensen and Petersen also showed that N-linked oligosaccharide consensus sequences in bovine milk OPN were unsubstituted.[15] FIGURE 4 also shows the conserved poly-Asp sequence. This stretch of several aspartic acids may be involved in binding OPN to hydroxyapatite in bone;[8,21] in this way OPN may influence the mineralization process.

Another highly conserved structure is a GRGDS sequence (FIG. 5) near the center of the molecule (residues 127–131 in rat OPN). This sequence has been implicated in binding to $\alpha_v\beta_3$ cell surface integrins on a number of cell types.[2–4] Thus, OPN will promote *in vitro* attachment and spreading of osteoblasts, fibroblasts, osteoclasts, epithelial cells, smooth muscle cells and other cells. This

	63 … 70 … 78
Rat	NESHDHMDDDDDDDDD-
Mouse	NESHDHMDDDDDDDDDD
Man	NESHDHMDDMDDEDDDD
Pig	NESPEQTDDVDDDDDED
Cow	NESPEQTDDLDDDDDNS
Chick	SKSQEAVDDDDDDDDNDS

FIGURE 4. Sequences in the OPN of six species showing the potential N-glycosylation site, conserved in five species, and the polyaspartic acid region.

activity is inhibited by inclusion of RGD peptides, but not RGE peptides in the assays. Appropriate antibodies against $\alpha_v\beta_3$ integrins also prevent attachment. The interaction of OPN with cells may also bring about intracellular signaling.[22] In addition, the GRGDS sequence may be involved in promoting migration of smooth muscle cells.[23,24]

The RS sequence at residues 137–138 (see FIG. 5) represents a thrombin cleavage site,[25] present in 5 of the 6 OPN species, and replaced by RA in chicken OPN. The cleavage of OPN by thrombin has been utilized in numerous studies to verify the authenticity of OPN. Thus, thrombin treatment generates two fragments that usually comigrate on gel electrophoresis in the Mr 28–35 kDa size range.[25] The interspecies retention of the thrombin cleavage site suggests that it has important biological relevance. The cleavage of circulating OPN by thrombin occurs after blood coagulation, again suggesting some function for this proteolysis. By comparing the activity of OPN with that of thrombin-cleaved OPN, Senger *et al.*[26] showed that thrombin markedly increased cell attachment and spreading. The exact nature of this enhancement needs to be elucidated.

We have reported that the COOH-terminal portion of rat OPN contains a non-RGD cell attachment component.[27] This activity was associated with a 28-kDa peptide isolated after cleavage of OPN at the R^{141}-S^{142} bond with Arg-C endoprotease (see FIG. 5). The 28-kDa peptide promoted attachment and spreading of

	130 … 140
Rat	PDGRGDSLAYGLRSKSRSFP
Mouse	PNGRGDSLAYGLRSKSRSFQ
Man	YDGRGDSVVYGLRSKSKKFR
Pig	NDGRGDSVVYGLRSKSKKFR
Cow	NDGRGDSVAYGLKSRSKKFR
Chick	NAGRGDSVAYGFRAKAHVVK

FIGURE 5. A highly conserved sequence in the central region of OPN is the GRGDS cell attachment sequence. Nearby is an RS sequence (residues 137–138; 139–140 in cow OPN) cleaved by thrombin.[25] In chicken OPN the sequence is RA. The RS sequence in rat OPN (residues 141–142) is also cleaved by Arg-C endoprotease.[27]

OPN *in vitro* almost as efficiently as OPN. The attachment was inhibited by RGD, but not RGE, peptides and by monoclonal antibody LM609, directed against $\alpha_v\beta_3$ integrins. The exact identity of this second cell attachment site and the significance of these observations are unknown at this time.

Secondary structure predictions for rat[28] and mouse[9] OPN indicate several stretches of α-helix and shorter segments of β-sheet structure. The GRGDS integrin binding sequence is in a β-turn and is flanked by two short β-sheet structures. This structure would probably be on the surface of the molecule and would "present" the RGD sequence for interaction with the cell surface receptors. This type of motif seems to be present in other RGD-containing cell attachment proteins, such as fibronectin. Prince[27] also predicted that rat bone OPN contains two heparin-binding sites and a calcium binding loop. Despite the predictions of significant levels of α-helix and β-sheet structures, circular dichroism measurements indicated that OPN extracted from bone displayed spectra indicating a random structure.[28] Presumably the harsh conditions utilized during demineralization of bone and purification of OPN resulted in denaturation. Secondary structure predictions must be verified with OPN carefully purified from tissues or cell culture media under nondenaturing conditions.

Another aspect of the structure of OPN relates to potential interactions with other extracellular matrix macromolecules to form higher order structures. Because OPN is secreted into a mineralizing matrix (osteoid) in bone,[29] OPN is likely to encounter other macromolecules, such as collagen. This consideration has led us to postulate that OPN interacts with other bone proteins to form some type of supramolecular structure; its biological activity would thus be vested in this structure and might differ from what we measure in *in vitro* experiments. To initially test this hypothesis, we used gel overlay experiments with bone proteins that were separated on a SDS polyacrylamide gel, probed with [^{125}I]OPN.[30] We were surprised to find that OPN formed a complex with osteocalcin. This interaction was prevented by adding excess osteocalcin in the reaction mixture with labeled OPN. In the reverse experiment, we showed that [^{125}I]osteocalcin formed a complex with OPN; again this complex was prevented with unlabeled osteocalcin. The interaction of OPN with osteocalcin was also demonstrated using solid-phase filter binding assays, which showed an equilibrium binding constant of about 10^{-8} M. The specificity of the binding in the solid-phase system was shown by including unlabeled osteocalcin and OPN with the reaction mixtures. Finally, the formation of a soluble complex between OPN and osteocalcin was shown by using a modified Hummel-Dreyer gel filtration assay. The exact nature of the OPN-osteocalcin interaction has not been elucidated: it could involve discrete regions of both proteins, but could also be conformation-dependent. Formation of a complex between OPN and osteocalcin is also consistent with their colocalization in developing bone, seen in ultrastructural immunolocalization experiments.[31]

We have also shown that OPN forms a complex when incubated with collagen types I, II, III, IV and V.[32] These experiments employed solid-phase assays and showed that the binding of OPN to the collagens (2 μg/well) was saturable at OPN concentrations above 100–120 nM. Equilibrium binding constants were 20–50 nM. The interactions of OPN with the collagen types were shown to be specific by concentration-dependent inhibition of binding of the radiolabeled OPN by unlabeled OPN. Some differences were noted between the binding to the different collagens. Taken together, these observations indicate that OPN could form a complex with collagen and with osteocalcin *in vivo* and that such interactions should be considered when projecting and studying the function of OPN.

REFERENCES

1. SENGER, D. R., D. F. WIRTH & R. O. HYNES. 1979. Cell **16:** 885–893.
2. BUTLER, W. T. 1989. Connect. Tissue Res. **23:** 123–136.
3. YOUNG, M. F., J. M. KERR, K. IBARAKI, A.-M. HEEGAARD & P. G. ROBEY. 1992. Clin. Orthop. Relat. Res. **281:** 275–294.
4. DENHARDT, D. T. & X. GUO. 1993. FASEB J. **7:** 1475–1482.
5. PATARCA, R., R. A. SAAVEDRA & H. CANTOR. 1993. Crit. Rev. Immunol. **13:** 225–246.
6. PRINCE, C. W., T. OOSAWA, W. T. BUTLER, M. TOMANA, A. S. BHOWN, M. BHOWN & R. E. SCHROHENLOHER. 1987. J. Biol. Chem. **262:** 2900–2907.
7. NAGATA, T., R. TODESCAN, H. A. GOLDBERG, Q. ZHANG & J. SODEK. 1989. Biochem. Biophys. Res. Commun. **165:** 234–240.
8. OLDBERG, A., A. FRANZEN & D. HEINEGÅRD. 1986. Proc. Natl. Acad. Sci. USA **83:** 8819–8823.
9. CRAIG, A. M., J. H. SMITH & D. T. DENHARDT. 1989. J. Biol. Chem. **264:** 9682–9689.
10. YOUNG, M. F., J. M. KERR, J. D. TERMINE, U. M. WEWER, M. G. WANG, O. W. MCBRIDE & L. W. FISHER. 1990. Genomics **7:** 491–502.
11. WRANA, J. L., Q. ZHANG & J. SODEK. 1989. Nucleic Acids Res. **17:** 10119.
12. KERR, J. M., L. W. FISHER, J. D. TERMINE & M. G. YOUNG. 1991. Gene **108:** 237–243.
13. MOORE, M. A., Y. GOTOH, K. RAFIDI & L. C. GERSTENFELD. 1991. Biochemistry **30:** 2501–2508.
14. NEAME, P. J. & W. T. BUTLER. Unpublished data.
15. SORENSEN, E. S. & T. E. PETERSEN. 1994. Biochem. Biophys. Res. Commun. **198:** 200–295.
16. FISHER, L. W. 1992. *In* Chemistry and Biology of Mineralized Tissues. H. Slavkin & P. Price, Eds. 177–186. Elsevier. New York.
17. NEMIR, M., M. W. DEVOUGE & B. B. MUKHERJEE. 1990. J. Biol. Chem. **264:** 18202–18208.
18. SINGH, K., M. W. DEVOUGE & B. B. MUKHERJEE. 1990. J. Biol. Chem. **265:** 18696–18701.
19. CHANG, P.-L. & C. W. PRINCE. 1993. Cancer Res. **53:** 2217–2220.
20. EK-RYLANDER, B., M. FLORES, M. WENDEL, D. HEINEGÅRD & G. ANDERSSON. 1994. J. Biol. Chem. **269:** 14853–14856.
21. GORSKI, J. P. 1992. Calcif. Tissue Int. **50:** 391–396.
22. MIYAUCHI, A., J. ALVAREZ, E. M. GREENFIELD, A. TETI, M. GRANO, S. COLUCCI, A. ZAMBONIN-ZALLONE, F. P. ROSS, Ş. L. TEITELBAUM, D. CHERESH & K. A. HRUSKA. 1991. J. Biol. Chem. **266:** 20369–20374.
23. LIAW, L., M. ALMEIDA, C. E. HART, S. M. SCHWARTZ & C. M. GIACHELLI. 1994. Circ. Res. **74:** 214–224.
24. YUE, T.-L., P. J. MCKENNA, E. H. OHLSTEIN, M. C. FARACH-CARSON, W. T. BUTLER, K. JOHANSON, P. MCDEVITT, G. Z. FEUERSTEIN & J. M. STADEL. 1994. Exp. Cell Res. **214:** 459–464.
25. SENGER, D. R., C. A. PERRUZZI, C. F. GRACEY, A. PAPADOPOULOS & D. G. TEMEN. 1989. Cancer Res. **48:** 5770–5774.
26. SENGER, D. R., C. A. PERRUZZI, A. PAPADOPOULOS-SERGIOU & L. VAN DE WATER. 1994. Mol. Biol. Cell **5:** 565–574.
27. VAN DIJK, S., J. A. D'ERRICO, M. J. SOMERMAN, M. C. FARACH-CARSON & W. T. BUTLER. 1993. J. Bone Miner. Res. **8:** 1499–1506.
28. PRINCE, C. W. 1989. Connect. Tissue Res. **21:** 345–350.
29. PINERO, G. J., M. C. FARACH-CARSON & W. T. BUTLER. 1995. Arch. Oral Biol. In press.
30. RITTER, N. M., M. C. FARACH-CARSON & W. T. BUTLER. 1992. J. Bone Miner. Res. **7:** 877–885.
31. MCKEE, M. D., M. C. FARACH-CARSON, W. T. BUTLER, P. V. HAUSCHKA & A. NANCI. 1993. J. Bone Miner. Res. **8:** 485–495.
32. RITTER, N. M. & W. T. BUTLER. Unpublished data.

Conformational Analyses on Soluble and Surface Bound Osteopontin

JEFFREY P. GORSKI,[a] EDWARD KREMER,
JAVIER RUIZ-PEREZ, GARY E. WISE,[b]
AND ANTONIO ARTIGUES

Division of Molecular Biology and Biochemistry
University of Missouri-Kansas City
Kansas City, Missouri 64110
and
[b]Department of Anatomy and Cell Biology
School of Veterinary Medicine
Louisiana State University
Baton Rouge, Louisiana 70803

INTRODUCTION

Osteopontin (OP) is an acidic phosphoprotein which represents a major component of bone matrix.[1-3] Structural homologies of OP with other metal binding proteins suggest a role in skeletal calcium metabolism.[4] A number of studies support such a role directly and indirectly. We[5] and Singh *et al.*[6] have demonstrated that osteopontin from bone, kidney, and human milk exhibits a large calcium binding capacity (about 50 atoms calcium/mole at saturation[5]). Interestingly, OP isolated from human urine (uropontin) acts as an inhibitor of *in vitro* calcium oxalate crystal growth formation.[7] Under certain *in vivo* conditions, however, OP is incorporated into urinary calcium oxalate stones where it can be extracted with EDTA.[8] In support for a causative role, the mRNA for OP is induced by glyoxylic acid in kidney distal tubular cells in a rat model of stone formation.[8] Hunter *et al.*[9] have found that OP is capable of inhibiting hydroxyapatite crystal formation by metastable solutions of calcium and phosphate, while both Boskey *et al.*[10] and Hunter *et al.*[9] have observed inhibition of crystal formation by OP in a steady state agarose gel system, although the effective concentration of OP was much higher in this model. Polyglutamic acid and dephosphorylated OP were also less effective as inhibitors of hydroxyapatite formation than OP itself. Immunolocalization studies at the EM level demonstrate that OP is concentrated within bone in cement lines and at mineralization fronts.[11-13] Since OP is synthesized both by osteoblasts and osteoclasts within bone,[14] the cellular source of OP deposited at these sites is not yet clear. Finally, OP has consistently been observed in calcifying atherosclerotic plaques from man and in animal models of atherosclerosis.[15]

Despite the above-mentioned studies and availability of a cloned cDNA sequence from a number of species, mechanistic questions remain regarding the function of OP *in vivo*. Several years ago we introduced the hypothesis that matrix proteins such as OP may possess several conformations, *i.e.*, one in semisolid

[a] Please address all correspondence and reprint requests to: Dr. Jeffrey P. Gorski, Room B109, School of Biological Sciences, University of Missouri-KC, Kansas City, MO 64110.

matrix and one in solution.[4] In this way, the functionality of OP may depend in part on the assumption of an '*in vivo* conformation.' Questions arising from a consideration of this hypothesis include: does binding of OP to hydroxyapatite induce a conformational change or select for a particular conformation, and does binding of calcium induce a conformational change in acidic phosphoproteins? We present results of several studies on OP to address these questions.

METHODS

Isolation of Osteopontin

Young rat calvaria were extracted with 4 M guanidine-HCl/0.5 M EDTA and osteopontin isolated as described by Gorski *et al.*[16] Purified osteopontin was radioiodinated with ^{125}I-iodine by the Iodobead method (Pierce Chem. Co.) in order to follow protein binding to hydroxyapatite, collagen and plastic surfaces. The concentration of OP stock solutions was determined by amino acid analysis.

Spectroscopy

Circular dichroism (CD) spectroscopy was carried out on OP solutions in 5 mM Tris-HCl buffer (pH 7.5–8.0), containing 20 to 150 mM sodium chloride and, with or without 1–5 mM calcium chloride, using a JASCO model J-720 spectropolarimeter. Spectra (2 nm bandwidth, sensitivity 20 mdeg, and scan speed of 10–20 nm/min) were collected at room temperature and are the result of averaging 10 repetitive scans. Baseline spectra were recorded for buffer alone and used to correct OP spectra. Manipulation of data files utilized software programs (version 1.20c) supplied by JASCO.

Fourier transformed infrared (FTIR) spectra were collected on OP samples with a Mattson Instruments Sirius 100 model spectrophotometer and analyzed with associated WinFirst software programs as previously described.[17] Calcium fluoride windows were used with a pathlength of 0.056 mm. Protein samples were concentrated prior to spectroscopy using Amicon devices with a 10,000 molecular weight membrane cut-off.

Binding Studies

The capacity of rat OP to bind to plastic, collagen, or hydroxyapatite surfaces was determined in 50 mM Tris-HCl buffer (pH 7.5), containing 150 mM sodium chloride and in the presence or absence of 1–10 mM calcium chloride. Negative controls were carried out similarly except that bovine serum albumin was substituted for osteopontin. Quantitation of bound OP was determined through gamma counting of a ^{125}I-labeled tracer or by an ELISA type protocol using monoclonal anti-osteopontin, peroxidase conjugated rabbit anti-mouse IgG antibodies, and 2,2′-azino-bis(3-ethylbenzthiazoline-6-sulfonic acid) substrate.

Immunoreactivity of Osteopontin on Surfaces

Monoclonal mouse anti-rat osteopontin antibody (clone MPIIIB10) from the NIH Developmental Studies Hybridoma Bank was used to detect osteopontin on surfaces. Monoclonal antibody binding was analyzed by use of a peroxidase-conjugated second antibody method, or by use of a ^{125}I-labeled second antibody and gamma counting.

Immunohistology

Sprague-Dawley rats, 1–5 days old postnatally, were sacrificed by cervical dislocation and the calvaria and adhering tissues were surgically removed. They were then frozen in liquid nitrogen, transferred to a cryostat, and embedded in Tissue-Tek. Frozen sections of 10 μm thickness were cut and fixed in 10% neutral buffered formalin for 10 min followed by rinsing in 0.01 M phosphate buffered saline (pH 7.4) (PBS).

One group of slides was demineralized with 0.25 M EDTA in 0.01 M PBS (pH 7.4) for 10 min and then washed with 0.01 M PBS. The other groups were not treated with EDTA. All slides were then incubated in 1% hydrogen peroxide for 10 min to eliminate endogenous peroxidase activity. To reduce nonspecific background staining, the slides were then incubated in a 1 : 5 dilution of colostrum-free bovine serum for 60 min. Monoclonal anti-osteopontin antibodies (in hybridoma culture supernatant fraction) were diluted 1 : 40 with 0.01 M PBS and then incubated with calvarial sections for 2 h at room temperature. Excess primary antibody was removed by rinsing with PBS and sections were then treated for 1 h with 1 : 100 diluted peroxidase-conjugated rabbit anti-mouse IgG (ICN Immunologicals, Lisle, IL). Slides were rinsed again and then incubated for 10 min with PBS containing 0.06% diaminobenzidine tetrahydrochloride) and 0.01% hydrogen peroxide. Following rinsing with water, slides were stained with hematoxylin and mounted in Permount. Negative controls included omission of monoclonal anti-osteopontin antibodies, incubation with mouse normal ascites fluid, and substitution of normal rabbit serum IgG for primary antibody.

RESULTS

Initial results with monoclonal anti-osteopontin antibody (clone MPIIIB10) and frozen neonatal rat calvarial sections demonstrated near background immunostaining (FIG. 1A). No evidence for reactivity was found in cells adjacent to mineralized regions in the absence of decalcification. In contrast, following treatment of sections with EDTA, strong staining of mineralizing osteoid areas and osteoblast-like cells was noted (FIG. 1B). Immunostaining for osteopontin appeared to be specific in that controls which substituted rabbit IgG or normal mouse ascites fluid for monoclonal antibody, or which omitted monoclonal antibody uniformly gave background results (FIG. 1C). On the other hand, immunoreactivity for bone acidic glycoprotein-75 was found to be strongest in calcified sections, reactivity which was largely lost upon decalcification (not shown). We considered two rationalizations for immunostaining results with MAb MPIIIB10. First, the OP epitope in bone may be covered up or inaccessible to the antibody. Alternatively, OP may undergo a conformational change when binding to hydroxyapatite leading to a loss of immunoreactivity with MAb MPIIIB10. In this latter case, treatment with EDTA would reverse this process through dissolution of hydroxyapatite. We sought to test these hypotheses through direct spectroscopic studies and, indirectly, through comparative analyses to determine if monoclonal antibody binding to OP is sensitive to the presence of calcium or to surface-induced changes in conformation.

A range of concentrations of OP were incubated with plastic microtiter wells and with freshly prepared hydroxyapatite crystals in the presence and absence of 1 mM calcium chloride. The amount of OP bound to these surfaces was then determined both by inclusion of a radioiodinated tracer and through immunoreactivity with MAb MPIIIB10. Negative controls substituted bovine serum albumin for OP. Results in FIGURE 2 illustrate that OP does bind weakly to plastic wells when monitored with a radioiodinated tracer. Binding reached an apparent maximum at about 1 pmole bound per well, representing about 1% of that initially present. In the absence of 1 mM calcium chloride binding to plastic is observed to be lowered by about 20% (FIG. 2A). Interestingly, OP adsorbed to plastic under these conditions is readily recognized by MAb MPIIIB10. Specific immunoreactivity following colorimetric detection with peroxidase-conjugated second antibody is plotted in FIGURE 2B. Although variability was increased with the multistep immune detection method over that with ^{125}I-labeled OP (FIG. 2A), it is clear that both methods detected slightly more OP bound to plastic surfaces in the presence of calcium than in the absence of divalent metal ion. Thus, MAb MPIIIB10 is able to recognize and bind to OP adsorbed on plastic surfaces and is able to detect as little as about 1 pmole of protein.

OP binding to hydroxyapatite crystals, as assessed with radioactive tracer, was much greater than to plastic (FIG. 3A). A first-order curve was obtained wherein binding averaged 20% of input OP over the range of concentrations examined (22–600 μM); binding was not saturable over this input concentration range. In several experiments, no significant difference was observed in the amount of OP bound to hydroxyapatite whether 1 mM calcium chloride was present or not. Interestingly, when hydroxyapatite crystals with adsorbed osteopontin were prepared identically, but without radioactive tracer, bound OP was not recognized by MAb MPIIIB10. Immunoreactivity of OP adsorbed to hydroxyapatite was assayed in two ways. *i.e.*, by assaying binding of ^{125}I-labeled MAb MPIIIB10 and by monitoring color development with peroxidase-conjugated second antibody. Results of detection with ^{125}I-labeled MAb MPIIIB10 are plotted in FIGURE 3B on the basis of a percentage of that obtained with bovine serum albumin negative controls. It is apparent that MAb MPIIIB10 recognition of OP- and albumin-treated hydroxyapatite crystals is indistinguishable over the entire range of protein input concentration tested. As predicted from the results shown in FIGURE 3A, the presence or absence of calcium had no effect on immunoreactivity with MAb MPIIIB10. Indirect immunoassays of MAb MPIIIB10 binding to OP-treated hydroxyapatite crystals are also plotted as a percentage of the optical density values of albumin-treated crystal negative controls (FIG. 3C). On the basis of this approach, OP-treated crystals gave values the same or lower than that of controls. We hypothesize that lower values may be a consequence of greater coverage of hydroxyapatite crystal surfaces by OP as compared with albumin, effectively reducing nonspecific binding of primary and secondary antibodies added subsequently (FIG. 3C). Since the amount of OP adsorbed to hydroxyapatite was at least 100-times greater than the limit of detection with MAb MPIIIB10 (FIG. 2B), we conclude that OP bound to hydroxyapatite exhibits a different conformation than when bound to plastic surfaces.

OP was analyzed by circular dichroism spectroscopy in an effort to examine directly the effect of millimolar calcium on its secondary structure. FIGURE 4 illustrates a representative near UV CD spectrum for OP at 150 μg/ml in 0.05 M Tris-HCl buffer (pH 7.5) containing 0.15 M sodium chloride. The spectrum, which

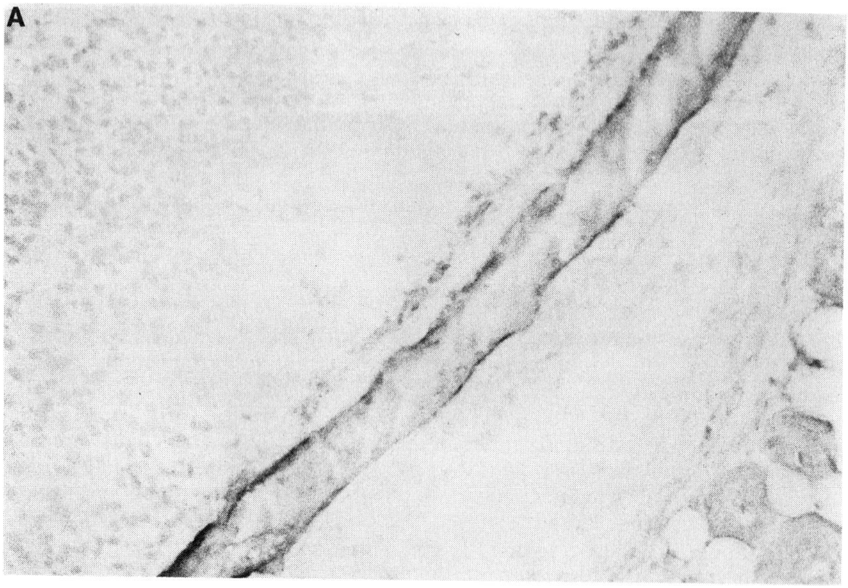

FIGURE 1. Pretreatment of calvarial sections with EDTA dramatically increases the expression of osteopontin antigen within mineralized regions. **(A)** Near background immunostaining for osteopontin is observed with monoclonal anti-osteopontin antibodies and mineralized calvarial sections. Culture supernatant containing anti-osteopontin antibodies was diluted 1 : 40 and used in immunostaining of undecalcified frozen sections as described in Methods. Magnification: 32×. **(B)** Antigenicity for osteopontin is greatly increased after decalcification. Frozen sections were fixed in 10% neutral buffered formalin and then demineralized with 0.25 M EDTA prior to immunostaining as in Methods. Magnification: 63×. **(C)** Negative control for immunostaining protocol. Calvarial section was incubated as above except that the primary antibody was omitted. Similar results were obtained when mouse ascites fluid or normal rabbit serum IgG was substituted for monoclonal anti-osteopontin antibodies. Magnification: 63×.

displays a single minimum at 198 nm, is in general representative of a random coil conformation. Addition of calcium chloride to a final concentration of 5 mM had little effect on the OP spectrum (FIG. 4). Next, the impact of increasing the OP concentration and decreasing the salt concentration to 0.02 M was tested; in order to carry out concurrent Fourier transformed infrared spectroscopic analysis, the water in OP samples was exchanged for D_2O. As shown in FIGURE 5, OP, at a concentration of 17.9 mg/ml and in low salt, exhibits a distinctly different spectrum which is now representative of a more organized secondary structure containing beta-pleated sheet. Interestingly, the single minimum at 204.5 nm reproducibly shifted to 202.5 nm following addition of calcium to 5 mM; in addition, the size of this absorption maximum increased by 17% upon addition of calcium. Since this change in OP conformation was dramatic and unexpected (compare FIGS. 4 and 5), we wished to determine whether the decrease in salt concentration or an approximate 100-fold increase in OP concentration was primarily responsible for the observed change in conformation. As a result, spectra for OP under conditions

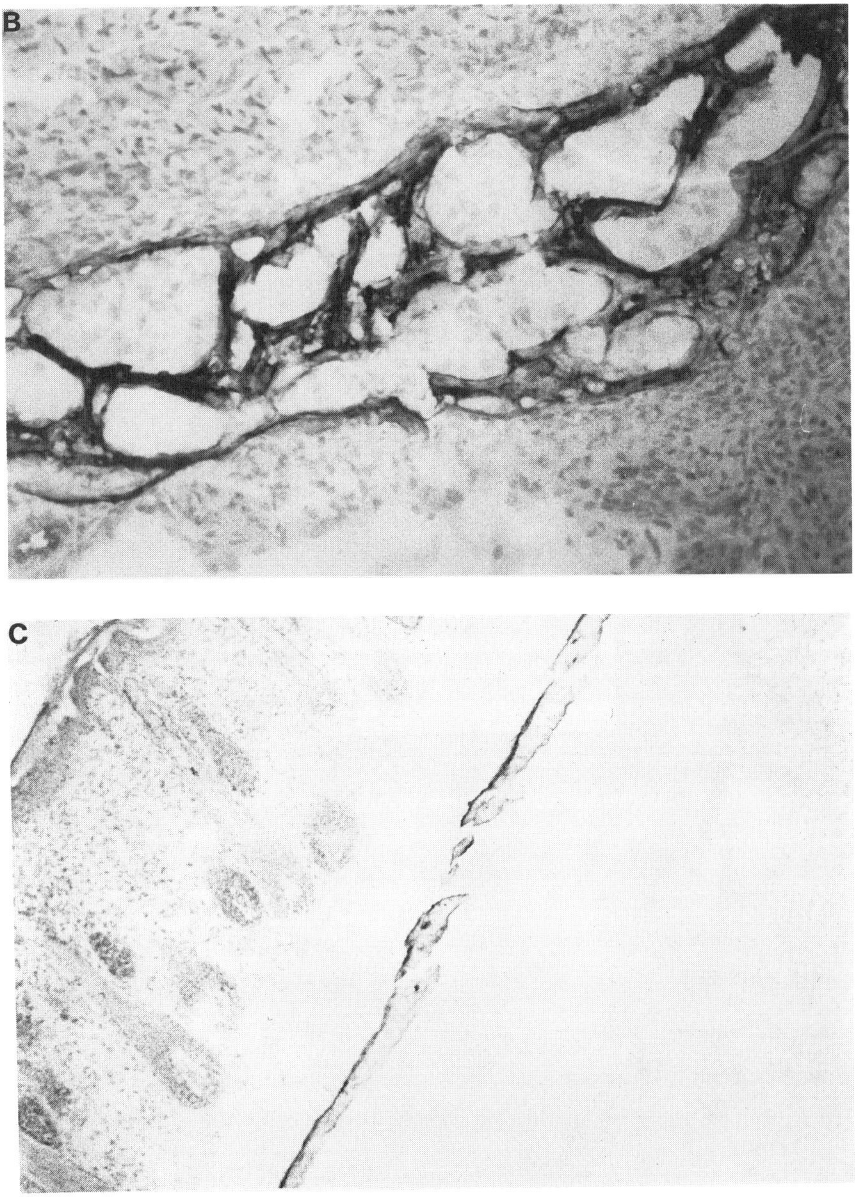

FIGURE 1 (*Continued*).

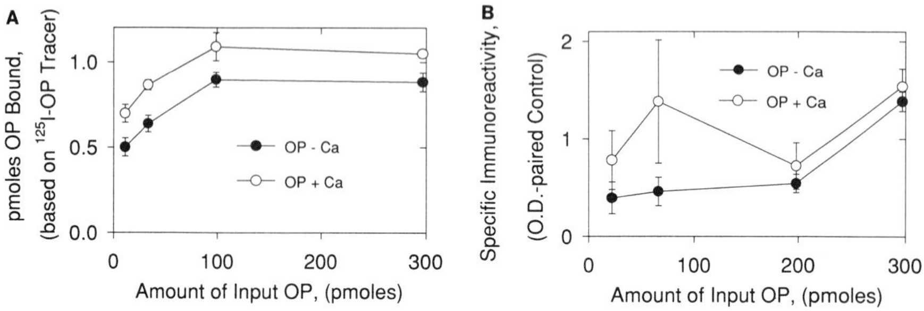

FIGURE 2. Osteopontin binding to plastic surfaces. **(A)** Binding as monitored by radioactive tracer. A mixture of ^{125}I-labeled tracer and cold OP was hydrated in 0.05 M Tris-HCl buffer (pH 7.5), containing 0.15 M sodium chloride with or without 1 mM calcium chloride, and, after dilution, adsorbed for 16 h at 37°C to plastic microtiter wells over a range of concentrations (3–0.1 μM). Wells were washed 5× with the appropriate buffer and processed for immunodetection with MAb MPIIIB10 and peroxidase-conjugated second antibody as described in Methods. Following optical density measurements, proteinase K was incubated in wells for 20 h at 55°C to release bound OP which was then counted. Bovine serum albumin (3–0.1 μM) was substituted for OP in control wells which were processed similarly to experimental wells. **(B)** Binding as assayed by colorimetric peroxidase conjugated second antibody method. Results plotted represent corrected optical density values for wells with bound OP (minus color values for paired albumin controls). *Error bars* on the figures refer to standard errors of the mean, SEM.

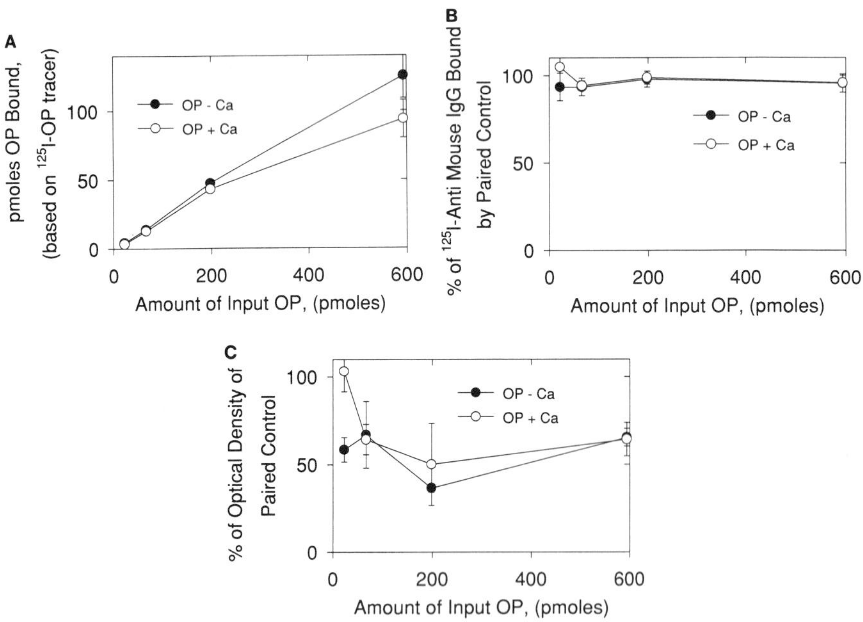

FIGURE 3. Osteopontin binding to hydroxyapatite crystals. **(A)** Amount of OP bound based on radioactive tracer. Mixtures of OP and ^{125}I-labeled tracer (6–0.22 μM) in 0.05 M Tris-HCl buffer (pH 7.5), containing 0.15 M sodium chloride with or without 5 mM calcium chloride were mixed with hydroxyapatite crystals for 4 h in siliconized microfuge tubes.

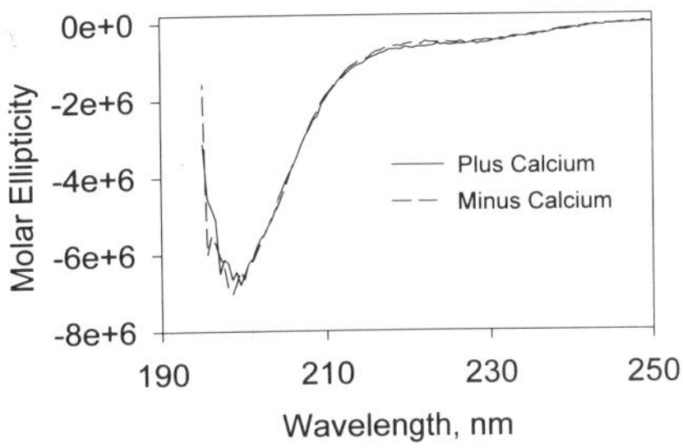

FIGURE 4. Circular dichroism spectroscopy of osteopontin under dilute conditions. Osteopontin was hydrated at 157 μg/ml in 0.05 M Tris-HCl buffer (pH 7.5), containing 0.15 M sodium chloride and spectra recorded before and after addition of calcium chloride to a final concentration of 5 mM. The plus calcium spectrum was normalized to account for this dilution of 5%.

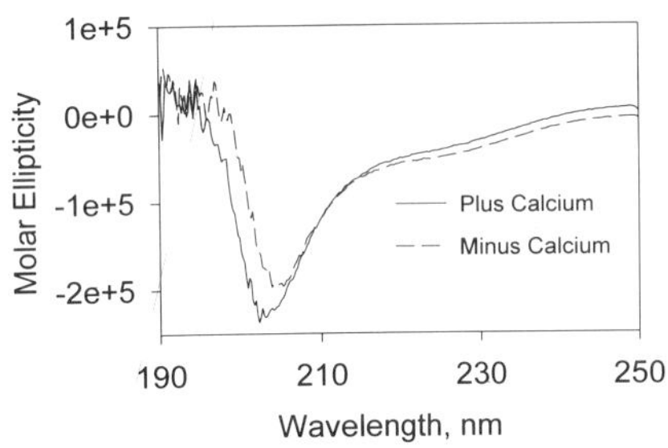

FIGURE 5. The effect of elevated concentration and low salt on the osteopontin spectrum. Osteopontin was hydrated with 0.005 M Tris-HCl buffer (pH 8.5), containing 0.02 M sodium chloride in D_2O, and, concentrated to 17.9 mg/ml as described in Methods. Circular dichroism spectra were collected for OP under these conditions before and after the addition of calcium chloride to a final concentration of 5 mM.

of dilute protein concentration and in 0.02 M sodium chloride were recorded with and without 5 mM calcium, and are presented in FIGURE 6. Under these conditions, the appearance of both spectra are again consistent with a randomly coiled polypeptide chain. A small decrease in the size of the absorption peak centered at 198 nm was observed, however, upon addition of calcium. Taken together these findings

---

Free OP was removed by four washes with the appropriate buffer. Controls substituted albumin for OP. Crystals with bound OP were incubated with MAb MPIIIB10 for 2 h, washed twice with buffer, and incubated for 1.5 h with peroxidase-conjugated second antibody prior to colorimetric development and absorption measurements on supernatant fractions. Hydroxyapatite crystals were then resuspended and counted in a gamma counter; raw counts were converted to pmoles of bound OP based on the specific activity of input ^{125}I-OP. Bound OP is plotted versus the amount of input OP. **(B)** Immunodetection of OP bound to hydroxyapatite crystals with ^{125}I-labeled second antibody. Hydroxyapatite crystals with bound OP or albumin (controls) were prepared as described in (A) above (except that ^{125}I-labeled OP tracer was omitted); samples were mixed with MAb MPIIIB10 for 2 h prior to washing steps; and bound antibody detected by incubating for 1.5 h with ^{125}I-labeled goat anti-mouse IgG. Results with OP crystals are plotted as a percentage of ^{125}I-labeled second antibody bound by paired negative controls versus the amount of input OP. **(C)** Measurement of OP bound to hydroxyapatite crystals with colorimetric immunoassay. Hydroxyapatite crystals with bound OP or albumin (controls) were prepared and subjected to immunoassay as described in (A) above. Absorption measurements of OP crystals are presented as a percentage of optical density values for paired albumin controls versus the amount of input OP. *Error bars* on the figures refer to standard errors of the mean, SEM.

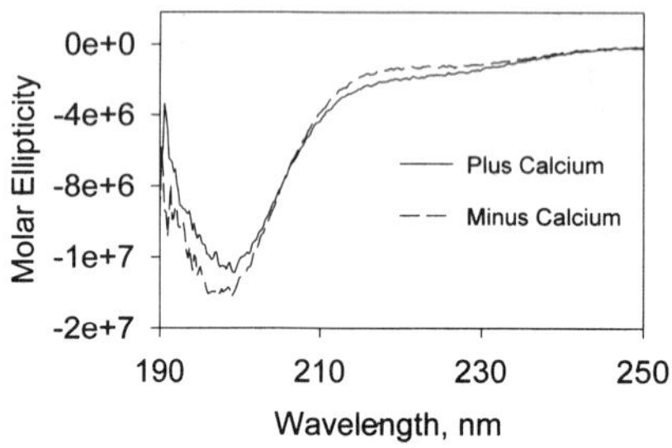

FIGURE 6. Circular dichroism spectroscopy of osteopontin in dilute conditions and in low salt. Osteopontin was hydrated at 157 μg/ml with 0.005 M Tris-HCl buffer (pH 8.2), containing 0.02 M sodium chloride in D_2O. Spectra were recorded both in the presence and absence of 5 mM calcium chloride.

indicate that the conformation of OP in solution is dependent upon its concentration and secondarily upon the salt and calcium concentration. Further support for this conclusion is provided by FTIR spectroscopic analyses carried out in deuterated water at 15 mg/ml OP in 0.005 M Tris-HCl buffer (pH 8.0), containing 0.02 M sodium chloride with or without 5 mM calcium chloride (TABLE 1). The amide I absorption band in both cases was subjected to a series of 3–4 separate deconvolution schemes and resultant quantitative estimates of the amount of different secondary structures were then averaged over all schemes. FTIR spectroscopic analysis indicates that OP under these concentrated conditions assumes an ordered structure composed of nearly equivalent contributions from beta turns, beta, alpha, and, random disordered conformations. A small, but detectable change was consistently observed upon addition of calcium to 5 mM, wherein a decrease of 10–15% in alpha helix coincided with a comparable increase in the amount of random structure (TABLE 1).

DISCUSSION

Direct binding studies demonstrate that osteopontin binds a large number of calcium atoms, estimated at 19–50 atoms per mole at saturation depending upon conditions. However, binding to hydroxyapatite crystals is apparently not affected by the presence of 1–10 mM calcium chloride in the binding buffer. Second, immunohistology of calvarial sections revealed that staining with monoclonal anti-osteopontin antibodies (clone MPIIIB10) is minimal unless sections are first treated with EDTA. In contrast, immunoreactivity towards bone acidic glycoprotein-75

TABLE 1. FTIR Spectroscopic Analysis of Osteopontin Plus and Minus 5 mM Calcium Chloride[a]

Without Calcium		With Calcium	
beta turns	17%	beta turns	22%
beta	24	beta	22
alpha	29	alpha	16
random	21	random	36
inter-beta	6	unknown	4
unknown	2		

[a] Results are averages of 3–4 separate deconvolution schemes and representative of results from two separate experiments.

was strongest in undecalcified sections. However, interpretation of these results is complicated by well documented problems of antibody accessibility in skeletal tissues. In an effort to circumvent such complications we examined the immunoreactivity of OP when adsorbed to plastic and hydroxyapatite surfaces, situations under which access of antibody to antigen should not be impaired. Although OP bound to plastic surfaces is reactive with MPIIIB10 antibodies, OP adsorbed to hydroxyapatite crystal surfaces is not recognized by these antibodies as assessed by two detection methods. Since it is anticipated that the detection methods used should have been able to resolve a 10% difference in reactivity as compared to negative controls, our results imply that most or all of the OP bound to hydroxyapatite (and in bone sections) exhibits a different conformation than when bound to plastic surfaces. Finally, solution circular dichroism and Fourier-transformed infrared spectroscopic studies indicate that the conformation of bone OP is dependent upon its concentration, and, secondarily to the presence or absence of calcium ion. OP used here was purified from bone in the presence of protein denaturants guanidine-HCl and urea. Detailed structural analyses of bone OP are justified since such analyses provide a baseline with which to compare with "native" preparations of OP, and, similar preparations from bone were used in most published calcium binding and cell adhesion studies.[18,19] At elevated OP concentrations, FTIR and CD methods detected a conformational shift upon addition of calcium ion to 5 mM. With both spectroscopic methods, addition of calcium appeared to increase the extent of disordered structure. An absence of disulfide bonds may make the conformation of OP particularly susceptible to environmental changes. We suggest that these findings support our hypothesis that bone matrix proteins exhibit a different conformation when adsorbed on hydroxyapatite crystal surfaces. Assumption of a more organized secondary structure in concentrated OP solutions (*i.e.*, 15 mg/ml) is consistent with these results in that local concentrations of OP within a semi-solid matrix may approach or exceed levels used here. However, the physiological significance of our findings must necessarily wait for confirmatory studies with native OP preparations. Prince[20] also observed that OP displayed a random coiled conformation at relatively dilute concentrations (0.75 mg/ml). Similarly, the presence of calcium ion concentrations up to 20 mM were without effect under these conditions. Sequence-based methods of secondary structure estimation predict the alpha helical and beta pleated sheet contents of OP as 41 and 12%, respectively,[20] values which are somewhat at odds with those estimated from FTIR studies here of isolated OP. It can be expected that OP's acidic character, polyacidic amino acid stretch, and posttranslational modifications may confound empirical computer prediction models based upon a distinctly different protein database.

SUMMARY

Immunohistology of calvarial sections revealed that staining with monoclonal anti-osteopontin antibodies (clone MPIIIB10) is minimal unless sections are first treated with EDTA. In contrast, following treatment of sections with EDTA, strong staining of mineralizing osteoid areas and osteoblast-like cells was noted (FIG. 1B). Immunostaining for osteopontin appeared to be specific in that controls which substituted rabbit IgG or normal mouse ascites fluid for monoclonal antibody, or which omitted monoclonal antibody uniformly gave background results (FIG. 1C). In an effort to circumvent problems of antibody accessibility we examined the immunoreactivity of OP when adsorbed to plastic and hydroxyapatite surfaces. Although OP bound to plastic surfaces is reactive with MPIIIB10 antibodies, OP adsorbed to hydroxyapatite crystal surfaces is not recognized by these antibodies as assessed by two detection methods. These results demonstrate that most or all of OP bound to hydroxyapatite exhibits a different conformation than when bound to plastic surfaces. On the basis of immunohistologic results with calvarial sections, we suggest that the conformation of native OP in bone and of isolated OP adsorbed to hydroxyapatite may be similar. Finally, solution circular dichroism and Fourier-transformed infrared spectroscopic studies indicate that the conformation of bone OP is dependent upon its concentration, and, secondarily to the presence or absence of calcium ion. With both spectroscopic methods, addition of calcium appeared to increase the extent of disordered structure. We suggest that these findings support our hypothesis that bone matrix proteins exhibit a different conformation when adsorbed on hydroxyapatite crystal surfaces. Assumption of a more organized secondary structure in concentrated OP solutions (*i.e.*, 15 mg/ml) is consistent with these results in that local concentrations of OP within a semisolid matrix may approach or exceed levels used here.

ACKNOWLEDGMENTS

The authors wish to thank Ms. Kathy M. Brown for her excellent technical and photographic assistance, and Ms. Karen Jensen, NIH Developmental Studies Hybridoma Bank, University of Iowa, for supplying anti-osteopontin antibody (clone MPIIIB10). JPG wants to especially thank Drs. Ron MacQuarrie and Marino Martinez-Carrion for their continued support of this research.

REFERENCES

1. FISHER, L. W., S. W. WHITSON, L. V. AVIOLI & J. D. TERMINE. 1983. Matrix sialoprotein of developing bone. J. Biol. Chem. **258:** 12723–12727.
2. FRANZEN, A. & D. HEINEGÅRD. 1985. Isolation and characterization of two sialoproteins present only in bone calcified matrix. Biochem. J. **232:** 715–724.
3. PRINCE, C. W., T. OOSAWA, W. T. BUTLER, M. TOMANA, A. S. BHOWN, M. BHOWN & R. E. SCHROHENLOHER. 1987. Isolation, characterization, and biosynthesis of a phosphorylated glycoprotein from rat bone. J. Biol. Chem. **262:** 2900–2907.
4. GORSKI, J. P. 1992. Acidic phosphoproteins from bone matrix: a structural rationalization of their role in biomineralization. Calcif. Tissue Int. **50:** 391–396.
5. CHEN, Y., B. S. BAL & J. P. GORSKI. 1992. Calcium and collagen binding properties of osteopontin, bone sialoprotein, and bone acidic glycoprotein-75 from bone. J. Biol. Chem. **267:** 24871–24878.
6. SINGH, K., D. DEIONARINE, V. SHANMUGAM, D. R. SENGER, A. B. MUKHERJEE, P.L. CHANG, C. W. PRINCE & B. B. MUKHERJEE. 1993. Calcium-binding properties of osteopontin derived from non-osteogenic sources. J. Biochem. **114:** 702–707.
7. SHIRAGA, H., W. MIN, W. J. VANDUSEN, M. D. CLAYMAN, D. MINER. C. H. TERRELL, J. R. SHERBOTIE, J. W. FOREMAN, C. PRZYSIECKI, E. G. NEILSON & J. R. HOYER. 1992. Inhibition of calcium oxalate crystal growth *in vitro* by uropontin: another member of the aspartic acid-rich protein superfamily. Proc. Natl. Acad. Sci. USA **89:** 426–430.
8. KOHRI, K., S. NOMURA, Y. KITAMURA, T. NAGATA, K. YOSHIOKA, M. IGUCHI, T. YAMATE, T. UMEKAWA, Y. SUZUKI, H. SINOHARA & T. KURITA. 1993. Structure and expression of mRNA encoding urinary stone protein (uropontin). J. Biol. Chem. **268:** 15180–15184.

9. HUNTER, G. K., C. L. KYLE & H. A. GOLDBERG. 1994. Modulation of crystal formation by bone phosphoproteins: structural specificity of the osteopontin-mediated inhibition of hydroxyapatite formation. Biochem. J. **300:** 723–728.
10. BOSKEY, A. L., M. MARESCA, W. ULLRICH, S. B. DOTY, W. T. BUTLER & C. W. PRINCE. 1993. Osteopontin-hydroxyapatite interactions *in vitro:* inhibition of hydroxyapatite formation and growth in a gelatin gel. Bone Miner. **22:** 147–159.
11. MCKEE, M. D., A. NANCI, W. J. LANDIS, Y. GOTOH, L. C. GERSTENFELD & M. J. GLIMCHER. 1990. Developmental appearance and ultrastructural immunolocalization of a major 66 kDa phosphoprotein in embryonic and post-natal chicken bone. Anat. Rec. **228:** 77–92.
12. HULTENBY, K., F. P. REINHOLT, A. OLDBERG & D. HEINEGÅRD. 1991. Ultrastructural immunolocalization of osteopontin in metaphyseal and cortical bone. Matrix **11:** 206–213.
13. MCKEE, M. D., M. C. FARACH-CARSON, W. T. BUTLER, P. V. HAUSCHKA & A. NANCI. 1993. Ultrastructural immunolocalization of non-collagenous (osteopontin and osteocalcin) and plasma (albumin and α_2 HS-glycoprotein) proteins in rat bone. J. Bone Miner. Res. **8:** 485–496.
14. MERRY, K., R. DODDS, A. LITTLEWOOD & M. GOWEN. 1993. Expression of osteopontin mRNA by osteoclasts and osteoblasts in modelling adult human bone. J. Cell Sci. **104:** 1013–1020.
15. GIACHELLI, C. M., N. BAE, M. ALMEIDA, D. T. DENHARDT, C. E. ALPERS & S. M. SCHWARTZ. 1993. Osteopontin is elevated during neointima formation in rat arteries and is a novel component of human atherosclerotic plaques. J. Clin. Invest. **92:** 1686–1696.
16. GORSKI, J. P., D. GRIFFIN, G. DUDLEY, C. STANFORD, R. THOMAS, C. HUANG, E. LAI, B. KARR & M. SOLURSH. 1990. Bone acidic glycoprotein-75 is a major synthetic product of osteoblastic cells and localized as 75- and/or 50-kDa forms in mineralized phases of bone and growth plate and in serum. J. Biol. Chem. **265:** 14956–14963.
17. ARTIGUES, A., A. IRIARTE & M. MARTINEZ-CARRION. 1994. Acid-induced reversible unfolding of mitochondrial aspartate aminotransferase. J. Biol. Chem. **269:** 21990–21999.
18. OLDBERG, A., A. FRANZEN & D. HEINEGÅRD. 1986. Cloning and sequence analysis of rat bone sialoprotein (osteopontin) cDNA reveals an Arg-Gly-Asp cell-binding sequence. Proc. Natl. Acad. Sci. USA **83:** 8819–8823.
19. OLDBERG, A., A. FRANZEN, D. HEINEGÅRD, M. PIERSCHBACHER & E. ROUSLAHTI. 1988. Identification of a bone sialoprotein receptor in osteosarcoma cells. J. Biol. Chem. **263:** 19433–19436.
20. PRINCE, C. W. 1989. Secondary structure predictions for rat osteopontin. Connect. Tissue Res. **21:** 15–20.

Effects of Okadaic Acid on Calcitriol- and Phorbol Ester-Induced Expression and Phosphorylation of Osteopontin in Mouse JB6 Epidermal Cells[a]

PI-LING CHANG, WEI-CHUNG YANG,
AND CHARLES W. PRINCE[b]

Department of Nutrition Sciences
326 Susan Mott Webb Nutrition Sciences Building
University of Alabama at Birmingham
1675 University Boulevard
Birmingham, Alabama 35294-3360

INTRODUCTION

The transformation of a normal cell into a tumor cell typically involves a process that can be summarized by three steps: initiation, promotion and progression. The JB6 family of clonal mouse epidermal cells developed by Colburn and co-workers includes clonal lines that are promotion-resistant (P−), promotion-sensitive (P+) or irreversibly transformed (Tx).[1,2] The P+ lines are initiated but not transformed cells and thus provide good models for studying the molecular events that occur during the promotion phase of tumorigenesis. In these cells the phorbol ester tumor promoter 12-*O*-tetradecanoylphorbol-13-acetate (TPA)-induces an irreversible tumorigenic transformation. Such transformed cells form tumors in nude mice and form large colonies when grown in soft agar.

Several years ago Smith and Denhardt[3] discovered an mRNA that was greatly induced by treating one of the P+ cell lines (JB6 Cl22) with TPA. Through the work of Craig, Denhardt and others this mRNA was found to code for the phosphoprotein osteopontin (OPN).[4] Expression of OPN, formerly called transformation-dependent secreted phosphoprotein by Senger,[5] is now known to be linked to tumorigenic transformation of many cell types. Furthermore, a causal role for OPN expression in not only tumorigenesis but also metastasis is suggested by two recent studies which showed that expression of OPN antisense could partially or completely abrogate the ability of normally metastatic cells to metastasize or form tumors *in vivo*.[6,7] Thus, OPN expression during promotion could be a necessary step in tumorigenesis. Our work reported here is directed at understanding how OPN expression and phosphorylation is regulated in the JB6 model of tumor promotion.

[a] The research reported here was supported by NIH grants DE06739 and DE00247 to CWP.
[b] Corresponding author.

Calcitriol Effects in JB6 Cells

We, as well as others, previously showed that phosphorylated OPN expression in bone cells can be enhanced by calcitriol, the hormonal form of vitamin D_3.[8] This increase in expression by bone cells appears to be mediated by activation of transcription of the OPN gene. The promoter of the OPN gene contains one and possibly two active vitamin D response elements (VDREs) that are responsible for regulating OPN expression in the presence of calcitriol.[9,10,13]

Because Hosoi *et al.*[11] showed that calcitriol, in a metabolite-specific, dose-dependent and receptor-mediated manner, could mimic the effects of TPA by transforming the parental, nonclonal JB6 cells, we asked if calcitriol could induce expression of phosphorylated OPN and, concurrently, transform one of the P+ clonal lines of JB6 cells (JB6 Cl41.5a).

Cell Proliferation

We first examined the ability of calcitriol to modulate cell proliferation.[12] Cells were seeded and grown for two days prior to the addition of calcitriol at concentrations of 0.1–100 ng/ml. Twenty-four hours of treatment did not significantly affect proliferation. However, 48 hours of treatment with more than 0.1 ng/ml of calcitriol showed a significant dose-dependent reduction in proliferation of these cells (FIG. 1).

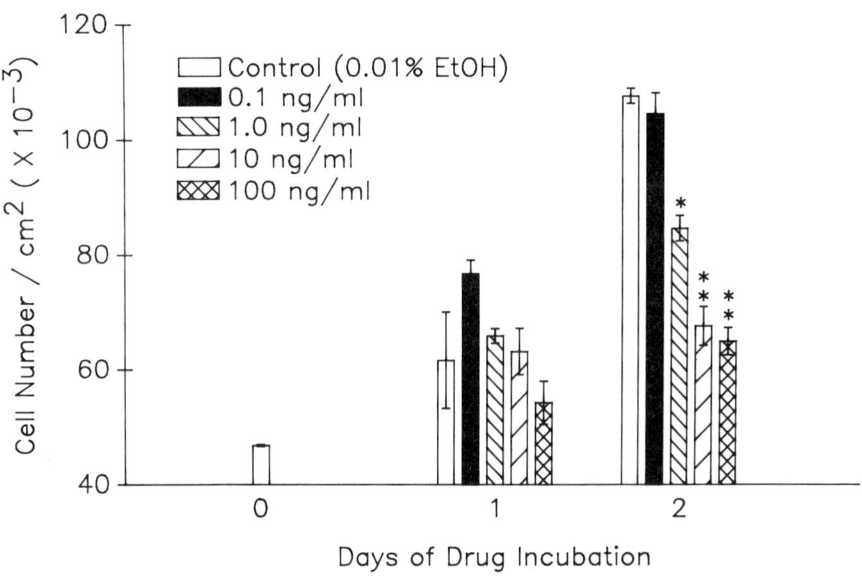

FIGURE 1. Growth of JB6 Cl41.5a cells treated with varying doses of calcitriol. Cells were plated at 20,000/cm^2 and grown for two days. Calcitriol was added, day 0, and the cells grown for two additional days. Cells were harvested by trypsin treatment and counted with a Coulter Counter. Values are means ± standard error of the mean. *, $p < 0.05$; **, $p < 0.01$ versus control at day 2. (From Chang & Prince.[12] Reprinted by permission from *Cancer Research.*)

Expression of OPN Protein and mRNA

To detect expression of phosphorylated OPN, cells were treated with 10 ng/ml of calcitriol or 10 ng/ml of TPA (a positive control) for 24 hours. After incubation in phosphate-free medium for 1 h, cells were incubated with $^{32}PO_4$ for 3 h. The medium was removed and proteins recovered by TCA precipitation and analyzed by 10% SDS-PAGE and autoradiography. No OPN could be detected except in the samples from TPA-treated cells (FIG. 2A, lanes 1–3). Immunoadsorption analyses of the same samples with a polyclonal antibody against rat bone OPN yielded similar results (data not shown).[12] These findings confirmed the results of Craig *et al.* that OPN is the major phosphoprotein secreted by TPA-treated JB6 cells.[4]

Similar experiments were done except proteins were incubated with ^{35}S-methionine to label newly synthesized proteins and OPN was identified by immunoadsorption, SDS-PAGE and fluorography. The immunoadsorption assays showed that calcitriol did in fact induce OPN expression (FIG. 2B, lanes 5–7) in a dose-dependent, metabolite-specific manner with a maximal increase of about 6-fold.[12]

To confirm these findings of an apparent increase in OPN expression, we examined the induction of OPN mRNA by calcitriol using Northern blot analyses. We found a 10-fold increase in OPN mRNA levels above the low basal levels seen in untreated cells[13] (FIG. 2C, lanes 1–3). Thus, calcitriol did induce OPN expression although the secreted protein was not phosphorylated, in contrast to the OPN induced by TPA.

Induction of Anchorage-Independent Growth

To evaluate the ability of calcitriol to transform the clonal JB6 Cl41.5a cells, we used an assay for anchorage-independent growth. Previous studies by Colburn and co-workers verified that growth of JB6 cells in soft agar is a reliable means for detecting their transformation.[1] Cells were incubated for 14 days in soft agar with calcitriol at doses of 0.05–500 ng/ml. The same concentrations of TPA were positive controls, and ethanol and DMSO were negative controls. As expected TPA exhibited a dose-dependent ability to induce colony formation—the maximal effect was seen at 5 ng/ml. Almost no colonies were seen in calcitriol-treated plates. In fact, these plates contained fewer colonies than the solvent control plates (FIG. 3).[12]

To summarize these studies, we found that calcitriol at up to 100 ng/ml inhibited the proliferation of JB6 Cl41.5a cells. Calcitriol induced OPN mRNA expression and secretion of nonphosphorylated OPN but did not induce anchorage-independent growth. This is in contrast to TPA, which induced cell proliferation, expression of OPN mRNA and phosphorylated OPN protein and also caused induction of anchorage-independent growth of these JB6 cells.

Effects of Calcitriol + TPA in JB6 Cells

Because of calcitriol's apparent inhibitory effects on JB6 cell proliferation and the fact that almost no colonies were formed when cells were incubated with calcitriol, we asked if calcitriol could inhibit the effects of TPA with respect to cell proliferation and transformation. We therefore assayed the combination of

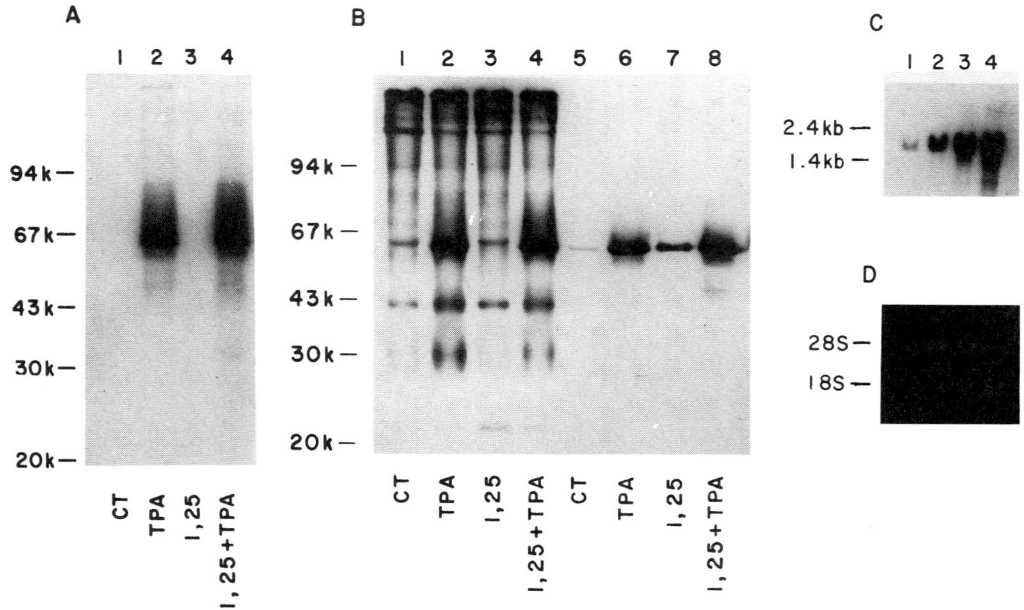

FIGURE 2. Expression of OPN protein and mRNA induced by TPA, calcitriol or TPA + calcitriol. **(A)** Autoradiogram of 10% SDS-PAGE of $^{32}PO_4$-labeled OPN secreted by JB6 Cl41.5a treated with TPA, calcitriol, or TPA with calcitriol for 24 h. TCA-precipitated proteins were run on 10% SDS-PAGE. *Lane 1,* 0.001% DMSO; *lane 2,* 10 ng/ml TPA; *lane 3,* 10 ng/ml calcitriol; *lane 4,* TPA and calcitriol both at 10 ng/ml. **(B)** Fluorogram of 10% SDS-PAGE of [^{35}S]methionine-labeled proteins secreted by JB6 Cl41.5a treated with TPA, calcitriol, or both drugs for 24 h. *Lanes 1–4* are TCA-precipitated proteins and *lanes 5–8* are immunoadsorbed OPN. 0.001% DMSO (*lanes 1 and 5*); 10 ng/ml TPA (*lanes 2 and 6*); 10 ng/ml calcitriol (*lanes 3 and 7*); TPA and calcitriol, both at 10 ng/ml (*lanes 4 and 8*). **(C)** Autoradiogram of Northern blot from JB6 Cl41.5a cells treated with TPA, calcitriol, or both drugs for 24 h. Cells were incubated with 0.001% DMSO (*lane 1*), 10 ng/ml calcitriol (*lane 2*), 10 ng/ml TPA (*lane 3*), or TPA and calcitriol, both at 10 ng/ml (*lane 4*). 10 μg of total RNA from each sample was run on a denatured agarose gel, transferred to Gene Screen *Plus* membrane, and probed with [Â-^{32}P]dCTP-labeled OPN cDNA (5.2×10^8 cpm/μg). **(D)** Ethidium bromide staining of ribosomal RNA from the membrane described in (C). (From Chang & Prince.[14] Reprinted by permission from *Cancer Research*.)

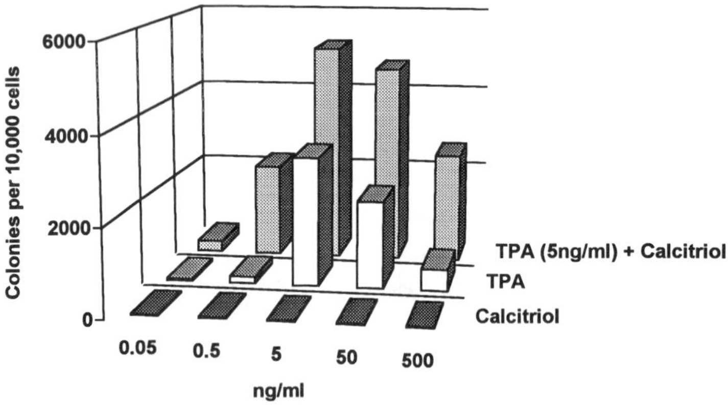

FIGURE 3. Anchorage-independent growth of JB6 Cl41.5a cells treated with calcitriol, TPA or TPA + calcitriol. 10,000 cells were suspended in 0.3% agar and layered on a base layer of 0.5% agar with both layers containing calcitriol (0.05–500 ng/ml), TPA (0.05–500 ng/ml) or TPA (5 ng/ml) plus calcitriol (0.05–500 ng/ml). Cells were incubated for 14 days at 37°C and then colonies of at least 10 cells were counted. Each treatment was done in triplicate.

both drugs for its ability to modulate cell proliferation, anchorage-independent growth and, also, OPN expression.

Cell Proliferation

Treatment of cells with 10 ng/ml of TPA or calcitriol for 2–3 days stimulated and inhibited, respectively, cell proliferation as expected. The combination of the two drugs resulted in less proliferation than in cells treated with TPA but more proliferation than in control cells. Thus, calcitriol did somewhat diminish the stimulatory effect of TPA on JB6 cells in monolayer culture.[14]

Induction of Anchorage-Independent Growth

Incubation of cells in soft agar for 14 days with 0.05–500 ng/ml of either TPA or calcitriol yielded the expected results, *i.e.*, dose-dependent induction of colonies by TPA and no colonies induced by calcitriol. The combination of 5 ng/ml of TPA and calcitriol at concentrations of 0.05–500 ng/ml did not result in inhibition of colony formation, as hypothesized, rather a 2–10-fold increase, depending on dosages, was seen (FIG. 3). Also, besides being more numerous, the colonies formed from the combination treatment were larger than those formed from just TPA treatment alone. Thus calcitriol significantly enhanced the ability of TPA to induce transformation of these cells.[14]

Expression of OPN Protein and mRNA

To determine if increased expression of OPN by JB6 cells treated with the combination of calcitriol and TPA coincided with this unexpected increase in colony formation, we used $^{32}PO_4$, ^{35}S-methionine labeling and Northern blot analyses (see FIG. 2A, lane 4; 2B, lanes 4, 8; 2C, lane 4, respectively). Twenty-four hour treatment with the combination of 10 ng/ml of TPA and calcitriol produced an increase in OPN mRNA and in total secreted OPN. OPN produced by cells treated with the combination of drugs was phosphorylated to a greater extent than OPN produced by cells treated with TPA alone. The increase in total secreted OPN and OPN mRNA was greater than additive (TABLE 1).[14] Thus, calcitriol not only enhanced TPA-induced colony formation, it also synergistically increased OPN expression. In fact, the extent of colony formation was well correlated with secretion of phosphorylated, but not nonphosphorylated, OPN (TABLE 1). Thus, we have found that calcitriol can enhance the tumorigenic transformation of JB6 Cl41.5a cells that is induced by TPA. Further, the extent of transformation correlates with the amount of phosphorylated OPN that is secreted by the JB6 cells.

Role of Okadaic Acid in OPN Expression and Phosphorylation

Since modulation OPN phosphorylation seemed to be a good predictor for the extent of transformation in this line of JB6 cells, we examined the role that phosphatases may have in regulating the phosphorylation of OPN and the extent of transformation of cells. The central question addressed was "will inhibition of protein serine/threonine phosphatases increase the extent to which OPN is phosphorylated in calcitriol- or TPA-treated cells and, if so, will the increased phosphorylation coincide with increased anchorage-independent growth?."

Phosphorylation and Expression of OPN

The main reagent used in these studies was okadaic acid (OKA), a polyether fatty acid that is a potent inhibitor of both protein phosphatase 1 (PP1, IC_{50} = 10 nM) and protein phosphatase 2A (PP2A, IC_{50} = 0.1 nM), two of the most common serine/threonine phosphatases in mammalian cells.[15] We plated JB6 cells, grew them for two days and then treated them with various combinations of OKA, calcitriol or TPA for 9–48 h. Nascent proteins were radiolabeled with $^{32}PO_4$, or ^{35}S-methionine and examined by SDS-PAGE on 10% polyacrylamide gels.

TABLE 1. Correlation of Phosphorylated OPN Expression with Induction of Anchorage-Independent Growth in Mouse JB6 Cells[a]

	Control	TPA	Calcitriol	TPA + Calcitriol
Relative level of OPN				
Protein	1	32	6	56
mRNA	1	38	10	70
Phosphorylation	—	+ +	—	+ + +
Extent of induction of anchorage-independent growth	<1%	30%	<1%	50%

[a] Cells were treated with TPA and/or calcitriol at 10 ng/ml; control cells were treated with 0.001% DMSO.

Preliminary studies were done to insure that the concentrations of OKA used in these studies were not toxic to the JB6 cells.

We consistently failed to observe significant increases in $^{32}PO_4$ incorporation into OPN when cells were treated with OKA alone (using from 5–50 nM) or in combination with calcitriol (10 ng/ml), suggesting that dephosphorylation of OPN by PP2A and/or PP1 is not involved in the secretion of the nonphosphorylated form of OPN induced by calcitriol. However, treatment of cells with the combination of 10 ng/ml TPA and 50 nM OKA for 16 h resulted in slightly increased $^{32}PO_4$ incorporation into OPN compared to that observed in cells treated with only TPA. No such increase was seen when cells were similarly treated for 9 h (compare FIG. 4A with 4B). However, increased OPN synthesis rather than increased phosphorylation seems to be responsible for these results. Comparable experiments using ^{35}S-methionine as the tracer also showed an increase in the amount of OPN secreted from cells treated with TPA + OKA compared to TPA-treated cells (FIG. 5A and 5B). Thus, OKA appeared to be inducing an increase in the amount of OPN protein present in TPA-treated cells. In JB6 cells treated for 48 h with 5 ng/ml TPA and 5 nM OKA versus 5 ng/ml TPA alone, similar increases were observed in $^{32}PO_4$ incorporation and ^{35}S-methionine-labeling of OPN. Since 5 nM OKA is only one-half of the IC_{50} for PP1 but is substantially greater than the IC_{50} of PP2A, this result suggests that PP2A rather than PP1 is somehow involved in regulating expression of the OPN gene in TPA-treated cells.

To confirm that OKA could elevate OPN expression, we did Northern blot analyses of OPN mRNA expression. We found that 24-h treatment of cells with 50 nM OKA + 10 ng/ml TPA increased OPN expression to a greater extent than TPA treatment alone (FIG. 6). In fact 50 nM OKA alone induced OPN mRNA expression about 4–5-fold over that of control cells. Examination of SDS-PAGE data (see FIGS. 4B and 5B) shows that this OPN is largely nonphosphorylated. This level of enhancement was the same as that seen with 10 ng/ml calcitriol treatment. The combination of OKA and calcitriol did not result in increased expression of OPN mRNA.

Thus, OKA appears to regulate the expression of OPN in JB6 Cl41.5a cells rather than the extent of OPN phosphorylation. When acting alone, OKA induces nonphosphorylated OPN, but when used in combination with TPA, the OPN is phosphorylated. These findings are very similar to the effects of calcitriol on OPN expression in these cells, though the mechanisms of induction are no doubt different. Since phosphorylated OPN is induced by TPA and nonphosphorylated OPN is induced by calcitriol and OKA, it seems likely that TPA also induces one or more kinases that phosphorylate OPN. The specific kinases and their subcellular localization are not known.

Induction of Anchorage-Independent Growth by Okadaic Acid

Fourteen-day incubations of JB6 Cl41.5a cells with 0.1–10 nM OKA failed to induce growth in soft agar. In monolayer cultures, these doses significantly reduced cell proliferation but did not appear to be cytotoxic. This growth inhibition could be overcome by the addition of 5 ng/ml of TPA (data not shown), so we examined the ability of various doses of OKA plus 5 ng/ml TPA to induce anchorage-independent growth. In three separate experiments, each done in triplicate, we found that 0.1, 0.5 and 1.0 nM OKA enhanced TPA-induced soft agar growth by 20–25%, but 5 and 10 nM OKA either had no effect or slightly inhibited TPA-induced soft agar growth by JB6 cells (see FIG. 7).

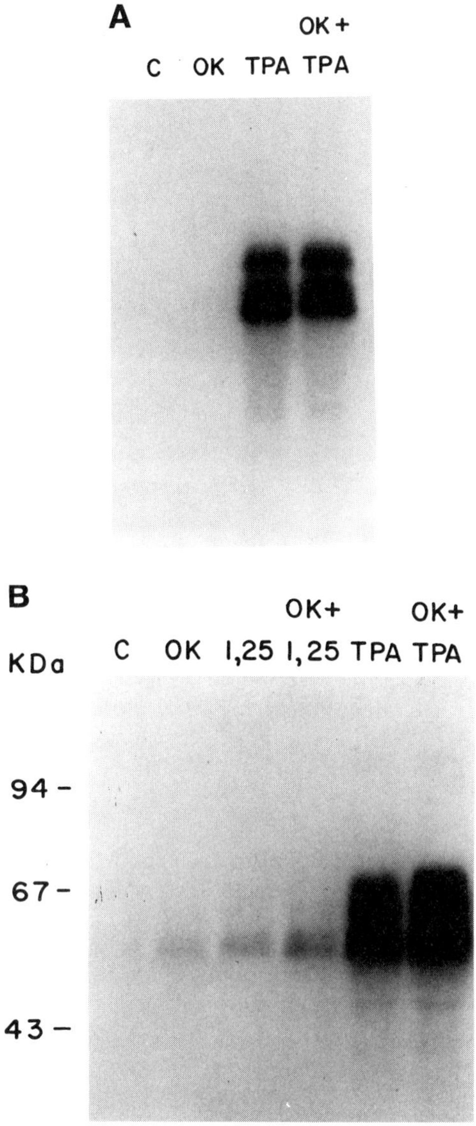

FIGURE 4. Synthesis of phosphorylated OPN by JB6 cells treated with OKA (50 nM), TPA (10 ng/ml), calcitriol (10 ng/ml), calcitriol + OKA or OKA + TPA. Shown are autoradiograms of 10% SDS-PAGE analyses of $^{32}PO_4$-labeled OPN isolated by immunoadsorption from cells treated with drugs **(A)** for 9 hours or **(B)** for 16 hours.

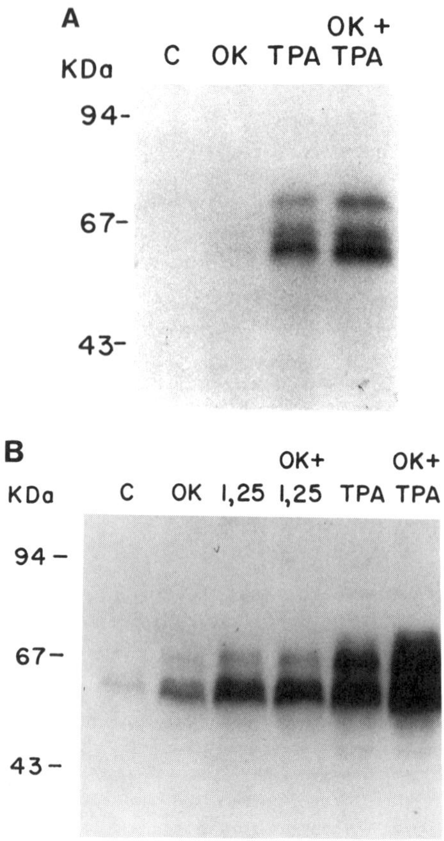

FIGURE 5. Synthesis of ^{35}S-methionine-labeled OPN by JB6 cells. Cells were treated as in FIGURE 4 except that they were labeled with ^{35}S-methionine and immunoadsorbed proteins were detected by fluorography. **(A)** and **(B)** represent drug treatment for 9 and 16 hours, respectively.

CONCLUSIONS

Colburn and co-workers have shown that activator protein 1 complex (AP-1) transcriptional activity is necessary for TPA-mediated transformation of JB6 cells. They found that P+ transfectants expressing a dominant negative *c-jun* mutant, TAM67, which encodes a protein product incapable of activating transcription, were inhibited from undergoing TPA-mediated transformation.[16] Thus, induction of *jun* and *fos* protein products is necessary for TPA to exert its transforming effect.

Furthermore, phosphorylation of *jun* proteins modulates their ability to activate transcription of a number of genes. Particularly phosphorylation of Ser63 and Ser73 appears to be necessary for *jun* proteins to possess transactivation capability.[17] Thus, one mechanism by which OKA-mediated regulation of OPN expression could occur is through modulation of the phosphorylation status of members of

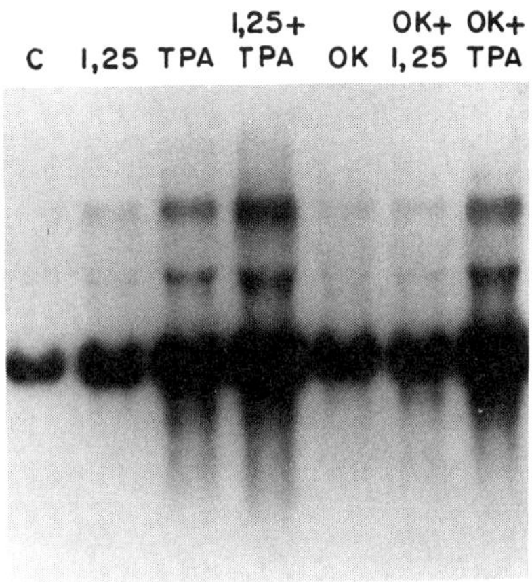

FIGURE 6. Northern blot analysis of OPN mRNA from JB6 cells treated with calcitriol (10 ng/ml), TPA (10 ng/ml), OKA (50 nM) or various combinations of the drugs. Cells were treated for 24 h and RNA isolated and analyzed as described in FIGURE 2C. Methylene blue staining of the membrane indicated equal loading of all lanes.

the *jun* and *fos* families of proteins and thereby regulating the induction of AP-1 activity in the OPN promoter. Perhaps OKA prevents dephosphorylation of Ser63 or Ser73 in the transcription activation domain of *c-jun* and thus enhances OPN expression. Other possibilities do exist, of course, but currently data are lacking to discriminate among them.

More recently Colburn and co-workers showed that induction of AP-1 transcriptional activity is necessary but not sufficient for induction of transformation of JB6 P+ cells.[18] They speculated that additional events, such as those related to cell adhesion or cytoskeletal organization, are also required for transformation of JB6 cells. Given OPN's role in tumorigenesis and its ability to promote adhesion of cells, it is possible that expression of OPN by JB6 P+ cells is also necessary for expression of the transformed phenotype.

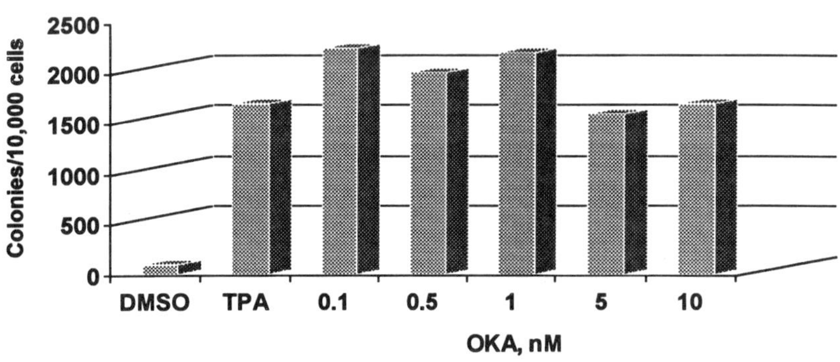

FIGURE 7. Anchorage-independent growth of JB6 Cl41.5a cells treated with TPA or TPA + OKA. 10,000 cells were suspended in 0.3% agar and layered on a base layer of 0.5% agar with both layers containing TPA (5 ng/ml) or TPA (5 ng/ml) plus OKA (0.1–10 nM). Cells were incubated for 14 days at 37°C and then colonies of at least 10 cells were counted. Each treatment was done in triplicate.

REFERENCES

1. Colburn, N. H., B. F. Former, K. A. Nelson & S. H. Yuspa. 1979. Tumour promoter induces anchorage independence irreversibly. Nature **281:** 589–591.
2. Colburn, N. H., B. A. Koehler & K. J. Nelson. 1980. A cell culture assay for tumor-promoter-dependent progression toward neoplastic phenotype: detection of tumor promoters and promotion inhibitors. Teratog. Carcinog. Mutagen. **1:** 87–96.
3. Smith, J. H. & D. T. Denhardt. 1987. Molecular cloning of tumor promoter-inducible mRNA found in JB6 mouse epidermal cells: induction is stable at high, but not at low, cell densities. J. Cell. Biochem. **34:** 13–22.
4. Craig, A. M., J. H. Smith & D. T. Denhardt. 1989. Osteopontin, a transformation-associated cell adhesion phosphoprotein, is induced by 12-O-tetradecanoylphorbol-13-acetate in mouse epidermis. J. Biol. Chem. **264:** 9682–9689.
5. Senger, D. R., C. A. Perruzzi & A. Papadopoulos. 1989. Elevated expression of secreted phosphoprotein I (osteopontin, 2ar) as a consequence of neoplastic transformation. Anticancer Res **9:** 1291–1300.
6. Behrend, E. I., A. M. Craig, S. M. Wilson, D. T. Denhardt & A. F. Chambers. 1994. Reduced malignancy of ras-transformed NIH 3T3 cells expressing antisense osteopontin RNA. Cancer Res. **54:** 832–837.
7. Gardner, H. A. R., B. Berse & D. R. Senger. 1994. Specific reduction in osteopontin synthesis by antisense RNA inhibits the tumorigenicity of transformed Rat1 fibroblasts. Oncogene **9:** 2321–2326.
8. Prince, C. W. & W. T. Butler. 1987. 1,25-Dihydroxyvitamin D_3 regulates the biosynthesis of osteopontin, a bone-derived cell attachment protein, in clonal osteoblast-like osteosarcoma cells. Collagen Relat. Res. **7:** 305–313.
9. Noda, M., R. L. Vogel, A. M. Craig, J. Prahl, H. F. DeLuca & D. T. Denhardt. 1990. Identification of a DNA sequence responsible for binding of the 1,25-dihydroxyvitamin D_3 receptor and 1,25-dihydroxyvitamin D_3 enhancement of mouse secreted phosphoprotein 1 (Spp-1 or osteopontin) gene expression. Proc. Natl. Acad. Sci. USA **87:** 9995–9999.
10. Craig, A. M. & D. T. Denhardt. 1991. The murine gene encoding secreted phosphoprotein 1 (osteopontin): promoter structure, activity, and induction *in vivo* by estrogen and progesterone. Gene **100:** 163–171.
11. Hosoi, J., E. Abe, T. Suda, N. H. Colburn & T. Kuroki. 1986. Induction of anchorage-independent growth of JB6 mouse epidermal cells by 1α,25-dihydroxyvitamin D_3. Cancer Res. **46:** 5582–5586.
12. Chang, P. L. & C. W. Prince. 1991. 1α,25-dihydroxyvitamin D_3 stimulates synthesis and secretion of nonphosphorylated osteopontin (secreted phosphoprotein 1) in mouse JB6 epidermal cells. Cancer Res **51:** 2144–2150.
13. Chang, P. L., A. L. Ridall & C. W. Prince. 1994. Calcitriol regulation of osteopontin expression in mouse epidermal cells. Endocrinology **135:** 863–869.
14. Chang, P. L. & C. W. Prince. 1993. 1α,25-dihydroxyvitamin D_3 enhances 12-O-tetradecanoylphorbol-13-acetate-induced tumorigenic transformation and osteopontin expression in mouse JB6 epidermal cells. Cancer Res. **53:** 2217–2220.
15. Cohen, P., C. F. B. Holmes & Y. Tsukitani. 1990. Okadaic acid: a new probe for the study of cellular regulation. Trends Biochem. Sci. **15:** 98–102.
16. Dong, Z., M. J. Birrer, R. G. Watts, L. M. Matrisian & N. H. Colburn. 1994. Blocking of tumor promoter-induced AP-1 activity inhibits induced transformation in JB6 mouse epidermal cells. Proc. Natl. Acad. Sci. USA **91:** 609–613.
17. Adler, V., C. C. Franklin & A. S. Kraft. 1992. Phorbol esters stimulate the phosphorylation of c-Jun but not v-Jun: regulation by the N-terminal δ domain. Proc. Natl. Acad. Sci. USA. **89:** 5341–5345.
18. Dong, Z., J. L. Cmarik, E. J. Wendel & N. H. Colburn. 1994. Differential transformation effiency but not AP-1 induction under anchorage-dependent and -independent conditions. Carcinogenesis **15:** 1001–1004.

Gene Expression and Phosphorylation of Mouse Osteopontin[a]

RAUL A. SAAVEDRA,[b,c,d] SEAN K. KIMBRO,[c,d]
DORIS N. STERN,[c,d,e] JAMIE SCHNUER,[d]
SAMY ASHKAR,[c,d] MELVIN J. GLIMCHER,[c,d]
AND CECILIA I. LJUBETIC[d]

[c]*Department of Orthopaedic Surgery*
Harvard Medical School
Boston, Massachusetts 02115

[d]*Laboratory for the Study of Skeletal Disorders*
and Rehabilitation
The Children's Hospital
Enders Pediatric Research Center
300 Longwood Avenue
Boston, Massachusetts 02115

[e]*Forsyth Dental Center*
Boston, Massachusetts 02115

INTRODUCTION

Osteopontin, a secreted glycosylated phosphoprotein of Mr ~66,000, is rich in acidic amino acids and sialic acid, and it is expressed in various tissues, including bone,[1-3] cartilage,[3,4] kidney,[2,5] decidua and placenta,[4,5] smooth muscle cells of arteries,[6] and cells of the immune system.[7-9] Osteopontin is also expressed by carcinomas and sarcomas[10-12] and by mouse epidermal cells treated with 12-O-tetradecanoylphorbol-13-acetate.[13,14] The primary sequence of osteopontin from various mammalian species exhibits several motifs, such as a cell surface-attachment sequence (*Arg-Gly-Asp*), numerous phosphorylation sites for various protein kinases, a polyaspartate motif, an N- and several O-glycosylation sites, three thrombin cleavage sites, and two heparin binding domains.[9,15]

The tissue localization, primary structure, and theoretical considerations have led to the notion that osteopontin is involved in various extracellular functions. These functions include the initiation and localization of apatite crystals in bone,[3,16,17] the regulation of calcium and phosphate,[9,16] the attachment of normal osteoclasts to the bone surface during the process of bone remodeling[18] and of neoplastic cells to the extracellular matrix,[11] and the immune response.[7-9]

The many cell types that express osteopontin and the various functions proposed for this protein raise the questions of whether there are multiple forms of

[a] This work was supported by grants IR29AR41046 and PO1AR34078 from the National Institutes of Health (NIH) to RAS and MJG, respectively, Grant RG24458A2 from the National Multiple Sclerosis Society, a grant from the Milton Fund (Harvard Medical School) to RAS, Training Grant T32ARO7112 to DNS, and NIH Fellowship F32AR08302 to KSK.

[b] Address correspondence to: Dr. Raul A. Saavedra, The Children's Hospital, Enders Pediatric Research Center, 300 Longwood Avenue, Boston, MA 02115.

this protein, and what are the mechanisms that the cells use to originate these forms. Two general mechanisms are used by cells to generate distinct but related forms of a protein: alternative splicing of transcripts and posttranslational modifications. In this article, we discuss studies on mouse osteopontin that address these two questions.

Expression of Osteopontin during Mouse Development

In order to better understand the function of osteopontin, we have initiated a project designed to determine the tissue localization of osteopontin during mouse development by immunocytochemical techniques. Polyclonal antibodies against mouse osteopontin were generated at the Animal Research Facility (Children's Hospital, Boston) using a recombinant form of the protein expressed in *Escherichia coli* (see below).[19] These antibodies are expected to recognize epitopes located on the polypeptide backbone of osteopontin. Swiss-Webster mouse embryos were removed at 14, 15, 16, and 18 days of gestation. A 1.5-day-old neonate mouse was also dissected. Embryos and neonate tissues were fixed with paraformaldehyde, or paraformaldehyde/glutaraldehyde, rinsed in phosphate buffered saline (PBS) solution, washed in PBS containing sucrose, and sectioned in a cryostat. The tissue sections were incubated first with the anti-osteopontin antibody and subsequently with an anti-rabbit IgG antibody labelled with horseradish peroxidase. We found osteopontin in developing bone, cartilage, kidney, and epithelial tissue, as reported by other researchers[3,4,20] (see Stern *et al.,* this volume). We also found osteopontin in the notochord of the 14-day-old mouse embryo (FIG. 1). At this stage of development, the notochord undergoes significant morphological changes which include the enlargement of those regions located between the future vertebrae, at the sites where the intravertebral disks will be formed. The enlarged regions of the notochord originate the future *nuclei pulposi* of the intervertebral disks. Our observation suggests that osteopontin is involved in the morphogenesis of the notochord in a manner that remains to be determined.

Analysis of RNA Transcripts from the Mouse Osteopontin Gene at Different Stages of Development and in Various Tissues

Mouse osteopontin is encoded by a single-copy gene located on chromosome 5.[9,15] We designed a set of experiments to determine whether various tissues and cell lines from mouse that express osteopontin use the mechanism of alternative splicing of transcripts from a single copy gene to generate distinct but related forms of the protein. Total RNA from mouse embryos at 8, 10, and 14 days, adult mouse kidney, placenta, and MC3T3-E1 osteoblast-like,[21] MGH-OGS osteosarcoma,[22] and NIH3T3 mouse cells were prepared. These RNAs were used in combination with primer BO2 (complementary to the 3′-untranslated region of exon VII, near the termination codon; TABLE 1)[14] and reverse transcriptase to generate the first strands of cDNAs. The resulting cDNAs were expected to contain the entire coding region of mouse osteopontin. These cDNAs were used as templates in combination with different pairs of opposite-orientation oligonucleotides complentary to various sequences located within or adjacent to the coding region of osteopontin (see TABLE 1) in polymerase chain reactions (PCRs). The oligonucleotides complementary to the 5′-side were BO3, BO4, or BO6, whereas the oligonucleotides complementary to the 3′-side were BO2, BO5, or BO7. The

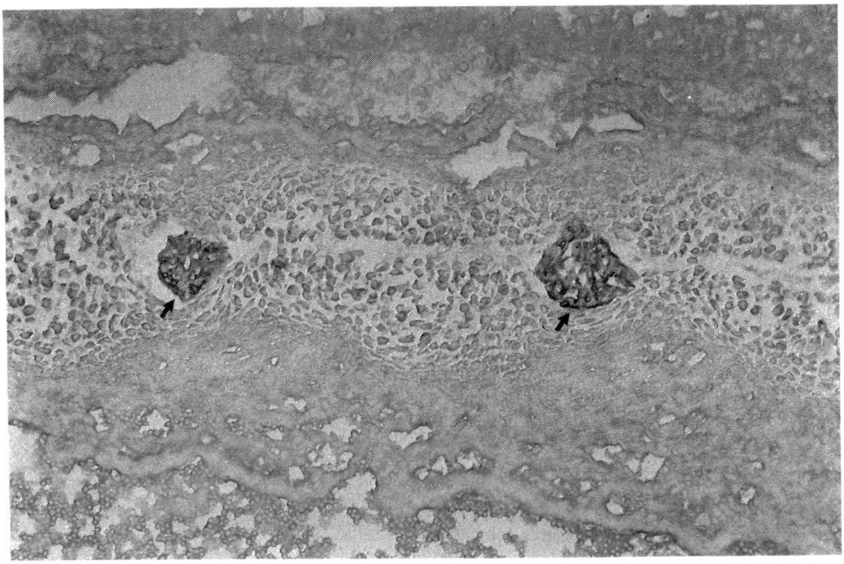

FIGURE 1. Osteopontin is found in the notochordal segments of a 14-day-old mouse embryo. Embryos were fixed in PBS, pH 7.4, containing paraformaldehyde, rinsed in PBS solution, followed by washes in PBS containing increasing concentrations (up to 30%) of sucrose. Fixed tissues were placed into TISSUE-TEK O.C.T., and eight- to ten-micrometer sections were cut on a cryostat and mounted on slides. The tissue sections were incubated first with a polyclonal antibody made in rabbit against mouse recombinant osteopontin, and then with an anti-rabbit IgG antibody labelled with horseradish peroxidase. Dorsal is *top* of micrograph. *Solid arrows* indicate the notochordal segments. Magnification = 200×.

PCRs were performed in the presence of ^{32}P-dCTP and the products analyzed in 4% high-resolution nondenaturing polyacrylamide gels. The results showed that all the tissues and cell lines expressed only transcripts which contained the complete coding region of mouse osteopontin (exons II through VII). FIGURE 2 shows results obtained with total RNA from MC3T3-E1 osteoblast-like, MGH-OGS osteosarcoma, and NIH3T3 mouse cells. Our results indicate that in all the tissues and cell lines from mouse examined, there is undetectable alternative splicing of

TABLE 1. Sequences of the Oligonucleotides Used in This Work[14]

Oligonucleotide	Sequence	Complementary to Exon
BO2	5′-CCACTGAACTGAGAAATGAGC-3′	VII
BO3	5′-GCGGCAGGCATTCTCGGAGGAAACC-3′	I
BO4	5′-CGACCATGAGATTGGCAGTGATTTG-3′	II
BO5	5′-CCAGGCTGGCTTTGGAACTTGC-3′	VII
BO6	5′-AGCCACAAGTTTCACAGCCACAAG-3′	VII
BO7	5′-GCTTCTGAGATGGGTCAGGCACCAG-3′	IV
PO1	5′-ATCACTGCCAATCTCATGGTCGTAG-3′	II

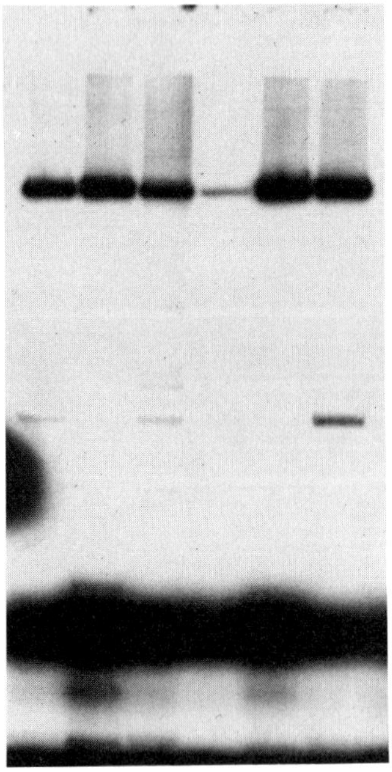

FIGURE 2. Only transcripts that contain the entire coding region (exons II through VII) of osteopontin are expressed by various cell lines from mouse. Total RNA and oligonucleotide BO2 (complementary to exon VII; see TABLE 1) were used to generate cDNAs. The cDNAs were used as templates in combination with oligonucleotides BO4 (complementary to exon II) and BO2 as primers, and ^{32}P dCTP in PCRs. 2ar, a plasmid that contained the entire coding region of mouse osteopontin,[14] was used as control template. The amplification products were separated by 4% polyacrylamide high-resolution gel electrophoresis. The expected size of the largest possible amplification product was 996 base pairs. The thick and diffused band seen in all lanes, near the *bottom* of the figure, was an artifact. *Lane 1,* MGH-OGS osteosarcoma mouse cells; *lane 2,* 2ar plasmid; *lane 3,* MGH-OGS cells treated with phorbol ester; *lane 4,* NIH3T3 mouse cells; *lane 5,* 2ar plasmid; *lane 6,* MC3T3-E1 osteoblast-like mouse cells. Similar results were obtained using RNAs from kidney, uterus, or placenta from mouse (not shown; see text).

coding exons. These results differ from those obtained in human, where two classes of osteopontin transcripts that result from the alternative splicing of exon V and code for two distinct but related forms of the protein have been detected.[23,24]

Two Classes of Osteopontin Transcripts that Differ at Their 5′-Untranslated Regions

We also investigated whether, in the mouse, there were osteopontin transcripts that differed from each other at sequences other than within the coding region. The osteopontin gene has a promoter sequence located immediately upstream from the untranslated exon I.[15] The promoter sequence and untranslated exon I are both located upstream from a sequence previously identified as the promoter of the mouse osteopontin gene,[8] and presently thought to be intron I of this gene.[15] We performed S1-nuclease protection experiments using a DNA probe of 312 base pairs that was complementary to a region which included the 3′-end of intron I and the 5′-end of exon II (the first coding exon) of mouse osteopontin, and RNAs from various cell lines and tissues described above. The results unequivocally indicated that the DNA probe was protected by all the RNAs from the tissues and cell lines tested, up to a nucleotide sequence similar to that reported earlier.[8] FIGURE 3 shows results obtained with total RNA from a 10-day-old mouse embryo. Our results and those of others[8,15] are consistent with the notion that, in the mouse, there are two classes of osteopontin transcripts that differ at their 5′-untranslated sequences. One class contains the untranslated exon I, whereas the other class contains sequences from the 3′-end of intron I. Both classes of transcripts, however, bear the complete coding region (exons II through VII) of osteopontin.

A Model to Study Posttranslational Modifications of Mouse Osteopontin

The relatively low diversity of osteopontin transcripts found in mouse (this article) and human[23,24] suggests that posttranslational modifications of the osteopontin polypeptide may be a major mechanism used by the cells to generate a group of distinct but related forms of this protein. Among the posttranslational modifications of osteopontin are phosphorylation,[1,25,26] glycosylation,[1,26] and sulfation.[27] We have focused our attention on the phosphorylation of osteopontin, since it has been shown that normal rat kidney cells secrete phosphorylated and dephosphorylated forms of osteopontin,[26] and osteoblasts secrete at least two

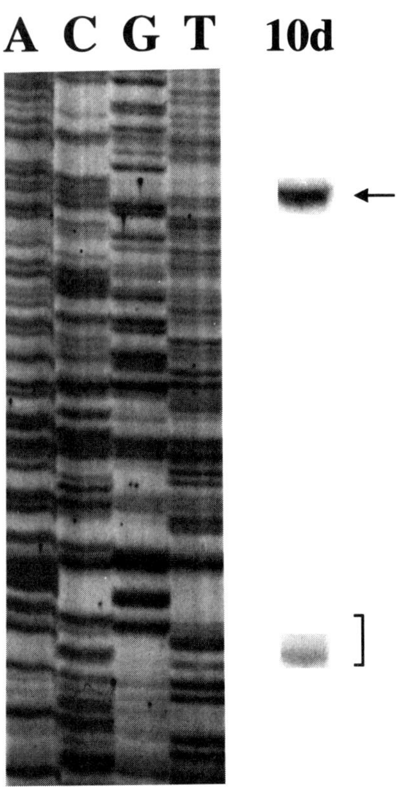

FIGURE 3. S-1 protection of a 312-base pair DNA probe complementary to the 3′-end of intron I and the 5′-end of exon II in the mouse osteopontin gene with total RNA from a 10-day-old mouse embryo. The DNA probe was generated by PCR using the cassette plasmid RSCL9 (R. Saavedra and C. Ljubetic, unpublished results) as a template and oligonucleotides OP7 (5′-GGTGTGGAGCTCTAGAGACATCAG-3′; complementary to the 3′-end of intron I) and PO1 (radioactively labelled with ^{32}P, and complementary to the 5′-end of exon II; see TABLE 1) as primers. Total RNA from a 10-day-old mouse embryo was incubated with the labelled DNA probe in 40 mM PIPES, 400 mM NaCl, 1 mM EDTA and 67% formamide at 55°C for 16 hours. The samples were diluted ten times in 300 mM NaCl, 30 mM NaAc and 3 mM $ZnCl_2$ and digested with S-1 nuclease at 37°C for 1 hour. The DNAs were separated on a 6% polyacrylamide sequencing gel. A sequencing reaction of plasmid RSCL9 with primer PO1 was used as a control. A, C, G, and T indicate adenine, cytosine, guanine, and thymidine, respectively. The full-length probe is indicated by a *solid arrow;* the protected region is marked by a *square bracket*. Similar results were obtained using total RNAs from various cell lines and tissues from mouse, and polyA RNA from mouse kidney (not shown; see text).

phosphorylated forms of the protein.[28,29] Furthermore, it has been found that osteopontin from bovine milk contains two classes of sites which are phosphorylated by casein kinase II and mammary gland casein kinase.[30] Phosphorylation of osteopontin, therefore, appears to play an essential role in its physiology.[31] To study the posttranslational modifications of mouse osteopontin, we expressed a recombinant form of this protein in *E. coli*.[19] Bacterial cells do not phosphorylate proteins in the same fashion as eukaryotic cells.[32] Consistent with this notion, mouse recombinant osteopontin contains no detectable phosphorylated residues; although its primary sequence exhibits more than fifty potential phosphorylation sites.[19] Therefore, recombinant osteopontin is an appropriate model to study the intricacies of phosphorylation.

Recombinant Osteopontin is Phosphorylated by Various Protein Kinases and Is Autophosphorylated on Tyrosine Residues

Mouse osteopontin expressed in *E. coli* is an appropriate substrate for phosphorylation by various protein kinases, including casein kinase II from mouse kidney and the catalytic subunit of cAMP-dependent protein kinase.[19] The phosphorylation of osteopontin may relate to the various proposed functions for this protein, such as mineralization[16,31] or cell attachment.[33]

We found that recombinant osteopontin, after incubation in the presence of ATP or GTP *in vitro,* became autophosphorylated on tyrosine residues.[34] Osteopontin extracted from mouse MC3T3-E1 cell cultures, rat tibiae, or human urine, however, cannot be autophosphorylated (S. Ahskar, unpublished results). It is possible that the tyrosine residues involved in the autophosphorylation of recombinant osteopontin are either sulfated (see Ref. 27) or inaccessible in most of the osteopontin molecules extracted from tissues or cell lines, or their autophosphorylation requires a yet undetermined condition. The functional significance of the autophosphorylation of osteopontin is presently not known. It may be related, however, to the specific interaction of osteopontin with other proteins of the extracellular milieu, as is the case in other systems. For instance, the specific interactions of growth factor receptors with intracellular proteins involve autophosphorylation of tyrosine residues.[35,36] After the binding of epidermal growth factor, the intracellular domain of its receptor becomes autophosphorylated on tyrosine residues to generate a site that serves as a recognition motif for the specific interaction with proteins which bear a src-homology 2 domain. This interaction results in the stimulation of a cascade of kinase activities that allows the transduction of signals from the extracellular milieu into the cell. Autophosphorylation of osteopontin could also generate sites for specific interactions with other proteins within the extracellular matrix and/or on the cell surface.[31] The interactions of autophosphorylated osteopontin with specific proteins could then trigger distinct biological responses, such as biomineralization of the extracellular matrix,[16] or cell attachment to the extracellular matrix[11,18] followed by the activation of specific intracellular signaling pathways.[15]

SUMMARY

Osteopontin is expressed in many different cell types and has been proposed to play several functions. Distinct forms of the protein have been detected. Various tissues and cell lines from mouse, however, exhibit two classes of transcripts with different 5′-untranslated ends but with an identical coding region (exons II through VII). These transcripts do not arise from the alternative splicing of coding exons. These results suggest that posttranslational modifications of osteopontin, such as phosphorylation, are a major mechanism to generate different forms of the protein. Mouse osteopontin was expressed in *E. coli* and used as a model to study its phosphorylation.

REFERENCES

1. Oldberg, A., A. Franzen & D. Heinegård. 1986. Cloning and sequence analysis of rat bone sialoprotein (osteopontin) cDNA reveals an Arg-Gly-Asp cell-binding sequence. Proc. Natl. Acad. Sci. USA **83:** 8819–8823.
2. Yoon, K., R. Buenaga & G. A. Rodan. 1987. Tissue specificity and developmental expression of rat osteopontin. Biochem. Biophys. Res. Commun. **148:** 1129–1136.
3. McKee, M. D., M. J. Glimcher & A. Nanci. 1992. High-resolution immunolocalization of osteopontin and osteocalcin in bone and cartilage during endochondral ossification in the chicken tibiae. Anat. Rec. **234:** 479–492.
4. Mark, M. P., W. Butler, C. W. Prince, R. D. Finkelman & J.-V. Ruch. 1988. Developmental expression of 44-kDa bone phosphoprotein (osteopontin) and bone γ-carboxyglutamic acid (Gla)-containing protein (osteocalcin) in calcifying tissue of rat. Differentiation **37:** 123–136.
5. Nomura, S., A. J. Wills, D. R. Edwards, J. K. Heath & B. L. M. Hogan. 1988. Developmental expression of 2ar (osteopontin), and SPARC (osteonectin) RNA as revealed by in situ hybridization. J. Cell Biol. **106:** 441–450.
6. Giachelli, C. M., N. Bae, M. Almeida, D. T. Denhardt, C. E. Alpers & S. M. Schartz. 1993. Osteopontin is elevated during neointima formation in rat arteries and is a novel component of human atherosclerotic plaques. J. Clin. Invest. **92:** 1686–1696.
7. Patarca, R., G. J. Freeman, R. P. Singh, F.-Y. Wei, T. Durfee, F. Blattner, D. C. Regnier, C. A. Kozak, B. A. Mock, H. C. Morse III, T. R. Jerrells & H. Cantor. 1989. Structural and functional studies of the early T-lymphocyte activation 1 (Eta-1) gene: definition of a novel T-cell dependent response associated with genetics resistance to bacterial infection. J. Exp. Med. **170:** 145–161.
8. Miyazaki, Y., M. Setoguchi, S. Yoshida, Y. Higuchi, S. Akizuki & S. Yamamoto. 1990. The mouse osteopontin gene: expression in monocytic lineages and complete nucleotide sequence. J. Biol. Chem. **265:** 14432–14438.
9. Patarca, R., R. A. Saavedra & H. Cantor. 1993. Molecular and cellular basis of genetic resistance to bacterial infection: the role of the early T-lymphocyte activation-1/osteopontin gene. Crit. Rev. Immunol. **13:** 225–246.
10. Senger, D. R., C. A. Perruzzi, C. F. Gracey, A. Papadopoulos & D. G. Tenen. 1988. Secreted phosphoproteins associated with neoplastic transformation: close homology with plasma proteins cleaved during blood coagulation. Cancer Res. **48:** 5770–5774.
11. Senger, D. R., C. A. Perruzzi & A. Papadopoulos. 1989. Elevated expression of secreted phosphoprotein I (osteopontin, 2ar) as a consequence of neoplastic transformation. Anticancer Res. **9:** 1291–1300.
12. Craig, A. M., G. T. Bowden, A. F. Chambers, M. A. Spearman, A. H. Greenberg, J. A. Wright, M. McLeod & D. T. Denhardt. 1991. Secreted phosphoprotein mRNA is induced in mouse skin and correlates with the metastatic potential of murine fibroblasts. Int. J. Cancer **46:** 133–137.
13. Smith, J. H. & D. T. Denhardt. 1987. Molecular cloning of a tumor promoter-inducible mRNA found in JB6 mouse epidermal cells: induction is stable at high, but not at low cell densities. J. Cell. Biochem. **34:** 13–22.
14. Craig, A. M., J. H. Smith & D. T. Denhardt. 1989. Osteopontin, a transformation-

associated cell adhesion phosphoprotein, is induced by 12-O-tetradecanoyl phorbol 13-acetate in mouse epidermis. J. Biol. Chem. **264:** 9682–9689.

15. DENHARDT, D. T. & X. GUO. 1993. Osteopontin: a protein with diverse functions. FASEB J. **7:** 1475–1482.
16. GLIMCHER, M. J. 1989. Mechanism of calcification: role of collagen fibrils and collagen-phosphoprotein complexes *in vitro* and *in vivo*. Anat. Rec. **224:** 139–153.
17. MCKEE, M. D., M. C. FARACH-CARSON, W. T. BUTLER, P. V. HAUSCHKA & A. NANCI. 1993. Ultrastructural immunolocalization of noncollagenous (osteopontin and osteocalcin) and plasma (albumin and α_2HS-glycoprotein) proteins in rat bone. J. Bone Miner. Res. **8:** 485–496.
18. REINHOLT, F. P., K. HULTENBY, A. OLDBERG & D. HEINEGÅRD. 1990. Osteopontin: a possible anchor of osteoblasts to bone. Proc. Natl. Acad. Sci. USA **87:** 4473–4475.
19. ASHKAR, S., D. TEPLOW, M. J. GLIMCHER & R. A. SAAVEDRA. 1993. *In vitro* phosphorylation of mouse osteopontin expressed in *E. coli*. Biochem. Biophys. Res. Commun. **191:** 126–133.
20. BROWN, L. F., B. BERSE, L. VAN DE WATER, A. PAPADOPOULUS-SERGIOU, C. A. PERRUZZI, E. J. MANSEAU, H. F. DVORAK & D. R. SENGER. 1992. Expression and distribution of osteopontin in human tissues: widespread association with luminal epithelial surfaces. Mol. Biol. Cell **3:** 1169–1180.
21. KODAMA, H., Y. AMAGAI, H. SUDO, S. KASAI & S. YAMANTO. 1981. Establishment of a clonal osteogenic cell line from newborn mouse calvaria. Jpn. J. Oral Biol. **23:** 899–901.
22. CHOI, C. H., R. S. SEDLACEK & H. D. SUIT. 1979. Radiation-induced osteogenic sarcoma of C3H mouse: effects of *Corynebacterium parvum* and WBI on its natural history and response to irradiation. Eur. J. Cancer **15:** 433–442.
23. KIEFER, M. C., D. M. BAUER & P. J. BARR. 1989. The cDNA and derived amino acid sequence for human osteopontin. Nucleic Acids Res. **17:** 3306.
24. YOUNG, M. F., J. M. KERR, J. D. TERMINE, U. M. WEBER, M. G. WANG, O. W. MCBRIDE & L. W. FISHER. 1990. cDNA cloning, mRNA distribution and heterogeneity, chromosomal localization and RFLP analysis of human osteopontin (OPN). Genomics **7:** 491–502.
25. NEMIR, M., M. W. DEVOUGE & B. B. MUKHERJEE. 1989. Normal rat kidney cells secrete both phosphorylated and nonphosphorylated forms of osteopontin showing different physiological properties. J. Biol. Chem. **264:** 18202–18208.
26. SINGH, K., M. W. DEVOUGE & B. B. MUKHERJEE. 1990. Physiological properties and differential glycosylation of phosphorylated and nonphosphorylated forms of osteopontin secreted by normal rat kidney cells. J. Biol. Chem. **265:** 18696–18701.
27. NAGATA, T., R. TODESCAN, H. A. GOLDBERG, Q. ZHANG & J. SODEK. 1989. Sulphation of secreted phosphoprotein I (SPPI, osteopontin) is associated with mineralized tissue formation. Biochem. Biophys. Res. Commun. **165:** 234–240.
28. KUBOTA, T., Q. ZHANG, J. L. WRANA, R. BER, J. E. AUBIN, W. T. BUTLER & J. SODEK. 1989. Multiple forms of SppI (secreted phosphoprotein, osteopontin) synthesized by normal and transformed rat bone cell populations: regulation by TGF-β. Biochem. Biophys. Res. Commun. **162:** 1453–1459.
29. KASUGAI, S., ZHANG, Q., C. M. OVERALL, J. L. WRANA, W. T. BUTLER & J. SODEK. 1991. Differential regulation of the 55 and 44 kDa forms of secreted phosphoprotein 1 (SPP-1, osteopontin) in normal and transformed rat bone cells by osteopetropic hormones, growth factors and a tumor promoter. Bone Miner. **13:** 235–250.
30. SORENSEN, E. S. & T. E. PETERSEN. 1994. Identification of two phosphorylation motifs in bovine osteopontin. Biochem. Biophys. Res. Commun. **198:** 200–205.
31. SAAVEDRA, R. A. 1994. The role of autophosphorylation and phosphorylation in the life of osteopontin. BioEssays **16:** 913–918.
32. GOLD, L. 1990. Expression of heterologous proteins in *Escherichia coli*. Methods Enzymol. **185:** 11–14.
33. PETTERSON, E., B. LUNING, H. MICKOS & D. HEINEGÅRD. 1991. Synthesis, NMR, and function of an O-phosphorylated peptide comprising the RGD-adhesion sequence of osteopontin. Acta Chem. Scand. **45:** 604–608.

34. ASHKAR, S., M. J. GLIMCHER & R. A. SAAVEDRA. 1993. Mouse osteopontin expressed in *E. coli* exhibits autophosphorylating activity of tyrosine residues. Biochem. Biophys. Res. Commun. **194:** 274–279.
35. PAWSON, T. 1992. Tyrosine kinases and their interactions with signaling proteins. Curr. Opin. Genet. Dev. **2:** 4–12.
36. MAYER, B. J. & D. BALTIMORE. 1993. Signalling through SH2 and SH3 domains. Trends Cell. Biol. **3:** 8–13.

Structure of the Osteopontin Gene and Its Promoter[a]

SHUNSUKE YAMAMOTO, NAOKI HIJIYA, MIHOKO SETOGUCHI, KEIKO MATSUURA, TETSUYA ISHIDA, YASUNORI HIGUCHI, AND SHIN'ICHIRO AKIZUKI

Department of Pathology
Oita Medical University
Hasama-machi
Oita 879-55, Japan

INTRODUCTION

Osteopontin (OPN), a sialoprotein of about 298 amino acids containing a Gly-Arg-Gly-Asp-Ser (GRGDS) is expressed in a variety of cell types, and shows diverse features of expression, although expression of OPN is not ubiquitous. Constitutive expression of OPN is found in bone, kidney, placenta, nerve cells, and macrophages, whereas inducible or enhanced expression of OPN is observed in epidermal and bone cells, T lymphocytes, and macrophages in response to a variety of agents.[1-11] These results suggest that the expression of OPN is controlled by complex regulatory systems which could differ between different cell types.

Previously, we reported the structure of the mouse OPN (mOPN) gene which consists of six exons, although another exon further upstream has also been reported.[6,12,13] A similar structure consisting of seven exons was also suggested for the porcine OPN (pOPN) gene,[14] although the sequences of exons 3 to 7 were not determined. Here, we present the complete nucleotide sequence of the human OPN (hOPN) gene and its 5′ upstream sequence, and compare its structure and potential regulatory sequences with those of mouse and porcine homologues. In addition, we analyzed the regulatory elements for the constitutive expression of the hOPN gene in the human monocytic cell line SCC-3, which strongly expresses the hOPN gene.

Structure of the hOPN Gene and Comparison with Those of the mOPN and pOPN Genes

We screened approximately 10^6 plaques of a human liver genomic DNA library with the hOPN cDNA as a probe and isolated one positive clone of 13 kb designated GOP-2. DNA from the clone was further analyzed by restriction enzyme mapping (see FIG. 2). This clone was then characterized by Southern blot analysis with

[a] This work was supported by grants from the Mitsuiseimei Kosei Jigyodan and the Ministry of Education, Science and Culture (No. 06281250).

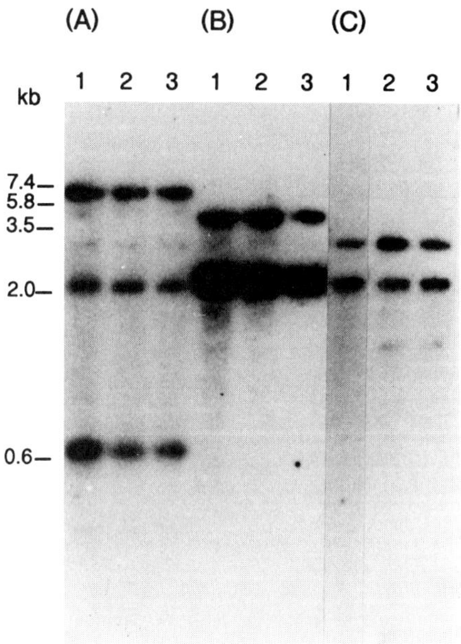

FIGURE 1. Southern blot analysis of the hOPN gene. Genomic DNA was extracted from cell lines, digested with EcoRI **(A)**, HindIII **(B)**, PstI **(C)**, electrophoresed on a 1% agarose gel, and blotted onto a nylon membrane. The membrane was hybridized with a ^{32}P-labeled hOPN cDNA probe. Lanes contained restriction enzyme-digested DNA from the following sources: *lane 1,* SCC-3; *lane 2,* Raji; *lane 3,* Molt 4.

distinct parts of the cDNA (FIG. 1). The SacI specific restriction fragments were subcloned into M13. Exon and intron sequences were determined by single-stranded sequencing.

As shown in FIGURE 2, the hOPN gene is organized into 7 exons spanning approximately 8 kb (FIG. 2, and see FIG. 4) similar to the murine homologue.[6,12,13] The combined coding sequence of the gene is almost identical to the published hOPN cDNA sequences.[15]

Sequences around the exon-intron junctions are shown in TABLE 1. Sizes of each exon and intron were roughly comparable with those for the mOPN gene,

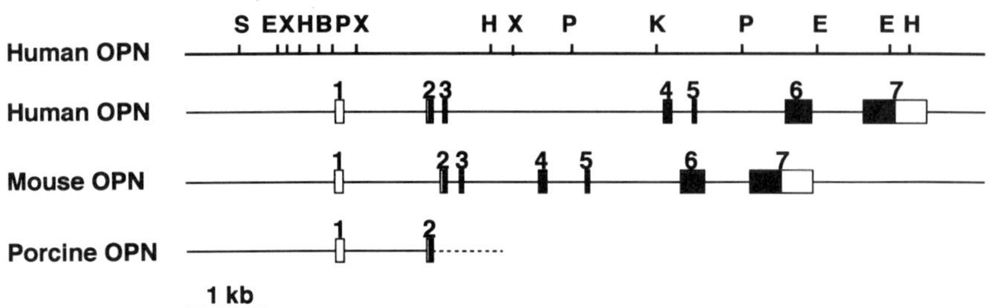

FIGURE 2. Exon and intron organization of the hOPN, mOPN and pOPN genes. Exons are *boxed*. *Filled boxes* are coding regions. *Open boxes* are untranslated regions. Abbreviations for restriction sites: E, EcoRI; S, SacI; X, XbaI; H, HindIII; B, BamHI: P, PstI; K, KpnI.

TABLE 1. Exon/Intron Organization of the hOPN Gene[a]

Exon No.	Exon Size (bp)	5′ Splice Site	Intron Size (bp)	3′ Splice Site
1	90	ACCAAGgtacag	1083	ttgcagGAAAAC
2	68	2 ATACCAgtgagt I P	109	3 tctaagGTTAAA V K
3	39	15 AAGCAGgtaagc K Q	2942	16 ttgtagCTTTAC L Y
4	81	42 CCACAGgtattt P Q	265	43 cttcagAATGCT N A
5	42	56 CAAGAGgtaagt Q E	1039	57 tttcagACCCTT T L
6	324	164 ATCCAGgtaaat I Q	692	165 attcagTACCCT Y P
7	829	TAAGAAtttggt		

[a] Codon numbers are shown above the nucleotide sequence.

except for intron 3, which was approximately 2.7 times longer than that of the mOPN gene. The difference in length was mainly due to the insertion of a 1750-bp sequence immediately before exon 4 in the hOPN gene. Exon 1 contained no coding region and was 1083 bp distant from exon 2, which included the region encoding 18 amino acids of which the 16 N-terminal residues appeared to be the signal sequence. The size of intron 1 was very close to that of the porcine homologue. Exons 3, 4, 5, and 6 encode 13, 27, 14, and 108 amino acid residues, respectively, in which exon 6 encodes the GRGDS sequence. Exon 7 encodes the C-terminal 134 amino acids and contains the 3′-untranslated region.

The transcription initiation site of the mOPN gene contains tandem AGC repeats,[12] although initiation begins with the second AGC tandem repeat. The nucleotide sequence of the hOPN gene also contained triple AGC repeats in exon 1. An S1 nuclease protection experiment was performed to identify the transcription initiation site of the hOPN gene. A 529-nucleotide BamHI-XbaI fragment containing the entire sequence of the first exon was end-labeled at the XbaI site and hybridized with SCC-3 RNA. Digestion with S1 nuclease resulted in major bands of 88, 90, 91, 93, and 97 nucleotide protected fragments (not shown). The strongest band, corresponding to the first nucleotide, A of the triple repeats of AGC in the 90-nucleotide species, was subsequently designated mRNA residue +1 (FIG. 3). To verify that these S1-protected products indeed represented the 5′ end of the hOPN mRNA, primer extention analysis was next performed. A 30-mer complementary to a sequence located near the 5′ end of the coding region of the mRNA was annealed to 50 μg of total cellular RNA derived from SCC-3 cells and extended. The primer extension product from this analysis revealed a major OPN message of 90 nucleotides and several minor bands. The main band (indicated in the figure by an *arrow*) corresponded to the transcription initiation site determined by S1 nuclease protection (FIG. 4).

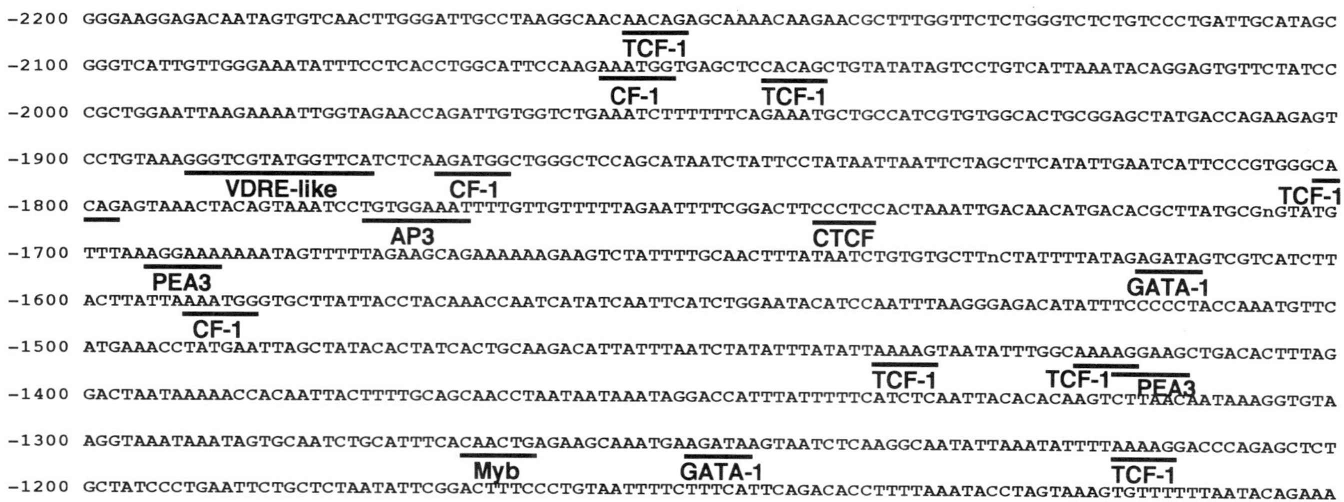

FIGURE 3. 5′ upstream region of the hOPN gene. Nucleotide positions are indicated in the *left margin*. Regulatory elements are indicated *below* the sequence. The *open square* shows a candidate cis region demonstrated by the present experiments.

FIGURE 3 (*Continued*).

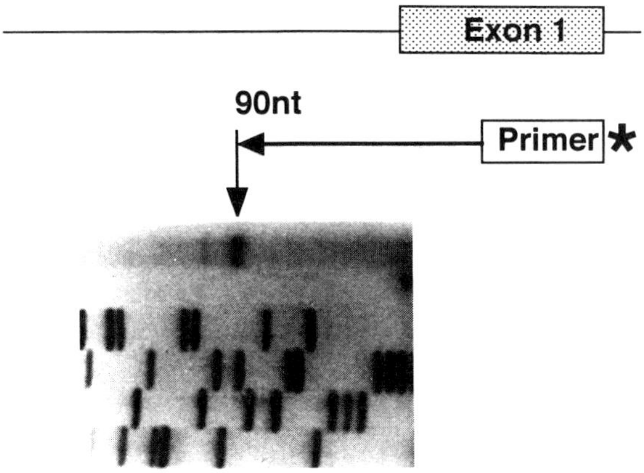

FIGURE 4. Primer extension analysis of hOPN mRNA. Total RNA from SCC-3 was annealed to a 30-base oligonucleotide in sequence to 5′ region specific to hOPN mRNA. The oligomers were extended and analyzed. Dideoxynucleotide sequencing reactions were electrophoresed in parallel as markers. *Arrow* indicates the transcriptional start site (nucleotide +1, see FIG. 5).

Comparison of Promoter Sequences with Those of the mOPN and pOPN Genes

The 5′ upstream sequence of the hOPN gene and further upstream sequence of the mOPN gene, part of which had been reported previously,[12] were sequenced, and compared with the corresponding sequences of the pOPN gene.[14] As shown in dot matrix comparison (FIG. 5), the 5′ upstream sequence of the hOPN gene up to −250 showed high homology with those of the mOPN and pOPN genes, whereas it was different from the mOPN gene from position −250 to −650.

The 5′ upstream sequence of the hOPN gene contained a number of potential regulatory sequences (FIG. 3). A TATA-like sequence, TTTAAA, which was similar to those in the mOPN and pOPN genes,[12,14] was found at position −27 to −22. Further upstream, there were a number of TATA-like sequences, although these might not play functional roles because of their distance from the cap site. Many CCAAT-like sequences[16,17] were scattered over the 5′ upstream region spanning from −73 to −2190, although they were situated further upstream than the usual CCAAT position, which resides within −50 to −100 bp. TCF-1 sequences were also abundant. Since the elements were found in T cells, they may be functional in these cells. The vitamin D-responsive element (VDRE)-like sequence,[18] which was similar to that in mouse,[12] was located at position −698 to −684. Another VDRE-like sequence, very similar to those in the mOPN and pOPN genes, was found at position −1892 to −1878.[14] Four NF-IL6 sequences were situated at positions −2091 to −2083, −1950 to −1942, −1778 to −1770, and −1007 to −999, two of these corresponding to those in the pOPN gene.[19] The erythrocyte, muscle, and fibroblast-specific motif, CCCTC and its inverse sequence GAGGG[20] were widely distributed. GATA-1(GF-1),[21,22] which show tissue specificity for erythroids, megakaryocytes, mast cells, and T cells, were found at positions −1616 to −1611, −1252 to −1247, −851 to −847, and −366 to −362, in which the sequence corresponding to that at position −851 to −847 was found in mouse and porcine homologues. The AP1 sequence TGACACA[23] was found at position −78 to −72. No AP1 sequence, however, was found at nucleotides −9 to −3, which accommodated the AP1 motif in the mouse sequence. An interferon regulatory factor-1 (IRF-1) binding sequence AACTGA[24,25] was

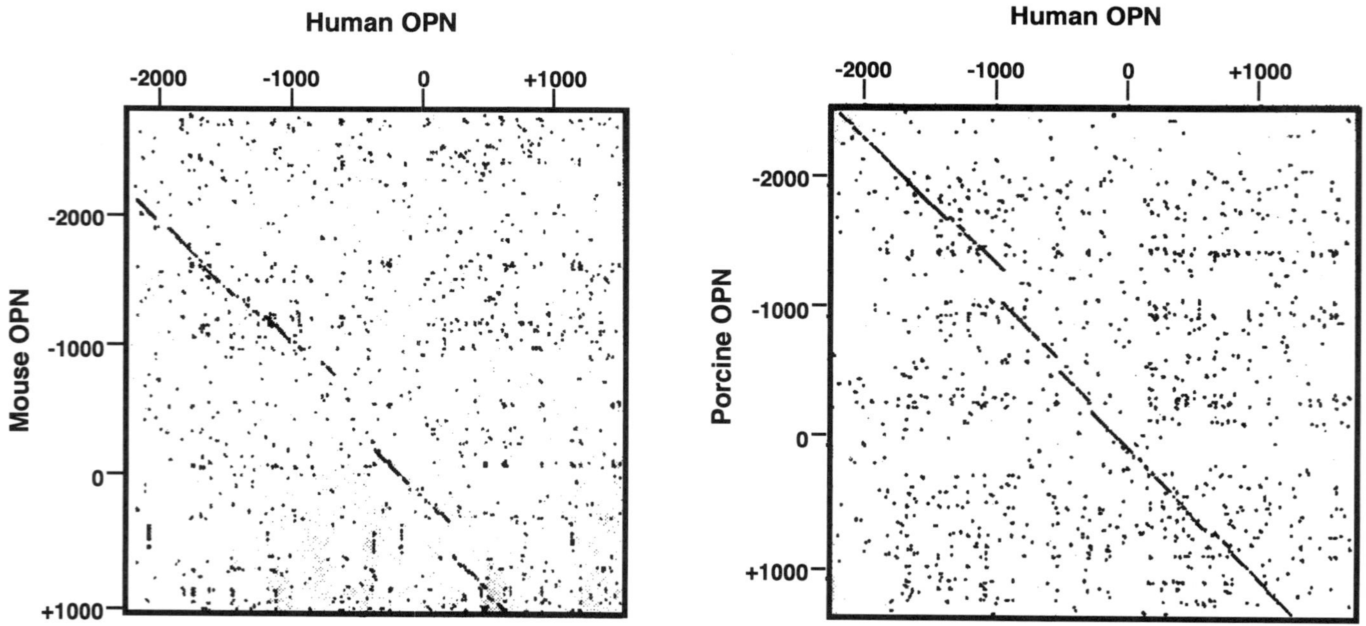

FIGURE 5. Dot matrix comparison of the 5′ upstream sequences of the hOPN, mOPN and pOPN genes. The plot was generated using a computer program with standard parameters (match >= 14/21).

found at position −1270 to −1264. Other ubiquitous or tissue-specific motifs including E2A,[26] E2BP,[27] Myb,[28] Sp1,[29] PPAR,[30] AP2,[31] SIF,[32] E4TF1,[33] Oct-1,[34] and Ets-1[35] were distributed mainly within 728 bp of the cap site. Three ubiquitous CF1 elements resided at positions −2059 to −2054, −1872 to −1867, and −1592 to −1587. PEA3, a primary target of signal transduction for a variety of factors such as TPA, EGF, serum, and oncogene products, were found at positions −1695 to −1690, and −1418 to −1413. Thus, the number of consensus sequences shared by OPN homologues from these three species were relatively few, although a number of consensus sequences in the far upstream region are shared by the hOPN and pOPN genes.

Cis-Acting Regulatory Elements in the hOPN Promoter Upstream Regions

To examine whether cis-acting elements were present in the 5′ upstream region of the hOPN gene, a panel of upstream hOPN gene fragments was inserted in front of the bacterial chloramphenicol acetyl transferase (CAT) reporter gene (FIG. 6). The constructs were transfected into SCC-3 cells which expressed hOPN. Electroporation was used for DNA transfection because SCC-3 cells are not adherent. Raji cells were used as a hOPN-negative background in which hOPN-specific promoters are not substantially activated.

The results obtained with the various constructs and deletions are presented as percentage of chloramphenicol conversion compared with that in plasmid pSV2CAT which contains the CAT gene transcribed from the Simian virus 40 early promoter. The 55-bp region was capable of directing a significant level of CAT synthesis in SCC-3 cells, although the region including 38 bp immediately upstream of the transcription start site showed no activity. The relative amount of reporter gene expression decreased by including additional 5′ sequences up to −79. Elongation of the 5′ upstream sequence up to −124, however, markedly increased the reporter gene expression in SCC-3 cells, although −107/+90CAT and −114/+90CAT showed no CAT activity. A further stretch of the upstream sequence diminished the reporter gene expression, although a marginal increase in expression was observed by elongating the upstream sequence up to −439. Further elongation of the 5′ upstream sequence diminished reporter gene expression. These results indicate that the 5′-flanking region at position −124 to −80 contain cis-acting regulatory sequences specific for SCC-3 cells. Deletion of the sequence between −124 to −80 localized the cis element within the sequences at positions −124 to −114, −104 to −95, and −94 to −80, although the sequence at position −104 to −95 might have no cis activity because the activity of −104/+90CAT did not exceed that of −94/+90CAT (FIG. 6). Levels of the CAT activities of these elements were high, reaching 200% to 400% of that in pSV2CAT.

The 5′ upstream regions of the mOPN gene responsible for constitutive and inducible expressions have been reported.[12] Regions at positions −777 to 79, −543 to 79, and −253 to 79 show promoter activities, of which the region at position −543 to 79 shows the strongest activity when tested in the murine epidermal cell line JB6. Activities directed by these regions are enhanced by TPA.[12] Since the mOPN gene is not normally expressed in JB6, the authors suggest that transcription from the mOPN promoter is normally repressed or controlled at the posttranscriptional level. CAT analysis for the promoter region of the pOPN gene revealed that a sequence at position −180 to 51 had remarkably high promoter activity in calvarial bone cells.[14] CAT analysis of the 5′ upstream region of hOPN

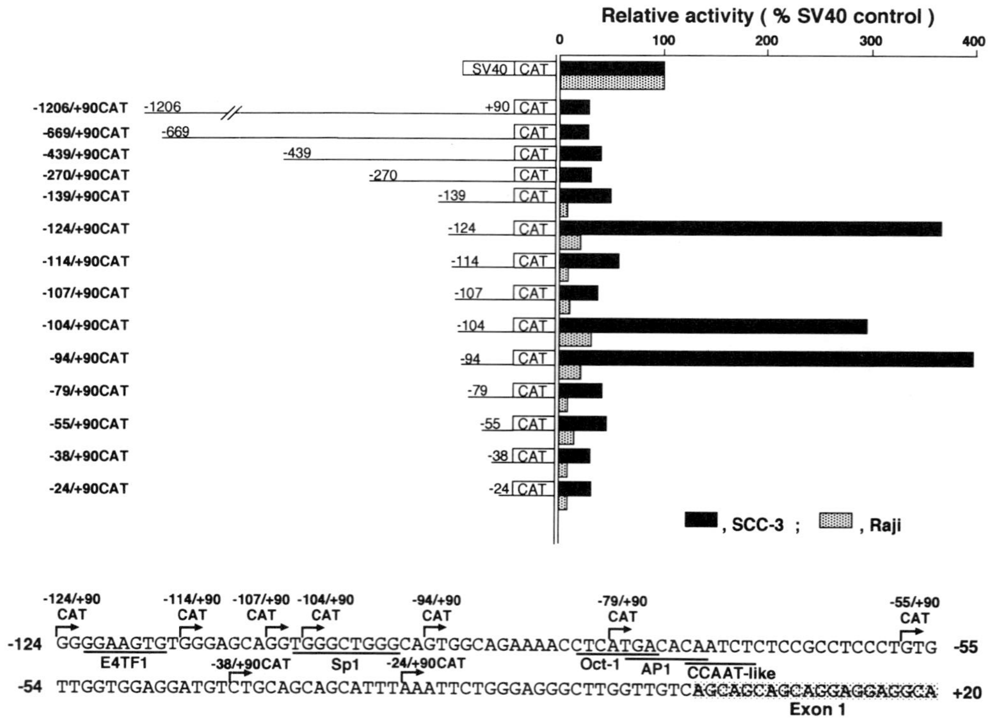

FIGURE 6. CAT activities of constructs containing regions 5′ upstream of the hOPN gene. Expression of 5′-deletion mutants of the hOPN/CAT fusion gene was tested in SCC-3 and Raji cells. A series of chimeric plasmids was constructed and transfected into SCC-3 cells. *Filled columns* are CAT activities in SCC-3 cells. *Shaded columns* are CAT activities in Raji cells. Nucleotide positions are indicated *above the underline*.

revealed that sequences at positions −124 to −115, and −94 to −80 accommodate strong cis-acting regulatory elements in SCC-3 cells, whereas that at position −439 to −409 contained only low levels of cis activity (data not shown).

The nucleotide sequences of the major cis-acting regions were highly conserved in hOPN, pOPN and mOPN genes. Therefore, we tested CAT activities of −124/+90CAT, −104/+90CAT and −94/+90CAT in a mouse macrophage cell line, HINS-B3, and found that these CAT clones showed significant CAT activities, although levels of the activities were low because of the differences in nucleotide sequences downstream of the cis regulatory regions between species (data not shown). These results suggest that the major cis-acting regions are also important in other species.

Interaction of Nuclear Protein with the cis-Regulatory Element

Having identified the functional regions of the hOPN promoter, we assessed whether these regions could serve as binding sites for specific nuclear proteins. To do so, we synthesized oligonucleotides corresponding in sequence to the major cis-acting regulatory regions and performed gel shift assays. Synthetic oligonucleotides were GATCGGGGAAGTGT, GATCGGGAGCAGGT, GATCGGGCTGGGCA, GATCGTGGCAGAAAACCTC, GATCGAAAACCTCATGA, and GATCCTCATGACACA corresponding to positions −124 to −115, −114 to −105, −104 to −95, −94 to −80, −88 to −76, and −82 to −72, designated A, B, C, D, E, and F, respectively. Each nucleotide formed a major band with nuclear extracts from SCC-3 cells. The band formed with probe A was competed out with corresponding unlabeled oligonucleotides (results not shown), but not unrelated ones (FIG. 7). The bands formed probes B and C were identical and also competed out with corresponding unlabeled oligonucleotides but not unrelated ones (not shown). The bands formed with probes D and E had identical electrophoretic mobility, although the band formed with probe D was weaker than that formed with probe E (FIG. 7). Probe F formed a very weak band. The major band formed with probe E was not competed with unlabeled F and A (not shown). These results suggested that the nucleotide position (G)AAAAC was involved in the formation of the major band. Each of the probes A, B, C, D, E and F formed a major band with nuclear extracts from Raji cells which showed little or no CAT activity. The electrophoretic mobilities of all these bands, however, were different from those with nuclear extracts from SCC-3 cells.

Probe A formed a band, which may be the band related to E4TF1. Probe B formed one band which could be that formed with a negative regulatory factor. Probe C formed a band which was possibly the band associated with Sp1, although the electrophoretic mobility of the band appeared identical with the band formed with probe B. However, the role of Sp1 is uncertain because −104/+90CAT did not confer additional CAT activity to −94/+90CAT, and −107/+90CAT abolished CAT activities. Thus, the region immediately downstream at position −107 to −105 is also likely to operate as a negative cis element. The bands formed with probe D were competed out with competitor E, and vice versa. These bands were not competed out with competitor F. Therefore, these results suggest that (G)AAAAC is important as a cis sequence. Since the signal intensities of the bands formed with F were less marked, involvement of Oct-1 may be minor, if any. Taken together, these data suggest that (G)AAAAC could play a major role

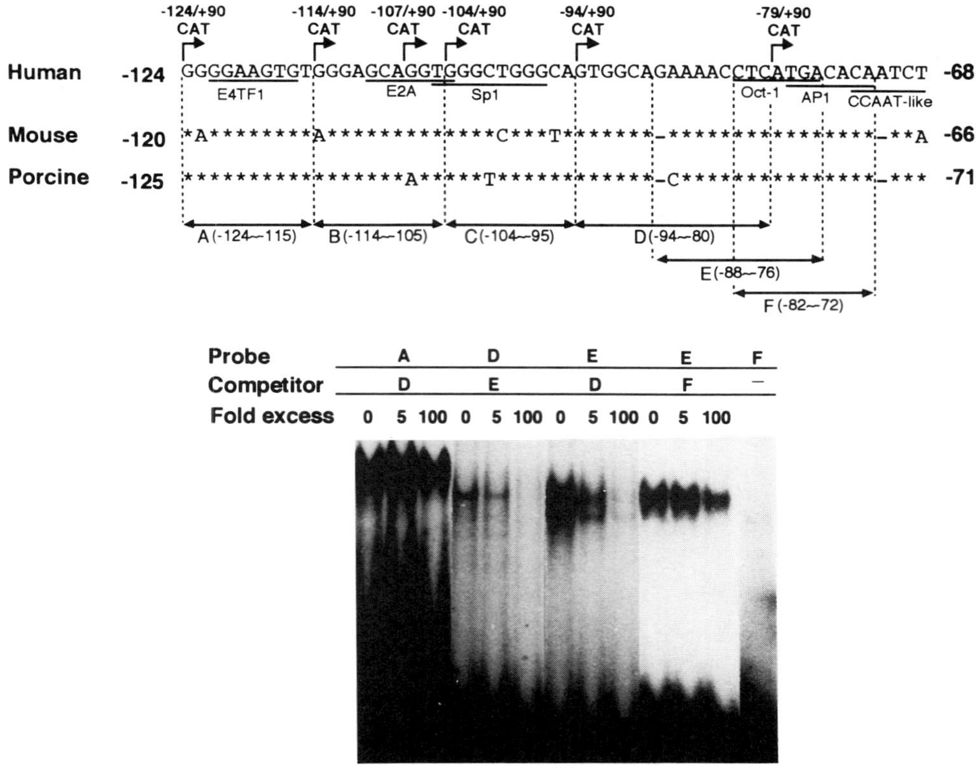

FIGURE 7. Gel shift and competition assays using A, D, E, and F fragments with SCC-3 nuclear extracts. Gel shift and competition assays were performed using synthetic oligonucleotides with SCC-3 nuclear extracts.

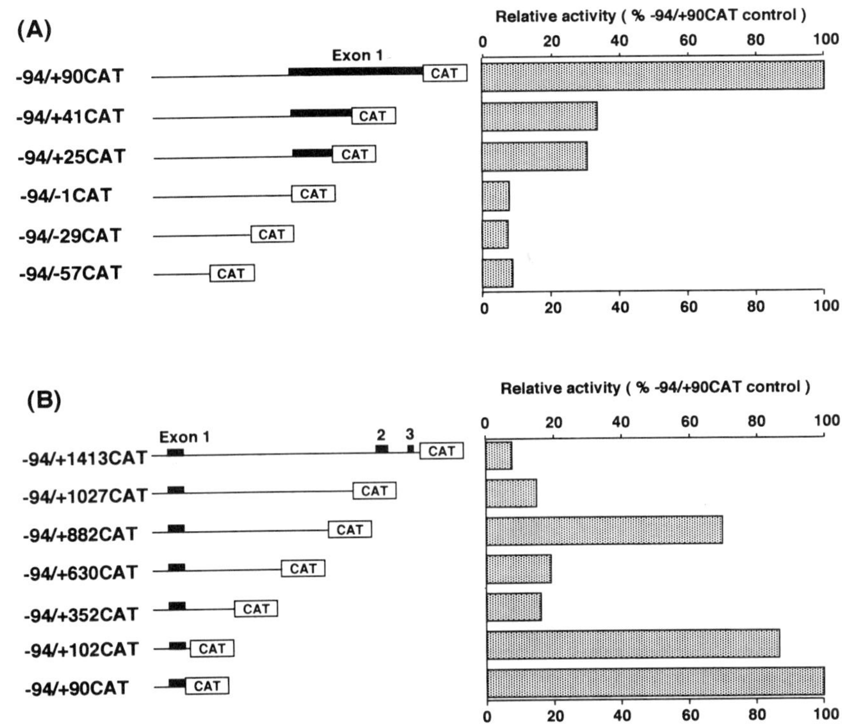

FIGURE 8. CAT activities of reporter constructs containing the first exon and introns. Expression of **(A)** 3′-deletion mutants of the first exon sequence of −94/+90CAT and **(B)** deletion mutants of intron 1, exon 2, intron 2, exon 3 and intron 3 of −94/+1413CAT were tested in SCC-3 cells. A series of chimeric plasmids was constructed and transfected into SCC-3 cells. *Shaded columns* are CAT activities in SCC-3 cells. Nucleotide positions are indicated *above the underline*.

in the promoter activity in the hOPN gene in SCC-3 cells. The region at position −124 to −115 containing the E4TF1 site may act to suppress the negative regulatory region between −107 and −105.

It has been reported from the results using the CAT assay, that TPA induces OPN gene expression in murine epithelial cells and that the region upstream of the mOPN gene contains TPA-responsive elements.[12] To test whether the TPA-responsive elements found in the major cis regulatory regions and the region immediately upstream of the hOPN gene are operative, we compared CAT activities using −124/+90CAT, −79/+90CAT, and −55/+90CAT clones in the presence of TPA (100 ng/ml). TPA, however, showed no effects in these CAT transfectants 3 and 16 h after stimulation. The AP1 site in the region at position −78 to −72 is conserved in all three species, although the AP1 site immediately upstream of the cap site found in the mOPN gene is not conserved in the pOPN and hOPN genes. The results of the present experiments demonstrated that the conserved AP1 site at position −78 to −72 was not responsible for enhancement of CAT activity in SCC-3, although it was bound with a nuclear protein. −124/+90CAT and −55/+90CAT also showed no TPA inducibility. VDRE at position −698 to −684 and −1892 to −1878, which have been reported to be functional in mouse and porcine homologues, respectively,[14,36] were found in hOPN. Whether these elements are functional in the hOPN gene awaits further study.

Role of the First Exon and Intron in Regulation of Expression of the hOPN Gene

We developed chimeric hOPN-CAT constructs to test the role of exons and introns in the regulation of expression of the hOPN gene. CAT clones with a deletion in the 3′ region of the first exon were constructed using −94/+90CAT, whereas those with deletion in the 3′ region of intron 1 were constructed using −94/+90CAT and a plasmid containing the region spanning the 5′ upstream sequence and intron 3 (−474/2032). −94/+90CAT showed the highest CAT activity. The CAT activity decreased as the 3′ region of the first exon became shorter, indicating that the full length of exon 1 was required for maximum CAT activity (FIG. 8). The original CAT clone −94/+1413CAT containing exon 1, intron 1, exon 2, intron 2 and exon 3 and a part of intron 3 showed no CAT activity. Deletion at the 3′ region of −94/+1413CAT resulted in an increase in CAT activity. −94/+882CAT containing a part of intron 1 showed a peak, while further deletion caused a marked drop in CAT activity. These results indicate that intron 1 contains both negative and positive cis-acting regions.

SUMMARY

We cloned the hOPN gene and its 5′ upstream region, and analyzed its exon-intron structure and potential regulatory sequences of the promoter region in comparison with those of mouse and porcine homologues. The hOPN gene consists of 7 exons that are similar to those of the mouse gene, although the hOPN gene is longer than the mouse homologue. This difference is attributable to an insertion of about 1750 bp immediately before exon 4 in the hOPN gene. A region of approximately 285 bp immediately upstream of the hOPN transcription initiation site was highly conserved and contained a number of potential cis regulatory consensus sequences. CAT analysis using SCC-3 cells demonstrated that nucleotides at positions −439 to −270, −124 to −80, and −55 to −39 contained cis-acting enhancing elements, in which the −124 to −80 element was much more active than the others. Deletion of the sequences between −474 and −270 localized the cis elements to the sequence at position −439 to −410, whereas the deletion between −124 to −80 localized it to −124 to −115, and −94 to −80 (data not shown). Gel shift analysis using synthesized double-stranded oligonucleotides corresponding to the 30 bp at position −439 to −410 (data not shown), and 10 and 15 bp regions at positions −124 to −115 and −94 to −80, respectively, as probes revealed that each probe formed one or two bands complexed with a nuclear protein prepared from SCC-3 cells. Inclusion of the full-length first exon in transfected reporter constructs resulted in maximum expression of CAT activity. Intron 1 contains both negative and positive regulatory cis regions.

REFERENCES

1. OLDBERG, A., A. FRANZEN & D. HEINEGÅRD. 1986. Proc. Natl. Acad. Sci. USA **83:** 8819–8823.
2. PATARCA, R., G. J. FREEMAN, R. P. SINGH, F.-Y. WEI, W. T. DURFEE, F. BLATTNER, D. C. REGNIER, C. A. KOZAK, B. A. MOCK, H. C. MORSE III, T. R. JERRELS & H. CANTOR. 1989. J. Exp. Med. **170:** 145–161.
3. YOON, K., R. BUENAGA & G. A. RODAN. 1987. Biochem. Biophys. Res. Commun. **13:** 1129–1136.
4. MARK, M. P., C. W. PRINCE, S. GAY, R. L. AUSTIN & W. T. BUTLER. 1988. Cell Tissue Res. **251:** 23–30.
5. NOMURA, S., A. J. WILLIS, D. R. EDWARDS, J. K. HEATH & B. L. M. HOGAN. 1988. J. Cell Biol. **106:** 441–450.
6. MIYAZAKI, Y., M. SETOGUCHI, S. YOSHIDA, Y. HIGUCHI, S. AKIZUKI & S. YAMAMOTO. 1990. J. Biol. Chem. **265:** 14432–14438.
7. RODAN, S. B., G. WESOLOWSKI, K. YOON & G. A. RODAN. 1989. J. Biol. Chem. **264:** 19934–19941.
8. NODA, M., K. YOON, C. W. PRINCE, W. T. BUTLER & G. A. RODAN. 1988. J. Biol. Chem. **263:** 13916–13921.
9. NODA, M., R. L. VOGEL, D. M. HASSON & G. A. RODAN. 1990. Endocrinol. **127:** 185–190.
10. NODA, M. & G. A. RODAN. 1989. J. Cell Biol. **108:** 713–718.
11. SMITH, J. H. & D. T. DENHARDT. 1987. J. Cell. Biochem. **34:** 13–22.
12. CRAIG, A. M. & D. T. DENHARDT. 1991. Gene **100:** 163–171.
13. BEHREND, E. I., A. F. CHAMBERS, S. M. WILSON & D. T. DENHARDT. 1993. J. Biol. Chem. **268:** 11172–11175.
14. ZHANG, Q., J. L. WRANA & J. SODEK. 1992. Eur. J. Biochem. **207:** 649–659.
15. KIEFER, M. C., D. M. BAUER & P. J. BARR. 1989. Nucleic Acids Res. **17:** 3306.
16. CHODOSH, L. A., A. S. BALDWIN, R. W. CARTHEW & P. A. SHARP. 1988. Cell **53:** 11–24.
17. SANTORO, C., N. MERMOD, P. C. ANDREWS & R. TJIAN. 1988. Nature **334:** 218–224.
18. SCHÜLE, R., K. UMESONO, D. J. MANGELSDORF, J. BOLADO, J. W. PIKE & R. M. EVANS. 1990. Cell **61:** 497–504.
19. AKIRA, S., H. ISSHIKI, T. SUGITA, O. TANABE, S. KINOSHITA, Y. NISHIO, T. NAKAJIMA, T. HIRANO & T. KISHIMOTO. 1990. EMBO J. **9:** 1897–1906.
20. LOBANEKOV, V. V., R. H. NICOLAS, V. V. ADLER, H. PATERSON, E. M. KLENOVA, A. V. POLOTSKAJA & G. H. GOODWIN. 1990. Oncogene **5:** 1743–1753.
21. ORKIN, S. H. 1990. Cell **63:** 665–672.
22. YAMAMOTO, M., L. J. KO, M. W. LEONARD, H. BEUG, S. H. ORKIN & J. D. ENGEL. 1990. Genes Dev. **4:** 1650–1662.
23. ANGEL, P., M. IMAGAWA, R. CHIU, B. STEIN, R. J. IMBRA, H. J. RAHMSDORF, C. JONAT, P. HERRICH & M. KARIN. 1987. Cell **49:** 729–739.
24. FUJITA, T., S. OHNO, H. YASUMITSU & T. TANIGUCHI. 1985. Cell **41:** 489–496.
25. KORBER, B., N. MERMOD, L. HOOD & I. STORYNOWSKI. 1988. Science **239:** 1302–1306.
26. MURRE, C., A. VORONOVA & D. BALTIMORE. 1991. Mol. Cell. Biol. **11:** 1156–1160.
27. KENNY, S. & R. V. GUNTAKA. 1990. Virology **176:** 483–493.
28. BIEDENKAPP, H., U. BORGMEYER, A. E. SIPPEL & K.-H. KLEMPNAUER. 1988. Nature **335:** 835–837.
29. SCHMIDT, M. C., Q. ZHOU & A. J. BERK. 1989. Mol. Cell. Biol. **9:** 3299–3307.
30. ISSEMANN, I. & S. GREEN. 1990. Nature **347:** 645–650.

31. WILLIAMS, T. & R. TJIAN. 1991. Genes Dev. **5:** 670–682.
32. WAGNER, B. J., T. E. HAYES, C. J. HOBAN & B. H. COCHRAN. 1990. EMBO J. **9:** 4477–4484.
33. WATANABE, H., T. WADA & H. HANDA. 1990. EMBO J. **9:** 841–847.
34. VERRIJZER, C. P., A. J. KAL & P. C. VAN DER VLIET. 1990. EMBO J. **9:** 1883–1888.
35. WASYLIK, B., C. WASYLIK, P. FLORES, A. BEGUE, D. LEPRINCE & D. STEHELIN. 1990. Nature **346:** 191–193.
36. NODA, M., R. L. VOGEL, A. M. CRAIG, J. PRAHL, H. F. DELUCA & D. T. DENHARDT. 1990. Proc. Natl. Acad. Sci. USA **87:** 9995–9999.

Characterization of the Rat Osteopontin Gene

Evidence for Two Vitamin D Response Elements[a]

AMY L. RIDALL,[b,c] ELIZABETH L. DAANE,[b]
DOUGLAS P. DICKINSON,[b] AND WILLIAM T. BUTLER[b]

Departments of [b]Basic Sciences and [c]Prosthodontics
Dental Branch
University of Texas Houston-Health Science Center
Houston, Texas 77030

INTRODUCTION

The seco-steroid vitamin D contributes to skeletal development and homeostasis by serving as a regulator to control several biological processes, including cell differentiation and calcium homeostasis. The molecular action of 1,25-dihydroxyvitamin D_3 ($1,25(OH)_2D_3$), the active vitamin D metabolite, has been shown to involve mechanisms similar to the classic steroid hormone action (for review see Darwish & DeLuca[1]). Binding of $1,25(OH)_2D_3$ to a nuclear heterodimeric receptor complex consisting of the vitamin D receptor (VDR) and one of the retinoid X receptors (RXRs)[2–4] within the target cell effects an increased affinity for specific DNA sequences located on the target gene. Binding to these DNA sequences, referred to as vitamin D response elements (VDREs), results in an enhancement of target gene transcription. A comparison of several VDRE sequences reveals a common motif consisting of two imperfect direct repeats of six-base-pair (bp) elements, separated by a three-bp spacer.

Osteopontin (OPN) expression is affected by $1,25(OH)_2D_3$ in osteoblastic cells at the level of transcription. Transcriptional regulation of OPN was first demonstrated by showing that the level of OPN protein secretion by rat osteoblast-like osteosarcoma cells (ROS 17/2.8) is increased approximately twofold by $1,25(OH)_2D_3$ in a concentration-dependent manner and that this effect is blocked by transcriptional inhibitors.[5] Subsequently, Noda *et al.*[6] identified a VDRE in the mouse *opn* gene that conferred responsiveness to $1,25(OH)_2D_3$ and bound the VDR. In this publication we report the isolation of a clone containing the rat *opn* gene and present evidence for the existence of two separate DNA sequences involved in the regulation at the transcriptional level by $1,25(OH)_2D_3$.

[a] Use of the GCG computer analysis package was supported by Grant CA-16672 from the National Cancer Institute. These studies were supported by an NIDR Dentist Scientist Award, K15 DE00218 (to ALR) and by NIH Grant RO1 AR39273 (WTB, Principal Investigator).

Isolation and Characterization of the Rat Osteopontin Gene

A rat spleen genomic library (Stratagene, La Jolla, CA) was screened using a probe corresponding to the rat OPN cDNA.[7] A combination of Southern blot, restriction analysis, and partial sequencing identified a clone (10.1.1) that contained the entire OPN coding sequences, as well as 7 kilobases (Kb) of upstream flanking sequences for rat OPN. Sequencing of the intron/exon boundaries confirmed the

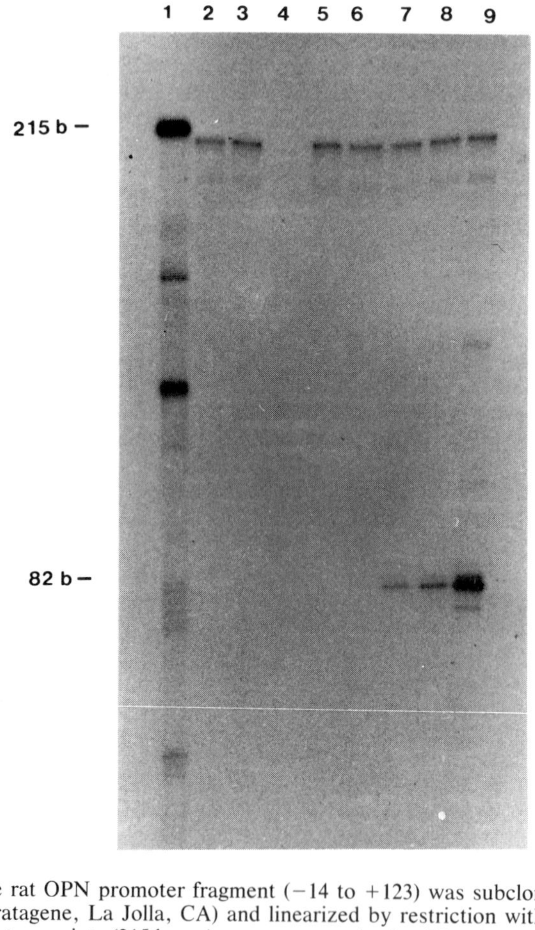

FIGURE 1. The rat OPN promoter fragment (−14 to +123) was subcloned into the pBS KS+ vector (Stratagene, La Jolla, CA) and linearized by restriction with AvaII enzyme. Full-length RNA transcripts (215 bases) were generated using T7 polymerse (*lane 1*). Transcripts were hybridized overnight at 42°C with either total RNA extracted from ROS 17/2.8 cells (*lanes 5–9*; increasing concentrations of RNA from 1–10 μg) or yeast RNA (*lanes 1–2*). All samples, except the control (*lane 1*), were digested with RNase cocktail (RNase A/RNase T1 mixture from Ambion, diluted 1 : 100). Fragments were electrophoresed on a 8% denaturing acrylamide gel. The size of the resulting protected fragment (82 b) was used to calculate the start site of transcription and to verify the position of the first intron/exon boundary.

existence of 7 exons, identical to the genomic structure reported for the mouse and chicken *opn* genes.[8-10] An RNase protection assay was used to identify the start site of transcription and to detect the potential existence of alternate mRNA species resulting from differential splicing (FIG. 1). The resultant 82-base protected fragment suggests that the transcription start site and position of the first intron is identical to that found in the mouse *opn* gene,[11] while detection of only one protected fragment suggested that no alternate 5′ end mRNA species are transcribed. Approximately 2.4 Kb of the 5′ upstream flanking region was sequenced and subjected to computer analysis to identify putative transcription factor binding sites (FIG. 2). Putative TATA and CAAT sequences were located appropriately 40–70 nucleotides (nt) upstream from the transcription start site. Deletion mutation constructs were fused to the firefly luciferase gene and transiently transfected into ROS 17/2.8 cells. These functional analysis studies indicated that deletion of nucleotides +9 to −106, which deleted the putative TATA and CAAT sequences, abolished sequences required for luciferase expression (FIG. 3).

Identification of Two Separate VDREs

DNA fragments containing vitamin D responsive sequences were identified by measuring luciferase activity of constructs of rat *opn* promoter sequences fused to the firefly luciferase reporter gene (LUC) after transient transfection in ROS/17/2.8 cells. The fusion constructs used in these experiments are diagrammed in FIGURE 4. An initial construct containing +140 to −701 nt showed no responsiveness. However, fusion of sequences from −701 to −2275 (1.7-EcoRI) to the SV40 promoted-LUC provided a 10-fold enhancement of LUC activity, following $1,25(OH)_2D_3$ induction. Additional experiments using the 1.7-EcoRI construct in the reverse orientation demonstrated that this enhancement is orientation independent. Specificity of the response to $1,25(OH)_2D_3$ was demonstrated by showing that reagents such as dexamethasone, phorbol esters (TPA) and all-trans retinoic acid (RA) did not alter luciferase activity (FIG. 5a).

To determine the location of the VDREs, unilateral deletions from the 5′ end were tested for $1,25(OH)_2D_3$ responsiveness. The results showed that two separate sequences conferred this response, although the amplitude of response differed substantially for each sequence (FIG. 5b). When both elements were combined in one construct, the overall enhancement was approximately 8–10-fold, whereas the distal (relative to the promoter) element showed a 6–8-fold increase and the proximal element a 2-fold increase when assayed individually. These data suggest that two functional VDREs are present in the rat *opn* gene and that their combined actions result in increased transcription rates that are additive. Sequence analysis showed that each fragment contained a DNA sequence that aligned well with the consensus VDRE sequence (TABLE 1): the proximal VDRE was located at −923 to −938 nt and the distal VDRE was located at −1920 to −1935 nt. Both VDRE elements were assayed in forward and reverse orientations. In each case the amplitude of responsiveness to $1,25(OH)_2D_3$ was identical in either orientation, indicating enhancer-like, orientation-independent activity. Additionally, the $1,25(OH)_2D_3$ effect on each VDRE was time and dose dependent, with enhanced luciferase activity detected as early as 6 h and a maximal response at 10 μM of $1,25(OH)_2D_3$ (data not shown).

Double-stranded oligonucleotides for the proximal and distal VDREs were synthesized and annealed for use in gel shifting assays. Preliminary data indicated that both VDREs are capable of competing the binding of the mouse VDRE

-2275 ....GGA

-2272 TGTCCTTCTCTGCTTTGCAGAACTGCGCCCAGCACACAGCGTCTCAACCTCAAGGTT

-2215 GCAGACACTGAAAGCCGGTGTGAGTGTAGGAAGCAGTCAGTCCTGTCGACCTGGGAT

-2158 CGCTGAAGGCATCAACAGAGCAGGACAGGGGTGCTTTGGTTCTCAGGGTCCTGGCCT

-2101 GGCTGCCAACTGGGTCATTGCTGAGAAATACTTCCTTAATCAGCATTCTGGGAAACG

-2044 GGGTGTATACAGCCTCTATCAGAAAGTCTTTGAAGCCTTTGACGGCTGCCTGTCACC

XbaI

-1987 CTTCTAGAAATGCTGCCACCGTGCTAGCTCAGGGGTTACAAAGAGTCCTGGAAGGTC

VDRE c-fos 5.2

-1930 ATATGGTTCAGCTCCGAGGTGGCGCCGTCCAGCAACATCTACTCCTATACCTCCATA

AP-1

-1873 ATTCGTGTTGAGTCATTCCTGTGGCTCAGGGTAAACTGCAGTGAACCCTGTGGACA

-1816 GTTCGTTACTTTAGGATTTGTGGAATTTCCCTGCACAGCGCTGAAAACAAACTCATG

-1759 TGGGTGTCGTGCTTAAAGGGCAGAAAGACTTTTTTTTTTATTAAAAAAAAAAGAAA

-1702 TAAGAAGCCTGTTTTTGCAACTTTATAACTTGTGTGATTTATATTCGATAGTCACAG

-1645 GTGTTGTCTTAATAAAGTGTGTACCAATGACCTTCAGAGTAATAGCACCAATTTACT

-1588 TAGAATGCATTCACCAAGAGATAACCCGAATGCTCCAAACAGCTTTGCACTATTCAG

10.I

-1531 CCTAACCCTGAAGAAGATTATTTAATCTATATTTATATTTAAAACAGAATTTTGTAG

-1474 GTGGCAACGGATGGCTAGTACAACAAAGGTTTGACTTAATTCTAATCTCAATTACTG

-1417 CGTGCCTCTCACACTGGGGATGTAAGGTAATAAGTCCTGTGTTCTCCATTTCAGGTA

7.4

-1360 ATGGAAGAAAGGAAGTAATTGCAGGGCAATACTAAATATTTAAAGCAGTTTGCTTCA

-1303 CGTATAATTTATCCCATTTGAATACCTCTGAACTATTCAGTAACGTTTTGTTAAGAC

-1246 AAATACTAAAACCTGCTTAAGTAGCCAACCTTATATATTCTAGGAAATAAAACTCAA

-1189 AAGAAGATAATTCGGTAATATAAGTTGTTTCTTTAAAGATATTCAGATACCCTTTAA

-1132 CGTGTACACCTTAAAGATCGATTCAAAGACGTTACAAATCACGGCTCCTTGCTCATG

-1075 CTGAAGATTTCATTCAGTAACAGTTAATACACACGTTCTTTAACATGTGGGGGAGGG

-1018 GGGCATTGCAACACGTGAGGTGCTGTTCACTGTACCAACGCTAGCTCATTGTCCTGC

VDRE

-961 AGATAGGCTAAGCGACTTGTGAAGGTCACACAGGGCACTGACTGCAGAAATCTGAAA

-904 GGCAGTCTGACTCCTGCAGTTTCCCCGCCCAGCAGGCATTCGGGCCTCACTCAAGCT

-847 GCTGTCTTCGCCGGCTCTCTCAGAAGCAACCACCACTTCTGTTCTATGTGGCCATAT

egr-1

-790 AGTTTAGATAGCGCGAGAACCATCACCCCCAGCTCCCCATCAGAGCAATGAGGTTCG

Eco RI

-733 TGTCTCTAAAGGTCAGTGGAGGCTCGAGAGGAATTCAGGGTCACGTGTGCTTTGTGC

-676 AGATGGCACATAGCGGCATAGTTCGAGTGTCAAAACCGAGGTGTTAGTTTGTTTTGC

-619 TTAGTTCCGAGGGGTTTTGTTTCTTTTGTGAGAAGGTCTCACTCTGTTGCCCATGCT

-562 GTCCTGAACACAGTACGTGCCCGAGGCTAGCCTCAAACTCACGGTGATCTTTTGAAG

-505 CACAAAACTGAGCCACCTATACCGTAAACAACCTAAGCTACTGAATACATGGCAAAA

-448 AAAAAAAAAAGAGGTTCTGAAATTTGTGCTGAGATAGCCGCTTTCAGTCTTCAGTTC

-391 AGCAAGTCTGAGAGAATCAAATGGTGTATCCAGTGTCCTTTATCTGTAACTTAAGAT

-334 ACGACAAGAGTCCATACCTTTCAGTTCTATTGATGTTTTCCCACTAATTCAGTAACA

ets

-277 ATAAAGAAAATCTCTCTTCCCTTCCTTATGGGTCCCTGATGCTCTTCCAGGATTCTA

-220 AATGCAGTCTATAAACGAAAAGGGTAGTTAATGACATTGTTCATAAGTAATGCTTTG

-163 GTGTGTTTCCTTCTCTTCCTTTTTTTTTTTAACCGCAAAACCAGAGTGGGGAGTGTA

CAAT

-106 GGAGCAGGTGGCCGGCCGTGGCAAAAACCTCATGACACATCACTCGCTCTGGTTGGT

TATA AvaII +1 Exon #1

-49 GGAGACCGTCTGAAACAGCGTTTAAATTCTGGGAGGTCCGAGCCACCAGCAGCAGGA

+9 CTGAAGGAGCTAAGCCTCAGCATCCTTGGCTTTGCAGTCTCCTGCGGCAAGCATTCT

Intron #1

+66 CGAGAAGCCAGCCAAGGTAAGCCTGCAGTTAAGCCTGCAGTGCCCGTGAGGCCGCAG

Eco RI

+123 TTCTCCTGGCTGAATTC

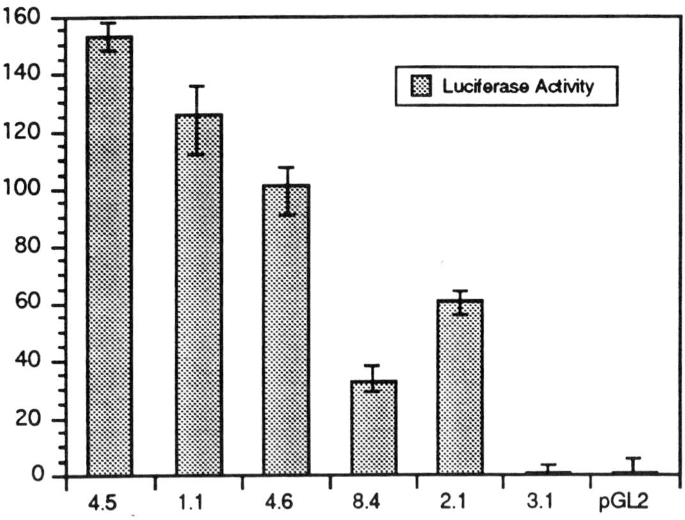

FIGURE 3. Luciferase expression of rat OPN promoter-luciferase constructs. The 4.5 rOPN-Luc construct was made by inserting a 4.5 XbaI fragment into the NheI site of the pGL2 vector (Promega, Madison, WI). Subsequent digestion with XhoI followed by religation provided a construct that contained from −1984 to +66 of the rat OPN promoter. Unilateral deletions from the 5′ end were accomplished by using the Promega Erase-a-base system. Deletion end points were determined by sequence analysis. ROS 17/2.8 cells were grown in Dulbecco's Modified Eagle's Medium-Ham's F12 (1 : 1) culture medium containing 10% fetal calf serum (DMEM-F12, 10% FCS). Cells were seeded at 40% confluency in 60-mm plates. Transfections were performed with 2 μg of test plasmid DNA per plate using 5 μl of lipofectin reagent (GIBCO-BRL, Gaithersburg, MD). Each plate was co-transfected with 2 μg of RSV-LacZ plasmid as a control. Transfections were performed in Opti-MEM media (GIBCO-BRL, Gaithersburg, MD) for 6 h after which the medium was replaced with DMEM-F12, 10% FCS. Cells were allowed to grow for 48 h after transfection, at which point they were washed twice with physiological buffered saline (PBS), then incubated in 400 μl of 1 × lysis buffer (Promega, Madison, WI) for 15 min. After centrifugation to remove cell debris, 40-μl aliquots were tested for luciferase activity using luciferase reagent (Promega) according to manufacturer's instructions and monitoring luminescence using a Turner Td-20e luminometer. β-Galactosidase activity was assayed in microtiter plates by addition of chlorophenol red-β-D-galactopyranoside (CPRG) and monitoring by absorbance at 560 nm. Data are from one of three experiments with similar results. Results represent triplicate samples.

FIGURE 2. 5′ flanking sequence of the rat *opn* gene from +139 to −2275 nt. GCG computer analysis revealed several putative transcription factor binding sites including sites for: early growth response (egr-1), ets, AP-1 and c-fos. The putative TATA and CAAT sequences are shown. The proximal and distal VDREs are shown and symbolized as *two direct repeating arrows*.

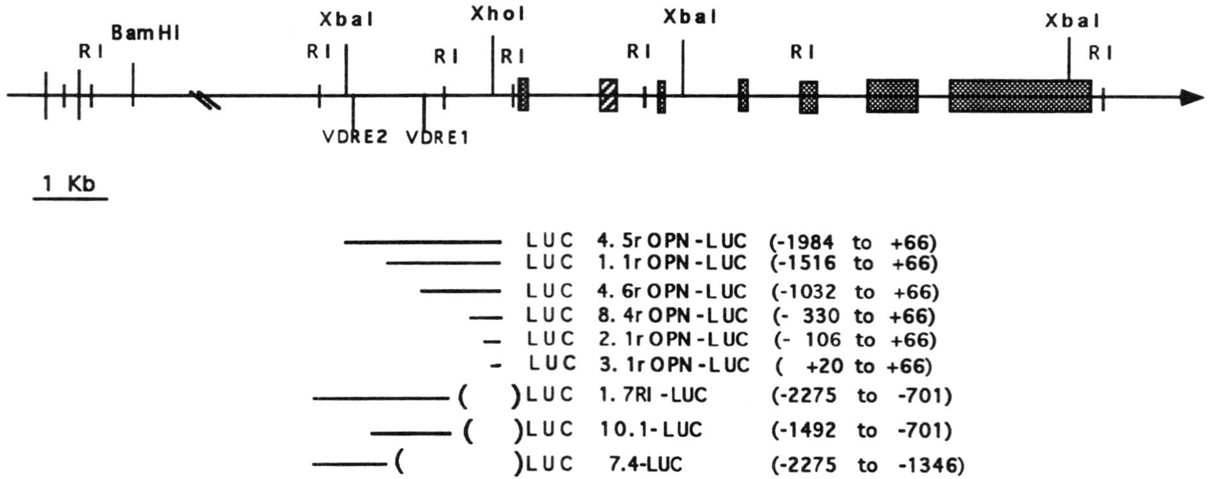

FIGURE 4. Partial restriction digest map of the rat *opn* gene showing exons (*hatched boxes*) and positions of both VDREs. Fusion constructs used in functional assays are diagrammed *below* the digest map.

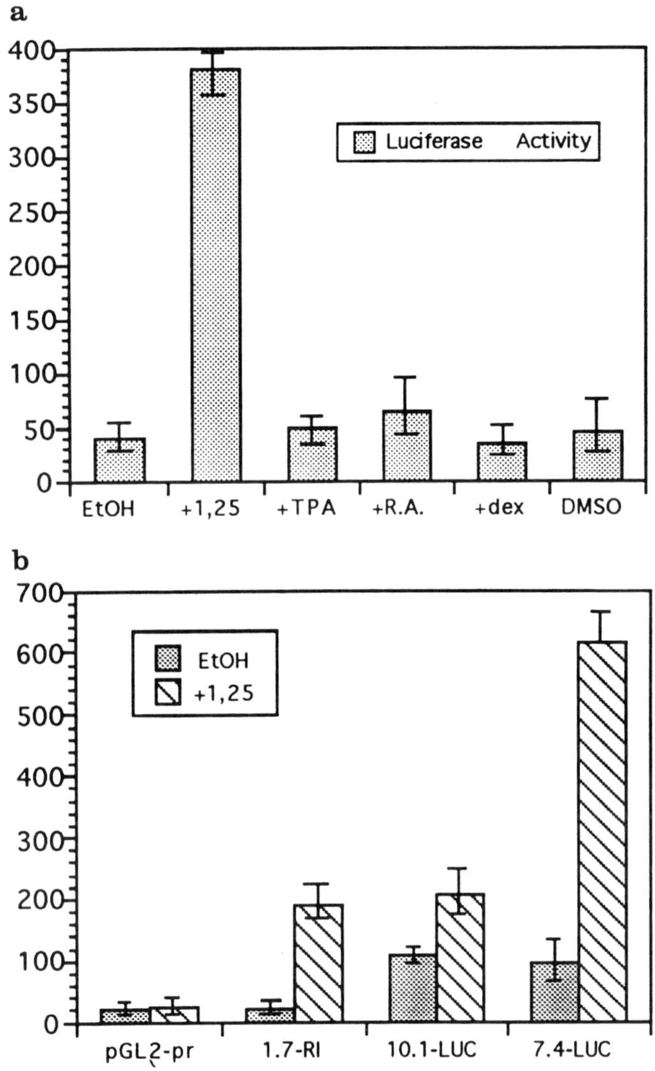

FIGURE 5. (a) Specificity of the 1.7-EcoRI construct to 1,25$(OH)_2D_3$. The 1.7-RI-LUC construct was made by subcloning the EcoRI fragment (−2275 to −701) into the pBS KS+ vector. The fragment was excised by a KpnI/BamHI digest and inserted into the KpnI/Bgl II sites of the pGL2-promoter vector which contains the SV40 promoter (Promega, Madison, WI). Transient transfections were performed as described previously. Cells were allowed to recover for 6–12 h before addition of 1,25$(OH)_2D_3$, dexamethasone (dex), all-trans retinoic acid (RA), phorbol esters (TPA) or vehicle control. The cells were incubated with test compound for 48 h prior to lysis. **(b)** Functional assays of constructs 10.1-LUC and 7.4-LUC. Constructs were made by Erase-a-base deletions of 1.7-RI-LUC and deletion end points were verified by sequence analysis. Transfections and 1,25$(OH)_2D_3$ induction were performed as above.

TABLE 1. Rat OPN VDREs Aligned against the VDRE Consensus Sequence and Several Reported VDREs

Gene	Nucleotide Sequence		
Calbindin D-9k	GGGTGG	CGG	AAGCCC
Rat osteocalcin	GGGTGA	ATG	AGGTCA
Mouse OPN	GGTTGA	CGA	GGTTCA
Human osteocalcin	GGGTGA	AGG	AGGTCA
Rat calcitriol 24-hydroxylase	GGGTGA	ACG	GGGGCA
Consensus	AGGTCA	AGG	AGGTCA
Rat OPN (proximal)	AGGTCA	CAC	AGGGCA
Rat OPN (distal)	AGGTCA	TAT	GGTTCA

with the VDR complex (Sara Peleg, personal communication). In preliminary experiments, we showed that a shifted complex is associated with each VDRE using nuclear extracts enhanced with the hVDR. Experiments are currently being performed to characterize the protein complexes.

ACKNOWLEDGMENTS

We thank Dr. Gideon Rodan (Merck, Sharpe & Dohme, West Point, PA) for the ROS 17/2.8 cells, Dr. J. Wesley Pike (Ligand Pharmaceutica Inc., San Diego, CA) for the pAd-hVDR plasmid, Dr. Mien Chie Hung (University of Texas M.D. Anderson Cancer Center, Houston, TX) for the RSV-LacZ plasmid and Dr. Kyong-geun Yoon for the ROP 3B-7 plasmid. In addition we would like to thank Dr. Sara Peleg (University of Texas M.D. Anderson Cancer Center, Houston, TX) for her assistance with the gel shifting assays.

REFERENCES

1. DARWISH, H. & H. F. DELUCA. 1993. Crit. Rev. Eukaryotic Gene Expression **3:** 89–116.
2. SONE, T., K. OZONO & J. W. PIKE. 1991. Mol. Endocrinol. **5:** 1578–1586.
3. KLIEWER, S. A., K. UMESONO, D. J. MANGELSDORF & R. M. EVANS. Nature **358:** 771–774.
4. MACDONALD, P. N., D. R. DOWD, S. NAKAJIMA, M. A. GALLIGAN, M. C. REEDER, C. A. HAUSSLER, K. OZATO & M. R. HAUSSLER. 1993. Mol. Cell. Biol. **13:** 5907–5917.
5. PRINCE, C. W. & W. T. BUTLER. 1987. Collagen Relat. Res. **7:** 305–313.
6. NODA, M., R. L. VOGEL, A. M. CRAIG, J. PRAHL, H. F. DELUCA & D. T. DENHARDT. 1990. Proc. Natl. Acad. Sci. USA **87:** 9995–9999.
7. YOON, K., R. BUENAGA & G. A. RODAN. 1987. Biochem. Biophys. Res. Commun. **148:** 1129–1136.
8. MIYAZAKI, Y., M. SETOGUCHI, S. TOSHIDA, Y. HIGUCHI, S. AKIZUKI & S. YAMAMOTO. 1990. J. Biol. Chem. **265:** 14432–14438.
9. BEHREND, E. I., A. F. CHAMBERS, S. M. WILSON & D. T. DENHARDT. 1993. J. Biol. Chem. **268:** 11172–11175.
10. RAFIDI, K., I. SIMKINA, E. JOHNSON, M. A. MOORE & L. C. GERSTENFELD. 1994. Gene **140:** 163–169.
11. CRAIG, A. M. & D. T. DENHARDT. 1991. Gene **100:** 163–171.

Regulation of Avian Osteopontin Pre- and Posttranscriptional Expression in Skeletal Tissues[a]

L. C. GERSTENFELD, T. UPOROVA, S. ASHKAR,
E. SALIH, Y. GOTOH, M. D. McKEE,[b] A. NANCI,[b]
AND M. J. GLIMCHER

Laboratory for the Study of Skeletal Disorders and Rehabilitation
Department of Orthopedic Surgery
Harvard Medical School and Children's Hospital
Boston, Massachusetts 02115
and
[b]Laboratory for Electron Microscopy
Department of Stomatology
Faculty of Dentistry
University of Montreal
Montreal, Quebec, Canada H3C 3J7

Osteopontin initially was characterized as one of the predominant noncollagenous proteins which is accumulated in the extracellular matrix (ECM) of bone in a wide variety of vertebrates.[6,21,29,40,41,45,68,71–73] Independent of studies on osteopontin within bone tissue, it was also identified as an early response gene during T-cell activation[43,57] and was induced by phorbol ester tumor promoter treatment in JB6 epidermal cells.[8,12,58,59] It was concurrently identified as a serum marker of malignant cell growth and was shown to be present within metastatic tumors from a wide variety of primary tissue origins in mice and humans.[7,11,23,53,54,59] More recently it was shown to be expressed in varying levels in many other vertebrate tissues.[4–6,26,28,34,36,40,55]

In studies from this laboratory which have primarily focused on characterizing the noncollagenous proteins of chicken bone, osteopontin was initially identified as a phosphorylated glycoprotein of ~66 kDa in the mineralized ECM of postnatal chicken bones.[21] Furthermore, osteopontin expression was found to be temporally concurrent with alkaline phosphatase, an early marker of osteoblast differentiation, during chicken osteoblast differentiation *in vitro*.[17] Osteopontin was shown to contribute approximately 2.0% to the total protein synthesized by osteoblasts,[21,35] and quantitative analysis of mRNA expression in mature osteoblast cultures demonstrated that it is expressed at ~100-fold lower levels than collagen type I and one half the levels seen for osteocalcin,[37] the most abundant noncollagenous protein expressed by osteoblasts. During chicken embryogenesis *in vivo*, osteopontin is expressed during bone development at much higher levels when skeletal development is well advanced, and maximal levels of expression are attained during postnatal development. These results demonstrate that osteopontin is quantitatively one of the most abundant noncollagenous proteins expressed by osteo-

[a] This work has been supported by NIH Grants HD22400 and AR34078.

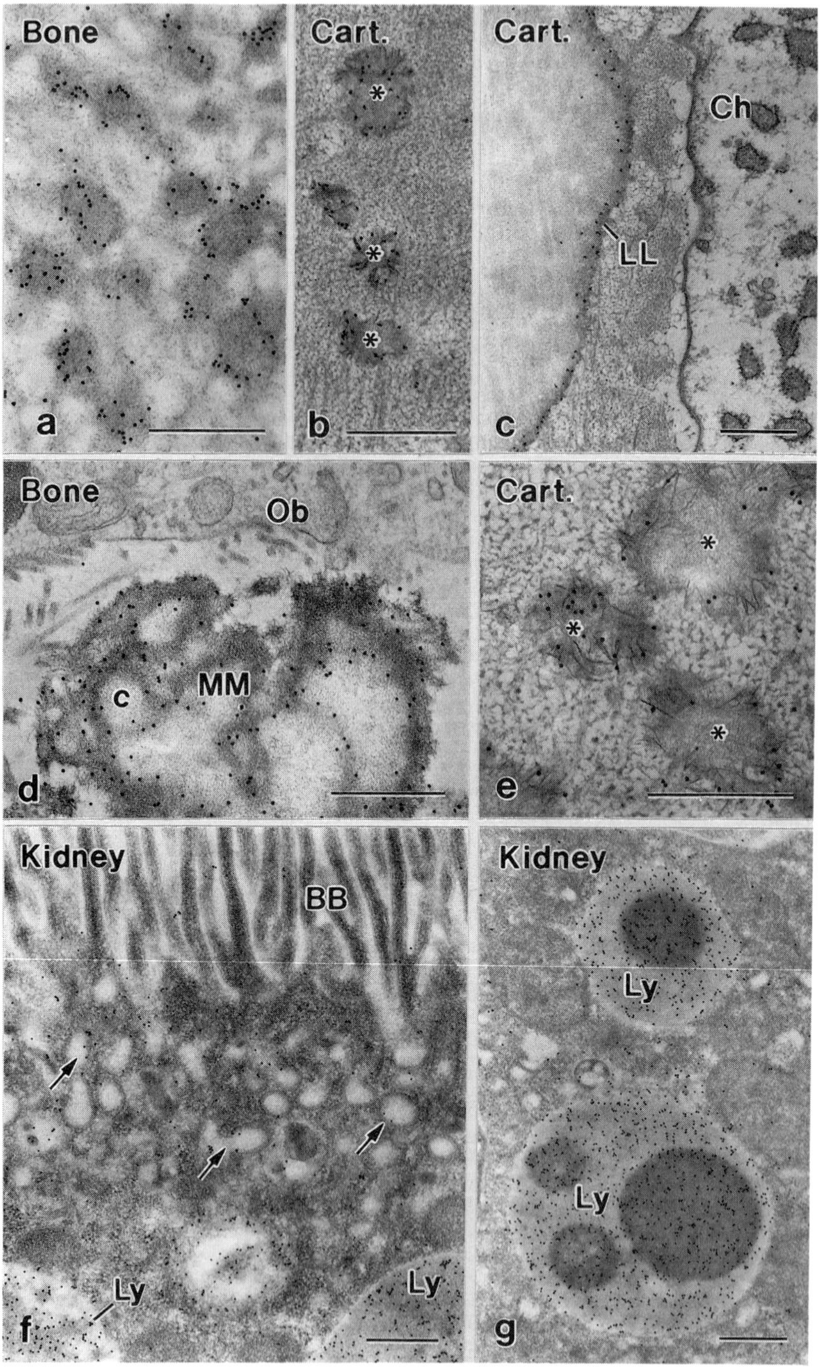
Bone
a
Cart.
*
*
*
b
Cart.
Ch
LL
c
Bone
Ob
c
MM
d
Cart.
*
*
*
e
Kidney
BB
Ly
Ly
f
Kidney
Ly
Ly
g

blasts, and its expression reaches its highest levels after the skeletal stem cell has committed to the osteoblast lineage.[35,70]

Immunolocalization of osteopontin at the histological level, within whole chicken embryos, demonstrated that the protein is first detected at stage 32 of embryonic development within the mineralized bone ECM throughout the animal's entire skeleton, and it continues to be expressed in these tissues throughout embryogenesis. Antibody reaction was also seen in the lining cells of the large intestine and the luminal lining cells of the embryonic kidney in stage-35 embryos.[5] Osteopontin was also shown to be expressed by chicken hypertrophic chondrocytes *in vitro*[27,37,67] and was detected *in vivo* in the mineralized matrices of growth plate cartilage undergoing endochondral ossification.[32] The results for the expression of osteopontin by osteoblasts and hypertrophic chondrocytes *in vitro,* its *in vivo* levels of expression during development and its tissue distribution within embryonic chickens are consistent with results seen for the expression of the osteopontin gene in mammals.[9,25,29,30,63,72] These data suggest, therefore, that osteopontin probably has generally maintained it basic functionality within the same tissues throughout vertebrate evolution.

Because osteopontin is an ECM protein, its ultrastructural distribution within the matrices of the various tissues in which it is expressed has been of considerable interest in understanding its possible functions. FIGURE 1 compares its ultrastructural localization in 17-day chicken embryonic bone with mineralized cartilage tissue *in vivo,* as well as with that seen in the ECM produced by cultured osteoblasts and hypertrophic chondrocytes *in vitro*. It is interesting to note that both cartilage and bone tissues produced *in vivo* and *in vitro* have the same localization of the protein primarily in electron-dense foci and irregular patches of matrix which have been shown to be rich in mineral.[30–32] The ability of osteoblasts and chondrocytes *in vitro* to synthesize and assemble osteopontin in their matrices in an ultrastructural fashion comparable to that seen *in vivo* suggests that these cells *in vitro* have maintained the phenotypic ability to synthesize protein which intrinsically assembles itself into the matrix and associates with the mineral phase. Within chicken cartilage, large areas of mineralized matrix are delimited by a lamina limitans that contains osteopontin. In comparing these results to those seen for the same tissues in mammals, near identical ultrastructural morphologies are observed.[30] While considerable data exist for the ultrastructural localization of this protein in mineralized tissues, much less is known for the kidney. In the embryonic chicken kidney *in vivo,* osteopontin shows a very specific association with polymorphic lysosome-

FIGURE 1. Electron micrographs of chicken bone, cartilage and kidney *in vivo* and *in vitro* after postembedding, colloidal-gold immunocytochemistry for osteopontin. During endochondral ossification in embryonic long bones, OPN is found primarily in electron-dense "patches" of organic matrix in mineralized bone **(a),** and in cartilage within small foci (*asterisks*) of calcification **(b)** and at the periphery of larger masses of calcified cartilage **(c)** delimited by a lamina limitans (LL). *In vitro,* chicken osteoblasts (Ob) **(d)** and chondrocyte **(e)** cultures also synthesize and assemble an extracellular matrix (MM, *asterisks*) that calcifies and shows intense labeling for OPN in a manner similar to that seen *in vivo*. In embryonic chicken kidney epithelial cells *in vivo* (**(f)** and **(g)**), osteopontin labeling is primarily associated with structures of the apical tubulovesicular system (*arrows*) and polymorphic lysosome-like elements (Ly) found dispersed throughout the cytoplasm. Brush borders (BB) are seen on the lumenal epithelial surface. Sections were obtained from undecalcified specimens and were conventionally stained with uranyl acetate and lead citrate. *Bars* = 0.5 μm.

like structures and elements of the tubulovesicular system. The epithelial cells over which the osteopontin is localized have a distinct brush border, providing evidence that embryonically these cells most likely are located within the proximal convoluted tubules. However, at this time, direct comparisons to mammalian kidney structures are not available and there are significant differences between both mammalian and avian kidney anatomy and physiology as well as between embryonic and postembryonic kidney functions.

Sequence analysis of osteopontin from various species has identified a number of conserved functional domains within this protein's primary structure (FIG. 2). These domains include: the signal peptide and the amino terminal end of the mature protein; several recognition sites for factor-independent casein kinase-like enzymes; several recognition sites for N- and O-linked glycosylation; a polyaspartate domain; and an RGD recognition site for integrin-mediated cell adhesion.[44] The amino terminal portion of this protein through the first 170 amino acids shows a much higher sequence similarity (~45%) than the carboxyl half of this protein (~25%) when comparing the osteopontins from various species.[35] The one exception to the lack of sequence conservation that was seen for the carboxyl portion of the protein was observed for two of the phosphorylation recognition sites (FIG. 3). It is interesting to note that for the available gene structures, the intron and exon organization has been shown to be conserved[48] with the following genomic organization: a 5′ exon which encodes untranslated sequence;[2] a signal peptide; a recognition sequence for phosphorylation by casein kinase II; a domain containing a linkage site for O-glycosylation; a second casein kinase II phosphorylation site; an exon containing three functional regions, the poly-Asp sequence, the RGD integrin recognition site and a potential N-glycosylation site; and a large C-terminal exon that contains both a potential site for N-glycosylation and three phosphorylation sites.

The identification of the RGD sequence in the osteopontin molecule has led to a major research effort directed towards identifying the functional characteristics of the integrin-binding domain. Numerous studies have investigated the ability of osteopontin to promote cellular adherence *in vitro* of a wide variety of cells including osteoblastic,[21,42] fibroblastic,[52,60,61] and osteoclastic[16,50] cell types. Available data indicates that this protein specifically interacts with the $\alpha_v\beta_3$ integrin isotype,[33,42,50] and a number of hypotheses have been put forward suggesting that osteopontin mediates osteoclast attachment to the mineralized ECM during bone resorption, thus playing an important role in this process.[49,50] An aspect of this protein which has been a major focus of the research from our laboratory is understanding the mechanisms by which osteopontin is phosphorylated and what the functional consequences of this aspect of the molecule's structure are. In our initial studies osteopontin was shown to be intracellularly labeled by ^{32}P in cultures of primary chicken embryo osteoblasts, and the label was isolated in both O-phosphoserine and O-phosphothreonine after protein hydrolysis and amino acid analysis.[21]

In more recent studies, the sites of phosphorylation of osteopontin were determined by metabolic labeling of the protein in cultured osteoblasts, isolation of the labeled protein, and protein sequence analysis of the phosphorylation sites.[1,51] A comparison of the results obtained from the labeling studies within cultured avian osteoblasts and that seen for osteopontin isolated from bovine milk[62] is presented in FIGURE 3. Of the sites which were identified, four out of six were conserved between the two different species of proteins and these sites were the same for protein synthesized by both mammary tissue or within bone. Whether the kinase enzyme(s) found within these different tissues are the same, or whether the nature

```
CHICKEN     -16   MKLAFLCLCFISIAAA/WPVSKSRQHAISASSEEK        18
RABBIT      -16   MRIAVICFCLLGMAYA/LPV----KHADSGSSEEK        14
BOVINE      -16   MRIAVICFCLLGIASA/LPV----KPTSSGSSEEK        14
HUMAN       -16   MRIAVICFCLLGITCA/IPV----KQADSGSSEEK        14
PORCINE     -16   MRIAVIAFCLWGFASA/LPV----KQTNSGSSEEK        14
RAT         -16   MRLAVVCLCLFGLASC/LPV----KVAEFGSSEEK        14
MOUSE       -16   MRLAVICFCLFGIASS/LPV----KVTDSGSSEEK        14
CONSENSUS         MrlAvicfCl-giasa/lPV----k-a-sgSSEEK

CHICKEN      44   QTQNDLASLQQTHYSSEENADVPEQPDFPDIPSKSQEAV---DDDDDGDNDS  92
RABBIT       49   QKQNLLT--PQNAMSSEEKDD----LKQETLPSKSIESHDHMDDIDEDEDDD  95
BOVINE       49   QKQTFLT--PQNSVSSEETDD----NKQNTLPSKSNESPEQTDDLDDDDDNS  95
HUMAN        34   QKQNLLA--PQNAVSSEETND----FKQETLPSKSNESHDHMDDMDDEDDDD  79
PORCINE      34   QKQTFLA--PQNTISSEETDD----FKQETLPSKSNESPEQTDDVDDDDDED  79
RAT          34   QKQNLLA--PQNSVSSEETDD----FKQETLPSNWNESHDHMDDDDDDDDDG  79
MOUSE        34   QKQNLLA--PQNAVSSEEKDD----FKQETLPSNSNESHDHMDDDDDDDDDD  79
CONSENSUS         QkQnlLA--pQn-vSSEEtdD----fkqetlPSKsnEshdhmDDdDDdDdDd

CHICKEN      96   DESDEVVTDFPTEAPV----TPFN-RGD-NAGRGDSVAY-GFRARSK       135
RABBIT      124   DESDEV-TVYPTEDAATIVETEVVPTYETYDGRGDSVAYRLKRSKSK      169
BOVINE      122   DESDEV--DFPTDIPTIAVFTPFIPTESANDGRGDSVAY-GLKSRSK      169
HUMAN       111   DESDELVTDFPTDLPATEVFTPVVPTVDTYDGRGDSVVY-GLRSKSK      158
PORCINE     107   DESDELVTDFPTDTPATDV-TPAVPTGDPNDGRGDSVVY-GLRSKSK      151
RAT         100   DESDESFT-ASTQADV---LTPIAPTVDVPDGRGDSLAY-GLRSKSR      141
MOUSE       100   DESDETVT-ASTQADT---FTPIVPTVDVPNGRGDSLAY-GLRSKSK      141
CONSENSUS         DESDE-vTdfpT-ap-----TP-vptvD--dGRGDSvaY-GLRSRSK
```

FIGURE 2. Comparison of the conserved domains within the osteopontin protein. Negative numbering of the sequences denotes the unprocessed signal peptide which is absent in the mature protein. Positive numbering begins from the first amino acid of the mature protein. The mature sequence is separated from the signal peptide sequence by a *slash*. Sequences are aligned in a linear fashion and absent amino acids are denoted by *hyphens*. Protein sequences from the different species were derived from the published literature.[12,25,35,41,66,69,72]

IDENTIFICATION OF PHOSPHORYLATION SITES

```
CHICK BONE              BOVINE MILK            POSITION OF SITE   ENZYME
 *  **                   ** **
ISASSEEKYD 27-35        TSSGSSEEKQ 22-29         SAME             CK II
   **                    * **
THYSSEENAD 71-80        NSVSSEETDD 59-68         SAME             CK II
   *                        *     *
DDATAEVGDS 168-177      PDATEEDFTS 160-169       SAME             CK II
                           *
                        HIESEEMHDA 170-179       ABSENT           CK II
      *                 *     *
SRPRFDSPEV 206-215      SKET-NSSEL 189-198       SAME             CK II
   *
SRESLASRSA 228-237                               ABSENT           UNKNOWN
                        *    * *
                        SHELDSASSE 264-273       ABSENT           UNKNOWN
```

COMPARISON OF AMINO TERMINAL SEQUENCES OF BSP AND OPN

```
       SIGNAL PEPTIDE                       CK II

BSP   MRTALL-LACLLATASAFSV-RSLLR-RARAGDSEE  NAVLKSRH-RYYLYRYAYPPLH
      :. :.: : : .. :.:..: .:  : .: .. :::  ..  .:.: .    ::     :
OPN   MKLAFLCL-CFISIAAAWPVSKS--RQHAISASSEEKYDP--RSHHTH----RY-----H

               CK II      CK II          CK II
BSP   RYKVGSDSSEE--EGDGSEEEEEGGAP-SHAGTQAAGE-GLTLGDVGP
         .: :.: .:  . .. ...   ..  .: ...  .. .    .  :
OPN   QDHVDSQS-QEHLQ-QT-QNDL-ASLQQTHYSSEENAD-VP---E-QP
```

FIGURE 3. Comparison of the amino terminal domains of two bone sialoproteins (BSP and OPN)[35,70] and identification of the phosphorylation sites of avian OPN in comparison to bovine OPN. In the *upper portion* of the figure, ten amino acid sequences containing the sites of phosphorylation determined by sequence analysis of metabolically (^{32}P)-labeled avian osteopontin synthesized by osteoblasts are compared to that derived from sequence analysis of osteopontin purified from bovine milk.[62] The positions of the phosphorylation sites within the primary sequences are denoted by *asterisks* and are numbered in the figure. The conservation of sequence and its enzyme recognition are denoted in the figure. In the *lower portion* of the figure, a comparison of the signal peptide and amino terminal sequences of avian OPN and BSP is depicted. *Double dots* denote exact amino acid conservation while *single dots* denote conservation of amino acid charge. Sequences are aligned in a linear fashion and absent amino acids are denoted by *hyphens*. The positions of the signal peptide and possible recognition sites for casein kinase II factor-independent kinase (CKII) are denoted in the figure.

of the sequence motif for phosphorylation is permissive in its recognition by multiple isoforms of enzymes which recognize it, must await purification and sequence analysis of the enzymes from bone and the other tissues in which the osteopontin gene is expressed. Variation in several of the phosphorylation sites that are present in the carboxyl ends of the molecule which do not fall in the class of sites recognized by the factor-independent class of protein kinases,[15,24,65] suggests the possibility that at least two classes of enzymes are capable of modifying the protein. Other data that are consistent with this possibility are results with JB6 cells that show variation in the extent of phosphorylation in the presence of either $1,25(OH)_2D_3$ or after induction of the protein with phorbol-12 myristate-13-acetate[8] and differences in the extent of phosphorylation seen in studies of cultured kidney cells.[35,36] At the present time, very little is known about the functional significance of the protein's phosphorylation. However, both the phosphate and carboxyl side chains may play a role in either the interaction of the protein with calcium[10,47] or with the mineral phase of bone, or it may have a function in controlling mineral initiation.[3,19,20]

Of considerable interest is the intracellular processing pathway by which osteopontin is modified and the intracellular compartment(s) in which the osteopontin is phosphorylated. It is to be noted in this context that bone sialoprotein (BSP), a second protein which also undergoes phosphorylation and, which is primarily expressed in skeletal tissue, contains the CK II recognition motif, and has a very similar amino terminal sequence when compared to osteopontin (FIG. 3 lower panel). These data suggest that the amino terminal structure for these two proteins may direct the intracellular routing of these proteins or confer the proper conformation for recognition by the factor-independent kinase enzyme within osteoblasts. In preliminary experiments, different pharmacological agents that block the transit for the protein at various stages of the secretory pathway in combination with radioactive pulse chase analysis provided biochemical evidence of the processing pathway of osteopontin. This deduced pathway for the posttranslational steps of modification of osteopontin is schematically summarized in FIGURE 4. These data indicate that the phosphorylation of osteopontin occurs intracellularly and further suggest that it is restricted to the Golgi apparatus.[1] The results from our studies are interesting since unlike casein kinase II which was previously shown to be either cytoplasmic or nuclear in origin[15] the enzyme that modifies osteopontin is intracellularly compartmentalized suggesting that either there is a unique mechanism by which casein kinase enzyme is localized to the Golgi or there is a unique form of the enzyme involved in the phosphorylation of osteopontin. The latter possibility is suggested since mammary gland casein kinase is unique and is found in the Golgi compartment.[24] Such a mechanism of compartmental phosphorylation could be used as a regulatory process by which different states of osteopontin phosphorylation could be generated.

In the last section of this review, a summary of the results pertaining to the studies examining the molecular mechanisms by which the avian osteopontin gene expression is regulated in skeletal cells is given. FIGURE 5 presents 1185 nt of the promoter sequence of the avian osteopontin gene while FIGURE 6 presents a comparison of the proximal 5′ promoter sequences for the gene from three different species. In comparing these different 5′ osteopontin promoters as well as comparing the avian 5′ osteopontin promoter to that of other avian genes, both species-unique and cross-species cis elements that possibly play a regulatory role were identified in the 5′ promoter sequence. Two species-specific elements were identified. A tandem repeat at −535 to −550 (TGATTCTGTGATTCTC) which is identified as a CR1 sequence. This sequence is a chicken-specific, middle repetitive sequence which was first identified in the ovomucoid gene and defines DNase hypersensitive sites for a number of transcriptionally active chicken-specific genes.[64] The second element is localized at −129 to −145 (CCACAAAGCCAAGAG). While this sequence is found in the proximal promoters of all of the osteopontin genes from different species, it is primarily seen in a number of other avian and *Xenopus laevis* promoter sequences. It is interesting to note that both the CR1 sequence and the sequence between −129 to −145 are also found in common in the promoter of the avian tenascin/cytotactin gene, which is a developmentally regulated adhesion glycoprotein.[22]

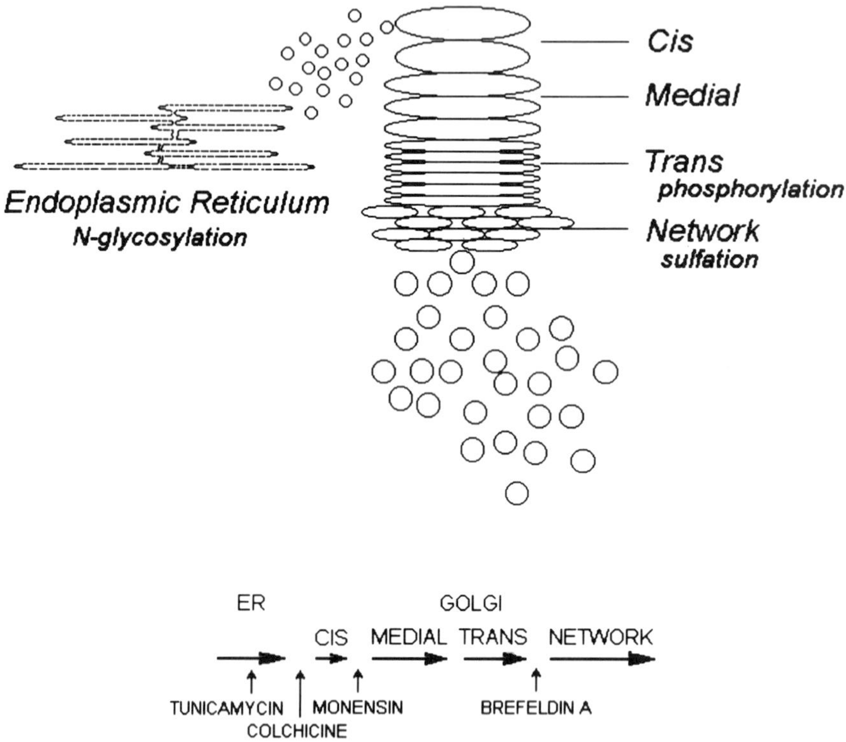

FIGURE 4. Diagrammatic representation of the processing pathway of osteopontin and experimentally determined sites of the intracellular compartments in which the various stages of posttranslational modification take place. *Arrows* for each of the metabolic inhibitors indicate the point up to where the inhibitor exerts its effect and where further posttranslational modifications are blocked. Based on experimental determination with the various metabolic inhibitors, the site of phosphorylation has been localized to the trans Golgi network.

The comparison of the first ~250 nt of the proximal promoters of the osteopontin genes from three different species show a very high degree of conservation of the elements found in this region of the promoter including TATA, CCAAT and AP1 sequences. An additional sequence at -180 to -205 was also observed and was found in all of the osteopontin gene promoters. In examining the functional characteristics of the avian osteopontin promoter, it was shown that the proximal promoter was very active both in mammalian osteoblasts[48] and in various avian skeletal cell populations. In either heterologous transfection assays or in assays of avian skeletal cell populations the proximal promoter showed at least $1.5\times$ levels of activity when compared to the SV40 promoter with its enhancer. In order to assess the specific nature of the promoter elements in skeletal cell populations, a comparison of the promoter activities in three different skeletal cell populations was carried out. These cell populations were: nonhypertrophic chondrocytes isolated from the caudal sterna of 17-day embryonic chicken in which the osteopontin gene is not expressed; immature or presumptive osteoblast cell populations isolated

5'

1185 GTGAGCAGTC ACTGAGCGAA TTTGGGTCTA TTTCTCCTTA TGTTTCTAAA CAGACAACAA AATGTAAGTA CAGGATGTCT

1105 GGTGGCAATG GCTGTGACAA AATGTTTATC TCATCTGTGC ACGATTAAGG ACAAACAGCC ATTCTCTCAC AAAGCGAACA

1025 CCTTCAAAAG CTGCAAAATC AGTAGCCAGA TATGTTGTGG ATGCTCTATG TGTGGAGTGC TCAAGGCCAG TTTGGATGAG

Δ946

955 GCCCTGGAGC TAGCCTGATC TGCTGGGTGG CAATCAGCCG CATGGGAAGG GGGCTGGGGT TGGGTAGGCT TTGAGGTCCC

875 TTCCAACTCA AACCATTCTG TGGTTCGATG ATCAAAATTA CAGAGAGAGA TAACAGCTTT AAGACAGCTA TGAACACTTT

795 TGCCTTCTAA AGACTGCAGG GAAGGATTTT CATGTGCTTT AATGACACGA CTCAATTGAA ATTTCAAAGA TACACGTGAG

715 GGAGAAAGGG TCTGGCACAG GCTTGCCCGG AGAGGTAGAT GGGAGTCACT GACCCTGGAG GTGTTCAAGA AATGTGGAGC

635 GTTGGGCTCA CTGAGGGACA TGGCTAGTGA ACGTGGTGGG GATGGCGCTG GGGTCAGAGT TGGTGAGCTT AGAGGTCTTT

CR1

555 TCCAA**TGATT CTGTGATTCT C**TGTGAGACT TGCAGCATGT CAGGTGCGGC CCCAAGATCC AGGCAAGGCA CAGCTGGAAC

475 AGACCCCCAT CAGCAGAGCT GACACTGCTA GATAGCAGAC ATAGGGTTTC TTCCTACGAT GTCCCTTGCT TCCCAAACCT

Δ368

395 CTTGGCACTC CTCACACTGC AGACATGACA CATCCCTACC CCTTACAGAA CAGGGCCCCT TACAGAACAG GGCGTGAAGG

315 GTTTCTGGGA ACCGGGCTCT GACTCCTTGC AAAAGGAGGA CGGGATGTGC CCCAAAGCCC TCCCTACACA CGATACCAAC

235 AACTGTGCAT TCAGCAGACC TGCTGCTGGT TAATGATCTT GCCCATAAGT AATGTTTTAA CCATAGAGTA GCTTTTGTTT

AP1 Δ77

155 GCTTCAGAAG CCACAAAGCC AAAGAGGAAG TGGATGAGAA CGTGGGCAGG ATGGTGGCCA ACAG**CCTCAT GACC**TCAGCT

CAT **TATA** 3' *tsp*

75 TGTTCCGCTG CACAGG**GATT GG**TGGCTGCT CTCTGGGACA GGGTT**TAAAT T**CCCCACGGG TGCAGATAGC CAGGC

FIGURE 5. Promoter sequence of the avian (chicken) *opn* gene. The 5′ promoter sequence encompassing-1195-+1 nt is presented.[48] The start positions of three promoter sequences used to drive the bacterial *cat* gene in transient transfection assays depicted in FIGURE 7 are denoted in the figure. The identification of several possible regulatory elements are denoted in the figure.

from 12-day embryonic chicken calvaria in which the gene is expressed at low levels; and mature osteoblasts isolated from 17-day embryos in which the gene is maximally expressed. A summary of these results is depicted in FIGURES 7A and B. Within populations of skeletal cells of increasing differentiation there was a progressive increase in the activity of the promoter, with the longest promoter construct showing the highest activity. Interestingly, even the shortest construct of −77 nt showed higher activity for the avian promoter in the most differentiated skeletal cells. Within chondrocytes, there was little difference between the promoter sequences of different lengths suggesting that in cell types in which the gene in not normally expressed, that the basal elements TATA and CCAAT are what is minimally needed for activity, but tissue specificity is probably imparted by the more 5′ sequences. Interestingly, in osteoblasts where the gene is maximally expressed there was no response to PMA. These results suggest that there are at least two separate mechanisms by which this gene is regulated. One set of mechanisms are operative in skeletal cells which produce maximal levels of gene expres-

```
MUS -212          GAAAAG--GGTA--G-TTAATGA-CATCGTTCATCAGTAATGCTTT--GTG
                  |||||   ||||  | ||||||| ||| || ||| ||||||| |||   ||
PIG -230    CA--C-GAAAAA--GCTA--G-TTAATGA-CATTGTACATAAGTAATGTTTTAACTG
                ||  | | | |   |||   | ||||||| | |||   |||||||||||||||||
CHK -227 GTGCATTCAGCAGACCTGCTGCTGGTTAATGATC-TTGCCCATAAGTAATGTTTTAAC--

MUS      T-GT-GTTTCCTTT-TCT-TC-CT-T-T-T-T-T-TT-TAA-CCACAAAACCAGAGGA
           |  |     | | | | |   | | | | | | | || ||| |||||||||||||||
PIG      CAGAGGGCGTGTGTGTGTGTGTGTGTGTGTGTGTTGTAA-CCACAAAACCAGAGGG
         || ||  |  |  |  | | | ||| | ||  | |  |    || ||||||| ||| || |
CHK      CATAGAG--TA-GC-T-T-T-TGT-T-TGC-T-TCAG----AAGCCACAAAGCCAAAGAG

                                                       AP 1
MUS      GGAAGTGTAG-GAGCAG-GTGGGCCGGGTA-G-TGGCAAAA-A-CCTCATGACA-CATCA
         |||||||| | |||||  |||||  ||| | | ||||| || | |||||||||| ||||
PIG      GGAAGTGTGG-GAGCAA-GTGGGTTGGGCA-G-TGGCACAA-A-CCTCATGACA-CATCT
         | ||||| |  ||| || |||||  ||  | | |||| ||| | |||||||||  || ||
CHK      G-AAGTG-GATGAG-AACGTGGGCAGG--ATGGTGGC-CAACAGCCTCATGACCTCAGCT

                           CAT                         TATA
MUS      C-T-CCACCTC-CTG--ATTGGTGGAGACTGTCTGG-ACCAGCATTTAAATTCT----GG
         | | ||  | |  ||   ||||||| |  ||| |||   |||| |||||||||     ||
PIG      C-T-CCGCCCC-CTGTG-TTGGTGGCGGATGTCTGGTG-CAGCCTTTAAATTC---A-GG
           | ||||  | | | | |||||||| | | ||||| | |||  |||||||||   | ||
CHK      TGTTCCGCTGCACAGGGATTGGTGGCTGCTCTCTGG-GACAGGGTTTAAATTCCCCACGG

                                +1
MUS      G---AG-TCCTTGCTTGGGTTTGCA
         |   || |||   ||        |
PIG      G-G-AGGTCCAGGCTGT--CAG-CA    3' tsp
         | | || |  || |     ||| |
CHK      GTGCAGAT--AG-C-----CAGGCA
```

FIGURE 6. Comparison of the sequences encoding the proximal promoter of chicken (CHK),[48] mouse (MUS),[13,38] and porcine (PIG)[74] osteopontin genes. The positions of the TATA, CCAAT, and AP1 elements are denoted in the figure. The sequence is presented beginning from the +1 nucleotide of the transcriptional start site (*tsp*). Overall sequence similarity is ~65% between the three sequences.

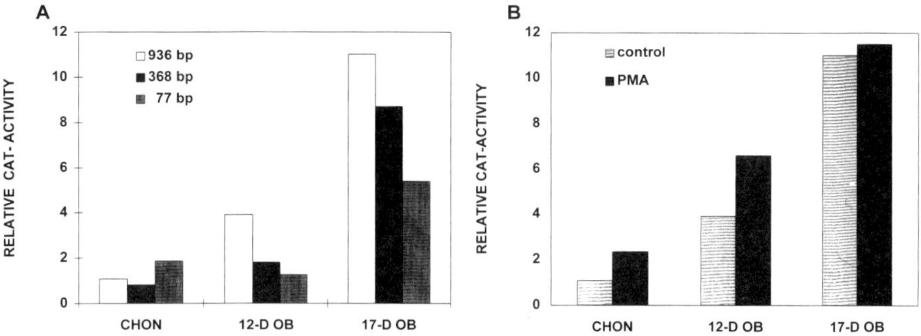

FIGURE 7. Differential activities of the avian osteopontin gene promoter in various skeletal cell populations. Osteopontin promoter activities were assayed by transient expression assay of the bacterial *cat* enzyme activity which was used as the reporter gene. Activities were measured as previously described[48] and are normalized to the ratio of the background activity of the CAT basic vector (Promega Inc., WI) minus promoter, to the activity of the CAT vector driven by the SV40 promoter and enhancer. This activity is set to 1. The *opn* promoter activities were measured in three primary skeletal cell populations: nonhypertrophic caudal sterna chondrocytes isolated from 17-day embryonic chicks (CHON); immature or nondifferentiated populations of presumptive osteoblasts isolated from the calvaria of 12 day chick embryo calvaria (12-D OB); and mature populations of committed osteoblasts isolated from 17 day chick embryonic calvaria (17-D OB). **(A)** The promoter activities of three separate promoter constructs containing sequences from −77, −368 and −936 nt 5′ to the *tsp*. **(B)** The response of the −936 nt promoter construct in three different primary skeletal cell lines to phorbol-12-myristate-13-acetate (PMA) treatment. PMA addition and concentration are as previously described.[48]

sion through either the presence of positive tissue specific factors or inactivation of negative factors that are operative independent of the AP1 site and are more 5′ to this element. A second set of mechanisms which are operative work through the AP1 site and probably reside within the proximal promoter.

In summary, the results presented here provide comparative data from a non-mammalian vertebrate in which conserved protein sequence and DNA regulatory elements were identified for osteopontin. Osteopontin is expressed during chicken embryonic development in a comparable fashion to that seen in mammals. It has maintained a large number of its sequence features including its RGD domain and its phosphorylation sites. Of considerable interest was the high degree of sequence conservation of the regulatory elements of the osteopontin genes' proximal promoter and the strength of the avian promoter when assayed in mammalian cells. Comparison of the regulation of the *opn* promoter in skeletal cell populations of varying degrees of differentiation provides evidence of two broad categories of mechanisms that regulate this gene activity. One set of mechanisms that are operative under normal developmental conditions which are tissue specific and are probably operative in tissues such as bone and kidney which produce maximal levels of expression. A second set of mechanisms are then operative that may be turned on in a ubiquitous fashion in response to exogenous activators occurs in early gene response such as what is seen in macrophages or T cells or during PMA induction or in response to factors such as TGFβ[39] or other cytokines.

ACKNOWLEDGMENTS

The authors wish to acknowledge all of our colleagues and technical staff who have worked on this research over the last several years. They include Drs. Mark Moore, Bill Landis, and Anne Todd, Mr. Erik Johnson, Ms. Kris Rafidi, and Ms. Irina Simkina. We wish to thank Lisa Lagasse for her editorial assistance with the manuscript.

REFERENCES

1. Ashkar, S., J. L. Schaffer, E. Salih, L. C. Gerstenfeld & M. J. Glimcher. 1994. Phosphorylation of osteopontin by Golgi kinases. Ann. N.Y. Acad. Sci. This volume.
2. Behrend, E. I., A. F. Chambers, S. M. Wilson & D. T. Denhardt. 1993. Comparative analysis of two alternative first exons reported for the mouse osteopontin gene. J. Biol. Chem. **268:** 11172–11175.
3. Boskey, A. L., M. Maresca, W. Ullrich, S. B. Doty, W. T. Butler & C. W. Prince. 1994. Osteopontin-hydroxyapatite interactions *in vitro:* inhibition of hydroxyapatite formation and growth in a gelatin-gel. Bone Miner. Res. In press.
4. Brown, L. F., B. Berse, L. Van De Water, A. Papadopoulos-Sergiou, C. A. Perruzzi, E. J. Manseau, H. F. Dvorak & D. R. Senger. 1992. Expression and distribution of osteopontin in human tissues; widespread association with luminal epithelial surfaces. Mol. Biol. Cell **3:** 1169–1180.
5. Bruder, S. P., A. I. Caplan, Y. Gotoh, L. C. Gerstenfeld & M. J. Glimcher. 1991. Immunohistochemical localization of a ~66kDa glycosylated phosphoprotein during development of the embryonic chick tibia. Calcif. Tissue Int. **48:** 429–437.
6. Butler, W. T. 1989. The nature and significance of osteopontin. Connect. Tissue Res. **23:** 123–136.
7. Chambers, A. F., E. I. Behrend, S. M. Wilson & D. T. Denhardt. 1992. Induction of expression of osteopontin (OPN; secreted phosphoprotein) in metastatic *ras*-transformed NIH3T3 cells. Anticancer Res. **12:** 43–48.
8. Chang, P. L. & C. W. Prince. 1991. 1α,25-Dihydroxyvitamin D_3 stimulates synthesis and secretion of nonphosphorylated osteopontin (secreted phosphoprotein 1) in mouse JB6 epidermal cells. Cancer Res. **51:** 2144–2150.
9. Chen, J., K. Singh, B. B. Mukherjee & J. Sodek. 1993. Developmental expression of osteopontin (OPN) mRNA in rat tissues: evidence for a role for OPN in bone formation and resorption. Matrix **13:** 113–123.
10. Chen, Y., B. S. Bal & J. P. Gorski. 1992. Calcium and collagen binding properties of osteopontin, bone sialoprotein, and bone acidic glycoprotein-75 from bone. J. Biol. Chem. **267:** 24871–24878.
11. Craig, A. M., G. T. Bowden, A. F. Chambers, M. A. Spearman, A. H. Greenberg, J. A. Wright, M. McLeod & D. T. Denhardt. 1990. Secreted phosphoprotein mRNA is induced during multistage carcinogenesis in mouse skin and correlates with the metastatic potential of murine fibroblasts. Int. J. Cancer **46:** 133–137.
12. Craig, A. M., J. H. Smith & D. T. Denhardt. 1989. Osteopontin, a transformation-associated cell adhesion phosphoprotein, is induced by 12-*O*-tetradecanoylphorbol 13-acetated in mouse epidermis. J. Biol. Chem. **264:** 9682–9689.
13. Craig, A. M. & D. T. Denhardt. 1991. The murine gene encoding secreted phosphoprotein (osteopontin): promoter structure, activity, and induction *in vivo* by estrogen and progesterone. Gene **100:** 163–171.
15. Edelman, A. M., D. K. Blumenthal & E. G. Krebs. 1987. Protein serine/threonine kinases. Annu. Rev. Biochem. **56:** 567–613.
16. Flores, M. E., M. Norgard, D. Heinegård, F. P. Reinholt & G. Andersson. 1992. RGD-directed attachment of isolated rat osteoclasts to osteopontin, bone sialoprotein and fibronectin. Exp. Cell Res. **201:** 526–530.
17. Gerstenfeld, L. C., Y. Gotoh, M. D. McKee, A. Nanci, W. J. Landis & M. J.

GLIMCHER. 1990. Expression and ultrastructural localization of a major 66 kDa phosphoprotein synthesized by chicken osteoblasts during *in vitro* mineralization. Anat. Rec. **228:** 93–103.

19. GLIMCHER, M. J. 1985. *In* Calcium in Biological Systems. R. P. Rubin, G. B. Weiss & J. W. Putney, Jr., Eds. 607–616. Plenum Publishing Co. New York.
20. GORSKI, J. P. 1992. Acidic phosphoproteins from bone matrix: a structural rationalization of their role in biomineralization. Calcif Tissue Int. **50:** 391–396.
21. GOTOH, Y., L. C. GERSTENFELD & M. J. GLIMCHER. 1990. Identification and characterization of the major bone specific phosphoprotein synthesized by cultured embryonic chick osteoblasts. Eur. J. Biochem. **187:** 49–58.
21. GOTOH, Y., M. D. PIERSCHBACHER, J. J. GRZESIAK, L. GERSTENFELD & M. J. GLIMCHER. 1990. Comparison of two phosphoproteins in chicken bone and their similarities to the mammalian bone proteins, osteopontin and bone sialoprotein II. Biochem. Biophys. Res. Commun. **173:** 471–479.
22. JONES, F. S., K. L. CROSSIN, B. A. CUNNINGHAM & G. M. EDELMAN. 1990. Identification and characterization of the promoter for the cytotactin gene. Proc. Natl. Acad. Sci. USA **87:** 6497–6501.
23. KASUGAI, S., Q. ZHANG, C. M. OVERALL, J. L. WRANA, W. T. BUTLER & J. SODEK. 1991. Differential regulation of the 55 and 44 kDa forms of secreted phosphoprotein (SPP-1, osteopontin) in normal and transformed rat bone cells by osteotropic hormones, growth factors and a tumor promoter. Bone Miner. **13:** 235–250.
24. KEMP, B. E. & R. B. PEARSON. 1990. Protein kinase recognition sequence motifs. TIBS **15:** 342–346.
25. KERR, J. M., L. W. FISHER, J. D. TERMINE & M. F. YOUNG. 1991. The cDNA cloning and RNA distribution of bovine osteopontin. Gene **108:** 237–243.
26. LAMPE, M. A., R. PATARCA, M. V. IREGU & H. CANTOR. 1991. Polyclonal B cell activation by the Eta-1 cytokine and the development of systemic autoimmune disease. J. Immunol. **147:** 2902–2906.
27. LIAN, J. B., M. D. MCKEE, A. M. TODD & L. C. GERSTENFELD. 1993. Induction of bone-related proteins, osteocalcin and osteopontin, and their matrix ultrastructural localization with development of chondrocyte hypertrophy *in vitro*. J. Cell. Biochem. **52:** 206–219.
28. LOPEZ, C. A., J. R. HOYER, P. D. WILSON, P. WATERHOUSE & D. T. DENHARDT. 1993. Heterogeneity of osteopontin expression among nephrons in mouse kidneys and enhanced expression in sclerotic glomeruli. Lab. Invest. In press.
29. MARK, M. P., W. T. BUTLER, C. W. PRINCE, R. O. FINKELMANN & J. V. RUCH. 1988. Developmental expression of 44 kDa bone phosphoprotein (osteopontin) and bone γ carboxyglutamic acid (Gla)-containing protein (osteocalcin) in calcifying tissues of rat. Differentiation **37:** 123–136.
30. MCKEE, M. D., M. C. FARACH-CARSON, W. T. BUTLER, P. V. HAUSCHKA & A. NANCI. 1993. Ultrastructural immunolocalization of noncollagenous (osteopontin and osteocalcin) and plasma (albumin and α_2HS-glycoprotein) proteins in rat bone. J. Bone Miner. Res. **8:** 485–496.
31. MCKEE, M. D., A. NANCI, W. J. LANDIS, Y. GOTOH, L. C. GERSTENFELD & M. J. GLIMCHER. 1990. Developmental appearance and ultrastructural immunolocalization of a major 66 kDa phosphoprotein in embryonic and post-natal chicken bone. Anat. Rec. **228:** 77–92.
32. MCKEE, M. D., M. J. GLIMCHER & A. NANCI. 1992. High-resolution immunolocalization of osteopontin and osteocalcin in bone and cartilage during endochondral ossification in the chicken tibia. Anat. Rec. **234:** 479–492.
33. MIYAUCHI, A., J. ALVAREZ, E. M. GREENFIELD, A. TETI, M. GRANO, S. COLUCCI, A. ZAMBONIN-ALLONE, F. P. ROSS, S. M. TEITELBAUM, D. CHERESH & K. A. HRUSKA. 1991. Recognition of osteopontin and related peptides by an $\alpha_v\beta_3$ integrin stimulates immediate cell signals in osteoclasts. J. Biol. Chem. **266:** 20369–20374.
34. MIYAZAKI, Y., M. SETOGUCHI, S. YOSHIDA, Y. HIGUCHI, S. AKIZUKI & S. YAMAMOTO. 1990. The mouse osteopontin gene: expression in monocytic lineages and complete nucleotide sequence. J. Biol. Chem. **265:** 14432–14438.

35. MOORE, M. A., Y. GOTOH, K. RAFIDI & L. C. GERSTENFELD. 1991. Characterization of a cDNA for chicken osteopontin: expression during bone development, osteoblast differentiation, and tissue distribution. Biochemistry **30:** 2501–2508.
36. NEMIR, M., M. W. DEVOUGE & B. B. MUKHERJEE. 1989. Normal rat kidney cells secrete both phosphorylated and nonphosphorylated forms of osteopontin showing different physiological properties. J. Biol. Chem. **264:** 18202–18208.
37. NEUGEBAUER, B. M., M. A. MOORE, M. BROESS, L. C. GERSTENFELD & P. V. HAUSCHKA. 1995. Characterization of structural sequences in the chicken osteocalcin gene: expression of osteocalcin by maturing osteoblasts and hypertrophic chondrocytes *in vitro*. J. Bone Miner. Res. **10:** 157–163.
38. NODA, M., R. L. VOGEL, A. M. CRAIG, J. PRAHL, H. F. DELUCA & D. T. DENHARDT. 1990. Identification of a DNA sequence responsible for binding of the 1,25-dihydroxyvitamin D_3 receptor and 1,25-dihydroxyvitamin D_3 enhancement of mouse secreted phosphoprotein 1 (Spp-1, or osteopontin) gene expression. Proc. Natl. Acad. Sci. USA **87:** 9995–9999.
39. NODA, M., K. YOON, C. W. PRINCE, W. T. BUTLER & G. A. RODAN. 1988. Transcriptional regulation of osteopontin production in rat osteosarcoma cells by type b transforming growth factor. J. Biol. Chem. **263:** 13916–13921.
40. NOMURA, S., A. J. WILLS, D. R. EDWARDS, J. K. HEATH & B. L. M. HOGAN. 1988. Developmental expression of 2ar (osteopontin) and SPARC (osteonectin) RNA as revealed by *in situ* hybridization. J. Cell Biol. **106:** 441–450.
41. OLDBERG, A., A. FRANZEN & D. HEINEGÅRD. 1986. Cloning and sequence analysis of rat bone sialoprotein (osteopontin) cDNA reveals an arg-gly-asp cell binding sequence. Proc. Natl. Acad. Sci. USA **83:** 8819–8823.
42. OLDBERG, A., A. FRANZEN, D. HEINEGÅRD, M. PIERSCHBACHER & E. RUOSLAHTI. 1988. Identification of a bone sialoprotein receptor in osteosarcoma cells. J. Biol. Chem. **263:** 19433–19436.
43. PATARCA, R., G. J. FREEMAN, R. P. SING, F. Y. WEI, T. DURFEE, F. BLATTNER, D. C. REGNIER, C. A. KOZAK, B. A. MOCK, H. C. MORESE, III, T. R. JERRELLS & H. CANTOR. 1989. Structural functional studies of the early T lymphocyte activation (Eta-1) gene. J. Exp. Med. **170:** 145–162.
44. PIERSCHBACHER, M. D. & E. RUOSLAHTI. 1984. Cell attachment activity of fibronectin can be duplicated by small synthetic fragments of the molecule. Nature **309:** 30–33.
45. PRINCE, C. W., T. OOSAWA, W. T. BUTLER, M. TOMANA, A. S. BHOWN, M. BHOWN & R. E. SCHROHENLOHER. 1987. Isolation, characterization and biosynthesis of a phosphorylated glycoprotein from rat bone. J. Biol. Chem. **262:** 2900–2907.
46. PRINCE, C. W. & W. T. BUTLER. 1987. 1,25-Dihydroxyvitamin D_3 regulates the biosynthesis of osteopontin, a bone-derived cell attachment protein, in clonal osteoblast-like osteosarcoma cells. Collagen Relat. Res. **7:** 305–313.
47. PRINCE, C. W. 1989. Secondary structure predictions for rat osteopontin. Connect. Tissue Res. **21:** 15–20.
48. RAFIDI, K., I. SIMKINA, E. JOHNSON, M. A. MOORE & L. C. GERSTENFELD. 1994. Characterization of the chicken osteopontin-encoding gene. Gene **140:** 163–169.
49. REINHOLT, F. P., K. HULTENBY, A. OLDBERG & D. HEINEGÅRD. 1990. Osteopontin: a possible anchor of osteoclasts to bone. Proc. Natl. Acad. Sci. USA **87:** 4473–4475.
50. ROSS, F. P., J. CHAPPEL, J. I. ALVAREZ, D. SANDER, W. T. BUTLER, M. C. FARACH-CARSON, K. A. MINTZ, P. G. ROBEY, S. L. TEITELBAUM & D. A. CHERESH. 1993. Interactions between the bone matrix proteins osteopontin and bone sialoprotein and the osteoclast integrin $\alpha_v\beta_3$ potentiate bone resorption. J. Biol. Chem. **268:** 9901–9907.
51. SALIH, E., S. ASHKAR, L. C. GERSTENFELD & M. J. GLIMCHER. 1994. Identification of the *in vivo* phosphorylated sites of secreted osteopontin from cultured chicken osteoblasts. Ann. N.Y. Acad. Sci. This volume.
52. SAUK, J. J., C. L. VAN KAMPEN, K. NORRIS, R. FOSTER & M. J. SOMERMAN. 1990. Persistent spreading of ligament cells on osteopontin/bone sialoprotein-1 or collagen enhances tolerance to heat shock. Exp. Cell Res. **188:** 105–110.
53. SENGER, D. R., C. A. PERRUZZI, C. F. GRACEY, A. PAPADOPOULOS & D. G. TENEN.

1988. Secreted phosphoproteins associated with neoplastic transformation, close homology with plasma proteins cleaved during blood coagulation. Cancer Res. **48:** 5770–5774.

54. SENGER, D. R., C. A. PERRUZZI & A. PAPADOPOULOS. 1989. Elevated expression of secreted phosphoprotein 1 (osteopontin, 2ar) as a consequence of neoplastic transformation. Anticancer Res. **9:** 1291–1300.
55. SHIRAGA, H., W. MIN, W. J. VANDUSEN, M. D. CLAYMAN, D. MINER, C. H. TERRELL, J. R. SHERBOTIE, J. W. FOREMAN, C. PRZYSIECKI, E. G. NEILSON & J. R. HOYER. 1992. Inhibition of calcium oxalate crystal growth *in vitro* by uropontin, a new member of the aspartic acid-rich protein superfamily. Proc. Natl. Acad. Sci. USA **89:** 426–430.
56. SINGH, K., M. W. DEVOUGE & B. B. MUKHERJEE. 1990. Physiological properties and differential glycosylation of phosphorylated and nonphosphorylated forms of osteopontin secreted by normal rat kidney. J. Biol. Chem. **265:** 18696–18701.
57. SINGH, R. P., R. PATARCA, J. SCHWARTZ, P. SINGH & H. CANTOR. 1991. Definition of a specific interaction between the early T lymphocyte activation 1 (Eta-1) protein and murine macrophages *in vitro* and its effect upon macrophages *in vivo*. J. Exp. Med. **171:** 1931–1942.
58. SMITH, J. H. & D. T. DENHARDT. 1989. Evidence for two pathways of protein kinase C induction of 2ar expression: correlation with mitogenesis. J. Cell. Physiol. **139:** 189–195.
59. SMITH, J. H. & D. T. DENHARDT. 1987. Molecular cloning of a tumor promoter-inducible mRNA found in JB6 mouse epidermal cells: induction is stable at high, but not at low, cell densities. J. Cell. Biochem. **34:** 13–22.
60. SOMERMAN, M. J., C. W. PRINCE, W. T. BUTLER, R. A. FOSTER, J. M. MOEHRING & J. J. SAUK. 1989. Cell attachment activity of the 44 kilodalton bone phosphoprotein is not restricted to bone cells. Matrix **9:** 49–54.
61. SOMERMAN, M. J., C. W. PRINCE, J. J. SAUK, R. A. FOSTER & W. T. BUTLER. 1987. Mechanism of fibroblast attachment to bone extracellular matrix: role of a 44 kilodalton bone phosphoprotein. J. Bone Miner. Res. **2:** 259–265.
62. SORENSEN, E. S. & T. E. PETERSEN. 1994. Identification of two phosphorylation motifs in bovine osteopontin. Biochem. Biophys. Res. Commun. **198:** 200–205.
63. STRAUSS, P. G., E. I. CLOSS, J. SCHMIDT & V. ERFLE. 1990. Gene expression during osteogenic differentiation in mandibular condyles *in vitro*. J. Cell Biol. **110:** 1369–1378.
64. STUMPH, W. E., C. P. HODGSON, M-J. TSAI & B. W. O'MALLEY. 1984. Genomic structure and possible retroviral origin of the chicken CR1 repetitive DNA sequence family. Proc. Natl. Acad. Sci. USA **81:** 6667–6671.
65. TAKAGAKI, Y. & M. J. GLIMCHER. 1990. Post translational processing of chicken bone phosphoproteins. Identification of bone (phospho) protein kinase. Biochem. J. **268:** 593–597.
66. TEZUKA, K., T. SATO, H. KAMIOKA, P. J. NIJWEIDE, K. TANAKA, T. MATSUO, M. OHTA, N. KURIHARA, Y. HAKEDA & M. KUMEGAWA. 1992. Identification of osteopontin in isolated rabbit osteoclasts. Biochem. Biophys. Res. Commun. **186:** 911–917.
67. TODD, A. & L. C. GERSTENFELD. 1993. Differential expression and regulation by $1{,}25(OH)_2$ D_3 of bone specific gene expression in chondrocytes during endochondral maturation. J. Bone Miner. Res. **8**(Suppl. 1): S222.
68. WEINREB, M., D. SHINAR & G. A. RODAN. 1990. Different pattern of alkaline phosphatase, osteopontin, and osteocalcin expression in developing bone visualized by *in situ* hybridization. J. Bone Miner. Res. **5:** 831–842.
69. WRANA, L. L., A. ZHANG & J. SODEK. 1989. Full length cDNA sequence of porcine secreted phosphoprotein-I (SPP-I osteopontin). Nucleic Acids Res. **17:** 3306.
70. YANG, R., Y. GOTOH, M. A. MOORE, K. RAFIDI & L. C. GERSTENFELD. 1994. Characterization of the avian bone sialoprotein (bsp) cDNA: comparisons to mammalian BSP and identification of conserved structural domains. J. Bone Miner. Res. In press.
71. YOON, K., R. BUENAGA & G. A. RODAN. 1987. Tissue specificity and expression of rat osteopontin. Biochem. Biophys. Res. Commun. **13:** 1129–1136.

72. YOUNG, M. F., J. M. KERR, J. D. TERMINE, U. M. WEWER, M. G. WANG, O. W. MCBRIDE & L. W. FISHER. 1990. cDNA cloning, mRNA distribution and heterogeneity, chromosomal location, and RFLP analysis of human osteopontin (OPN). Genomics **7:** 491–502.
73. ZHANG, Q., C. DOMENICUCCI, H. A. GOLDBERG, J. L. WRANA & J. SODEK. 1990. Characterization of fetal porcine bone sialoproteins, secreted phosphoprotein I (osteopontin), bone sialoprotein, and a 23-kDa glycoprotein. J. Biol. Chem. **265:** 7583–7589.
74. ZHANG, Q., J. L. WRANA & J. SODEK. 1992. Characterization of the promoter region of the porcine OPN (osteopontin, secreted phosphoprotein 1) gene: identification of positive and negative regulatory elements and a "silent" second promoter. Eur. J. Biochem. **207:** 649–659.

Osteopontin at the Tumor/Host Interface

Functional Regulation by Thrombin-Cleavage and Consequences for Cell Adhesion[a]

DONALD R. SENGER,[b] LAWRENCE F. BROWN, CAROLE A. PERRUZZI, AGELIKI PAPADOPOULOS-SERGIOU, AND LIVINGSTON VAN DE WATER

Departments of Pathology
Beth Israel Hospital
and
Harvard Medical School
Boston, Massachusetts 02215

INTRODUCTION

A variety of observations have suggested that osteopontin (OPN) expression may be altered significantly in cancer and that OPN is important for tumorigenesis and metastasis. For example, OPN expression is increased substantially in experimental cell lines following neoplastic transformation by several agents,[1-6] and OPN expression is induced in mouse epidermis with tumor promoter.[7,8] In addition, OPN concentrations are substantially elevated in the blood of patients with metastatic cancer in comparison with the relatively low levels of OPN which circulate normally,[9] and OPN expression correlates with the metastatic potential of mouse cell lines.[7] Furthermore, transfection of tumor cell lines for expression of OPN antisense RNA yields cell populations with reduced malignant potential.[10,11]

Collectively, the foregoing observations raise the possibility that OPN functions importantly in cancer. However, OPN expression in human tumors had not been examined directly. Therefore, we undertook a series of analyses to quantitate OPN mRNA and to identify sites of synthesis in human carcinomas. Because OPN is known to have a GRGDS-containing cell binding domain,[12] we also investigated adhesive interactions between tumor cell lines and OPN, including analyses of the cell surface integrins that mediate such interactions. Moreover, since thrombin cleaves OPN in close proximity to the GRGDS sequence[4,13] and cleavage of OPN by thrombin likely occurs in tumors, we examined the consequences of cleavage by thrombin for OPN adhesive properties and for interactions between OPN and cell surface receptors.

[a] Supported by USPHS Grant CA34025, awarded by the National Cancer Institute, Department of Health and Human Services, and by a grant from the Beth Israel Hospital Pathology Foundation, Inc.

[b] Correspondence: Donald R. Senger, Department of Pathology, Research North, Beth Israel Hospital, 330 Brookline Avenue, Boston, MA 02215.

Expression of OPN mRNA in Human Carcinomas

We began our studies with Northern analyses of total RNAs isolated from tumor and adjacent normal tissues; all 12 primary carcinomas examined exhibited very substantial increases in OPN mRNA compared to adjacent normal tissue when normalized for β-actin mRNA and ribosomal RNA. Markedly high levels of OPN mRNA were also present in the two colon carcinoma metastases that we analyzed (FIG. 1, panels A and B). In sharp contrast to the carcinomas, benign tumors (colonic adenomas from a patient with familial polyposis and a uterine smooth muscle leiomyoma) showed no increase in OPN mRNA as compared with corresponding control tissues (FIG. 1, panels B and F).

Next, to ascertain which cells were the primary sources of OPN mRNA in human carcinomas, we performed *in situ* hybridization with antisense and control sense OPN riboprobes. In tumors of the colon (14 cases examined), OPN mRNA was not detected in tumor cells with antisense probe, but intense and specific labeling was observed in a population of cells distinct from tumor cells (FIG. 2). The cells containing readily detectable levels of OPN transcript were most abundant at the advancing edge of tumors (FIG. 2A, B) and near areas of necrosis (FIG. 2C, D). Moreover, cells strongly labeled for OPN mRNA were detected not only in primary tumors but also in metastases to lymph nodes (FIG. 2G, H). By routine microscopy the cells expressing OPN mRNA appeared to be macrophages, and their identity was confirmed with combined *in situ* hybridization for OPN mRNA and immunostaining with antibody to the macrophage marker KP1.[14]

Similar to our findings with colon carcinomas, neither tumor cells of ductal carcinoma *in situ* of the breast nor tumor cells of infiltrating ductal carcinomas of the breast labeled detectably with OPN antisense riboprobe (14 cases total). However, in each case, tumor macrophages labeled strongly and a similar pattern of labeling was observed in breast carcinomas metastatic to lymph nodes (not shown). We also analyzed OPN mRNA expression in carcinomas of the lung (FIG. 3E, F), bladder, prostate, ovary, thyroid, and metastatic melanoma by *in situ* hybridization (22 cases total). As in carcinomas of the colon and breast, OPN mRNA was detected in tumor macrophages but not tumor cells, indicating that macrophages were the predominant sources of OPN transcript in these carcinomas. However, and in contrast to the foregoing examples, we observed strong and specific labeling of both tumor cells and tumor-associated macrophages in carcinomas of the kidney (FIG. 3A, B) (13 out of 14 cases) and endometrium (FIG. 3C, D) (4 out of 8 cases), indicating that in these human cancers, tumor cells as well as tumor macrophages were the principal sources of OPN transcript.

To summarize, Northern analyses and *in situ* hybridization demonstrated substantially increased levels of OPN mRNA in primary autochthonous human carcinomas and metastases in comparison with corresponding normal tissues. In most carcinomas, macrophages intimately associated with tumor cells were the major source of OPN mRNA. However, in carcinomas of the kidney and endometrium, comparably high levels of OPN mRNA were detected in both tumor cells and tumor-associated macrophages. Comparisons performed previously showed that transformed cells overexpress OPN in comparison with corresponding normal cells,[1-7] and consequently it was somewhat surprising that macrophages, and not tumor cells, were the major sources of OPN mRNA in most of the carcinomas we examined. However, there are several possible explanations. For example, most of the cell lines previously examined were fibroblastic, and the human tumors we examined included only carcinomas. Moreover, previous studies focused primarily on OPN expression by cells in culture or tumor cells which had been

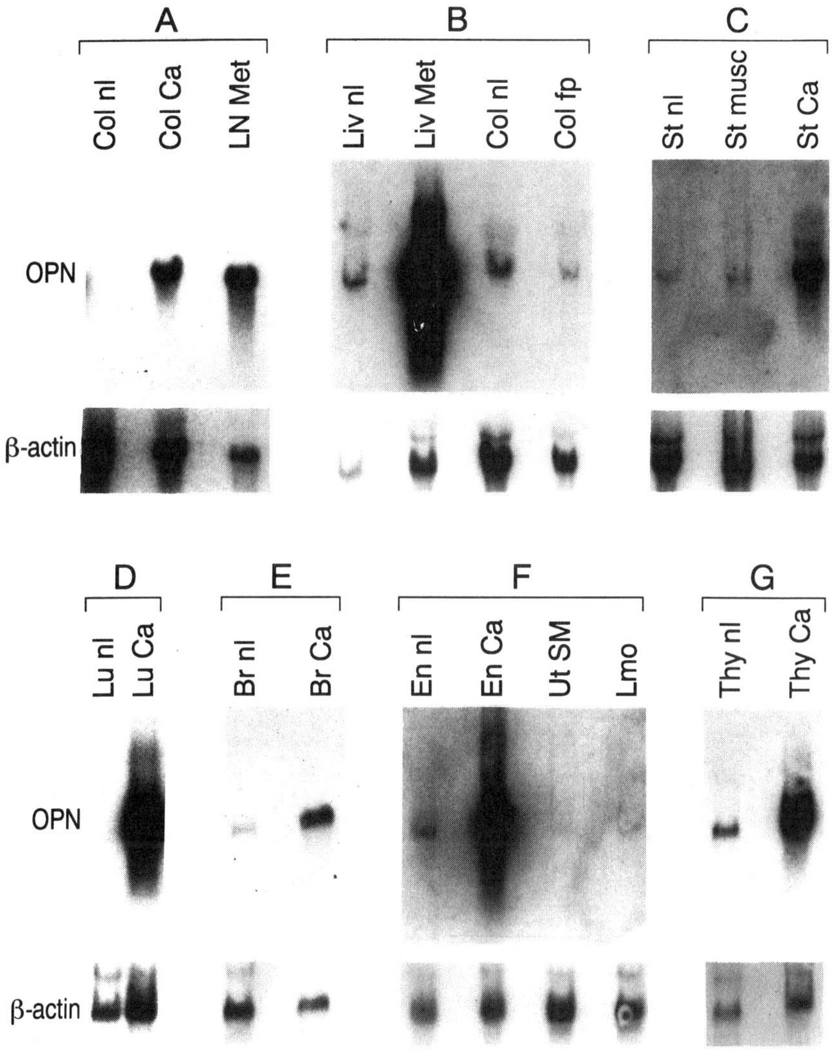

FIGURE 1. OPN mRNA in human tumors in comparison with corresponding normal tissues, as determined by Northern analyses. Total RNA was isolated from each of the specimens and equal quantities (20 μg) were subjected to electrophoresis and blotting. Blots were hybridized with a cDNA probe for OPN (1.5 kb), subsequently stripped, and rehybridized with a cDNA probe for β-actin mRNA (1.9 kb), as indicated. **(A)** (Col nl) normal colon, (Col Ca) colon carcinoma, (LN Met) colon carcinoma metastasis to lymph node; **(B)** (Liv nl) normal liver, (Liv Met) colon carcinoma metastasis to liver, (Col nl) normal colon, (Col fp) benign colonic adenomas from patient with familial polyposis; **(C)** (St nl) normal stomach mucosa, (St musc) normal stomach muscle, (St Ca) stomach carcinoma; **(D)** (Lu nl) normal lung, (Lu Ca) lung adenocarcinoma; **(E)** (Br nl) normal breast, (Br Ca) breast carcinoma; **(F)** (En nl) normal endometrium, (En Ca) endometrial carcinoma, (Ut SM) normal uterine smooth muscle, (Lmo) benign uterine smooth muscle leiomyoma; **(G)** (Thy nl) normal thyroid, (Thy Ca) papillary carcinoma of the thyroid. (From Brown *et al.*[14] Reprinted by permission from the *American Journal of Pathology*.)

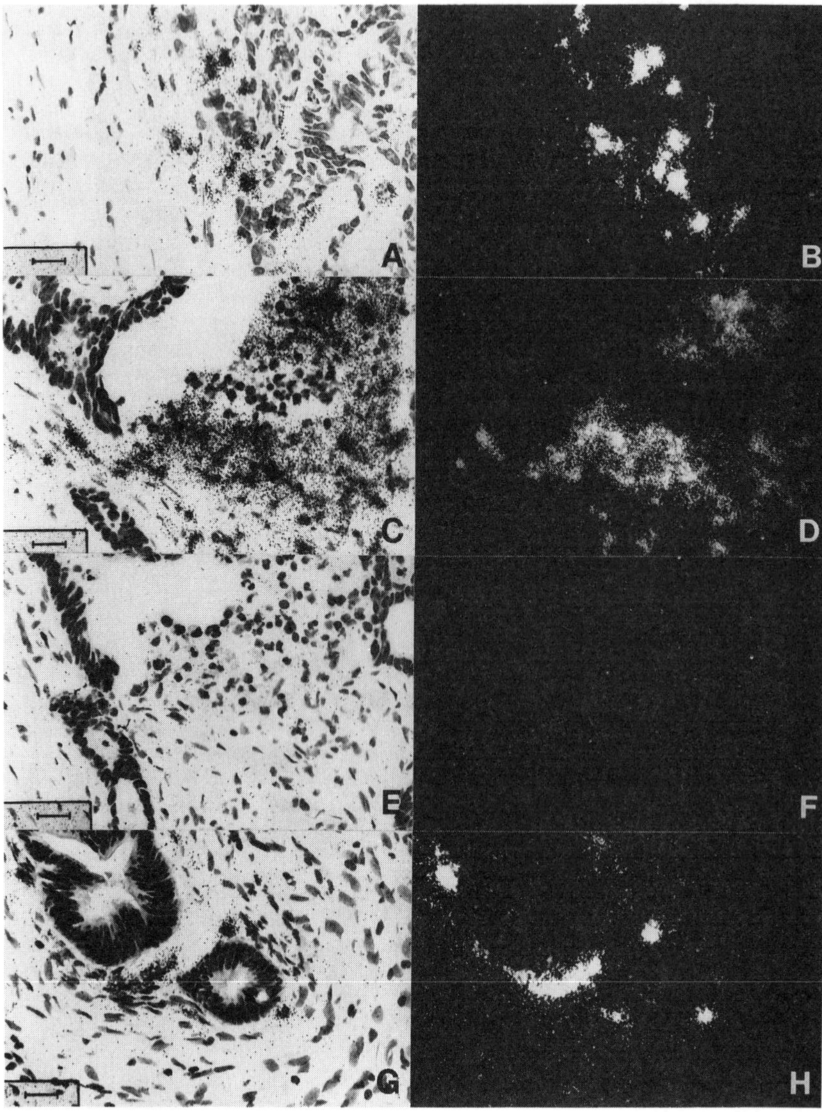

FIGURE 2. *In situ* hybridization with OPN riboprobes. Bright **(A)** and darkfield **(B)** photomicrographs of the same microscopic field with colonic tumor to the *right* and host stromal tissue to the *left*. Note intense labeling of cells in tumor stroma for OPN mRNA at the invasive edge of tumor (*center*). **(C, D)** Colonic carcinoma (*left*) with focal necrosis leading to disruption of glandular epithelium and release of necrotic contents into surrounding tissue (*right*). Note intense labeling of cells in tumor stroma (*right*) for OPN mRNA in the area of disruption. **(E, F)** Serial section of the site of glandular disruption depicted in (C) and (D); low background and no specific labeling with control sense probe. **(G, H)** Colon carcinoma metastatic to a lymph node with strong labeling for OPN mRNA in stromal cells immediately adjacent to malignant glands at *upper left* and *center* of field. (A–H): *bar* represents 25 μm. (From Brown *et al.*[14] Reprinted by permission from the *American Journal of Pathology*.)

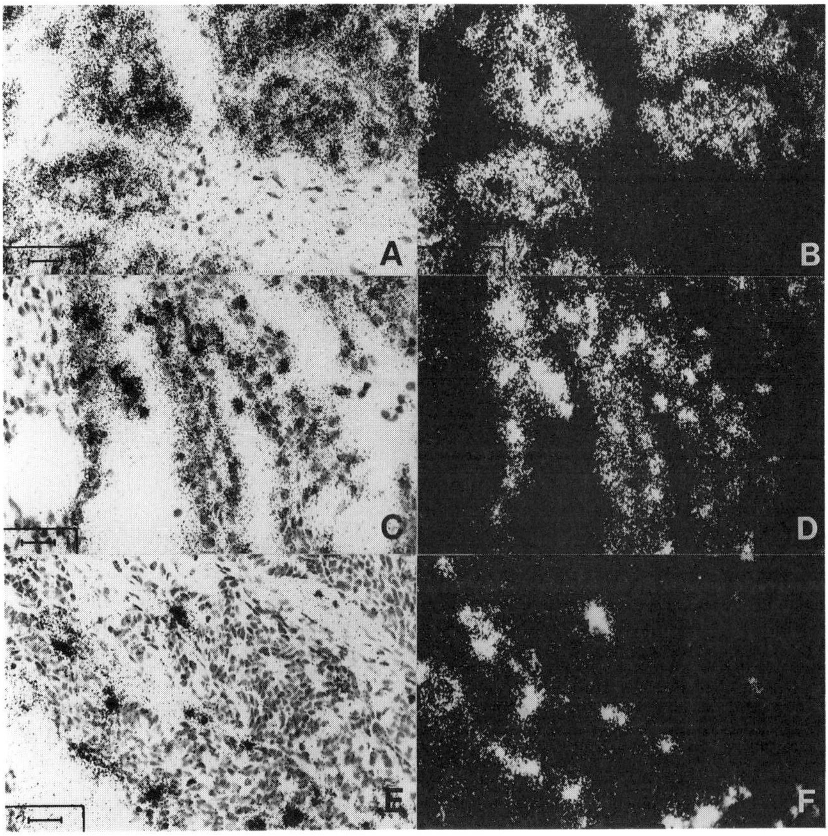

FIGURE 3. *In situ* hybridization of renal, endometrial, and lung carcinomas with OPN riboprobes. Bright **(A)** and darkfield **(B)** photomicrographs of the same microscopic fields of renal cell carcinoma. Note strong labeling of nests of tumor cells for OPN mRNA. **(C, D)** Endometrial carcinoma with strong labeling for OPN mRNA in malignant glandular epithelium as well as in macrophages (appearing as individual cells within gland lumens). **(E, F)** Poorly differentiated lung carcinoma; scattered macrophages label strongly for OPN mRNA while tumor cells do not label. (A–F): *bar* represents 25 μm. (From Brown *et al.*[14] Reprinted by permission from the *American Journal of Pathology*.)

serially passaged as tumors in animals, whereas this study examined OPN mRNA expression by spontaneously arising human carcinomas *in vivo*. Regardless, the collective evidence derived both from spontaneously arising human tumors as well as experimental tumor cell lines is consistent with the hypothesis that overexpression of OPN in tumors is important for the pathogenesis of cancer.

OPN Adhesive Function in Tumors and Regulation by Thrombin-Cleavage

It is particularly intriguing that macrophage synthesis of OPN mRNA was found to be prominent at the invasive edge of human carcinomas, and immunohistochemical staining of these tumors for OPN protein confirmed that OPN protein was present at the tumor/host interface, often in association with tumor cells as well as macrophages.[14] Moreover, OPN was also found in the tumor extracellular matrix[14] where, for example, it may be immobilized through covalent or noncovalent interactions with fibronectin.[15,16] These observations suggest that OPN participates in adhesive interactions in tumors, particularly at the tumor/host interface, and therefore we investigated interactions between OPN and tumor cells *in vitro*. In particular, we investigated the consequences of cleavage of OPN by thrombin for OPN adhesive function. As demonstrated previously, thrombin cleaves OPN at a site which is six amino acid residues carboxy-terminal to the GRGDS cell binding domain,[4,13] and this cleavage occurs under physiological conditions, *e.g.*, in the course of blood coagulation.[9] Moreover, the existence of activated thrombin in tumors is predicted by the presence of tumor-associated fibrin deposits (reviewed in Ref. 17, 18). Therefore, cleavage of OPN by thrombin most likely occurs in tumors and, moreover, in other pathologies where the coagulation pathway is active and OPN is present.

Cell Attachment and Spreading on Thrombin-Cleaved OPN in Comparison with Native OPN

To compare cell adhesion to native and thrombin-cleaved OPN, we performed cell attachment and spreading assays with both human and rat OPN and a variety of tumor as well as nontumor cell lines. For comparisons of OPN with thrombin-cleaved OPN, assays were performed according to two protocols: by immobilizing equal quantities of native or thrombin-cleaved OPN on substrata, followed by incubation with bovine serum albumin (BSA) to block remaining protein binding sites; or by coating substrata with OPN, incubating with BSA, and subsequently cleaving substratum-bound OPN *in situ* with added thrombin.[19] Adhesion assays performed according to both protocols gave indistinguishable results; typical experiments are presented in FIGURE 4 and TABLE 1. With all the cell lines and with both adhesion assay protocols, and regardless of whether OPN was covalently bound to substratum or passively adsorbed, thrombin-cleaved OPN promoted more extensive cell attachment and spreading than uncleaved OPN. Moreover, HT1080 cells, which did not attach detectably to native OPN in these assays, consistently demonstrated attachment to thrombin-cleaved OPN. As a control and to eliminate possible complications due to cell-secreted proteins, in several experiments we preincubated and assayed cells in the presence of cycloheximide, at concentrations of the drug sufficient to abolish protein synthesis, and found that these cells also attached more rapidly to thrombin-cleaved OPN in comparison with native OPN (TABLE 1, Experiment B).

Findings presented in FIGURE 4 and TABLE 1 cannot be explained by the presence of contaminating thrombin or, alternatively, by more efficient coating of substrata by thrombin-cleaved OPN in comparison with coating of substrata by native OPN. For assays employing OPN which was cleaved by thrombin prior to coating of substrata (FIG. 4, TABLE 1), all detectable thrombin was removed by repurification of OPN.[19] Furthermore, we investigated the possibility that thrombin might promote cell attachment and spreading and found for the cell types employed here and at thrombin concentrations used, that thrombin markedly inhibited cell attachment to cleaved OPN such that attachment was less than that observed on native OPN alone.[19] Moreover, we compared radioiodinated OPN before and after cleavage by thrombin and found native

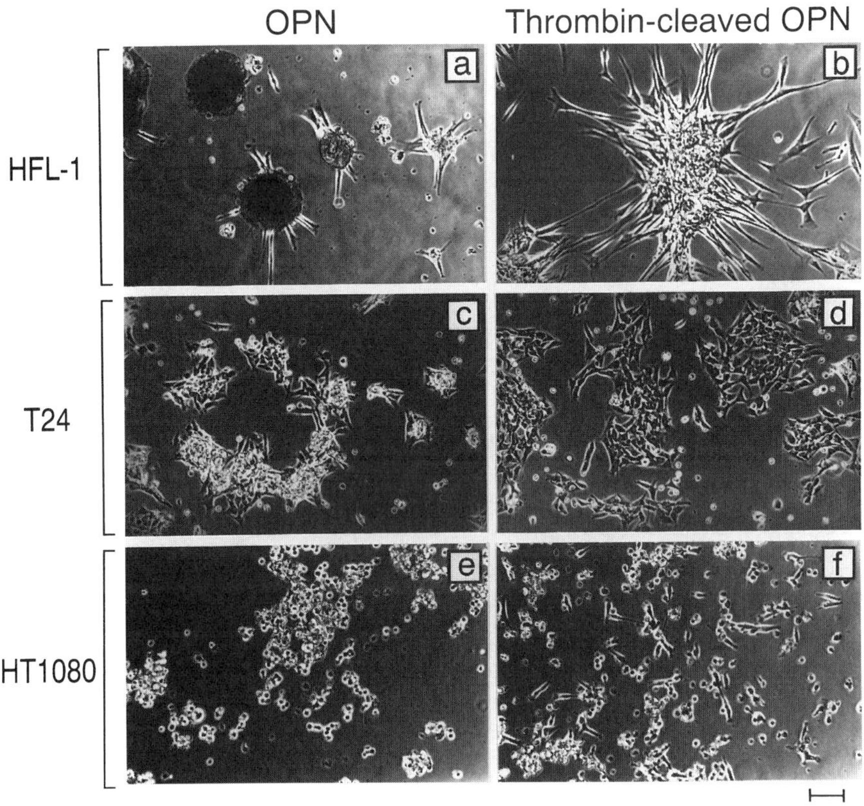

FIGURE 4. Attachment and spreading of HFL-1, T24, and HT1080 cells on uncleaved OPN and thrombin-cleaved OPN. For experiments with HFL-1 cells shown here, amino-derivatized glass coverslips[41] were coated covalently with human OPN (6 μg/ml) followed by BSA (10 mg/ml), and then incubated with saline alone **(a)** or saline with thrombin (2 U/ml) **(b)** followed by extensive washing; cells then were added. For experiments with T24 and HT1080 cells, glass coverslips were coated covalently with human OPN (6 μg/ml) followed by BSA **(c, e)** or with thrombin-cleaved human OPN (6 μg/ml) from which thrombin had been removed **(d, f)**.[19] Increased cell attachment and spreading on thrombin-cleaved OPN in comparison with uncleaved OPN was observed consistently in experiments performed according to both protocols. We observed no cell attachment to substrata coated with BSA alone (not shown). HFL-1 cells were photographed at 23 hr, T24 cells at 18 hr, and HT1080 cells at 5 hr after plating. *Bar* = 24 μm. Cell types: HFL-1 (human lung fibroblast), T24 (human bladder carcinoma), HT1080 (human fibrosarcoma). (From Senger *et al.*[19] Reprinted by permission from *Molecular Biology of the Cell.*)

and cleaved OPN-coated substrata with very similar efficiencies.[19] Finally, we consistently observed much more extensive cell attachment and spreading on substrata which was coated first with native OPN, next incubated with BSA to block residual protein binding sites, and subsequently incubated with thrombin to cleave substratum-bound OPN (FIG. 4, TABLE 1). Therefore, we conclude

TABLE 1. Cell Attachment to OPN in Comparison with Attachment to Thrombin-Cleaved OPN[a]

Experiment	Cell Line	# of Assays	Time	% Cell Binding to: BSA Alone	OPN	Thrombin-Cleaved OPN
A. Human OPN covalently coupled to substrata	HT1080	(3)	5 hr	4.0 ± 1.7	3.7 ± 2.1	20.3 ± 8.5
	HISM	(4)	4 hr	2.3 ± 0.5	39.8 ± 10.1	80.8 ± 11.0
	T24	(5)	18 hr	4.0 ± 1.2	16.8 ± 3.7	54.4 ± 4.6
	HFL-1	(6)	1.5 hr	3.5 ± 0.6	34.3 ± 8.0	75.2 ± 5.9
	SK MEL-2	(3)	1.5 hr	5.6 ± 2.4	9.0 ± 2.7	63.7 ± 2.8
B. Human OPN covalently coupled to substrata; cells treated with cycloheximide	T24	(3)	1.5 hr	2.4 ± 0.6	27.0 ± 3.5	94.1 ± 6.3
	HFL-1	(3)	1.5 hr	6.7 ± 2.5	22.3 ± 1.0	77.7 ± 3.5
C. Rat OPN passively adsorbed to substrata	T24	(6)	1.5 hr	2.7 ± 0.3	16.0 ± 7.6	79.3 ± 15.5
	HFL-1	(5)	1.0 hr	1.8 ± 0.2	36.9 ± 3.2	70.8 ± 7.7

[a] Attachment assays were performed with ^{35}S-labeled cells on substrata prepared by coating with either 6 μg/ml human OPN, 6 μg/ml thrombin-cleaved human OPN, or 10 μg/ml rat OPN, followed by BSA to block remaining protein binding sites, as follows. Experiments (A) with HT1080 cells were performed with amino-derivatized glass coverslips[41] to which equal quantities of human OPN or thrombin-cleaved human OPN were coupled covalently. Experiments (A and B) with HISM, T24, HFL-1, and SK MEL-2 cells were performed with amino-derivatized plastic wells (Costar Corp.) to which human OPN had been coupled covalently; where indicated, OPN was subsequently cleaved *in situ* by thrombin. Also, experiments (B) were performed with cells pretreated with the protein synthesis inhibitor cycloheximide (20 μg/ml) for 1 hr and maintained in the inhibitor during the assay (1.5 hr). Experiments (C) were performed with plastic passively coated with rat OPN (*i.e.*, noncovalent adsorption); where indicated rat OPN was subsequently cleaved *in situ* by thrombin. Cell types (all of human origin): HT1080 (fibrosarcoma), HISM (intestinal smooth muscle), T24 (bladder carcinoma), HFL-1 (lung fibroblast), SK MEL-2 (melanoma). (Modified from Senger *et al.*[19])

that markedly increased cell adhesion to thrombin-cleaved OPN in comparison with native OPN, as demonstrated with these assays, is a direct consequence of enzymatic cleavage of OPN by thrombin.

Structure of Thrombin-Cleaved OPN: Evidence for Dissociation of the Two Derived Polypeptides

As noted above, we determined previously with amino acid sequence analyses that OPN contains a single major thrombin-cleavage site and that, in both rat and human OPN, this site is only six amino acid residues from the carboxy-terminal side of the GRGDS cell-binding domain.[4,13] The two thrombin-generated fragments appear to be of very similar size because they are not easily resolved from each other in polyacrylamide gels containing sodium dodecyl sulphate (SDS); under such conditions both fragments derived from human OPN have electrophoretic mobilities corresponding to a molecular weight (M_r) of ~35,000.[13] OPN contains no cysteine residues which might contribute to intrachain bonding; and consequently, the OPN fragments migrate similarly in SDS polyacrylamide gels, whether or not reducing agents are present. However, to investigate the possibility that the two fragments remain associated through noncovalent interactions, we subjected thrombin-cleaved OPN to electrophoresis in the absence of SDS and other denaturing agents.[19] As shown in FIGURE 5, thrombin-cleaved OPN migrated in nondenaturing polyacrylamide gels as a single band of unresolved polypeptides corresponding to ~M_r 30,000. If the two OPN fragments had remained associated, mobility of thrombin-cleaved OPN under these electrophoretic conditions would be similar to that of uncleaved OPN which corresponded to ~M_r 67,000 in this system (FIG.

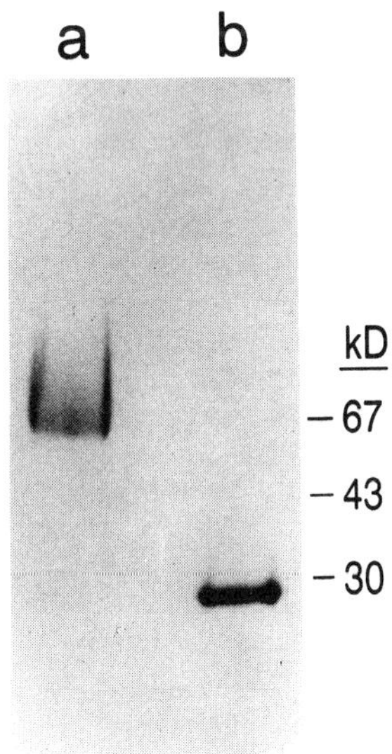

FIGURE 5. Electrophoresis of human OPN and thrombin-cleaved human OPN in nondenaturing conditions. Proteins were subjected to electrophoresis on a 4–20% gradient polyacrylamide gel in Tris-glycine buffer, pH 8.3, towards the anode. **(a)** 4 μg native OPN, **(b)** 4 μg thrombin-cleaved OPN; molecular weight markers = bovine serum albumin (67,000), chicken ovalbumin (43,000), and bovine erythrocyte carbonic anhydrase (30,000). Gel was stained with Coomassie Brilliant Blue R-250. (From Senger *et al.*[19] Reprinted by permission from *Molecular Biology of the Cell.*)

5). Thus, electrophoretic analysis of thrombin-cleaved OPN under nondenaturing conditions indicated that the two polypeptides generated by thrombin-cleavage did not remain associated.

Functional Activity of the GRGDS Cell-Binding Domain in Thrombin-Cleaved OPN

Cell attachment to native OPN was shown previously to be inhibited by soluble GRGDSP peptides but not inhibited by control GRGESP peptides[12] thus providing evidence for the importance of its GRGDS domain in promoting cell attachment. Similarly, and as expected if the GRGDS domain remained functional in thrombin-cleaved OPN, we observed complete inhibition of HT1080 cell attachment to thrombin-cleaved OPN in the presence of soluble GRGDSP peptide (300 μM) but little or no inhibition in the presence of an equal concentration of control GRADSP or GRGESP peptide. To investigate further the importance of the GRGDS domain for cell binding to thrombin-cleaved OPN, we raised a rabbit antibody to a 20-amino-acid synthetic peptide representing the GRGDS domain of human OPN (peptide sequence: VPTVDTYDGRGDSV-VYGLRC, with terminal cysteine added for conjugation).[19] As determined by enzyme-linked immunoadsorbent assays (ELISAs), antibody to this peptide was specific for OPN and did not bind detectably to other GRGDS-containing proteins. When employed in adhesion assays, this antibody markedly inhibited cell attachment and spreading on thrombin-cleaved OPN (FIG. 6), providing direct evidence for the importance of the GRGDS region in mediating cell attachment to thrombin-cleaved OPN. Similarly, this antibody also substantially blocked cell attachment to native OPN.[19]

Binding of Both Native OPN and Thrombin-Cleaved OPN to the Cell Surface Integrin $\alpha_v\beta_3$

It has been demonstrated that the LM609 blocking monoclonal antibody to integrin $\alpha_v\beta_3$ inhibits interactions between osteoclasts and OPN, thus implicating $\alpha_v\beta_3$ as an important osteoclast receptor for OPN.[20] However, OPN binding to $\alpha_v\beta_3$ had not been demonstrated with affinity chromatography. Therefore we sought to determine if $\alpha_v\beta_3$ binding to OPN could be demonstrated with affinity chromatography, and, moreover, if the thrombin-cleaved form of OPN also bound $\alpha_v\beta_3$. Affinity supports were constructed through covalent coupling of either human OPN or thrombin-cleaved human OPN to beads of crosslinked agarose. We found that cells attached to such beads but not to control beads, indicating that OPN still promoted cell adhesion following covalent coupling to the agarose support. Detergent extracts of human placenta were incubated with the OPN-agarose beads, followed by extensive washing of the beads, and elution of the bound proteins with buffer containing 20 mM EDTA. Subsequently, eluted proteins were analyzed for α_v and β_3 integrin subunits by electrophoresis and immunoblotting; a typical experiment is shown in FIGURE 7. As determined with antibody staining, both α_v and β_3 subunits bound and were eluted from columns covalently coupled to GRGDSPK peptide (RGD in FIG. 7), the 20-amino-acid OPN peptide corresponding to the GRGDS domain of OPN (OPN-RGD), OPN, and thrombin-cleaved OPN (OPN cl.). Moreover, the only major protein bands seen on silver-stained polyacrylamide gels were coincident with the α_v and β_3 bands identified on immu-

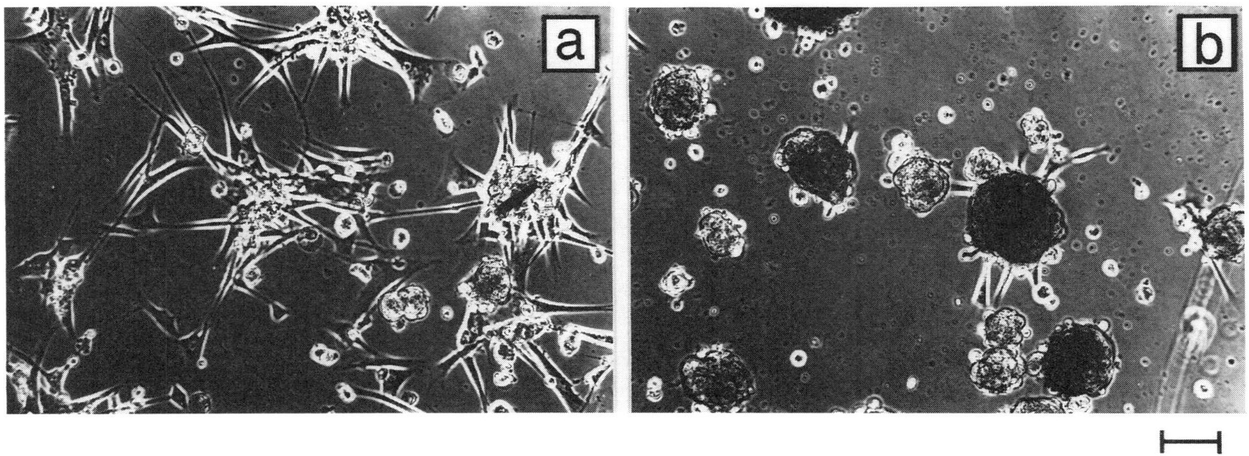

FIGURE 6. Inhibition of HFL-1 cell attachment to thrombin-cleaved OPN by antibody raised to a 20-amino-acid peptide containing the GRGDS cell binding domain of human OPN. HFL-1 cells were plated on thrombin-cleaved OPN as in FIGURE 4 in the presence of either: **(a)** 30 μg/ml control rabbit IgG or **(b)** 30 μg/ml affinity-purified rabbit antibody to a 20-amino-acid synthetic peptide corresponding to the GRGDS domain of OPN. This peptide antibody substantially inhibits cell attachment and spreading on both thrombin-cleaved OPN (b) and uncleaved OPN (not shown). Cells were photographed 18 hr after plating. *Bar* = 24 μm. (From Senger *et al.*[19] Reprinted by permission from *Molecular Biology of the Cell.*)

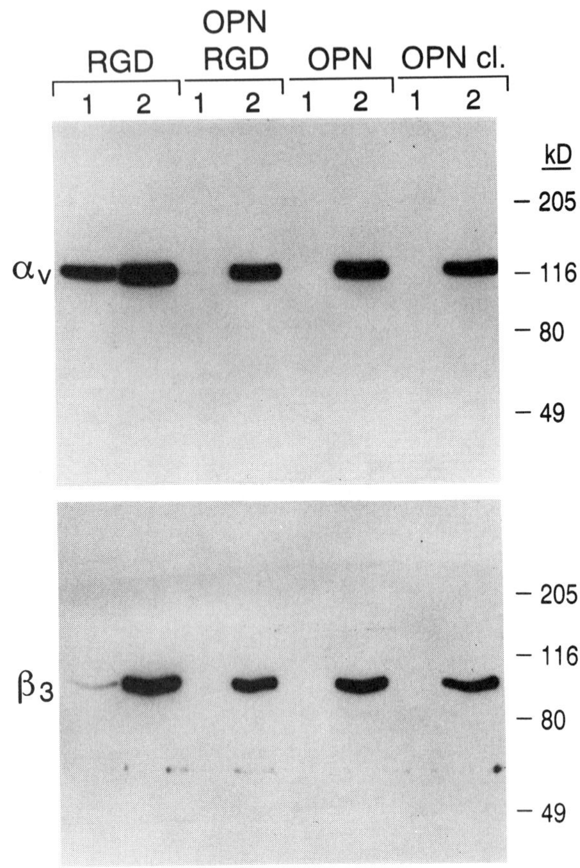

FIGURE 7. Immunoblots of placental integrins eluted from affinity columns. 1.5 ml columns were constructed with covalent coupling of the following peptides or proteins to cross-linked agarose beads: RGD = control RGDSP(K) peptide (10 mg); OPN RGD = 20-amino-acid human OPN peptide containing GRGDS (20 mg); OPN = uncleaved human OPN (10 mg); OPN cl. = thrombin-cleaved human OPN (10 mg). *Lanes 1* and *2* represent 80 μl volumes of successive 1.0 ml fractions collected from the indicated columns after extensive washing and immediately following addition of elution buffer containing 20 mM EDTA. Electrophoresis was carried out under reducing conditions on 7.5% polyacrylamide gels containing SDS, electroblotted, and stained with antibody to α_v or β_3, as indicated. α_v and β_3 were also present in fractions eluted from all of the columns with buffer containing soluble GRGDS peptide (1 mg/ml), although in lower amounts than in fractions eluted with 20 mM EDTA (not shown). (From Senger *et al.*[19] Reprinted by permission from *Molecular Biology of the Cell.*)

noblots. To demonstrate direct association between α_v and β_3 subunits, which were bound and eluted from OPN and thrombin-cleaved OPN columns, we performed immunoprecipitations with β_3 antibody and subjected the immunoprecipitates to immunoblotting with α_v and β_3 antibodies. β_3 antibody precipitates of eluted proteins contained both α_v and β_3 subunits (as detected by immunoblotting) indicating that $\alpha_v\beta_3$ heterodimers bound to OPN and thrombin-cleaved OPN affinity columns.[19]

To investigate directly the importance of $\alpha_v\beta_3$ for tumor cell attachment and spreading on OPN and thrombin-cleaved OPN, we performed adhesion assays with human tumor cell lines in the presence of $\alpha_v\beta_3$ blocking monoclonal antibody LM609[21] (kindly provided by Dr. David Cheresh, Scripps Clinic and Research Foundation, La Jolla, CA). As shown in TABLE 2, this $\alpha_v\beta_3$ antibody almost completely abolished tumor cell attachment on thrombin-cleaved OPN and also inhibited substantially or abolished tumor cell attachment on uncleaved OPN. In addition, we observed substantial inhibition of T24 bladder carcinoma spreading on thrombin-cleaved and uncleaved OPN with this antibody (not shown); there was no inhibition by control IgG in any of the experiments. Thus, our data indicate for the four human tumor cell lines examined (derived from a glioblastoma, a melanoma, a fibrosarcoma, and a bladder carcinoma) that $\alpha_v\beta_3$ is a major functional receptor for both native and thrombin-cleaved OPN.

Finally, we also considered the possibility that members of the β_1 integrin family might also be receptors for OPN and/or thrombin-cleaved OPN. To address these questions we performed cell attachment assays in the presence of a functionally blocking monoclonal antibody to β_1 (clone P4C10 from Gibco/BRL).[22] Although we observed some inhibition (~25%) of U87MG glioblastoma, HT1080 fibrosarcoma, and J82 bladder carcinoma cell attachment to OPN and/or thrombin-cleaved OPN with this antibody, in each case we also observed a similar degree of inhibition with a control nonblocking β_1 monoclonal antibody (clone LM534 from Chemicon).[23,24] Thus, we did not find any convincing evidence that β_1 integrins function in tumor cell attachment to OPN, and based on the data derived from both affinity chromatography and cell attachment assays, it seems probable that $\alpha_v\beta_3$ is the principal, if not the only receptor for both native and thrombin-cleaved OPN.

CONCLUSIONS

Given the foregoing information, we are led to several conclusions regarding the potential importance of OPN in human cancers and regulation of OPN-mediated cell adhesion by thrombin-cleavage. First, OPN mRNA and OPN protein are readily detectable in spontaneously arising human carcinomas and, overall, OPN is substantially overexpressed in tumors in comparison with corresponding normal tissues. Much of OPN expression in tumors is attributable to invading macrophages closely associated with tumor cells, although tumor cells of kidney and endometrial carcinomas also express relatively substantial levels of OPN mRNA. Most significantly, OPN expression is particularly prominent at the invasive edge of tumors, *i.e.*, the tumor/host interface, where as a component of the extracellular matrix it could strongly influence tumor cell adhesion and consequently the process of tumor cell invasion into the surrounding stroma.

Second, thrombin-cleavage of OPN serves to further activate the adhesive properties of this molecule and such activation likely occurs in the tumor microenvironment. Cleavage of OPN within six-amino-acid residues of the GRGDS sequence suggests a major conformational change in the molecule which probably affects the conformation of the GRGDS domain. Moreover, antibody raised to a 20-amino-acid synthetic peptide corresponding to the GRGDS domain of human OPN also substantially blocked cell attachment and spreading on thrombin-cleaved

TABLE 2. Inhibition of Cell Attachment to OPN and Thrombin-Cleaved OPN by $\alpha_v\beta_3$ Blocking Ab (LM609)[a]

			% Cell Attachment in the Presence of:		
Cell Line and Substrata	# of Assays	Time	Control IgG (10 μg/ml)	LM609 IgG (10 μg/ml)	Average % Inhibition by LM609 Ab
U87MG on OPN	3	5 hr	19.6 ± 7.9	8.3 ± 5.8	58%
U87MG on thrombin-cleaved OPN	3	5 hr	44.3 ± 7.3	3.4 ± 5.0	92%
SK MEL-2 on OPN	3	1.5 hr	11.4 ± 7.9	0.4 ± 0.6	97%
SK MEl-2 on thrombin-cleaved OPN	3	1.5 hr	77.2 ± 8.7	0	100%
HT1080 on thrombin-cleaved OPN	4	24 hr	23.5 ± 6.6	3.5 ± 0.6	85%

[a] Cell types (all of human origin): U87MG (glioblastoma), SK MEL-2 (melanoma), HT1080 (fibrosarcoma). Assays were performed with human OPN, covalently coupled to substrata as in TABLE 1.

OPN, providing direct evidence for the importance of the GRGDS region in mediating the increased cell attachment and spreading observed on thrombin-cleaved OPN. An intriguing hypothesis which could explain increased cell attachment and spreading on thrombin-cleaved OPN is that the resulting structural change in the molecule unmasks the GRGDS sequence, allowing greater accessibility to cell surface receptors. Alternatively, receptors may bind both uncleaved and thrombin-cleaved OPN equally, but receptor interactions with thrombin-cleaved OPN may result in signaling which promotes greater adhesion. Additional studies will be required to distinguish among the possibilities.

As described here, functional regulation of OPN by thrombin is relevant to OPN from both human and rat sources, and such regulation is likely relevant to OPN expressed in other species. Sequences derived from OPN cDNAs predict that the thrombin-sensitive arg-ser peptide bond, six residues carboxyterminal to the GRGDS sequence and present in human and rat OPN,[4,13] also is conserved in OPNs from mouse and pig.[8,25] A similar site is also predicted to exist seven and eight residues carboxyterminal to the GRGDS sequence present in rabbit and bovine OPNs, respectively;[26,27] and chicken OPN is cleaved by thrombin,[28] most probably at an arg-ala bond six residues carboxyterminal to the GRGDS sequence.[29,30] In addition, many of the posttranslationally modified forms of OPN have been shown to be sensitive to thrombin-cleavage[31,32] as have OPNs from a variety of sources including blood plasma,[9] milk,[13] bone cells[32] as well as recombinant OPN expressed in *E. coli*.[33] Thus, thrombin-cleavage is likely relevant to various forms of OPN from most or all species.

Finally, we have demonstrated that the cell surface integrin $\alpha_v\beta_3$ is the major protein in placental extracts which binds thrombin-cleaved OPN as well as native OPN, and moreover that LM609 blocking monoclonal antibody to $\alpha_v\beta_3$[21] substantially inhibits the attachment of human glioblastoma, melanoma, fibrosarcoma, and carcinoma cell lines to both forms of OPN. Consequently, it is likely that $\alpha_v\beta_3$ is the principal tumor cell receptor for OPN whether or not OPN is cleaved by thrombin. Interestingly, there are several lines of evidence that the $\alpha_v\beta_3$ integrin is likely important for tumor growth and invasion. For example, increased $\alpha_v\beta_3$ expression has been shown to correlate with melanoma vertical growth phase and metastasis[34–36] and with an aggressive class of glioblastoma.[37] Moreover, melanoma cells selected for low level expression of $\alpha_v\beta_3$ are less tumorigenic than parental cells and tumorigenicity can be restored following transfection for $\alpha_v\beta_3$ expression.[38,39] Furthermore, a monoclonal antibody that blocks $\alpha_v\beta_3$ function has been reported to inhibit melanoma tumor growth in nude mice.[40] Our findings that the more highly adhesive, thrombin-cleaved form of OPN as well as native OPN are both ligands for tumor cell $\alpha_v\beta_3$ and that OPN is overexpressed in carcinomas, particularly at the tumor/host interface, likely relate to $\alpha_v\beta_3$ function in tumors and moreover indicate that OPN is well positioned to participate in adhesive interactions directly involved in tumor invasion and metastasis.

REFERENCES

1. SENGER, D. R., D. F. WIRTH & R. O. HYNES. 1980. Transformation-specific secreted phosophoproteins. Nature **286:** 619–621.
2. SENGER, D. R., B. B. ASCH, B. D. SMITH, C. A. PERRUZZI & H. F. DVORAK. 1983. A secreted phosphoprotein marker for neoplastic transformation of both epithelial and fibroblastic cells. Nature **302:** 714–715.
3. CRAIG, A. M., M. NEMIR, B. B. MUKHERJEE, A. F. CHAMBERS & D. T. DENHARDT. 1988. Identification of the major phosphoprotein secreted by many rodent cell lines

as 2ar/osteopontin: enhanced expression in H-ras-transformed 3T3 cells. Biochem. Biophys. Res. Commun. **157:** 166–173.

4. SENGER, D. R., C. A. PERRUZZI & A. PAPADOPOULOS. 1989. Elevated expression of secreted phosphoprotein I (osteopontin, 2ar) as a consequence of neoplastic transformation. Anticancer Res. **9:** 1291–1299.
5. TUCK, A. B., S. M. WILSON, R. KHOKHA & A. F. CHAMBERS. 1991. Different patterns of gene expression in ras-resistant and ras-sensitive cells. J. Natl. Cancer Inst. **83:** 485–491.
6. CHAMBERS, A. F., E. I. BEHREND, S. M. WILSON & D. T. DENHARDT. 1992. Induction of expression of osteopontin (OPN; secreted phosphoprotein) in metastatic, ras-transformed NIH 3T3 cells. Anticancer Res. **12:** 43–47.
7. CRAIG, A. M., G. T. BOWDEN, A. F. CHAMBERS, M. A. SPEARMAN, A. H. GREENBERG, J. A. WRIGHT, M. MCLEOD & D. T. DENHARDT. 1990. Secreted phosphoprotein mRNA is induced during multi-stage carcinogenesis in mouse skin and correlates with the metastatic potential of murine fibroblasts. Int. J. Cancer **46:** 133–137.
8. CRAIG, A. M., J. H. SMITH & D. T. DENHARDT. 1989. Osteopontin, a transformation-associated cell adhesion phosphoprotein, is induced by 12-O-tetradecanoylphorbol 13-acetate in mouse epidermis. J. Biol. Chem. **264:** 9682–9689.
9. SENGER, D. R., C. A. PERRUZZI, C. F. GRACEY, A. PAPADOPOULOS & D. G. TENEN. 1988. Secreted phosphoproteins associated with neoplastic transformation: close homology with plasma proteins cleaved during blood coagulation. Cancer Res. **48:** 5770–5774.
10. BEHREND, E. I., A. M. CRAIG, S. M. WILSON, D. T. DENHARDT & A. F. CHAMBERS. 1994. Reduced malignancy of ras-transformed NIH 3T3 cells expressing antisense osteopontin RNA. Cancer Res. **54:** 832–837.
11. GARDNER, H. A. R., B. BERSE & D. R. SENGER. 1994. Specific reduction in osteopontin synthesis by antisense RNA inhibits the tumorigenicity of transformed Rat1 fibroblasts. Oncogene **9:** 2321–2326.
12. OLDBERG, A., A. FRANZEN & D. HEINEGÅRD. 1986. Cloning and sequence analysis of rat bone sialoprotein (osteopontin) cDNA reveals an Arg-Gly-Asp cell-binding sequence. Proc. Natl. Acad. Sci. USA **83:** 8819–8823.
13. SENGER, D. R., C. A. PERRUZZI, A. PAPADOPOULOS & D. G. TENEN. 1989. Purification of a human milk protein closely similar to tumor-secreted phosphoproteins and osteopontin. Biochim. Biophys. Acta **996:** 43–48.
14. BROWN, L. F., A. PAPADOPOULOS-SERGIOU, B. BERSE, E. J. MANSEAU, K. TOGNAZZI, C. A. PERRUZZI, H. F. DVORAK & D. R. SENGER. 1994. Osteopontin expression and distribution in human carcinomas. Am. J. Pathol. **145:** 610–623.
15. BENINATI, S., D. R. SENGER, E. CORDELLA-MIELE, A. B. MUKHERJEE, I. CHACKALAPARAMPIL, V. SHANMUGAM, K. SINGH & B. B. MUKHERJEE. 1994. Osteopontin: its transglutaminase-catalyzed posttranslational modifications and cross-linking to fibronectin. J. Biochem. **115:** 675–682.
16. SINGH, K., M. W. DEVOUGE & B. B. MUKHERJEE. 1990. Physiological properties and differential glycosylation of phosphorylated and nonphosphorylated forms of osteopontin secreted by normal rat kidney cells. J. Biol. Chem. **265:** 18696–18701.
17. DVORAK, H. F. 1986. Tumors: wounds that do not heal. Similarities between tumor stroma generation and wound healing. N. Engl. J. Med. **315:** 1650–1659.
18. DVORAK, H. F., J. A. NAGY, B. BERSE, L. F. BROWN, K.-T. YEO, T.-K. YEO, A. M. DVORAK, L. VAN DE WATER, T. M. SIOUSSAT & D. R. SENGER. 1992. Vascular permeability factor, fibrin, and the pathogenesis of tumor stroma formation. Ann. N.Y. Acad. Sci. **667:** 101–111.
19. SENGER, D. R., C. A. PERRUZZI, A. PAPADOPOULOS-SERGIOU & L. VAN DE WATER. 1994. Adhesive properties of osteopontin: regulation by a naturally occurring thrombin-cleavage in close proximity to the GRGDS cell-binding domain. Mol. Biol. Cell **5:** 565–574.
20. ROSS, F. P., J. CHAPPEL, J. I. ALVAREZ, D. SANDER, W. T. BUTLER, M. C. FARACH-CARSON, K. A. MINTZ, P. G. ROBEY, S. L. TEITELBAUM & D. A. CHERESH. 1993. Interactions between the bone matrix proteins osteopontin and bone sialoprotein

and the osteoclast integrin $\alpha_v\beta_3$ potentiate bone resorption. J. Biol. Chem. **268:** 9901–9907.
21. Cheresh, D. A. & R. C. Spiro. 1987. Biosynthetic and functional properties of an Arg-Gly-Asp-directed receptor involved in human melanoma cell attachment to vitronectin, fibrinogen, and von Willebrand factor. J. Biol. Chem. **262:** 17703–17711.
22. Carter, W. G., E. A. Wayner, T. S. Bouchard & P. Kaur. 1990. The role of integrins $\alpha2\beta1$ and $\alpha3\beta1$ in cell-cell and cell-substrate adhesion of human epidermal cells. J. Cell Biol. **110:** 1387–1404.
23. Wayner, E. A., R. A. Orlando & D. A. Cheresh. 1991. Integrins alpha v beta 3 and alpha v beta 5 contribute to cell attachment to vitronectin but differentially distribute on the cell surface. J. Cell Biol. **113:** 919–929.
24. Takada, Y. & W. Puzon. 1993. Identification of a regulatory region of integrin $\beta1$ subunit using activating and inhibiting antibodies. J. Biol. Chem. **268:** 17597–17601.
25. Wrana, J. L., Q. Zhang & J. Sodek. 1989. Full-length cDNA sequence of porcine secreted phosphoprotein-I (SPP-1, osteopontin). Nucleic Acids Res. **17:** 10119.
26. Tezuka, K., T. Sato, H. Kamioka, P. J. Nijweide, K. Tanaka, T. Matsuo, M. Ohta, N. Kurihara, Y. Hakeda & M. Kumegawa. 1992. Identification of osteopontin in isolated rabbit osteoclasts. Biochem. Biophys. Res. Commun. **186:** 911–917.
27. Kerr, J. M., L. W. Fisher, J. D. Termine & M. F. Young. 1991. The cDNA cloning and RNA distribution of bovine osteopontin. Gene **108:** 237–243.
28. Gotoh, Y., M. D. Pierschbacker, J. J. Grzesiak, L. Gerstenfeld & M. J. Glimcher. 1990. Comparison of two phosphoproteins in chicken bone and their similarities to the mammalian bone proteins, osteopontin and bone sialoprotein II. Biochem. Biophys. Res. Commun. **173:** 471–479.
29. Moore, M. A., Y. Gotoh, K. Rafidi & L. C. Gerstenfeld. 1991. Characterization of cDNA for chicken osteopontin: expression during bone development, osteoblast differentiation, and tissue distribution. Biochemistry **30:** 2501–2508.
30. Castagnola, P., P. Bet, R. Quarto, M. Gennari & R. Cancedda. 1991. cDNA cloning and gene expression of chicken osteopontin. J. Biol. Chem. **266:** 9944–9949.
31. Nagata, T., R. Todescan, H. A. Goldberg, Q. Zhang & J. Sodek. 1989. Sulphation of secreted phosphoprotein I (SPPI, osteopontin) is associated with mineralized tissue formation. Biochem. Biophys. Res. Commun. **165:** 234–240.
32. Kubota, T., Q. Zhang, J. L. Wrana, R. Ber, J. E. Aubin, W. T. Butler & J. Sodek. 1989. Multiple forms of SPPI (secreted phosphoprotein, osteopontin) synthesized by normal and transformed rat bone cell populations: regulation by TGF-β. Biochem. Biophys. Res. Commun. **162:** 1453–1459.
33. Ashkar, S., D. B. Teplow, M. J. Glimcher & R. A. Saavedra. 1993. *In vitro* phosphorylation of mouse osteopontin expressed in *E. coli*. Biochem. Biophys. Res. Commun. **191:** 126–133.
34. Albelda, S. M., S. A. Mette, D. E. Elder, R. Stewart, L. Damjanovich, M. Herlyn & C. A. Buck. 1990. Integrin distribution in malignant melanoma: association of the beta 3 subunit with tumor progression. Cancer Res. **50:** 6757–6764.
35. Kramer, R. H., M. Vu, Y.-F. Cheng & D. M. Ramos. 1991. Integrin expression in malignant melanoma. Cancer Metastasis Rev. **10:** 49–59.
36. Hart, I. R., M. Birch & J. F. Marshall. 1991. Cell adhesion receptor expression during melanoma progression and metastasis. Cancer Metastasis Rev. **10:** 115–128.
37. Gladson, C. L. & D. A. Cheresh. 1991. Glioblastoma expression of vitronectin and the $\alpha_v\beta_3$ integrin. Adhesion mechanism for transformed glial cells. J. Clin. Invest. **88:** 1924–1932.
38. Felding-Habermann, B., B. M. Mueller, C. A. Romerdahl & D. A. Cheresh. 1992. Involvement of integrin alpha V gene expression in human melanoma tumorigenicity. J. Clin. Invest. **89:** 2018–2022.
39. Sanders, L. C., B. Felding-Habermann, B. M. Mueller & D. A. Cheresh. 1992. Role of alpha v integrins and vitronectin in human melanoma cell growth. Cold Spring Harbor Sym. Quant. Biol. **57:** 233–240.

40. Boukerche, H., O. Berthier-Vergnes, M. Bailly, J. F. Dore, L. L. K. Leung & J. L. McGregor. 1989. A monoclonal antibody (LYP18) directed against the blood platelet glycoprotein IIb/IIIa complex inhibits human melanoma growth *in vivo*. Blood **74:** 909–912.
41. Werb, Z., P. M. Tremble, O. Behrendtsen, E. Crowley & C. H. Damsky. 1989. Signal transduction through the fibronectin receptor induces collagenase and stromelysin gene expression. J. Cell Biol. **109:** 877–889.

Regulation and Function of Osteopontin in Ras-Transformed Cells[a]

ANN F. CHAMBERS[b]

London Regional Cancer Centre
and
Oncology Department
University of Western Ontario
London, Ontario N6A 4L6, Canada

INTRODUCTION

Tumor metastasis, or the spread of cancer cells from a primary tumor to distant sites, is a major cause of treatment failure in cancer patients. Cancer that has not metastasized is more treatable than cancer that has seeded distant, and often undetected, metastases. In order to metastasize, a cancer cell must complete a series of steps, including growth and vascularization at the primary site, intravasation into vascular or lymphatic vessels, arrest in capillaries of a new organ, extravasation and migration into tissue of this organ, and finally growth and vascularization in this new organ environment (reviews, Refs. 1, 2). An understanding of the molecular mechanisms of these steps will help in the identification of molecular targets for future therapeutic intervention to prevent or minimize the effect of metastasis.

We have used *ras*-transformed murine NIH 3T3 cells as a model for studying molecular mechanisms of metastasis.[3–7] NIH 3T3 cells are an immortalized, nontumorigenic and nonmetastatic fibroblast cell line. NIH 3T3 cells that have been transformed with the activated, T24 H-*ras* gene (from a human bladder cancer) are both tumorigenic and metastatic, in proportion to their levels of Ras p21 protein expression.[4,5] Transformation by an activated *ras* oncogene results in constitutive activation of signal transduction pathways (which are the subject of much current research[8–11]), the effect of which is to regulate a set of genes whose expression we postulate can affect steps in metastasis (reviews, Refs. 12, 13). Many human tumors carry activated *ras* genes; however, the pathways that result in increased metastatic ability can be activated by various mechanisms (including oncogenes other than *ras*, by activating aspects of the same signal transduction pathways), suggesting that genes and functions identified by this strategy may have relevance to human cancers with, as well as without, activated *ras* oncogenes.

Using this approach, we have identified several genes whose expression is altered in *ras*-transformed cells and whose gene products appear to play a func-

[a] Research in the author's laboratory is supported by grants from the National Cancer Institute of Canada, as well as from London Regional Cancer Centre Regionally Designated Funds, the Victoria Hospital Research Development Fund, the Ontario Thoracic Society, and the Breast Cancer Society of Canada. The author is a Career Scientist of the Ontario Cancer Treatment and Research Foundation.

[b] Address for correspondence: Dr. Ann F. Chambers, London Regional Cancer Centre, 790 Commissioners Road East, London, Ontario N6A 4L6, Canada.

tional role in metastasis (reviews, Refs. 12, 13). These include: calcyclin, a calcium-binding cytoplasmic protein,[14] which has been linked to progression in melanoma;[15] cathepsins L and B (cysteine proteinases) and their inhibitors[16] and metalloproteinases and their inhibitors,[7] which have been postulated to contribute to tumor invasion and growth (reviews, Refs. 17, 18); and osteopontin (OPN), a secreted integrin-binding protein.[7,19–22] The regulation of OPN by the *ras* oncogene, the contribution of OPN to malignancy, and the effect of site-directed mutagenesis of the RGD (arginine-glycine-aspartic acid) amino acid sequence on OPN function, will be discussed in this review.

Regulation of OPN in ras*-Transformed Cells*

Expression of OPN (RNA and secreted protein) is increased in a series of *ras*-transformed NIH 3T3 cell lines, in proportion to their levels of Ras expression and metastatic ability.[19–21] OPN expression has been shown to be responsive to a number of factors, including estrogen and progesterone, calcitriol (vitamin D_3), and tumor promoters, as well as the *ras* oncogene (reviews, Refs. 23–25). The murine OPN gene promoter contains a number of putative transcription factor binding sites, including sequences that are responsive to Ras expression (*e.g.*, AP1), suggesting that the regulation of OPN expression is subject to complex control.[26,27]

Recent work in collaboration with David Denhardt's laboratory, by Guo *et al.*,[26] has clarified the regulation of OPN expression in *ras*-transformed cells. We found that the increase in OPN expression in response to Ras is due, at least in part, to increased transcription of OPN in *ras*-transformed cells. In experiments using fragments of the OPN promoter attached to reporter gene constructs, two regions of the promoter were found to contain positive Ras-responsive regulatory elements, with additional regions containing negative regulatory elements. Guo *et al.*[26] have identified a novel Ras-activated enhancer in the murine OPN promoter, whose sequence is distinct from other known Ras-response elements. This sequence, GGAGGCAGG, located at −725 to −717 in the OPN promoter, was deduced by methylation interference studies, and resulted in increased transcription activity in Ras-expressing cells. Several nuclear proteins were found to be able to bind to this sequence. Based on the finding that one of these proteins appeared to be associated with metastatic ability in a series of cell lines, we have designated this protein "metastasis-associated transcription factor," or MATF. This protein may be a novel member of the ETS family of transcription regulators, based on the sequence to which it binds and the inability of several other members of this family to compete with MATF for interaction with this sequence. These results suggest that activation of the Ras signal transduction pathway results in increased expression of OPN, via regulation of OPN transcription. Further, they suggest that cells that are metastatic may express regulatory protein(s) that results in activation of transcription of OPN.

Contribution of OPN to Malignancy

OPN was identified as a tumor-associated protein (review, Ref. 28), as well as a protein of bone and other mineralized tissues (review, Ref. 23). Craig *et al.*[19] showed that these were in fact the same protein. OPN is produced by many transformed cell lines in culture (review, Ref. 23), although in our experience this is not universal and some transformed cells do not produce OPN. Expression of OPN also has been shown to increase progressively in tumor promoter-induced multistage carcinogenesis in mouse skin, being lowest in normal epidermis, somewhat increased in papillomas, and further increased in carcinomas.[20] However, the association of OPN with the transformed phenotype did not indicate whether OPN contributed to the malignant properties of cancer cells or was a protein produced as a consequence of malignant transformation.

We were thus interested to learn if OPN was able to contribute functionally to malignancy. We transfected PAP2 cells (metastatic, *ras*-transformed NIH 3T3) with a mammalian expression vector containing the murine OPN cDNA in an antisense orientation.[29] Two independent clones that expressed antisense OPN RNA were selected. Both of these clones had markedly reduced malignant properties, including reduced experimental metastatic ability and reduced tumorigenic growth rate in nude mice.[29] In contrast, control clones did not show reduced malignancy. One of the antisense OPN-expressing clones failed to form tumors after subcutaneous injection in nude mice, while the other clone formed tumors after an increased lag period; interestingly, the tumors that did arise from this clone had uniformly lost expression of the antisense OPN RNA, in all mice, suggesting that expression of antisense OPN RNA *in vivo* was inconsistent with tumor growth.[29] Additional details of this study are presented elsewhere in this volume (Behrend *et al.*). Similar findings have recently been reported by Gardner *et al.*,[30] who found that B77-Rat1 (*src* oncogene-transformed) cells that expressed antisense OPN RNA had reduced abilities to form experimental metastases and to form colonies in soft agar. Together these studies suggest that OPN production by cancer cells is indeed able to contribute functionally to the malignant properties of the cells.

The molecular mechanisms by which OPN production by cancer cells can contribute to their malignancy, including primary tumor growth and metastatic ability, remain to be clarified. Many functions have been proposed for OPN, in its various normal and pathological contexts (reviews, Refs. 23–25). A possible new role for OPN in malignancy is suggested by the recent finding, from Denhardt's laboratory, that OPN can block nitric oxide synthesis induced by γ-interferon or lipopolysaccharide (LPS) in kidney proximal tubule cells.[31] This suggested that OPN could also play a role in regulation of nitric oxide produced by other tissues. Interactions of cancer cells with the vascular system are important at various stages of metastasis, including intravasation, extravasation, and tumor growth. We have extended the idea of OPN regulation of nitric oxide production to vascular tissue,[32] in preliminary studies in which we showed that recombinant OPN downregulates LPS-induced nitric oxide synthase (iNOS) activity, in rat thoracic aorta tissue. Nitric oxide has been suggested to play a role in limiting metastases.[33] We incorporated these ideas in an hypothesis that proposes that OPN production by cancer cells could contribute to malignancy by protecting cancer cells from oxidative attack by various host tissues, including macrophages and cells of the vascular system.[34] Given the many functions that have been proposed for different regions of the OPN molecule, it seems likely that OPN may well contribute to malignancy by more than one mechanism.

The details of the contribution, if any, of OPN to specific human cancers remain to be ascertained. We have found that OPN is expressed in many human tumors, and is not produced in others (O'Malley, Casson, Engel, Harris, and Chambers, unpublished observations). In addition, we have found that tumor-associated OPN, in a given tumor, can be due to expression of OPN by the tumor cells, by tumor-associated macrophages and other inflammatory cells, or both.

The clinical significance of OPN expression by various subsets of cells in tumors remains to be clarified.

Site-Directed Mutagenesis of the RGD Sequence of OPN

OPN proteins from all species studied have a conserved GRGDS (glycine-arginine-glycine-aspartic acid-serine) amino acid sequence (diagrammed in FIG. 6, Bautista *et al.*[35]), typical of integrin-binding cell adhesion proteins. We and many others have found that OPN has integrin-mediated cell adhesive function for a variety of cell types. Binding of cells to extracellular OPN (*e.g.*, in developing bone, immune responses, or cancer) might play a role in fixing cells in a particular location in the body. In addition to this proposed "attachment" function, however, evidence is accumulating for other integrin-binding proteins to suggest that binding of RGD-containing proteins to appropriate cell surface integrins can transmit a signal to the cell (reviews, Refs. 36, 37). This integrin-mediated signal transduction could result in altered cell behavior and function, via changes in gene expression, rendering cells responsive to changes in the local environment. Some evidence suggests that binding of OPN to cells may be able to initiate responses from the cells. As examples, binding of OPN to osteoclasts has been shown to result in cytosolic calcium fluxes,[38] and attachment of cells to OPN-coated surfaces can induce heat shock proteins.[39,40] Such studies imply that OPN can transmit a signal on binding to cells. The nature and effect of OPN-mediated signal transduction, as well as the range of possible target cells in normal and pathological situations, remain to be discovered.

We have prepared recombinant OPN protein (murine and human), as GST (glutathione-S-transferase)-fusion proteins, and showed that these proteins were functional in both cell adhesion and chemotaxis assays.[41,42] This finding indicated that sequences in the protein backbone of OPN are sufficient for these functions (at least as measured *in vitro*), leaving the role of posttranslational modifications, such as glycosylation and phosphorylation, to be clarified. The ability of OPN to induce directed migration in MDA-MB-435 human breast cancer cells,[42] as well as in normal human gingival fibroblasts and *ras*-transformed fibroblasts, suggests that OPN could play such a role in malignancy, perhaps in the targeting of metastatic cells to OPN-rich tissues. Interestingly, it was recently reported that breast cancer cells that had metastasized to bone had increased expression of the $\alpha v\beta 3$ integrin, to which OPN can bind.[43]

We then asked whether the conserved RGD sequence in OPN was required for the adhesive and migration-inducing functions of the protein. We directly tested the hypothesis that the RGD amino acid sequence of OPN is required for cell adhesion and chemotaxis.[42] We generated mutated recombinant OPN proteins, using site-directed mutagenesis, in which the RGD sequence was either deleted completely or mutated to RGE (arginine-glycine-glutamic acid), which binds poorly to integrins (FIG. 1). We found that both the mutant proteins were nonfunctional in cell adhesion and chemotaxis assays with a variety of human and mouse cell lines (TABLE 1), suggesting that an intact RGD sequence in OPN is required for these functions.[42] This conclusion is also supported by our recent development of monoclonal antibodies to human OPN, in collaboration with J. Harris.[35] One of these antibodies (Mab53) apparently binds to an epitope of OPN that contains the RGD sequence, and this antibody blocks cell adhesion to OPN.[35] In contrast, another monoclonal antibody (Mab87-B), whose epitope maps elsewhere in OPN, does not interfere with the cell adhesion function of the protein.[34] Details of that

GST-mOPN: GST......VPTVDVPNG**RGD**SLAYGLRSKSRSFQVSDEQY....

"delRGD": GST......VPTVDVPNG - - - SLAYGLRSKSRSFQVSDEQY....

"RGE": GST......VPTVDVPNG**RGE**SLAYGLRSKSRSFQVSDEQY....

FIGURE 1. Site-directed mutagenesis was used to produce recombinant murine OPN proteins in which the RGD sequence was either deleted ("delRGD") or mutated to RGE ("RGE"). Full-length proteins were produced as GST-fusion proteins, as described by Xuan *et al.*;[41,42] only the region around the RGD sequence is diagrammed.

study are presented elsewhere in this volume, by Bautista *et al.* Taken together these findings indicate that the majority of OPN's adhesive and chemotactic functions, as well as any signal transduction activity generated as a consequence of binding of OPN to cells, depend on the RGD sequence.

CONCLUSION

Many malignant cells produce OPN. Recent studies using antisense OPN RNA support the idea that OPN produced by tumor cells can contribute functionally to the malignant behavior of the cells. Expression of the H-*ras* oncogene can induce expression of OPN by transcriptional regulation, and can also induce tumorigenic and metastatic behavior in NIH 3T3 cells. We have presented evidence that other metastatic cells also may upregulate transcription of OPN, perhaps by expression of a nuclear protein that is a new member of the ETS family of transcriptional regulators. Site-directed mutagenesis of OPN to destroy the integrin-binding RGD amino acid sequence of the protein resulted in OPN that was unable to support cell adhesion or directed migration of various transformed and normal cells. OPN could promote malignancy by inducing integrin-mediated

TABLE 1. Mutation of OPN's RGD Sequence by Site-Directed Mutagenesis Destroys Cell Adhesion and Chemotaxis Function[a]

	GST-mOPN	"delRGD"	"RGE"
Cell adhesion			
PAP2	yes	no	no
MDA-MB-435	yes	no	no
Gingival fibroblasts	yes	no	no
Chemotaxis			
PAP2	yes	no	no
MDA-MB-435	yes	no	no
Gingival fibroblasts	yes	no	no

[a] The RGD sequence was either deleted ("delRGD") or was mutated to the sequence RGE (arginine-glycine-glutamic acid) ("RGE"), as shown in FIGURE 1. Recombinant proteins (control and two mutant proteins), as GST-OPN fusion proteins, were tested for ability to support cell adhesion and directed cell migration in Transwell chambers. Cells are PAP2 (*ras*-transformed NIH 3T3), MDA-MB-435 human mammary carcinoma cells, and human gingival fibroblasts. Data and additional experimental details are presented in Xuan *et al.*[42]

responses in host or tumor cells. Many questions still remain in understanding the details of the role and regulation of OPN in malignancy.

SUMMARY

Transformation of nontumorigenic NIH 3T3 fibroblasts with an activated *ras* oncogene produces malignant cells that can metastasize. This induction of malignant behavior is due to changes in gene expression induced by Ras-mediated signal transduction. Osteopontin expression is induced in response to Ras, due to increased osteopontin transcription mediated by Ras-responsive regions in the osteopontin gene promoter. The increased expression of osteopontin contributes functionally to the malignant ability of the cells. *Ras*-transformed cells that express antisense osteopontin RNA show markedly reduced ability to form tumors and to metastasize in experimental animals. Increased osteopontin expression in human tumors thus may also contribute to increased malignancy. Site-directed mutagenesis of recombinant osteopontin protein indicates that an intact RGD sequence is required for cell adhesion and induction of chemotaxis, consistent with the idea that integrin-mediated signal transduction is a consequence of osteopontin binding to cells. Osteopontin may contribute to malignancy by inducing responses in host and/or tumor cells.

ACKNOWLEDGMENTS

The valuable contributions of laboratory members (including Elke Behrend, Charulata Hota, Alan Tuck, Sylvia Wilson, and Jian-wu Xuan) and collaborators (David Denhardt, Ann Marie Craig, Xiaojia Guo; Charles Prince; John Harris, Diosdado Bautista; Martha Somerman, John D'Errico, Yoichiro Shigeyama; Frances O'Malley; Alan Casson; Jay Engel) to the studies reviewed here are gratefully acknowledged.

REFERENCES

1. Nicolson, G. L. 1993. Cancer progression and growth: relationship of paracrine and autocrine growth mechanisms to organ preference of metastasis. Exp. Cell Res. **204:** 171–180.
2. MacDonald, N. J. & P. S. Steeg. 1993. Molecular basis of tumour metastasis. Cancer Surv. **16:** 175–199.
3. Bondy, G. P., S. Wilson & A. F. Chambers. 1985. Experimental metastatic ability of H-*ras*-transformed NIH 3T3 cells. Cancer Res. **45:** 6005–6009.
4. Hill, S. A., S. Wilson & A. F. Chambers. 1988. Clonal heterogeneity, experimental metastatic ability, and p21 expression in H-*ras*-transformed NIH 3T3 cells. J. Natl. Cancer Inst. **80:** 484–490.
5. Chambers, A. F., G. H. Denhardt & S. M. Wilson. 1990. *Ras*-transformed NIH 3T3 cell lines, selected for metastatic ability in chick embryos, have increased proportions of p21-expressing cells and are metastatic in nude mice. Invasion & Metastasis **10:** 225–240.
6. Tuck, A. B., S. M. Wilson & A. F. Chambers. 1990. *Ras* transfection and expression does not induce progression from tumorigenicity to metastatic ability in mouse LTA cells. Clin Exp. Metastasis **8:** 417–431.

7. TUCK, A. B., S. M. WILSON, R. KHOKHA & A. F. CHAMBERS. 1991. Different patterns of gene expression in ras-resistant and ras-sensitive cells. J. Natl. Cancer Inst. **83:** 485–491.
8. BOGUSKI, M. S. & F. MCCORMICK. 1993. Proteins regulating Ras and its relatives. Nature **366:** 643–654.
9. HALL, A. 1994. A biochemical function for Ras—at last. Science **264:** 1413–1414.
10. LEEVERS, S. J., H. F. PATERSON & C. J. MARSHALL. 1994. Requirement for Ras in Raf activation is overcome by targeting Raf to the plasma membrane. Nature **369:** 411–414.
11. KHOSRAVI-FAR, M. & C. DER. 1994. The Ras signal transduction pathway. Cancer Metastasis Rev. **13:** 67–89.
12. CHAMBERS, A. F. 1992. Mechanisms of oncogene-mediated alterations in metastatic ability. Biochem. Cell Biol. **70:** 817–821.
13. CHAMBERS, A. F. & A. B. TUCK. 1993. *Ras*-responsive genes and tumor metastasis. Crit. Rev. Oncog. **4:** 95–114.
14. GUO, X., A. F. CHAMBERS, C. L. J. PARFETT, P. WATERHOUSE, L. C. MURPHY, R. E. REID, A. M. CRAIG, D. R. EDWARDS & D. T. DENHARDT. 1990. Identification of a serum-inducible mRNA (5B10) as the mouse homolog of calcyclin: tissue distribution and expression in metastatic, *ras*-transformed NIH 3T3 cells. Cell Growth Differ. **1:** 333–338.
15. WETERMAN, M. A. J., G. N. P. VAN MUIJEN, H. P. J. BLOEMERS & D. J. RUITER. 1993. Expression of calcyclin in human melanocytic lesions. Cancer Res. **53:** 6061–6066.
16. CHAMBERS, A. F., R. COLELLA, D. T. DENHARDT & S. M. WILSON. 1992. Increased expression of cathepsins L and B and decreased activity of their inhibitors in metastatic, *ras*-transformed NIH 3T3 cells. Mol. Carcinog. **5:** 238–245.
17. COLELLA, R., A. F. CHAMBERS & D. T. DENHARDT. 1993. Anticarcinogenic activities of naturally occurring cysteine proteinase inhibitors. *In* Protease Inhibitors as Cancer Chemopreventative Agents. W. Troll & A. R. Kennedy, Eds. 199–216. Plenum Publishing Co. New York.
18. LIOTTA, L. A., P. S. STEEG & W. G. STETLER-STEVENSON. 1991. Cancer metastasis and angiogenesis: an imbalance of positive and negative regulation. Cell **64:** 327–336.
19. CRAIG, A. M., M. NEMIR, B. B. MUKHERJEE, A. F. CHAMBERS & D. T. DENHARDT. 1988. Identification of the major phosphoprotein secreted by many rodent cell lines as 2ar/osteopontin: enhanced expression in H-*ras*-transformed 3T3 cells. Biochem. Biophys. Res. Commun. **157:** 166–173.
20. CRAIG, A. M., G. T. BOWDEN, A. F. CHAMBERS, M. A. SPEARMAN, A. H. GREENBERG, J. A. WRIGHT, M. MCLEOD & D. T. DENHARDT. 1990. Secreted phosphoprotein mRNA is induced during multi-stage carcinogenesis in mouse skin and correlates with the metastatic potential of murine fibroblasts. Int. J. Cancer **46:** 133–137.
21. CHAMBERS, A. F., E. I. BEHREND, S. M. WILSON & D. T. DENHARDT. 1992. Induction of expression of osteopontin (OPN; secreted phosphoprotein) in metastatic, *ras*-transformed NIH 3T3 cells. Anticancer Res. **12:** 43–48.
22. CHAMBERS, A. F., C. HOTA & C. W. PRINCE. 1993. Adhesion of metastatic, ras-transformed NIH 3T3 cells to osteopontin, fibronectin, and laminin. Cancer Res. **53:** 701–706.
23. BUTLER, W. T. 1989. The nature and significance of osteopontin. Connect. Tissue Res. **23:** 123–136.
24. DENHARDT, D. T. & X. GUO. 1993. Osteopontin: a protein with diverse functions. FASEB J. **7:** 1475–1482.
25. PATARCA, R., R. A. SAAVEDRA & H. CANTOR. 1993. Molecular and cellular basis of genetic resistance to bacterial infection: the role of the early T-lymphocyte activation-1/osteopontin gene. Crit. Rev. Immunol. **13:** 225–246.
26. GUO, X., Y. P. ZHANG, D. A. MITCHELL, D. T. DENHARDT & A. F. CHAMBERS. Identification of a ras-activated enhancer in the mouse osteopontin promoter and its interaction with a putative ETS-related transcription factor whose activity correlates with metastatic potential. Mol. Cell. Biol. In press.

27. CRAIG, A. M. & D. T. DENHARDT. 1991. The murine gene encoding secreted phosphoprotein 1 (osteopontin): promoter structure, activity, and induction *in vivo* by estrogen and progesterone. Gene **100:** 163–171.
28. SENGER, D. R., C. A. PERRUZZI & A. PAPADOPOULOS. 1989. Elevated expression of secreted phosphoprotein I (osteopontin, 2ar) as a consequence of neoplastic transformation. Anticancer Res. **9:** 1291–1299.
29. BEHREND, E. I., A. M. CRAIG, S. M. WILSON, D. T. DENHARDT & A. F. CHAMBERS. 1994. Reduced malignancy of ras-transformed NIH 3T3 cells expressing antisense osteopontin RNA. Cancer Res. **54:** 832–837.
30. GARDNER, H. A. R., B. BERSE & D. R. SENGER. 1994. Specific reduction in osteopontin synthesis by antisense RNA inhibits the tumorigenicity of transformed Rat1 fibroblasts. Oncogene **9:** 2321–2326.
31. HWANG, S. M., C. A. LOPEZ, D. E. HECK, C. R. GARDNER, D. L. LASKIN, J. D. LASKIN & D. T. DENHARDT. 1994. Osteopontin inhibits induction of nitric oxide synthase activity by inflammatory mediators in mouse kidney epithelial cells. J. Biol. Chem. **269:** 711–715.
32. SCOTT, J. A., D. G. MCCORMACK, J. W. XUAN & A. F. CHAMBERS. 1994. Osteopontin inhibits induction of nitric oxide synthase in rat vascular tissue by lipopolysaccharide. Clin. Exp. Metastasis **12:** 27 (abstract 65).
33. DONG, Z., A. H. STAROSELSKY, X. QI, K. XIE & I. J. FIDLER. 1994. Inverse correlation between expression of inducible nitric oxide synthase activity and production of metastases in K-1735 murine melanoma cells. Cancer Res. **54:** 789–793.
34. DENHARDT, D. T. & A. F. CHAMBERS. 1994. Overcoming obstacles to metastasis—defenses against host defenses: osteopontin (OPN) as a shield against attack by cytotoxic host cells. J. Cell. Biochem. **56:** 48–51.
35. BAUTISTA, D. S., J. W. XUAN, C. HOTA, A. F. CHAMBERS & J. F. HARRIS. 1994. Inhibition of RGD-mediated cell adhesion to osteopontin by a monoclonal antibody against osteopontin. J. Biol. Chem. **269:** 23280–23285.
36. HYNES, R. O. 1992. Integrins: versatility, modulation, and signaling in cell adhesion. Cell **69:** 11–25.
37. JULIANO, R. L. 1994. Signal transduction by integrins and its role in the regulation of tumor growth. Cancer Metastasis Rev. **13:** 25–30.
38. MIYAUCHI, A., J. ALVAREZ, E. M. GREENFIELD, A. TETI, M. GRANO, S. COLUCCI, A. ZAMBONIN-ZALLONE, F. P. ROSE, S. L. TEITELBAUM, D. CHERESH & K. A. HRUSKA. 1991. Recognition of osteopontin and related peptides by an $\alpha_v\beta_3$ integrin stimulates immediate cell signals in osteoclasts. J. Biol. Chem. **266:** 20369–20374.
39. SAUK, J. J., C. L. VANKAMPEN, K. NORRIS, J. MOEHRING, R. A. FOSTER & M. J. SOMERMAN. 1990. Persistent spreading of ligament cells on osteopontin/bone sialoprotein-I or collagen enhances tolerance to heat shock. Exp. Cell Res. **188:** 105–110.
40. SAUK, J. J., C. L. VANKAMPEN, K. NORRIS, R. FOSTER & M. J. SOMERMAN. 1990. Expression of constitutive and inducible HSP70 and HSP47 is enhanced in cells persistently spread on OPN or collagen. Biochem. Biophys. Res. Commun. **172:** 135–142.
41. XUAN, J. W., C. HOTA & A. F. CHAMBERS. 1994. Recombinant GST-human osteopontin fusion protein is functional in RGD-dependent cell adhesion. J. Cell. Biochem. **54:** 247–255.
42. XUAN, J. W., C. HOTA, Y. SHIGEYAMA, J. A. D'ERRICO, M. J. SOMERMAN & A. F. CHAMBERS. Site-directed mutagenesis of the arginine-glycine-aspartic acid sequence in osteopontin destroys cell adhesion and migration functions. J. Cell. Biochem. In press.
43. LIAPIS, H., S. L. TEITELBAUM, S. KITAZAWA, J. W. SMITH, K. MERKEL & F. P. ROSS. 1994. Selective expression of the integrin $\alpha_v\beta_3$ by bone-residing breast cancer metastases. J. Bone Mineral. Res. **9**(Suppl. 1): S146 (abstract 102).

Osteopontin Expression in Cardiovascular Diseases[a]

CECILIA M. GIACHELLI, LUCY LIAW,
CHARLES E. MURRY, STEPHEN M. SCHWARTZ,
AND MANUELA ALMEIDA

Department of Pathology SJ-60
University of Washington
School of Medicine
Seattle, WA 98195

INTRODUCTION

Specific adhesive interactions are increasingly recognized as prerequisite for cellular differentiation, growth and movement. These basic cell processes are thought to be essential for normal cardiovascular development and to contribute significantly to the remodeling processes of neointima formation, angiogenesis, and inflammation associated with several cardiovascular diseases including atherosclerosis and restenosis. Thus the role of adhesive proteins in cardiovascular function is an area of considerable interest. One of the more recently discovered components of the remodeling vascular milieu is osteopontin. The studies described below relate to our ongoing effort to examine the role of adhesive proteins, particularly osteopontin and its integrin-type receptors, in these clinically significant processes.

Cloning of Osteopontin as a Gene Differentially Expressed following Arterial Injury

In vivo and *in vitro* studies have shown that at least two smooth muscle cell phenotypes can be identified in the aorta of the rat: one common to young (pup) rats and the injured artery, and the second predominantly observed in the adult media.[1–3] The development of the injury-related phenotype appears to be critical for arterial neointima formation, a common vascular lesion seen in several vascular diseases including atherosclerosis and restenosis.[3–5] To identify markers for the injury phenotype, differential cloning was performed between phenotypically distinct subsets of cultured smooth muscle cells derived at different stages of vascular development.[6] One of the genes identified in this manner was named 2B7. As shown in FIGURE 1A, 2B7 mRNA levels were consistently higher in rat pup-derived cells than adult aortic smooth muscle cells. In addition, 2B7 mRNA levels were upregulated in the rat carotid artery in response to injury by balloon catheter angioplasty *in vivo* (FIG. 1B). Sequence analysis of this clone identified it as the rat cDNA for osteopontin.[6,7] Western blot analysis utilizing a monoclonal antibody

[a] Supported in part by NIH Grants HL03174, HL07312, HL47151, HL18645, and DK47659.

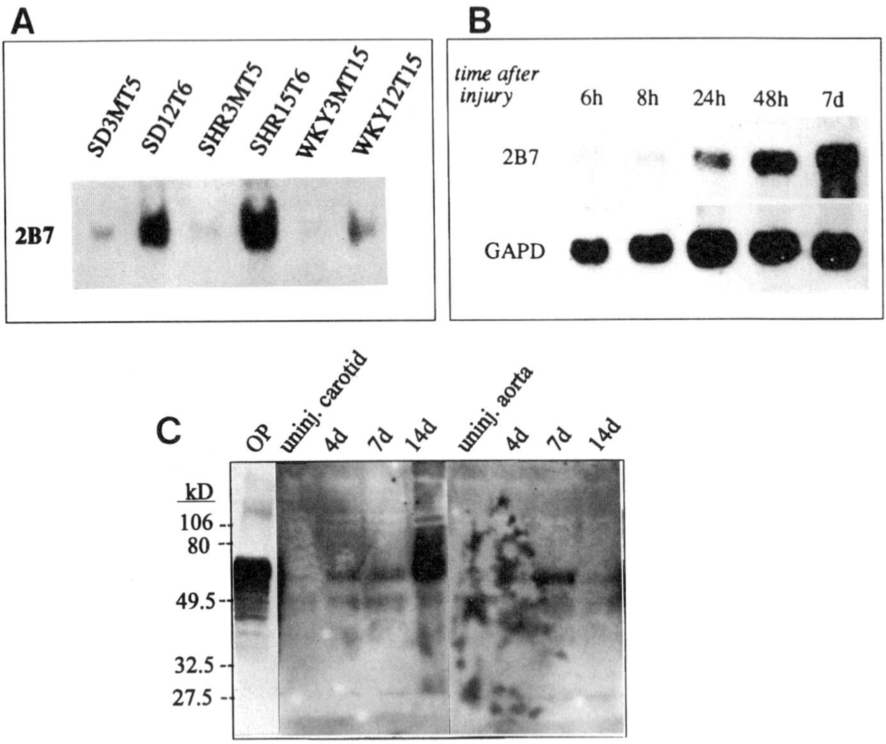

FIGURE 1. Osteopontin is upregulated in phenotypically distinct smooth muscle cells and in the injured rat carotid artery. **(A)** Northern blot probed with osteopontin probe, 2B7. SD12T6, SHR15T6, and WKY12T15 are different pup-derived smooth muscle cell isolates, and SD3MT5, SHR3MT5, and WKY3MT15 are different adult-derived smooth muscle cell isolates. **(B)** Northern blot probed with 2B7 (*upper*) or glyceraldehyde-6-phosphate-dehydrogenase probe, GAPD (*lower*). RNA was derived from rat carotid arteries at the indicated times following balloon angioplasty. **(C)** Western blot using anti-osteopontin antibody (OP199) of protein samples (25 μg) derived from uninjured arteries or from arteries at indicated times following balloon angioplasty. An identical gel stained with Coomassie indicated equivalent loading of the samples (not shown). OP = 1 μg purified rat smooth muscle-derived osteopontin. (A and B from C. M. Giachelli *et al.*[6] Reprinted by permission from *Biochemical and Biophysical Research Communications.*)

directed against rat bone osteopontin confirmed the elevation of a protein of about 64–66 kD in the injured rat carotid and aorta antigenically related to osteopontin (FIGURE 1C). This size is consistent with that of purified rat smooth muscle-derived osteopontin (see below).

To determine the localization of osteopontin within the rat artery, *in situ* hybridization and immunohistochemistry were performed. As shown in FIGURE 2, osteopontin mRNA levels were low in uninjured rat arteries, but strikingly elevated in the injured vessel. Similar results were obtained for osteopontin protein following antibody staining.[7] Of particular significance, smooth muscle cells form-

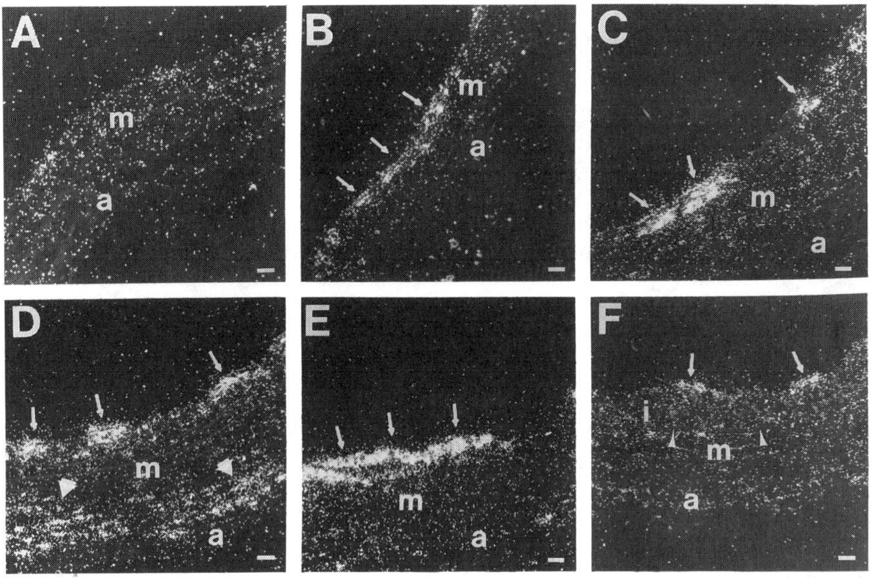

FIGURE 2. Osteopontin mRNA localizes to the neointima in rat arteries following balloon angioplasty. Rat arterial sections were hybridized to the antisense (A–F) riboprobes generated from transcription of the 2B7 rat osteopontin plasmid. **(A)** Uninjured carotid; **(B)** 2-d ballooned carotid showing increased grains over cells adjacent to the lumen (*arrows*); **(C)** 2-d ballooned aorta showing a similar pattern of high intensity signal localized to the luminal smooth muscle cell (*arrows*); **(D)** 4-d ballooned aorta showing high levels of grains over the neointimal (*arrows*) smooth muscle cells, and weaker signal over the outer medial smooth muscle cells (*arrowheads*); **(E)** 7-d ballooned aorta showing a 2–3-cell-layer-thick accumulation of neotimal smooth muscle cells (*arrows*) strongly positive for osteopontin mRNA; **(F)** 3-wk ballooned aorta showing elevated grains over a subpopulation of neointimal cells closest to the lumen (*arrows*), but levels in the remainder of the intima (i) similar to those in the underlying media (m). *Arrowheads* point to the internal elastic lamina. a, adventitial; i, neointima; m, media. *Bar,* 10 μM. (From Giachelli *et al.*[7] Reprinted by permission from the *Journal of Clinical Investigation.*)

ing the neointima (*i.e.*, those phenotypically distinct from the medial smooth muscle cells) were the primary producers of osteopontin, validating our cloning strategy. Osteopontin protein and mRNA were expressed transiently, with the highest levels seen between days 2 and 14, when cellular replication, migration and extracellular matrix deposition were highest.[7] In our most recent studies, we found that, in addition to neointimal smooth muscle cells, endothelial cells regenerating at the arterial wound edge also elaborate osteopontin (Liaw and Giachelli, unpublished observations). These data confirm the expression of osteopontin in phenotypically distinct smooth muscle cells, and suggest a role for osteopontin in the remodeling processes critical for neointima formation.

Upregulation of Osteopontin in Cardiovascular Disease States

The striking elevation of osteopontin during neointima formation in injured rat arteries suggested a potential role for this protein in human cardiovascular diseases characterized by thickened intima formation such as atherosclerosis and restenosis. Indeed, immunostaining of atherosclerotic human carotid arteries showed osteopontin protein accumulations in both cellular and acellular regions of the plaque, and in close association with calcium deposits (see below). This was in contrast to the absence of osteopontin staining in normal vessels.[7] The major producer of osteopontin in plaques was shown by *in situ* hybridization to be the macrophage, but distinct smooth muscle and endothelial populations also expressed osteopontin mRNA and protein. In particular, homogeneous areas of stellate smooth muscle cells and endothelial cells composing neovessels were strongly positive. Both primary and restenotic (those occurring following unsuccessful angioplasty) plaques showed similar osteopontin distributions.[8] This intimate association of osteopontin with diseased human intimas strongly implicates an involvement of osteopontin in atherosclerotic processes including smooth muscle migration, inflammation and angiogenesis.

Vascular remodeling is also a key feature of hypertensive vessels, and we recently obtained evidence for a potential involvement of osteopontin in this process. Angiotensin II (AII) infusion in rats elevates blood pressure and induces microvascular hyperplasia of type I mesenteric vessels. This lesion, resembling the type seen in individuals with malignant hypertension, is tightly correlated with osteopontin expression (Schwartz and Lombardi, unpublished observations). In addition to microvascular effects, AII-induced hypertension also leads to tubulointerstitial disease characterized by a macrophage infiltrate and a pronounced fibrotic response in the renal tubular interstitium. Of particular interest, a dramatic increase in renal cortical mRNA levels was noted which preceded pathologic changes in tubular and interstitial cells (FIG. 3). Immunostaining of adjacent sections confirmed upregulation of tubular osteopontin protein in selected distal tubule cells and revealed that sites with elevated osteopontin protein correlated with regions of macrophage accumulation.[9] In addition, three other rat models of tubulointerstitial disease were found to be associated with elevated cortical osteopontin expression.[10] These observations, coupled with recent findings that osteopontin may be a proinflammatory molecule[11] suggest that osteopontin overexpression may facilitate macrophage accumulation at sites of renal tubulointerstitial injury.

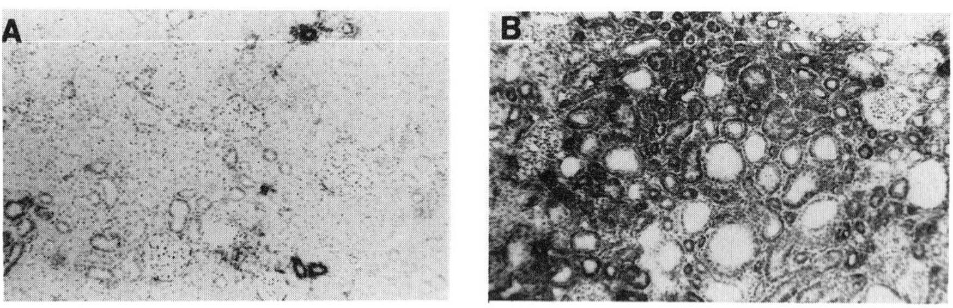

FIGURE 3. Osteopontin mRNA is induced in AII-treated rat kidneys. Rats were infused for 14 days with vehicle alone **(A)** or AII **(B)**, and *in situ* hybridization was performed using a 35S-labeled 2B7 riboprobe. Photomicrographs of the resulting hybridization signals were obtained at 200× magnification on the same field using bright field illumination. Original magnification = 200×. (From Giachelli *et al.*[9] Reprinted by permission from *Kidney International.*)

As mentioned above, osteopontin was found closely associated with dystrophic calcification in our inspections of carotid endarterectomy-derived atherosclerotic plaque tissue.[7] We and others subsequently confirmed and extended these observations. Extracellular accumulations of osteopontin were found to be tightly correlated with sites of calcium deposition in primary and restenotic coronary atherectomy specimens,[8] aortic plaques,[12,13] and coronary autopsy specimens.[14] Furthermore, highly significant associations were found between the degree of osteopontin expression and the degrees of both calcification and inflammation in lesions of aortic valvular stenosis (K. O'Brien, unpublished observations). These data suggest that, rather than representing an unmodifiable process, the dystrophic calcification seen in aortic stenosis and atherosclerosis may be an actively regulated process open to therapeutic intervention (K. O'Brien, unpublished observations).

Macrophages Express Osteopontin after Cardiac Injury

The observations that osteopontin promoted adhesion and migration of cultured vascular cells (see below), and was upregulated following balloon angioplasty *in vivo,* prompted us to explore other injury models. We hypothesized that osteopontin might be upregulated in the repair of experimental cardiac injury or myocardial infarction in humans.[15] To test this hypothesis, rat hearts were injured by freeze-thaw across an intact diaphragm as previously described.[16] This model produces a reproducible zone of coagulation necrosis, which then heals in a manner very similar to acute infarcts. The hearts were studied from one day to four weeks after injury, providing a time course of lesion healing. No osteopontin was detected in normal myocardium by immunostaining, Western blotting, or *in situ* hybridization. On days 1 and 2 after injury, osteopontin mRNA and protein were expressed at high levels by macrophages infiltrating necrotic myocardium (FIG. 4A, B). Double-labeling with the macrophage marker ED1, however, demonstrated that only a subset of macrophages contained osteopontin protein. Western blot analysis showed a single 66-kD band in injured myocardium, which was absent from control tissue (FIG. 4C).

Although macrophages remained abundant cellular components of the ensuing granulation tissue and scar tissue, the expression of osteopontin was diminished at day 4 and markedly downregulated at one and four weeks after injury, with only rare cells expressing the message or protein. The downregulation of osteopontin expression paralleled the removal of necrotic tissue. In fact, in some hearts the only macrophages making osteopontin were those surrounding residual necrotic muscle. In a human heart removed for transplantation from a patient eight days after a myocardial infarction, Northern blotting showed marked upregulation of osteopontin mRNA in areas of necrosis and granulation. Immunostaining localized osteopontin protein to cells morphologically consistent with macrophages. Transient upregulation of osteopontin was not unique to cardiac injury, because a subset of macrophages infiltrating injured lung, skin and skeletal muscle also showed a similar time course of osteopontin expression.[15]

These studies demonstrated that macrophages synthesize osteopontin in the early inflammatory response to necrotizing injury in the heart, lung, skeletal muscle, and skin. Thus, synthesis of osteopontin by macrophages appears to be a generalized response in the reaction to tissue injury. Although macrophages persist in these lesions, osteopontin is dramatically downregulated as healing proceeds. These results provide the first evidence that osteopontin may be important in

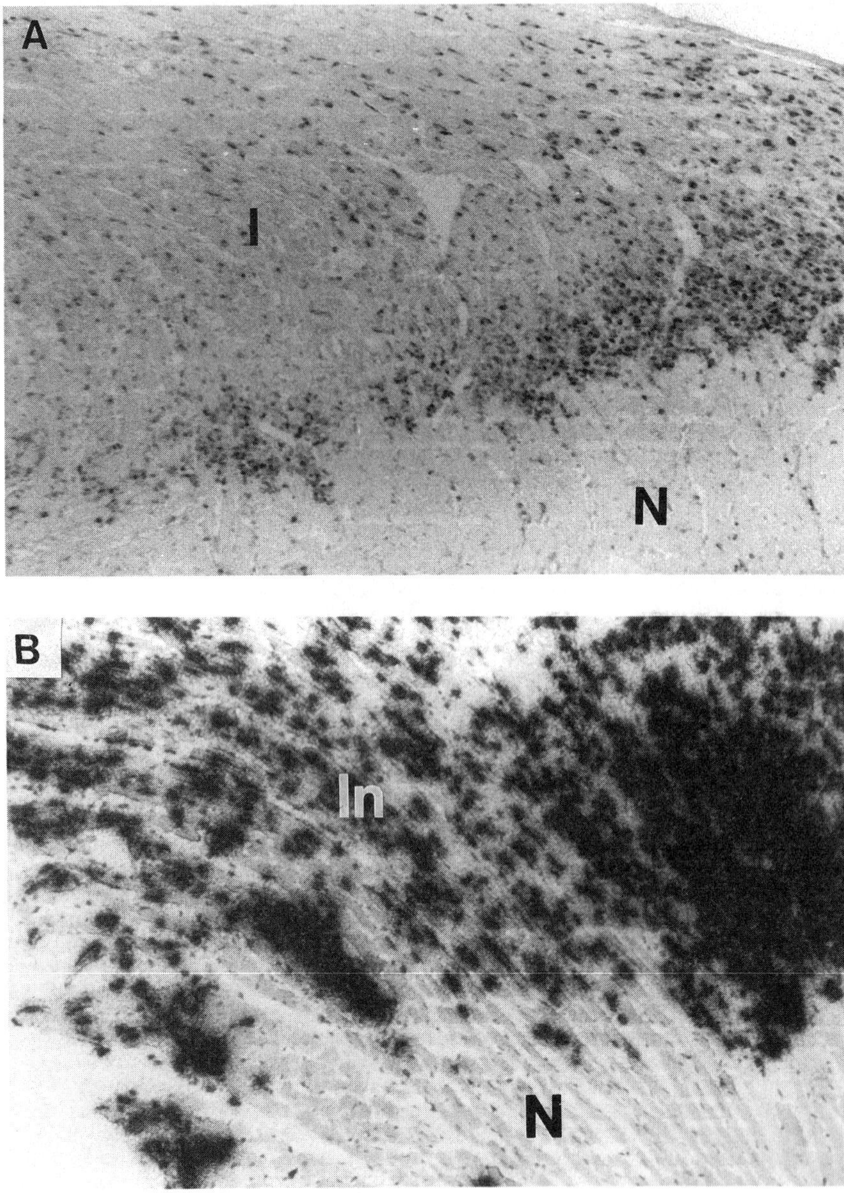
A
I
N
B
In
N

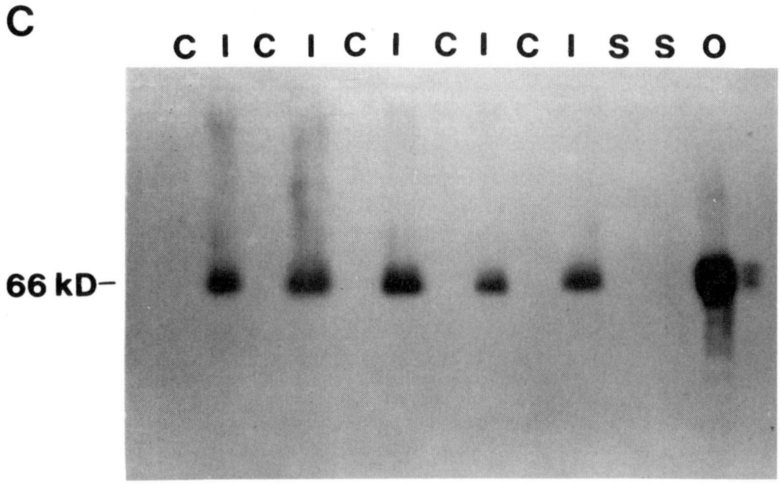

FIGURE 4 (*Continued*).

healing after tissue injury, possibly in cellular adhesion, chemotaxis and/or phagocytosis.

Purification and Characterization of Osteopontin Derived from Rat Smooth Muscle Cells

Based on the compelling expression patterns of osteopontin in multiple cardiovascular diseases and tissues injuries, experiments aimed at determining the function of osteopontin in vascular cells were undertaken. Rat pup smooth muscle cells were chosen as a source of osteopontin based on their high levels of expression of osteopontin mRNA and secreted protein compared to adult smooth muscle cells (FIG. 1).[7] Immunostaining of the respective cultures was consistent with these findings, and showed strong Golgi-like staining for osteopontin in cultured

FIGURE 4. Macrophages express osteopontin following cardiac injury. Immunostaining for macrophage marker, ED1 **(A)** or *in situ* hybridization for osteopontin **(B)** in sections of rat heart one day following cardiac freeze-thaw injury. Osteopontin signal colocalizes with macrophage infiltrating the wound. In = area of injury; N = normal myocardium. **(C)** Western blot analysis of protein samples obtained from control (C), sham-operated (S) or injured (I) hearts one day following freeze-thaw injury. The apparent size of purified rat smooth muscle-derived osteopontin (*lane 0*) is given in kilodaltons (kD). (From Murry *et al.*[15] Reprinted by permission from the *American Journal of Pathology*.)

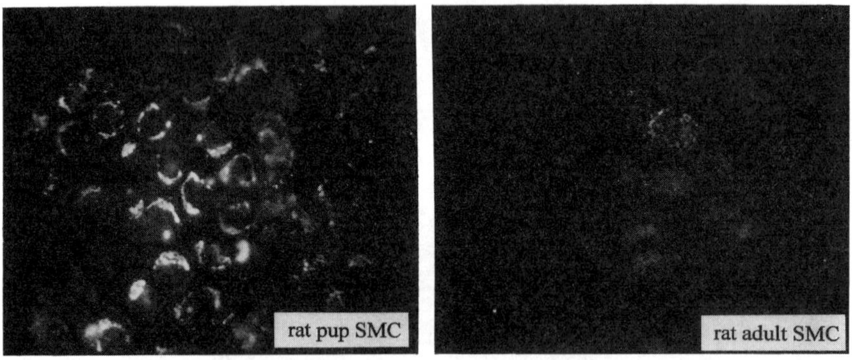

FIGURE 5. Smooth muscle cells derived from rat pup synthesize high levels of osteopontin. Rat pup or adult aortic smooth muscle cells were immunolabeled with a fluorescently labeled anti-osteopontin antibody (OP199[18]). Original magnification = 200×.

pup, but not adult, smooth muscle cells (FIG. 5). Serum-free media conditioned for 48 hours was collected, dialyzed, and fractionated over a DEAE-sepharose column with a linear salt gradient (FIG. 6A). Fractions containing osteopontin (eluting between 0.4 and 0.6 mol/L NaCl) were identified by immunoblotting with anti-osteopontin antibody and pooled. Osteopontin was adsorbed to barium citrate as previously described[17] and eluted with 0.2 M sodium citrate. The osteopontin preparation obtained in this manner was judged to be >95% homogeneous as determined by Coumassie staining and Western blotting (FIG. 6B, C) and N-terminal sequencing.[18] Two major proteins of apparent molecular weight 64 and 66 kD were observed on 9% SDS-PAGE gels, with traces of a lower (50-kD) band and high molecular weight bands (potentially aggregates).

The biochemical properties of rat smooth muscle-derived osteopontin obtained by proteolytic cleavage, *in vitro* phosphorylation studies, and lectin carbohydrate analyses are shown in TABLE 1. According to these data, rat smooth muscle-derived osteopontin is phosphorylated, contains sialic acid, and can be cleaved by thrombin, properties similar to those reported for bone-, urine- and milk-derived osteopontin.[19] Osteopontin was also identified as enterokinase substrate (Ling and Giachelli, unpublished observation).

Functional Properties of Rat Vascular Osteopontin

Both endothelial and smooth muscle cells adhered well to dishes coated with osteopontin (FIG. 7). However, half-maximal adhesion was observed in endothelial cells at osteopontin concentrations fivefold lower than that required for smooth muscle cell adhesion (7 nM versus 30 nM). Cells attached, spread and formed focal contacts on osteopontin, suggesting that osteopontin could initiate cytoskeletal organization in both cell types. Interaction with smooth muscle and endothelial cells was RGD- and divalent cation-dependent, and a neutralizing antibody against the $\alpha_v\beta_3$ integrin could partially block cell adhesion to osteopontin.[18] Thus, two major cell types involved in vascular remodeling not only express osteopontin, but can interact with it.

TABLE 1. Biochemical Properties of Smooth Muscle-Derived Osteopontin[a]

Protein Characteristics	
molecular weight	64 + 66 kd doublet
high molecular weight forms	yes
thrombin cleavage	yes
enterokinase cleavage	yes
Glycosylation Pattern	
Lectin: Specificity	Content
GNA: terminal mannose	–
SNA: sialic acid linked α(2-6) to galactose	+ + +
MAA: sialic acid linked α(2-3) to galactose	+ +
PNA: core disaccharide galactose β(1-3) N-acetylgalactosamine	–
DSA: Galβ(1-4)GlcNAc	+ +
Phosphorylation Pattern	
Incorporation of $[^{32}P]H_3PO_4$	64 kd + 66 kd doublet

[a] Protein characteristics determined by Western blot analysis on purified osteopontin; glycosylation pattern determined by the use of lectins on purified osteopontin; phosphorylation pattern determined by phosphate labeling of smooth muscle cells, followed by immunoprecipitation with a specific antibody. *Abbreviations:* GNA, *Galanthus nivalis* agglutinin; SNA, *Sambucus nigra* agglutinin; MAA, *Maackia amurensis* agglutinin; PNA, peanut agglutinin; DSA, *Datura stramonium* agglutinin.

Since it has been proposed that osteopontin might act as a migration factor,[20,21] this property was examined using the modified Boyden chamber chemotaxis assay. In this assay, test substances are separated from cells by a porous polycarbonate membrane, and migration is determined by quantitating cell movement across the membrane. Osteopontin stimulated dose-dependent migration of rat vascular smooth muscle cells as shown in FIGURE 8A. This activity was chemotactic, since maximal effects were only noted when a gradient of osteopontin was provided to the cells (FIG. 8B). This direct demonstration of cellular migration stimulated by osteopontin has implications not only for the cell movements critical to vascular lesion formation, but to other cell movements associated with elevated osteopontin including macrophage infiltration[15] and tumor metastasis.[20,21]

Vascular Receptors for Osteopontin and Their Functions

As mentioned above, vascular cell adhesion to osteopontin was RGD- and divalent cation-dependent, suggesting that receptors for osteopontin might include integrins. Indeed, a neutralizing antibody directed against $\alpha_v\beta_3$ integrin was shown to block about half of osteopontin's adhesive activity, consistent with previous studies implicating $\alpha_v\beta_3$ as an osteopontin receptor.[22] However, smooth muscle cells with low $\alpha_v\beta_3$ surface expression could also interact with osteopontin in an $\alpha_v\beta_3$-independent but RGD-dependent manner. Thus, we reasoned that other integrins might also be capable of interacting with osteopontin.

To identify these, a battery of neutralizing antibodies targeting various integrin complexes or subunits were tested for their ability to interfere with vascular cell adhesion to osteopontin. As shown in FIGURE 9A, anti-$\alpha_v\beta_5$ and anti-β_1 antibodies both inhibited $\alpha_v\beta_3$-deficient cell adhesion to osteopontin, whereas antibodies against α_2, α_3 or α_5 subunits had little effect. A combination of both antibodies

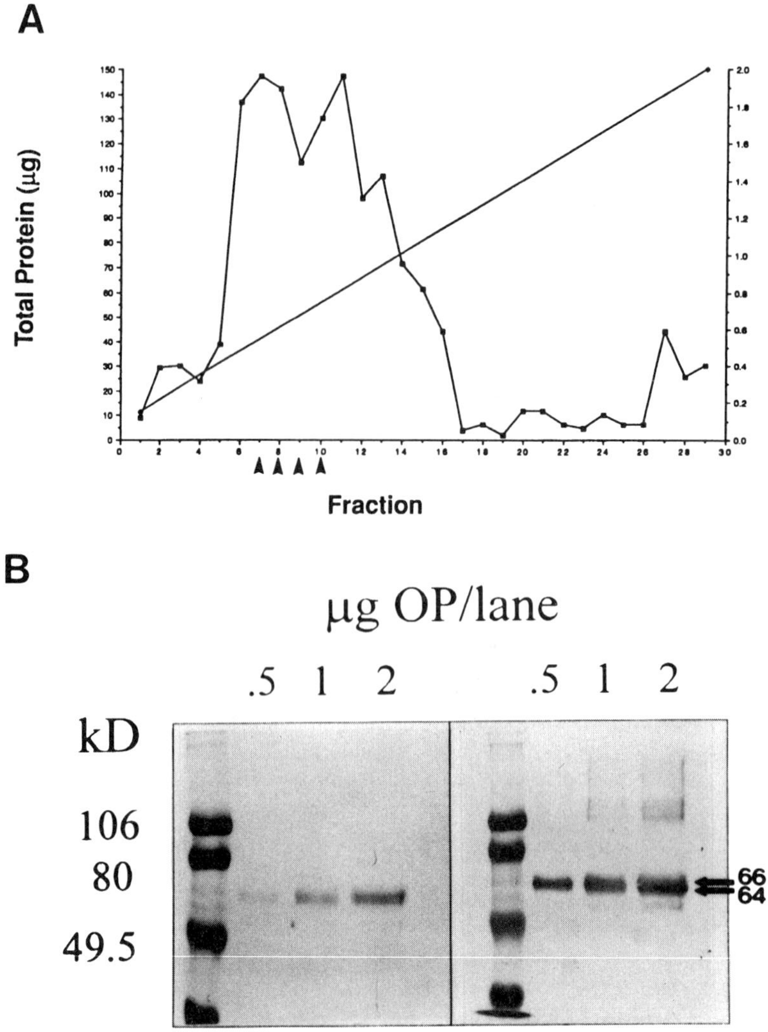

FIGURE 6. Purified osteopontin derived from rat smooth muscle cells. **(A)** Elution profile of dialyzed pup-conditioned media from DEAE-sepharose. Fractions containing osteopontin were identified by immunodotblotting (indicated with an *arrow*), pooled and then absorbed to barium citrate, washed and eluted into 0.2 M sodium citrate. **(B)** Purified osteopontin was electrophoresed on 9% polyacrylamide gels and either stained with Coomassie blue (*left panel*) or electroblotted onto a nylon membrane (*right panel*), and Western analysis was performed using an anti-osteopontin antibody. ((B) from Liaw *et al.*[18] Reprinted by permission from *Circulation Research*.)

completely blocked these cells' ability to adhere to osteopontin. Likewise, a combination of anti-$\alpha_v\beta_3$, β_1 and $\alpha_v\beta_5$ antibodies completely blocked adhesion in $\alpha_v\beta_3$-containing cells (FIG. 9B).[22] $\alpha_v\beta_1$ is most likely the β_1 integrin involved, since the $\alpha_v\beta_1$ complex could be immunoprecipitated from vascular smooth muscle cells, and an antibody directed against the α_v subunit completely blocked adhesion in cells containing or lacking $\alpha_v\beta_3$. Thus, in addition to $\alpha_v\beta_3$, osteopontin can interact with $\alpha_v\beta_1$ and $\alpha_v\beta_5$ integrins.[22] An important functional consequence of multiple receptors was noted when we tested whether cell surface $\alpha_v\beta_3$ affected vascular smooth muscle migration to osteopontin. As shown in FIGURE 10A, only cells containing $\alpha_v\beta_3$ were able to migrate towards an osteopontin stimulus. This migration was abolished when cells were preincubated with a neutralizing anti-$\alpha_v\beta_3$ antibody (FIG. 10B). Thus, while all three receptors facilitate adhesion to osteopontin, $\alpha_v\beta_3$ is required for migration.

DISCUSSION

Our studies have clearly implicated osteopontin in a number of fibrotic and inflammatory processes, including the cardiovascular pathologies of atherosclerosis and restenosis, the response of tissue to injury, and wound repair. In almost all cases, osteopontin expression correlated best with early tissue remodeling and was usually associated with cells that were actively migrating. Cell types that normally express little osteopontin gain this function in response to tissue remodeling and include macrophages, smooth muscle and endothelial cells.

Based on its adhesive and migratory functions in vascular smooth muscle cells and endothelial cells, we propose a model for the regulation and role of osteopontin in arterial remodeling. An injurious stimulus initiates the release of vasoactive molecules, including growth factors, from smooth muscle cells, endothelial cells, and platelets. These factors increase expression of genes including osteopontin, integrins and proteases required for smooth muscle and endothelial cell adhesion and migration. By stimulating both smooth muscle cells and endothelial cell adhesion and migration, osteopontin might favor intimal formation and reendothelialization, both required for complete wound repair. This model is supported by recent studies showing that a cyclic peptide, PenRGD, which is selective for α_v-containing receptors, inhibited neointimal thickening in rabbit arteries following angioplasty by ~60%.[23] This same peptide is an effective and discriminating inhibitor of smooth muscle cell adhesion to osteopontin in our hands.[22]

Our data have also demonstrated that osteopontin can interact with three different integrin-type receptors: $\alpha_v\beta_3$, $\alpha_v\beta_1$ and $\alpha_v\beta_5$. Interestingly, these three receptors appear to mediate different functions of osteopontin. As summarized in FIGURE 11, while $\alpha_v\beta_3$, $\alpha_v\beta_1$ and $\alpha_v\beta_5$ can all mediate adhesive interactions with osteopontin, only the $\alpha_v\beta_3$ mediates migration of smooth muscle cells to osteopontin. The mechanism responsible for stimulation of migration by osteopontin is intriguing, since it occurs at relatively low concentrations (nM levels) and appears to be independent of its adhesive activity, since cells containing the adhesive receptors for osteopontin, $\alpha_v\beta_1$ and $\alpha_v\beta_5$, failed to migrate. This suggests that a specific signaling cascade may be initiated by osteopontin through $\alpha_v\beta_3$ that results in cellular migration, and perhaps other as-yet unknown events.

One possible mechanism of unique signaling pathways may be sequestration of integrins into different compartments of the adherent cell, and we have observed

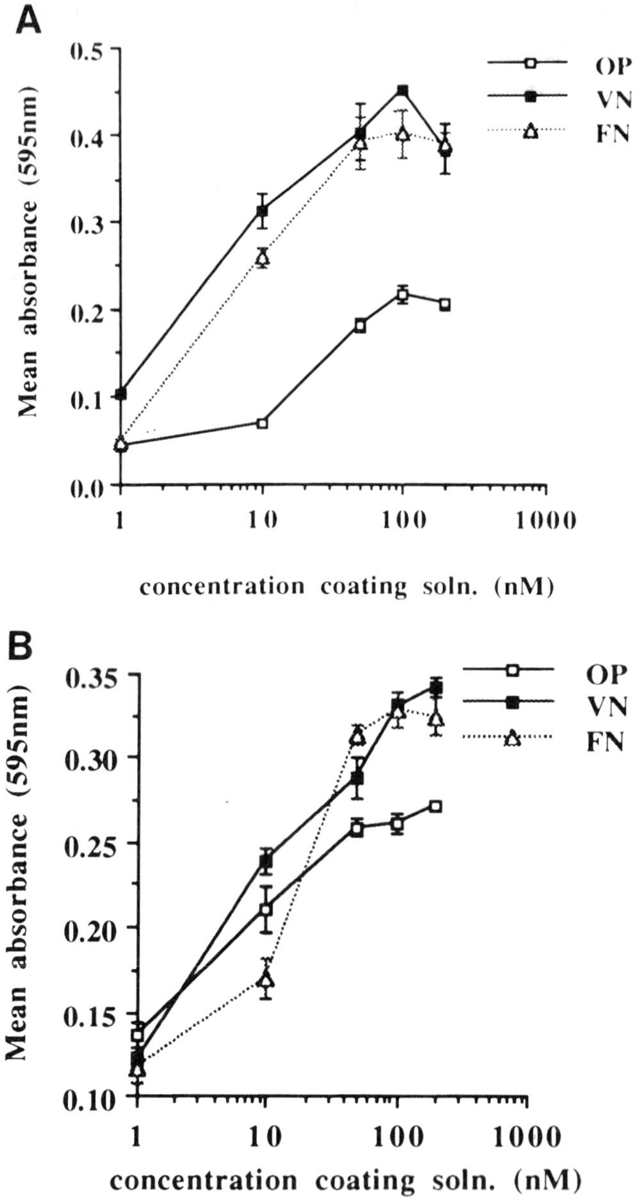

FIGURE 7. Endothelial and smooth muscle cells adhere to smooth muscle-derived osteopontin. **(A)** Adhesion of adult rat aortic smooth muscle cells to wells coated with increasing concentrations of osteopontin used to coat wells. **(B)** Adhesion of bovine aortic endothelial cells to wells coated with increasing concentrations of osteopontin. Mean absorbance is proportional to cell number in these assays.[18] (From Liaw *et al.*[18] Reprinted by permission from *Circulation Research*.)

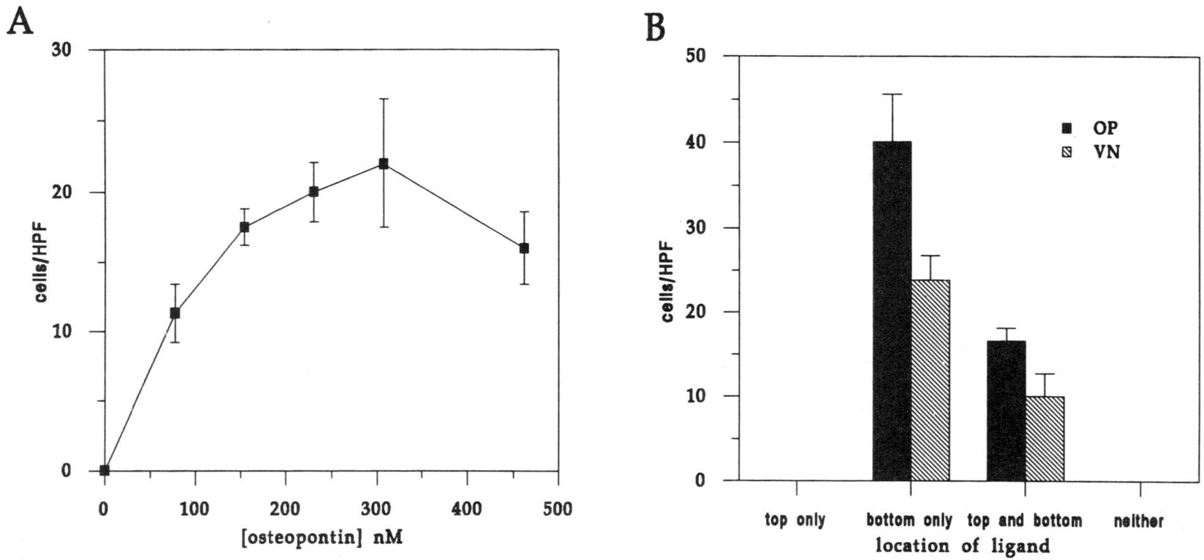

FIGURE 8. Osteopontin stimulates directed migration of rat aortic smooth muscle cells. **(A)** Cells were allowed to migrate to increasing concentrations of osteopontin in a modified Boyden chamber chemotaxis chamber. **(B)** Requirement of an osteopontin gradient for smooth muscle migration to osteopontin. Cells/HPF = cell migrated per high power (400×) field. (From Liaw *et al.*[18] Reprinted by permission from *Circulation Research*.)

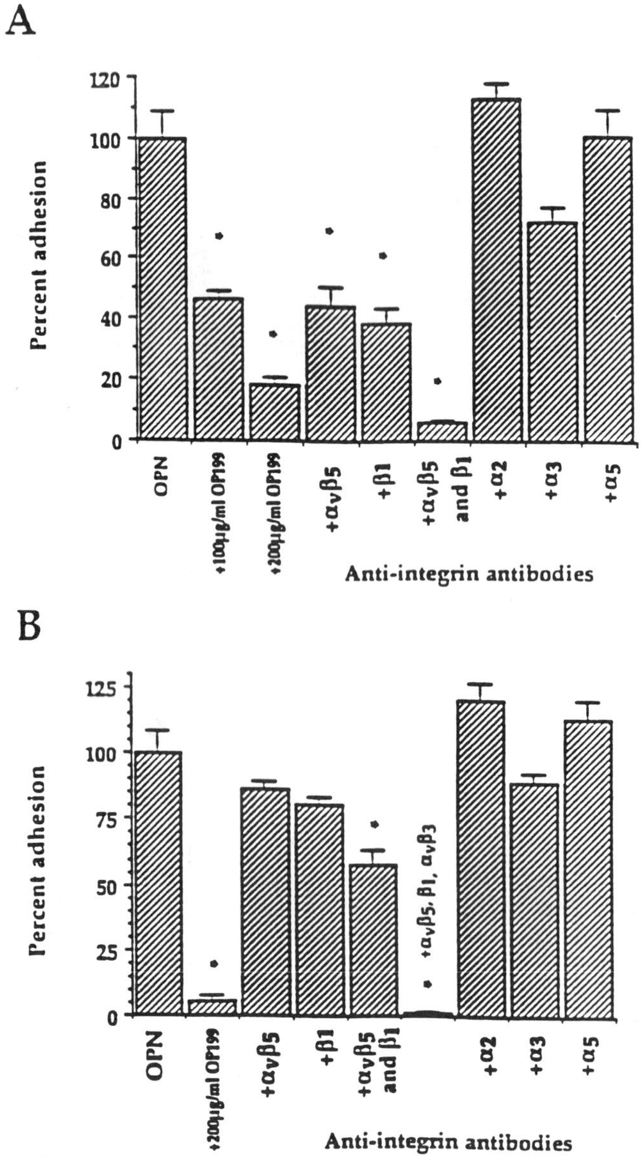

FIGURE 9. Three distinct integrins interact with osteopontin on vascular smooth muscle cells. **(A)** Effect of neutralizing anti-integrin antibodies or anti-osteopontin antibody (OP199) on adhesion of smooth muscle cells lacking $\alpha_V\beta_3$ integrin. **(B)** Effect of neutralizing anti-integrin antibodies or anti-osteopontin antibody (OP199) on adhesion of smooth muscle cells containing $\alpha_V\beta_3$. (From Liaw *et al.*[22] Reprinted by permission from the *Journal of Clinical Investigation.*)

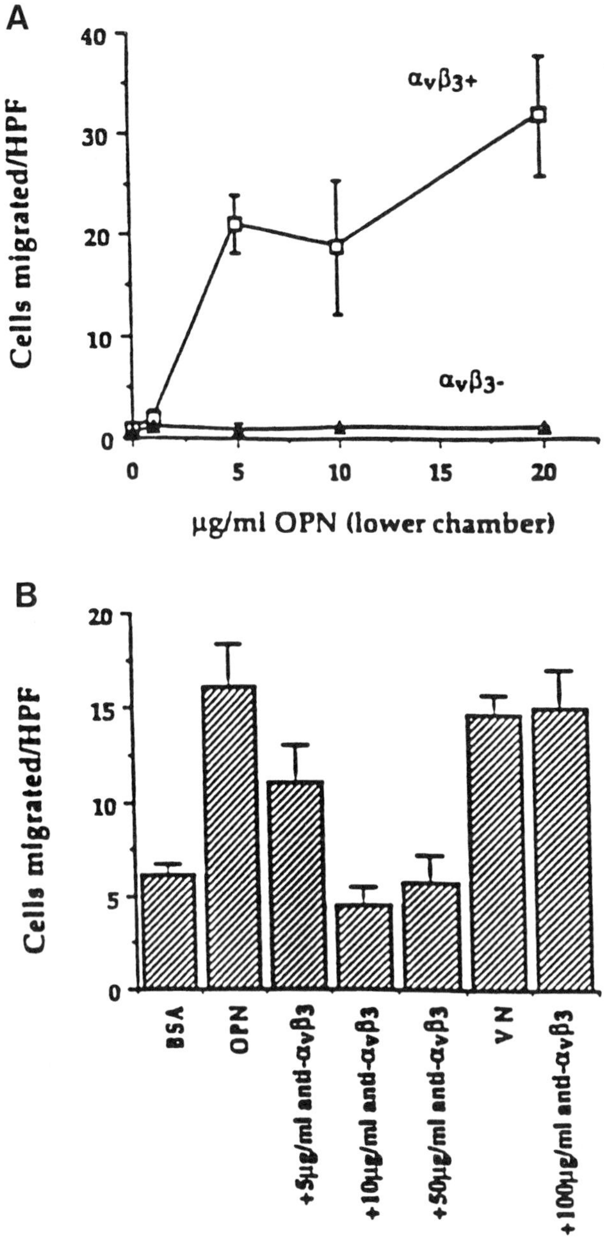

FIGURE 10. $\alpha_v\beta_3$ is required for smooth muscle cell migration to osteopontin. **(A)** Migration of smooth muscle cells either containing ($\alpha_v\beta_3+$) or lacking $\alpha_v\beta_3$ ($\alpha_v\beta_3-$) to osteopontin. **(B)** Effect of neutralizing anti-$\alpha_v\beta_3$ antibody (LM609) on migration of $\alpha_v\beta_3+$ cells. (From Liaw *et al.*[22] Reprinted by permission from the *Journal of Clinical Investigation*.)

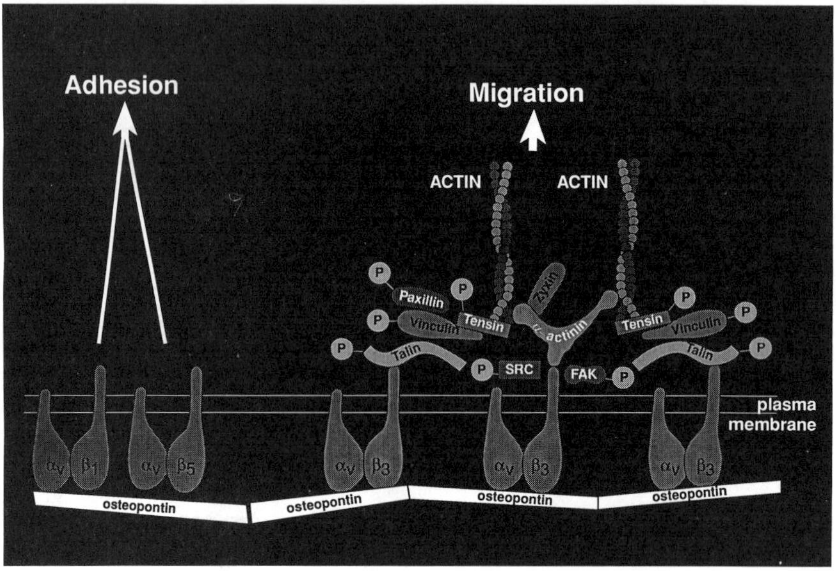

FIGURE 11. Distinct receptors mediate the adhesive and migratory effects of osteopontin in vascular smooth muscle cells.

such differential localization of integrins (FIG. 11). Once cells are adherent and spread on osteopontin, $\alpha_v\beta_3$ is detected in focal contact regions by immunocytochemistry in both smooth muscle and endothelial cells.[22] However, $\alpha_v\beta_5$ remains diffuse on the cell surface, and β_1 additionally localizes to regions of cell borders and cell-contacts. These data are consistent with previous findings comparing the localization of $\alpha_v\beta_3$ and $\alpha_v\beta_5$ in cells adherent to vitronectin.[24]

It is interesting to speculate that sequestration of $\alpha_v\beta_3$ in focal contact regions includes the integrin in a signaling complex that is known to include pp60c-src, pp62c-yes, phosphoinositide 3-kinase, and protein kinase C.[25] The formation of such a signaling complex may allow a cascade that eventuates in cell migration. This hypothesis is supported by several reports of $\alpha_v\beta_3$ distribution and importance in cell migration.[26,27] The existence of multiple downstream events signaled through $\alpha_v\beta_3$ interaction with osteopontin is further supported by studies in osteoclasts showing a clear signaling role for osteopontin through the $\alpha_v\beta_3$ receptor in the induction of bone resorption.[28] Our future studies will be aimed at understanding the signaling pathways involved in osteopontin-stimulated migration in vascular cells.

SUMMARY

Adhesive interactions are recognized requirements for cellular proliferation, migration and differentiation during normal morphogenesis as well as disease. By differential cloning, osteopontin was identified as an adhesive protein upregulated during vascular remodeling and neointima formation in both rat models and human vascular diseases including atherosclerosis and restenosis. In functional studies, purified osteopontin promoted adhesion, focal contact formation, and migration of vascular smooth muscle and endothelial cells. Utilizing neutralizing antibodies, three integrin-type receptors, $\alpha_v\beta_3$, $\alpha_v\beta_1$, and $\alpha_v\beta_5$ were found to support cellular adhesion to osteopontin. In contrast, only cells containing the $\alpha_v\beta_3$ integrin could migrate towards an osteopontin gradient, demonstrating for the first time that different functions of osteopontin are mediated via distinct receptors. These results suggest a model whereby osteopontin, via its integrin-type receptors, contributes to vascular remodeling during development and disease by facilitating smooth muscle migration and simultaneously promoting endothelial coverage of the affected area.

REFERENCES

1. Majesky, M. W., C. M. Giachelli, M. A. Reidy & S. M. Schwartz. 1992. Rat carotid neointimal smooth muscle cells re-express a developmentally regulated phenotype during repair of arterial injury. Circ. Res. **71:** 759–768.
2. Giachelli, C. M., M. W. Majesky & S. M. Schwartz. 1991. Developmentally regulated cytochrome P450IA1 expression in cultured rat vascular smooth muscle cells. J. Biol. Chem. **266:** 3981–3986.
3. Kocher, O., O. Skalli, D. Cerutti, F. Gabbiani & G. Gabbiani. 1985. Cytoskeletal features of rat aortic cells during development. An electron microscopic, immunohistochemical and biochemical study. Circ. Res. **56:** 829–838.
4. Campbell, G. R. & J. H. Campbell. 1990. The phenotypes of smooth muscle expressed in human atheroma. Ann. N.Y. Acad. Sci. **598:** 143–158.
5. Schwartz, S. M., R. L. Heimark & M. W. Majesky. 1990. Developmental mechanisms underlying pathology of arteries. Phys. Rev. **70:** 1177–1209.
6. Giachelli, C. M., N. Bae, D. M. Lombardi, M. W. Majesky & S. M. Schwartz. 1991. Molecular cloning and characterization of 2B7, a rat mRNA which distinguishes smooth muscle cell phenotypes *in vitro* and is identical to osteopontin (secreted phosphoprotein I, 2aR). Biochem. Biophys. Res. Commun. **177:** 867–873.
7. Giachelli, C., N. Bae, M. Almeida, D. Denhardt, C. E. Alpers & S. M. Schwartz. 1993. Osteopontin expression is elevated during neointima formation in rat arteries and in human atherosclerotic plaques. J. Clin. Invest. **92:** 1686–1696.
8. O'Brien, E. R., M. R. Garvin, D. K. Stewart, T. Hinohara, J. B. Simpson, S. M. Schwartz & C. M. Giachelli. 1994. Osteopontin is synthesized by macrophage, smooth muscle and endothelial cells in primary and restenotic human coronary atherosclerotic plaques. Arterioscler. Thromb. In press.
9. Giachelli, C., R. Pichler, D. Lombardi, D. Denhardt, C. E. Alpers, S. M. Schwartz & R. J. Johnson. 1994. Osteopontin expression in angiotensin II-induced tubulointerstitial nephritis. Kidney Int. **45:** 515–524.
10. Pichler, R., C. M. Giachelli, D. Lombardi, J. Pippin, K. Gordon, C. E. Alpers, S. M. Schwartz & R. J. Johnson. 1994. Tubulointerstitial disease in glomerulonephritis: potential role of osteopontin (uropontin). Am. J. Pathol. **144:** 915–926.
11. Singh, K., A. B. Mukherjee, M. W. DeVouge & B. B. Mukherjee. 1992. Differential processing of osteopontin transcripts in rat kidney- and osteoblast-derived cell lines. J. Biol. Chem. **267:** 23847–23851.
12. Hirota, S., M. Imakita, K. Kohri, A. Ito, E. Morii, S. Adachi, H. Kim, Y. Kitamura, C. Yutani & S. Nomura. 1993. Expression of osteopontin messenger RNA by macrophages in atherosclerotic plaques. Am. J. Pathol. **143:** 1003–1008.
13. Ikeda, T., T. Shirasawa, Y. Esaki, S. Yoshiki & K. Hirokawa. 1993. Osteopontin mRNA is expressed by smooth muscle-derived foam cells in human atherosclerotic lesions of the aorta. J. Clin. Invest. **92:** 2814–2820.
14. Shanahan, C. M., N. R. Cary, J. C. Metcalfe & P. L. Weissberg. 1994. High

expression of genes for calcification-regulating proteins in human atherosclerotic plaques. J. Clin. Invest. **93:** 2393–2402.

15. MURRY, C. E., C. M. GIACHELLI, S. M. SCHWARTZ & R. VRACKO. 1994. Macrophages express osteopontin during repair of myocardial necrosis. Am. J. Pathol. **145:** 1450–1462.
16. VRACKO, R., D. CUNNINGHAM, R. G. FREDERICKSON & D. THORNING. 1988. Basal lamina of rat myocardium. Its fate after death of cardiac myocytes. Lab. Invest. **58:** 77–87.
17. SENGER, D. R., C. A. PERRUZZI, A. PAPADOPOULOS & D. G. TENEN. 1989. Purification of a human milk protein closely similar to tumor-secreted phosphoproteins and osteopontin. Biochim. Biophys. Acta **996:** 43–48.
18. LIAW, L., M. ALMEIDA, W. DOWNEY, C. E. HART, S. M. SCHWARTZ & C. M. GIACHELLI. 1994. Osteopontin promotes vascular cell adhesion and spreading and is chemotactic for smooth muscle cells *in vitro*. Circ. Res. **74:** 214–224.
19. DENHARDT, D. & X. GUO. 1993. Osteopontin—a protein with diverse functions. FASEB J. **7:** 1475–1482.
20. SENGER, D. R., C. A. PERRUZZI & A. PAPADOPOULOS. 1989. Elevated expression of secreted phosphoprotein I (osteopontin, 2ar) as a consequence of neoplastic transformation. Anticancer Res. **9:** 1291–1299.
21. CRAIG, A. M., G. T. BOWDEN, A. F. CHAMBERS, M. A. SPEARMAN, A. H. GREENBERG, J. A. WRIGHT, M. MCLEOD & D. T. DENHARDT. 1990. Secreted phosphoprotein mRNA is induced during multi-stage carcinogenesis in mouse skin and correlates with the metastatic potential of murine fibroblasts. Int. J. Cancer **46:** 133–137.
22. LIAW, L., M. P. SKINNER, E. W. RAINES, R. ROSS, D. A. CHERESH, S. M. SCHWARTZ & C. M. GIACHELLI. 1994. The adhesive and migratory effects of osteopontin are mediated via distinct cell surface integrins: Role of $\alpha_v\beta_3$ in smooth muscle migration to osteopontin. J. Clin. Invest. In press.
23. CHOI, E. T., L. ENGEL, A. D. CALLOW, S. SUN, J. TRACHTENBERG, S. SANTORO & U. S. RYAN. 1994. Inhibition of neointimal hyperplasia by blocking $\alpha_v\beta_3$ integrin with a small peptide antagonist G*pen*GRGDSPCA. J. Vasc. Surg. **19:** 125–134.
24. WAYNER, E. A., R. A. ORLANDO & D. A. CHERESH. 1991. Integrins $\alpha_v\beta_3$ and $\alpha_v\beta_5$ contribute to cell attachment to vitronectin but differentially distribute on the cell surface. J. Cell Biol. **113:** 919–929.
25. FOX, J. E. 1994. Transmembrane signaling across the platelet integrin glycoprotein IIb-IIIa. Ann. N.Y. Acad. Sci. **714:** 75–87.
26. CLYMAN, R. I., F. MAURAY & R. H. KRAMER. 1992. β-1 and β-3 integrins have different roles in the adhesion and migration of vascular smooth muscle cells on extracellular matrix. Exp. Cell Res. **200:** 272–284.
27. LEAVESLEY, D. I., M. A. SCHWARTZ, M. ROSENFELD & D. A. CHERESH. 1993. Integrin β-1 and β-3-mediated endothelial cell migration is triggered through distinct signaling mechanisms. J. Cell Biol. **121:** 163–170.
28. ROSS, F. P., J. CHAPPEL, J. I. ALVAREZ, D. SANDER, W. T. BUTLER, M. C. FARACH-CARSON, K. A. MINTZ, P. G. ROBEY, S. L. TEITELBAUM & D. A. CHERESH. 1993. Interactions between the bone matrix proteins osteopontin and bone sialoprotein and the osteoclast integrin $\alpha_v\beta_3$ potentiate bone resorption. J. Biol. Chem. **268:** 9901–9907.

Osteopontin-Induced Modifications of Cellular Functions[a]

DAVID T. DENHARDT, CECILIA A. LOPEZ,
ELLEN E. ROLLO, SHIAW-MIN HWANG,[b]
XIAO-RONG AN, AND SUSAN E. WALTHER

Department of Biological Sciences
Rutgers University
P.O. Box 1059
Piscataway, New Jersey 08855

INTRODUCTION

OPN is an extracellular phosphoprotein found in certain tissues and apparently all body fluids; its level is frequently elevated under pathological conditions (reviews[1,2]). It binds calcium and interacts with the vitronectin receptor, the $\alpha_v\beta_3$ integrin. OPN expression is especially abundant in normal bone, kidney, decidua, and various epithelial surfaces.[3] Several groups have characterized OPN found in bone (review[4]). The protein has a GRGDS sequence capable of mediating attachment and spreading of osteosarcoma cells, periodontal ligament cells and fibroblasts on plastic[5,6] possibly via the vitronectin receptor.[7–9] Although cell attachment almost certainly is an important function of the protein, this does not seem to be its only function. The phosphorylated, bone-derived version of OPN is capable of binding multiple Ca^{++} ions with relatively low K_d such that roughly 25 of 50 potential calcium binding sites are occupied at a calcium concentration of 1–2 mM in 60 mM KCl.[10] Miyauchi *et al.*[11] have reported that OPN and certain RGD-containing peptides acting through the $\alpha_v\beta_3$ integrin cause chicken osteoclasts to secrete calcium via a calcium-dependent ATPase in a calmodulin-dependent fashion. Zimolo *et al.*[12] found in contrast in rodent osteoclasts that OPN and other $\alpha_v\beta_3$ ligands induced an increase in $[Ca^{++}]_i$, apparently by stimulating Ca^{++} release from internal stores.

Shiraga *et al.*[13] isolated an aspartic acid-rich protein ("uropontin") from human urine and identified it as OPN. Their demonstration, and also that of Worcester *et al.*[14] that the protein inhibited the growth of calcium oxalate crystals suggests that one role for OPN may be to suppress the precipitation of calcium salts in the urine. Lopez *et al.*[15] detected high levels of the protein associated with the apical surface of cells lining the thick ascending limb of the loop of Henle, and further noted an accumulation in sclerotic glomeruli. The protein is also found in the endometrial glands of nonpregnant secretory phase uterus.[16] Northern blot analysis of RNA preparations from pregnant and lactating mice revealed that the gene was expressed in both dorsal and ventral skin (dermis plus epidermis),[17] consistent

[a] This research was supported by Grant DC01295 from the National Institutes of Health, the Bureau of Biological Research of Rutgers University, the Cancer Federation, and the Palczuk Fund.

[b] Present address: Culture Collection and Research Center, PO Box 246, Hsinchu 30099, Taiwan, R.O.C.

with the detection by Senger *et al.*[18] of OPN in milk. Waterhouse *et al.*[19] have characterized OPN expression in the decidua and placenta, particularly in marrow-derived granulated metrial gland cells during pregnancy, and suggested that it may mediate transport of calcium from the maternal circulation to the developing embryo.

Mouse *opn* was first identified in our laboratory on the basis of the inducibility of its mRNA in cultured JB6 cells by the tumor-promoting phorbol ester TPA.[20] The mouse gene has been cloned, the transcriptional start site identified, and elements in the promoter/enhancer responding to 1,25-dihydroxyvitamin D_3 and possibly to female sex hormones have been located.[17,21,22] We have found that expression of the gene is in part regulated by *ras*[23–25] perhaps accounting for its increased expression in metastatic disease.[26] Quite independently, Patarca *et al.*[27] cloned OPN as the "early T lymphocyte activation 1" (*Eta-1*) gene. These investigators suggested that expression of the gene in T cells was enhanced in response to infection as a defense mechanism acting to attract macrophages to extravascular sites.[28] Consistent with an antiinfectious activity are the findings that *opn* is allelic with a locus controlling susceptibility to rickettsia infection[29] and that elevated concentrations of OPN are found in the blood of patients suffering gram-negative sepsis.[30] OPN binds to macrophages via an RGD motif and has been reported to enhance secretion of IgM and IgG by mixtures of macrophages and B cells.[31] Senger *et al.*[30] detected the protein at low levels in normal human plasma, and at considerably elevated levels in the plasma of patients with gram-negative sepsis.

MATERIALS AND METHODS

Cell Culture and Cytotoxicity Determinations

Primary mouse kidney proximal tubule epithelial (PTE) cells were grown as described.[32,33] The mouse monocyte-macrophage cell line RAW264.7 (ATCC TIB 71) and the mouse mastocytoma cell line P815 (ATCC TIB 64) were grown in DMEM (Dulbecco's modified essential medium) supplemented with 10% fetal bovine serum, 50 μg/ml penicillin and streptomycin to 80% confluence in a standard CO_2 incubator. Stria vascularis epithelial (SVE) cells are primary cultures grown from explants of the stria vascularis (Lopez and Davis, in preparation) for 2–3 months. They were maintained in DMEM supplemented with 10% fetal bovine serum in 5% CO_2 at 37°C. Rat hepatoma cells (ATCC CRL 1600) were grown in SWIMS s-77 medium. P815 cells were prelabeled with ^{51}Cr and used as the target cell for RAW264.7 macrophage-mediated cell killing as described.[34] They were mixed with the activated macrophages at an effector-to-target ratio of 3 : 1 in wells of 96-well plates at a concentration of 8×10^4 cells/well total. After 16 h incubation 50 μl of medium was removed and the ^{51}Cr released into the medium quantified in a scintillation counter. Specific ^{51}Cr release was calculated using the formula

$$\frac{\text{experimental release} - \text{spontaneous release}}{\text{maximum release} - \text{spontaneous release}} \times 100 = \text{percent specific } ^{51}\text{Cr release}$$

Nuclear Run-on Analysis

PTE cells were stimulated with LPS and γIFN in the presence or absence of 0.1 nM OPN for 12 h. After removal of the medium, the cells were washed twice with PBS at 4°C and harvested into 5 ml of PBS by scraping. The cells were collected by centrifugation at 500 g for 5 min and then lysed by vortexing for 10 sec in 4 ml of NP-40 lysis buffer (10 mM Tris HCl, pH 7.4, 10 mM NaCl, 3 mM $MgCl_2$, and 0.5% NP-40) at 0°C. After 5 min of incubation on ice the nuclei were collected by centrifugation for 5 min at 500 g, 4°C, and washed once with the NP-40 lysis buffer. The final pellet was resuspended in 200 μl glycerol storage buffer (50 mM Tris HCl, pH 8.3, 40% glycerol, 5 mM $MgCl_2$, 0.1 mM EDTA) and stored in liquid nitrogen. For run-on reactions, freshly thawed nuclei (10–20 $\times$ 10^6) were incubated in transcription buffer (5 mM Tris HCl, pH 8.0, 25 mM $MgCl_2$, 150 mM KCl, 0.5 mM each of ATP, CTP, GTP and 2 mM dithiothreitol) containing 10 u/ml RNasin ribonuclease inhibitor (Promega Corp., Madison, WI) and 100 μCi [α^{32}P]UTP (800 Ci/mmol, Amersham Corp., Arlington Heights, IL) at 30°C for 10 min. The labeled transcripts were isolated by the addition of 5 vol of the TRI reagent following the RNA isolation protocol. Equal amounts of [^{32}P]RNA probes were hybridized to mouse macrophage iNOS cDNA, rat brain cNOS cDNA, mouse β-actin cDNA and mouse GAPD cDNA which had been slot-blotted onto a nylon membrane (500 ng/slot, GeneScreen Plus). Prehybridization (55°C, 12 h) and hybridization (55°C, 48 h) were performed in glass scintillation vials with the same solution (50% formamide, 0.8 M NaCl, 20 mM Na_2HPO_4, pH 6.5, 1 mM EDTA, 5 $\times$ Denhardt's solution, 1% SDS, and 250 μg/ml sheared salmon sperm DNA). Blots were washed 3 $\times$ with 50 mM NaCl, 20 mM Na_2HPO_4, pH 6.5, 1 mM EDTA and 0.1% SDS at 60°C for 15 min and exposed to X-ray film at −70°C with intensifying screens. The relative intensities of the resulting autoradiograms were quantified by scanning densitometry.

Preparation and Purification of Recombinant Human OPN

Human OPN cDNA (OP10, a generous gift from M. F. Young and L. W. Fisher, Bone Research Branch, NIH) was cleaved with *Xba*I and *Xho*I from the pBLUESCRIPT SK+/− vector, releasing a 1,492-bp insert that was subsequently cloned into the *Bam*HI site of the mammalian expression vector pNMH.[35] This expression vector is under the control of the mouse metallothionein-1 promoter and allows selection with G418. One clone (NO1L) was isolated from a transfected human kidney cell line (MH2) that contained multiple copies of the expression cassette integrated into the chromosome and could be induced with 0.5 μM Cd^{++} to express a high level of OPN. NO1L cells were plated into forty 150-mm tissue culture dishes at 1 $\times$ 10^6 cells/dish and allowed to reach confluency. The cells were refed with fresh medium containing 10% FBS a day before they were serum-starved. Upon Cd^{++} induction for 18 h, conditioned medium from serum-starved NO1L cells was collected and cellular debris was removed by centrifugation at 1,000 rpm for 5 min. Clarified conditioned medium was collected and fractionated on a DEAE-Sepharose column that had been equilibrated with 0.05 mM Tris HCl, pH 7.5. The column was washed with two column volumes of starting buffer and bound proteins were eluted with a 0.1–0.6 M NaCl gradient. The fractions were monitored by absorbance at 280 nm, pooled appropriately, dialyzed against water, and lyophilized. Each pool was analyzed on SDS-PAGE using 12% polyacrylamide mini-gels and blotted onto a nitrocellulose membrane. The blotted proteins were probed with anti-human OPN antibody (LF-7). The pooled fractions containing osteopontin were subjected to gel filtration on a Sepharose G-100 column equilibrated with 0.01 M Tris HCl, pH 7.5. Fractions were pooled according to peak absorbance, dialyzed against water, freeze-dried, and analyzed by Western blotting

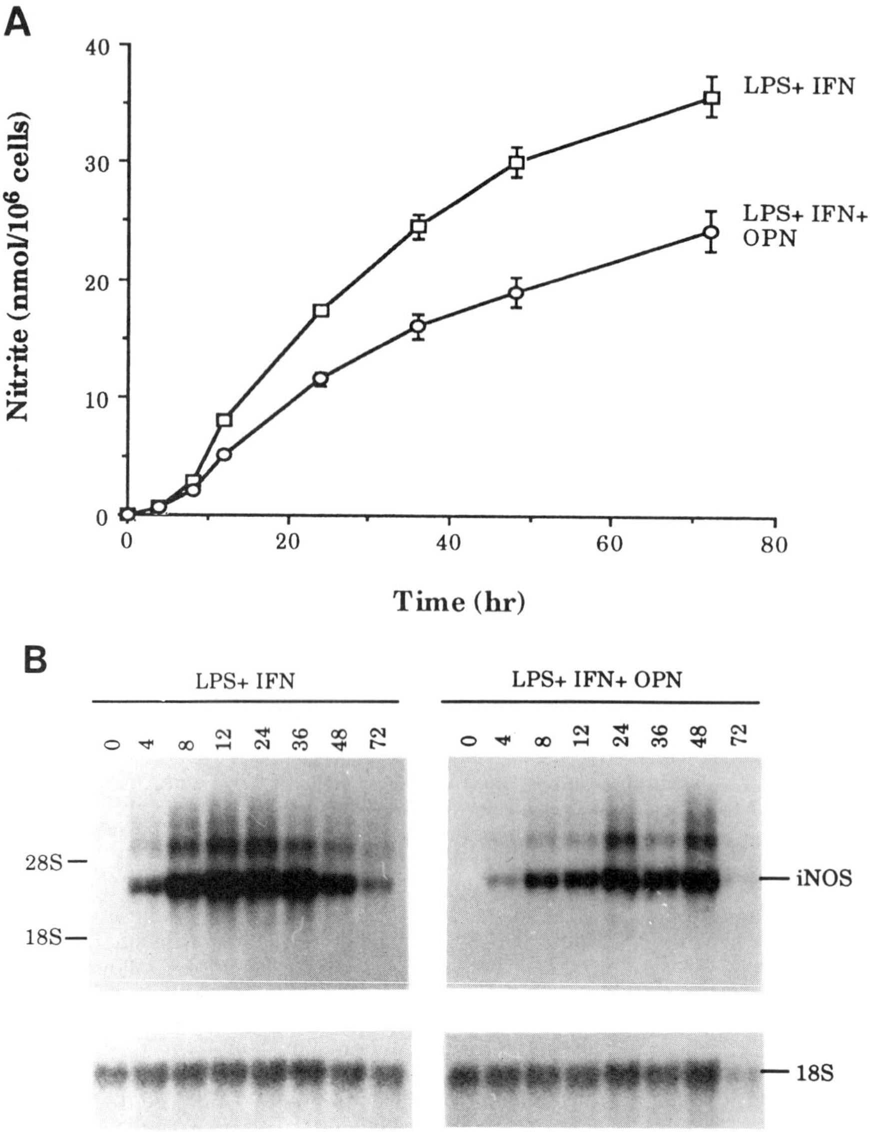

FIGURE 1. OPN can inhibit induction of iNOS at the transcriptional level. **(A–C)** Time course of the induction of NO production and iNOS mRNA by LPS and γIFN in the presence and absence of OPN. Primary mouse PTE cells were stimulated with 100 ng/ml LPS and 100 u/ml γIFN in the presence and absence of 0.1 nM human recombinant OPN made in mammalian cells. **(A)** Nitrite concentration was determined in the culture supernatants at the indicated time as described.[32] The values are the mean ± standard error from three separate plates. **(B)** Total RNA was purified from the cells at the indicated times, electrophoresed and blotted, and probed for both the iNOS mRNA and 18S rRNA as described.[32]

C

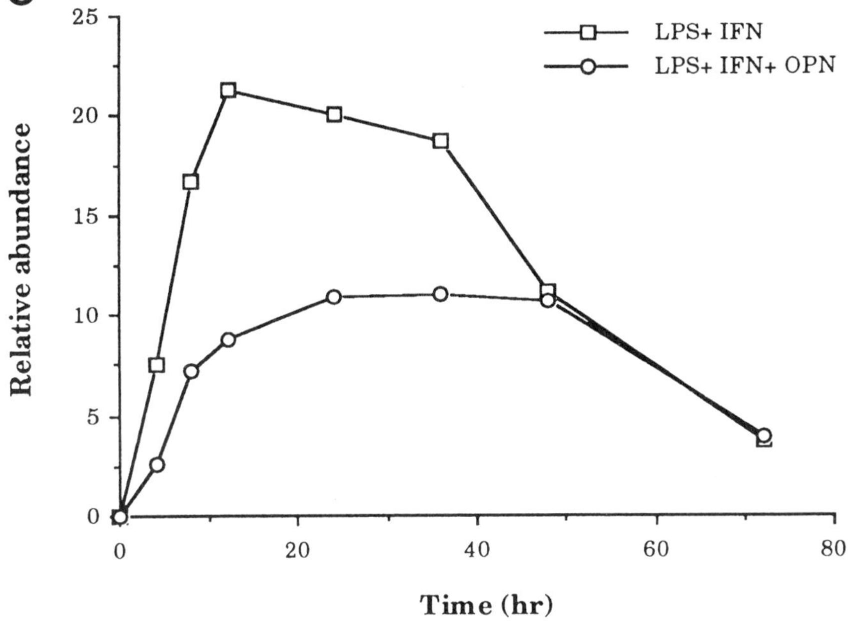

D

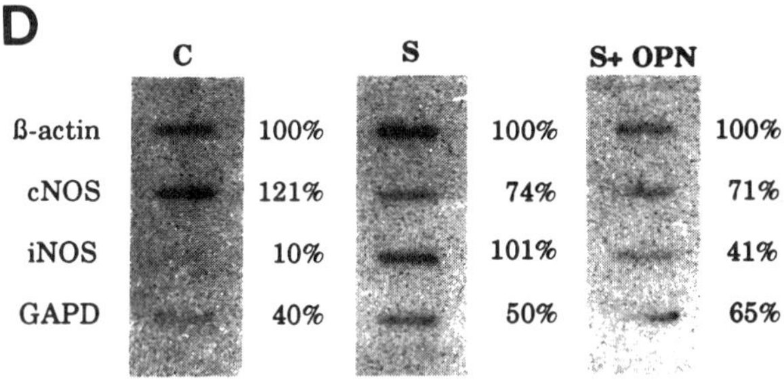

FIGURE 1 (*Continued*). **(C)** The relative abundance of the iNOS mRNA was quantified and normalized against the 18S rRNA signal by scanning densitometry. **(D)** Nuclear run-on analysis of cNOS and iNOS expression in PTE cells stimulated for 12 h with LPS plus γIFN alone or in combination with 0.1 nM OPN. The levels of transcription of iNOS, cNOS, β-actin and glyceraldehyde phosphate dehydrogenase (GAPD) were determined as described in Materials and Methods. *Lanes* C, S, and S + OPN are respectively control, stimulated with γIFN and LPS, and stimulated similarly but in the presence of OPN (0.1 nM).

(Lopez and Denhardt, in preparation). For the purposes of this research we have taken the MW of the glycosylated, phosphorylated protein made in mammalian cells to be 50,000.

Expression of OPN as a soluble protein (predicted MW 32,350) in *E. coli* was accomplished using the pET8C vector induced with IPTG for 3–4 h at 30°C. OPN was purified from a 20–40% ammonium sulfate protein fraction by hydroxyapatite chromatography and SDS-PAGE. OPN was eluted from the appropriate portion of the polyacrylamide gel with a solution containing 0.1% SDS and 10 mM sodium acetate, pH 7.4, precipitated with 90% methanol at −70°C overnight, and redissolved in water (Rollo and Denhardt, in preparation). Protein concentrations were determined using the Pierce BCA protein assay reagent.

Immunoprecipitation of Phosphorylated Proteins

SVE cells were plated in 100-mm petri dishes at 5×10^5 cells/dish. After 3 days, semiconfluent and quiescent cultures were washed twice in phosphate-buffered saline and incubated in 5 ml serum-free DMEM for 18 h. On the day of the experiment, plates were treated with 50 ng/ml (~1 nM) purified OPN for 3 min. The cells were quickly washed with ice-cold PBS containing 100 μM sodium orthovanadate and lysed at 4°C in 750 μl of modified RIPA buffer. Cell lysates were sedimented at 14,000 rpm for 10 min. Total proteins (200 μg), as analyzed by the BCA protein assay were preincubated with 40 μl of protein A-Sepharose for 3 h at 4°C. Cell lysates were precleared with protein A-Sepharose and then

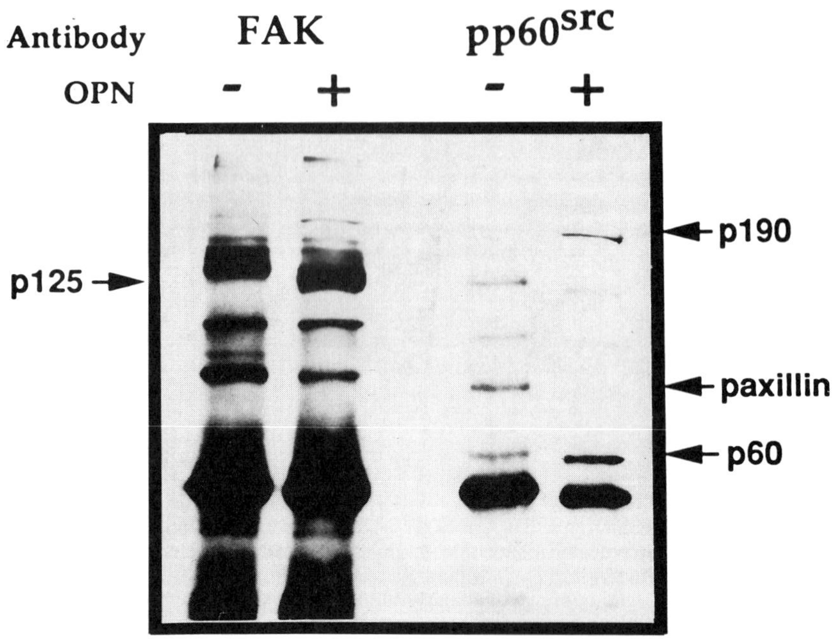

FIGURE 2. OPN can alter intracellular protein tyrosine phosphorylation. Tyrosine-phosphorylated proteins were detected with an antiphosphotyrosine antibody (PY-54) among proteins immunoprecipitated from SVE (stria vascularis epithelial) cells with antiserum to focal adhesion kinase, FAK,[39] or to pp60^{src}.[40,41] An increase in tyrosine phosphorylation of a protein of approximately 190 kD and pp60^{src} was observed in 3 min after OPN (1 nM) addition. Furthermore, it was accompanied by a dephosphorylation p125$^{\underline{fak}}$ and possibly of paxillin.

immunoprecipitated with specific antibodies overnight at 4°C. The antibodies used were monoclonal antibodies to pp125 focal adhesion kinase (a gift from Dr. J. T. Parsons, University of Virginia, Charlottesville, VA) and anti-pp60$^{\underline{src}}$ (a gift from Dr. D. J. Fujita, University of Calgary, Calgary, AL). To precipitate the antigen-antibody complex, 40 μl of protein A-Sepharose conjugated to goat anti-mouse IgG was added to the samples and mixing continued for another 2 h. The immunoprecipitates were pelleted and washed three times in ice-cold lysis buffer and electrophoresed on an SDS-7.5% polyacrylamide gel. After electrophoresis, the proteins were transferred to a nitrocellulose membrane. For Western blotting, the membranes were blocked for 1 h in TTBSB (10 mM Tris HCl, pH 7.5, 0.9% NaCl, 0.1% Tween 20, 5% bovine serum albumin) and probed with monoclonal antibody to phosphotyrosine (PY-54, 1 μg/ml, Oncogene Science, Uniondale, NY) for 1 h. The blots were then washed with three changes of TTBS (TTBSB without the BSA). This was followed by incubation with a secondary antibody conjugated to horseradish peroxidase (goat anti-mouse IgG at 1 : 3000 dilution, Bio-Rad, Hercules, CA). The blots were once again washed with three changes of TTBS and developed using enhanced chemiluminescence.

Hypoxia/Reoxygenation and Viability

Hypoxia and reoxygenation treatments were performed as described.[33] PTE cells were cultured 2–3 weeks to an 80–90% confluence in 25-cm^2 flasks. The medium was changed to 2 ml serum-free medium for 2 h prior to exposure to hypoxic conditions. The culture flasks were flushed with humidified 95% N_2 and 5% CO_2 and then maintained in that atmosphere in a 37°C gas chamber for the time required. After this hypoxic exposure, flasks were reoxygenated and placed in a normal cell culture incubator with 95% air and 5% CO_2 for 1–2 h.

RESULTS

Working with primary cultures of mouse kidney proximal tubule epithelial cells, Hwang *et al.*[32] discovered that OPN could inhibit the induction of iNOS (inducible nitric oxide synthase) mRNA by LPS (lipopolysaccharide) and γIFN (γ-interferon). The time course of this effect and its underlying basis are revealed by the data in FIGURE 1. OPN reduced the accumulation of nitrite (a measure of NO production) in the medium (FIG. 1A) and lessened the induction of the iNOS mRNA in cells stimulated with LPS and γIFN (FIG. 1B,C); further, there was a shift in the time of maximal induction to a later time. The nuclear run-on data presented in FIG. 1D indicate that at least part of the inhibition was at the level of transcription. When normalized to β-actin, transcription of the iNOS gene was reduced by about half; transcription of the cNOS gene on the other hand was inhibited by LPS + γIFN, and OPN had no additional effect. Measurements of the half-life of the iNOS mRNA in the presence and absence of OPN yield a half-life of 87 min in both cases when measured starting at 20 h after stimulation with LPS and βIFN in the presence and absence of OPN.[36] This ability of OPN to inhibit iNOS induction was inhibitable by antibody to OPN and by the peptide GRGDS, suggesting that the effect is mediated via an integrin, presumably the $\alpha_v\beta_3$ integrin.[32]

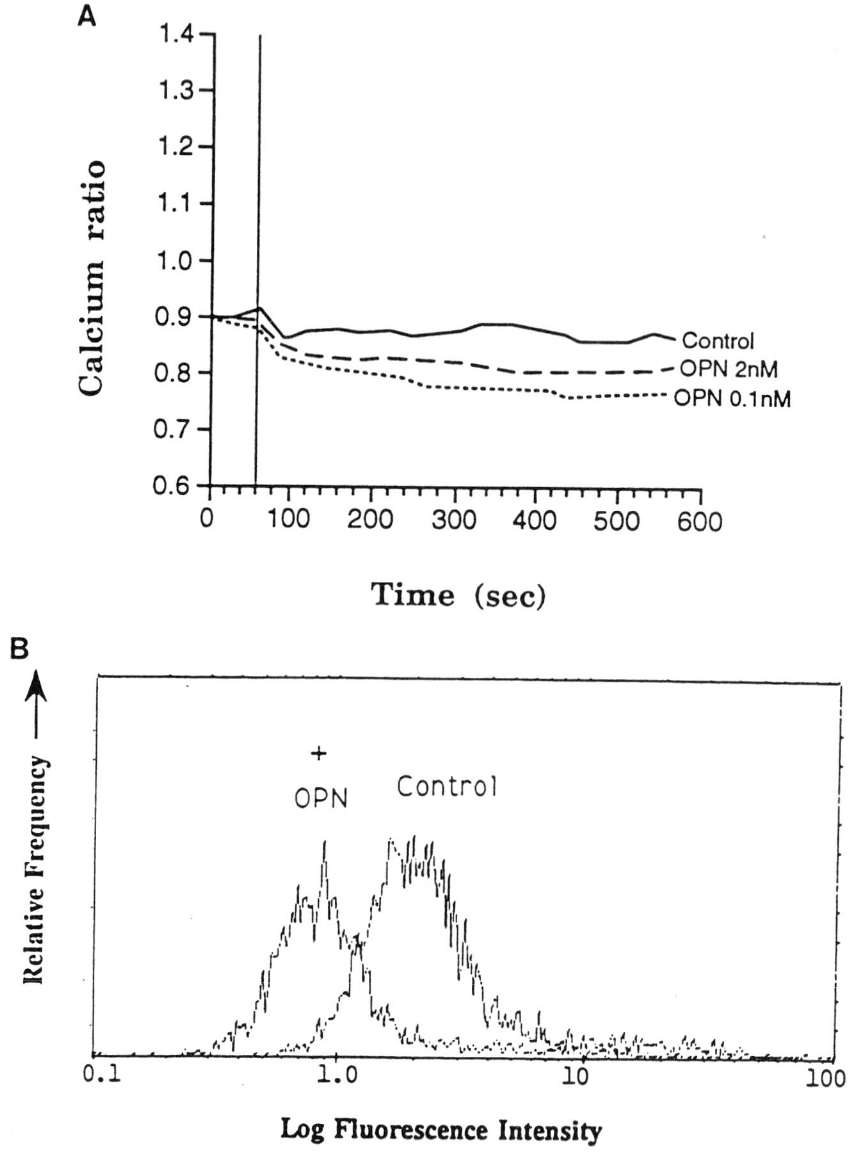

FIGURE 3. OPN can modulate intracellular calcium and hydrogen peroxide levels. **(A)** Reduction in $[Ca^{++}]_i$ in PTE cells exposed to OPN. PTE cells were incubated with 1 μM Indo-1AM [1-(2-amino-5-(6-carboxyindol-2-yl)phenoxy)-2-(2′-amino-5′methylphenoxy) ethane-N,N,N′,N′-tetraacetic acid)] and analyzed on a Meridian 570 Anchored Cell Analysis System (ACAS) equipped with an Innova 90 5-W laser adjusted to an output of 100 mW. Calcium levels were quantified by scanning single 150 × 150-μm fields containing about 10 cells every 30 sec for a total of 10 min and quantifying the emission at 500 nm (calcium-free Indo-1) and 400 nm (calcium-bound Indo-1). After 1 min, the cells were stimulated with

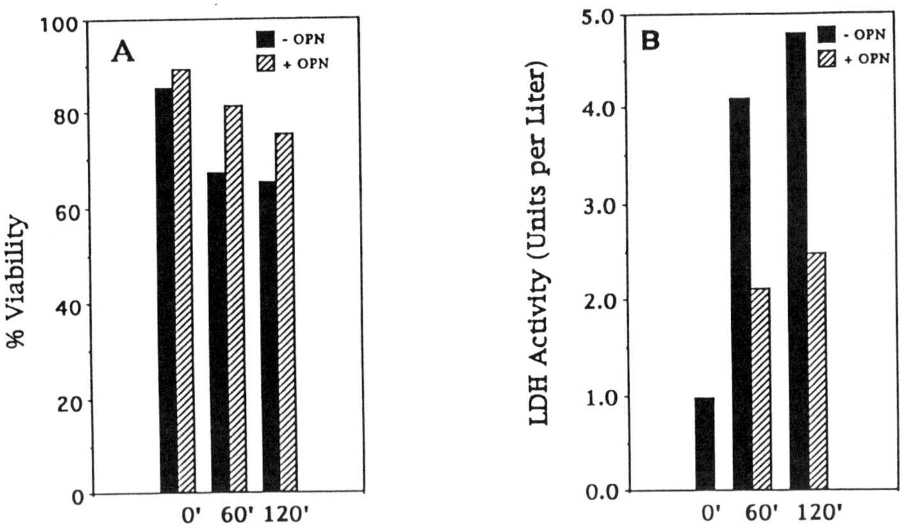

FIGURE 4. OPN can protect PTE cells from killing by hypoxia/reoxygenation. Subconfluent PTE cells were subjected to 60 or 120 min of hypoxia followed by 60 min reoxygenation in the presence or absence of 1 nM recombinant OPN.[33] **(A)** Cell viability as measured by trypan blue exclusion. **(B)** Cell viability as measured by lactate dehydrogenase release.[42]

The fact that OPN can alter gene expression at the transcriptional level implies the existence of a signal transduction cascade that leads from a cell surface receptor to nuclear proteins involved in regulating mRNA transcription. Research with various receptors has uncovered the existence of a signalling network functioning via the phosphorylation, and dephosphorylation, of proteins on tyrosine, serine and threonine residues. To determine if such a signalling network were operational here, we asked whether there were changes in the phosphorylation of specific cellular proteins in response to OPN stimulation. As can be seen in FIG. 2, this is indeed the case. This figure shows enhanced phosphorylation, 3 min after addition of OPN, and $pp60^{src}$ and p190; there was a concomitant reduction (at this 3-min time point) of $p125^{FAK}$ and paxillin phosphorylation. Some of the changes in protein phosphorylation we have detected are transient, lasting only a few minutes (*e.g.*, $pp60^{src}$) whereas others

---

2 nM or 0.1 nM OPN; the *solid line* is the control untreated cells. We have often observed, both in this kind of experiment and in others that lower concentrations of OPN can be more effective than higher concentrations. **(B)** Reduction in $[H_2O_2]_i$ in rat hepatoma cells exposed to OPN. Cultured cells were preloaded with dichlorofluorescein-diacetate, which yields a signal proportional to the H_2O_2 content of the cell. The cells were then exposed to approximately 1 μg/ml OPN for 30 min and then examined for cell-associated fluorescence by flow cytometry. A significant decline in the intracellular concentration of hydrogen peroxide was detected.

A

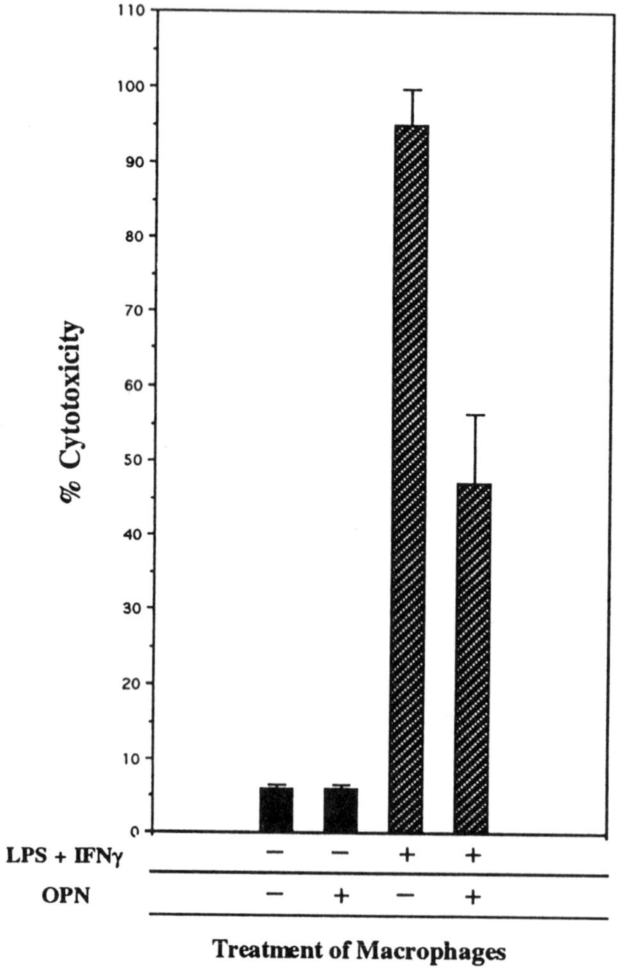

FIGURE 5. OPN can inhibit NO production by and the cytotoxic action of activated RAW264.7 macrophage-like cells. **(A)** P815 tumor cells were labeled with ^{51}Cr and then cocultured for 16 h with RAW264.7 cells treated as indicated in the figure. Stimulation was with 100 ng/ml LPS and 100 U/ml γIFN in the presence or absence of 3 pg/ml (0.1 pM) recombinant OPN (made in *E. coli*). The effector : target ratio was 3 : 1. **(B)** Accumulation of nitrite in the medium of the cocultures described in (A).

can be longer lasting ($p125^{FAK}$ and paxillin); there are also differences among cell types regarding which proteins are observed to be phosphorylated or dephosphorylated (Lopez and Denhardt, in preparation). Additional details can be found elsewhere in this volume in the contribution by Lopez, Davis, Mou, and Denhardt.

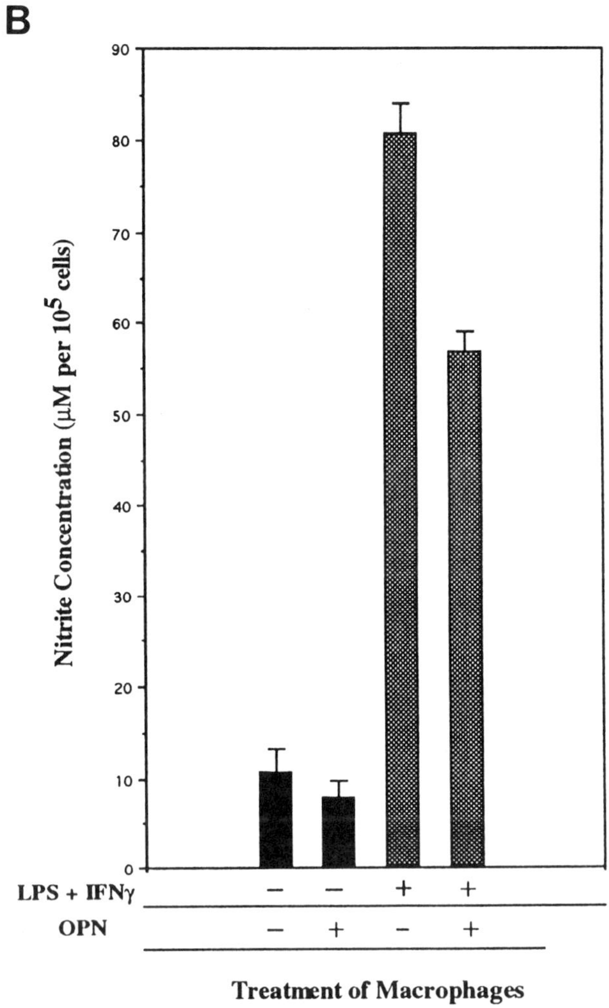

FIGURE 5 (*Continued*).

Other actions of OPN on cells include its ability to modify intracellular Ca^{++} and peroxide levels. In osteoclasts, both a decrease (in avian cells[11]) and an increase (in rodent cells[12]) in Ca^{++} have been reported. In the studies on the mouse PTE cells presented in FIGURE 3A, OPN caused a reduction in intracellular Ca^{++} levels. With regard to the level of hydrogen peroxide in the cell, we observed a decrease in response to the addition of osteopontin to H4 hepatoma cells (FIG. 3B).

Hypoxia followed by reoxygenation can lead to the death of sensitive cells, possibly by the influx of Ca^{++} ions as the result of a loss of membrane impermeability or by the increase in intracellular oxidants and free radicals that accompanies reoxygenation. As shown in FIGURE 4, the addition of OPN to the medium of the PTE cells during the reoxygenation phase appeared to enhance their survival as assessed both by trypan blue exclusion and by lactate dehydrogenase release (Walther and Denhardt, in preparation).

Perhaps by a related mechanism having to do with reducing oxidative damage, OPN can also make tumor cells more resistant to the cytotoxic action of activated macrophages. This is illustrated in FIGURE 5, which shows a rough parallel between the amount of nitrite (a measure of NO production) produced by activated RAW264.7 cells and their ability to kill p815 tumor cells (Rollo and Denhardt, in preparation).

DISCUSSION

The impact of osteopontin on cell physiology is impressive for an integrin-binding protein. Although originally implicated largely in cell adhesion, it has become apparent recently that OPN is also a potent signalling molecule, capable of changing the phosphorylation status of a number of cell proteins. This is perhaps not unexpected for two reasons. One is that it has been apparent for some time that when cells adhere to a substrate, extracellular matrix for example, there are substantial changes in protein phosphorylation and gene expression. The second line of evidence revealing a signalling function for OPN comes from its cloning as *Eta-1*, early T cell activation gene 1. Cantor and his colleagues[2,31] have made a good argument for considering OPN as a cytokine, a molecule capable of signalling the cell. The interesting question that we now face is that of which of the many potential functions of OPN are most important to the physiology of the organism. It seems that this might best be elucidated by studies on a knockout mouse.

In the survey of our research presented in this brief review we have mixed in experiments with recombinant OPN made in mammalian cells and hence fully modified (FIGS. 1–4) and in *E. coli,* and hence not phosphorylated or glycosylated (FIG. 5). For most of the studies presented, both OPN made in both hosts has been studied, and in our hands we have not observed significant differences between the two types of purified proteins. However, we have not done any adhesion experiments with them, nor have we looked at interactions with bone matrix or hydroxyapatite. Phosphorylation may well be important for adhesion to hydroxyapaptite, whereas glycosylation may function to increase the half-life of the protein in the plasma.

The ability to modulate Ca^{++} secretion, which may reflect a role of OPN in bone metabolism, may also be important in minimizing the injurious effects of the influx of Ca^{++} that occurs whenever the integrity of the plasma membrane is compromised—during hypoxia/reperfusion, for example. Alternatively, the protective action of OPN during hypoxia/reperfusion, and during the attack of activated cytotoxic cells, may reflect an ability to reduce intracellular oxidant levels. Considerable additional research will be necessary before we are likely to have a coherent understanding of the mechanisms and significance of these actions. The suggestion has been made that OPN upregulation in tumor cells serves as a defense against host defenses[37]—in other words, OPN promotes the survival of the tumor cell in the host organism, perhaps by blocking apoptosis. As shown in

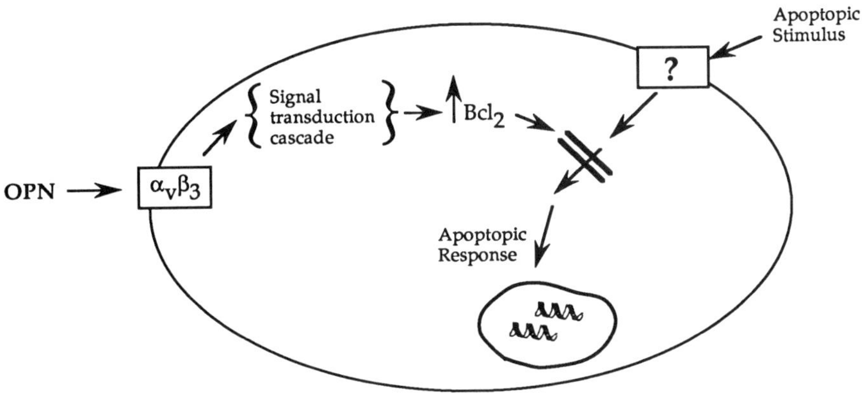

FIGURE 6. Model to explain how OPN could block apoptosis. We suggest that OPN, acting through the $\alpha_v\beta_3$ integrin, stimulates an increase in *bcl*2 activity and thereby opposes an alternative signal that would lead to an apoptotic response.

the model in FIGURE 6, we suggest that OPN acts to upregulate *bcl*2 expression and thus block a signal transduction pathway that would otherwise lead to the apoptotic response.[38]

SUMMARY

Osteopontin (OPN) serves both a cell attachment function and a cell signalling function via the $\alpha_v\beta_3$ integrin. We have investigated the action on mammalian cells of recombinant OPN made both in *E. coli* and in human cells. In its cell signalling capacity it initiates a signal transduction cascade that includes changes in the intracellular calcium ion levels and the tyrosine phosphorylation status of several proteins including pp60^{src} and components of focal adhesion complexes. Effects on gene expression include suppression of the induction of nitric oxide synthase by inflammatory mediators. OPN can also reduce cell peroxide levels, promote the survival of cells exposed to hypoxia, and inhibit the killing of tumor cells by activated macrophages.

ACKNOWLEDGMENTS

We thank Drs. Debra Laskin, Edward Yurkow, and Carol Gardner for their help in obtaining the data in FIGURE 3.

REFERENCES

1. DENHARDT, D. T. & X. GUO. 1993. Osteopontin: a protein with diverse functions. FASEB J. **7:** 1475–1482.
2. PATARCA, R., R. A. SAAVEDRA & H. CANTOR. 1993. Molecular and cellular basis of

genetic resistance to bacterial infection: the role of the early T-lymphocyte activation-1/osteopontin gene. Crit. Rev. Immunol. **13:** 225–246.

3. Brown, L. F., B. Berse, L. I. Van De Water, A. Papadopoulos-Sergiou, C. A. Perruzzi, E. J. Manseau, H. R. Dvorak & D. R. Senger. 1992. Expression and distribution of osteopontin in human tissues: widespread association with luminal epithelial surfaces. Mol. Biol. Cell **3:** 1169–1180.
4. Butler, W. T. 1989. The nature and significance of osteopontin. Connect. Tissue Res. **23:** 123–136.
5. Somerman, M. J., R. A. Foster, G. M. Imm, J. J. Sauk & S. Y. Archer. 1989. Periodontal ligament cells and gingival fibroblasts respond differently to attachment factors *in vitro*. J. Periodontol. **60:** 73–77.
6. Somerman, M. J., B. Shroff, W. S. Agraves, G. Morrison, A. M. Craig, D. T. Denhardt, R. A. Foster & J. J. Sauk. 1990. Expression of attachment proteins during cementogenesis. J. Biol. Buccale **18:** 207–214.
7. Horton, M. A. & J. Davies. 1989. Adhesion receptors in bone. J. Bone Miner. Res. **4:** 803–808.
8. Heinegård, D. & A. Oldberg. 1989. Structure and biology of cartilage and bone matrix noncollagenous macromolecules. FASEB J. **3:** 2042–2051.
9. Ruoslahti, E. & F. G. Giancotti. 1989. Integrins and tumor cell dissemination. Cancer Cells **1:** 119–126.
10. Chen, Y., B. S. Bal & J. P. Gorski. 1992. Calcium and collagen binding properties of osteopontin, bone sialoprotein, and bone acidic glycoprotein-75 from bone. J. Biol. Chem. **267:** 24871–24878.
11. Miyauchi, A., J. Alvarez, E. M. Greenfield, A. Teti, M. Grano, S. Colucci, A. Zambonin-Zallone, F. P. Ross, S. L. Teitelbaum, D. Cheresh & K. A. Hruska. 1991. Recognition of osteopontin and related peptides by an $\alpha_v\beta_3$ integrin stimulates immediate cell signals in osteoclasts. J. Biol. Chem. **266:** 20369–20374.
12. Zimolo, Z., G. Wesolowski, H. Tanaka, J. Hyman, J. R. Hoyer & G. A. Rodan. 1993. Soluble $\alpha_v\beta_3$ integrin ligands increase intracellular [Ca^{++}] in rat osteoclasts and mouse derived osteoclast-like cells. Am. J. Physiol. **35:** C376–C381.
13. Shiraga, H., W. Min, W. J. VanDusen, M. D. Clayman, D. Miner, C. H. Terrell, J. R. Sherbotie, J. W. Foreman, C. Przysiecki, E. G. Neilson & J. R. Hoyer. 1992. Inhibition of calcium oxalate crystal growth *in vitro* by uropontin: a new member of the aspartic acid-rich protein superfamily. Proc. Natl. Acad. Sci. USA **89:** 426–430.
14. Worcester, E. M., S. S. Blumenthal, A. M. Beshensky & D. L. Lewand. 1992. The calcium oxalate crystal growth inhibitor protein produced by mouse kidney cortical cells in culture is osteopontin. J. Bone Miner. Res. **7:** 1029–1036.
15. Lopez, C. A., J. R. Hoyer, P. D. Wilson, P. Waterhouse & D. T. Denhardt. 1993. Heterogeneity of osteopontin expression among nephrons in mouse kidneys and enhanced expression in sclerotic glomeruli. Lab. Invest. **69:** 355–363.
16. Young, M. F., J. M. Kerr, J. D. Termine, U. M. Wewer, M. G. Wang, O. W. McBride & L. W. Fisher. 1990. cDNA cloning, mRNA distribution and heterogeneity, chromosomal location, and RFLP analysis of human osteopontin (OPN). Genomics **7:** 491–502.
17. Craig, A. M. & D. T. Denhardt. 1991. The murine gene encoding secreted phosphoprotein (osteopontin): promoter structure, activity and induction *in vivo* by estrogen and progesterone. Gene **100:** 163–171.
18. Senger, D. R., C. A. Perruzzi, A. Papadopoulos & D. G. Tenen. 1989. Purification of a human milk protein closely similar to tumor-secreted phosphoproteins and osteopontin. Biochim. Biophys. Acta **996:** 43–48.
19. Waterhouse, P., R. S. Parhar, X. Guo, P. K. Lala & D. T. Denhardt. 1992. Regulated temporal and spatial expression of the calcium-binding proteins calcyclin and OPN (osteopontin) in mouse tissues during pregnancy. Mol. Reprod. Dev. **32:** 315–323.
20. Smith, J. H. & D. T. Denhardt. 1987. Molecular cloning of a tumor promotor-inducible mRNA found in JB6 mouse epidermal cells; induction is stable at high, but not at low densities. J. Cell. Biochem. **34:** 12–22.

21. Craig, A. M., J. H. Smith & D. T. Denhardt. 1989. Osteopontin, a transformation-associated cell adhesion phosphoprotein, is induced by 12-*O*-tetradecanoylphorbol-13-acetate in mouse epidermis. J. Biol. Chem. **264:** 9682–9689.
22. Noda, M., R. L. Vogel, A. M. Craig, J. Prahl, H. F. DeLuca & D. T. Denhardt. 1990. Identification of a DNA sequence responsible for binding of the 1,25-dihydroxy-vitamin D_3 receptor and 1,25-dihydroxyvitmain D_3 enhancement of mouse secreted phosphoprotein 1 (*Spp-1,* or osteopontin) gene expression. Proc. Natl. Acad. Sci. USA **87:** 9995–9999.
23. Craig, A. M., G. T. Bowden, A. F. Chambers, M. Spearman, A. H. Greenberg, J. Wright, M. McLeod & D. T. Denhardt. 1990. Secreted phosphoprotein mRNA is induced during multistage carcinogenesis in mouse skin and correlates with the metastatic potential of murine fibroblasts. Int. J. Cancer **46:** 133–137.
24. Chambers, A. F., E. I. Behrend, S. M. Wilson & D. T. Denhardt. 1992. Induction of expression of osteopontin (OPN; secreted phosphoprotein) in metastatic, *ras*-transformed NIH3T3 cells. Anticancer Res. **12:** 43–48.
25. Guo, X., Y. P. Zhang, D. Mitchell, D. T. Denhardt & A. F. Chambers. 1995. Identification of a *ras*-activated enhancer in the mouse osteopontin promoter and its interaction with a putative ETS-related transcription factor whose activity correlates with the metastatic potential of the cell. Mol. Cell. Biol. **15**. In press.
26. Senger, D. R., C. A. Perruzzi & A. Papadopoulos. 1989. Elevated expression of secreted phosphoprotein 1 (osteopontin, 2ar) as a consequence of neoplastic transformation. Anticancer Res. **9:** 1291–1300.
27. Patarca, R., G. J. Freeman, R. P. Singh, F.-Y. Wei, T. Durfee, F. Blattner, D. C. Regnier, C. A. Kozak, B. A. Mock, H. C. Morse III, T. R. Jerrells & H. Cantor. 1989. Structural and functional studies of the early T lymphocyte activation (Eta-1) gene. J. Exp. Med. **170:** 145–161.
28. Singh, R. P., R. Patarca, J. Schwartz, P. Singh & H. Cantor. 1990. Definition of a specific interaction between the early T-lymphocyte activation 1 (Eta-1) protein and murine macrophages *in vitro* and its effect upon macrophages *in vivo*. J. Exp. Med. **171:** 1931–1942.
29. Fet, V., M. E. Dickinson & B. L. M. Hogan. 1989. Localization of the mouse gene for secreted phosphoprotein 1 (Spp-1) (2ar, osteopontin, bone sialoprotein 1, 44-kDa bone phosphoprotein, tumor-secreted phosphoprotein) to chromosome 5, closely linked to *ric* (*Rickettsia resistance*). Genomics **5:** 375–377.
30. Senger, D. R., C. A. Perruzzi, C. F. Gracey, A. Papadopoulos & D. G. Tenen. 1988. Secreted phosphoproteins associated with neoplastic transformation: close homology with plasma proteins cleaved during blood coagulation. Cancer Res. **48:** 5770–5774.
31. Patarca, R., F.-Y. Wei, P. Singh, M. I. Morasso & H. Cantor. 1990. Dysregulated expression of the T cell cytokine Eta-1 in $CD4^-8^-$ lymphocytes during the development of murine autoimmune disease. J. Exp. Med. **172:** 1177–1183.
32. Hwang, S.-M., C. A. Lopez, D. E. Heck, C. R. Gardner, D. L. Laskin, J. D. Laskin & D. T. Denhardt. 1994. Osteopontin inhibits induction of nitric oxide synthase gene expression by inflammatory mediators in mouse kidney epithelial cells. J. Biol. Chem. **269:** 711–715.
33. Hwang, S.-M., P. D. Wilson, J. D. Laskin & D. T. Denhardt. 1994. Age and development-related changes in osteopontin and nitric oxide synthase mRNA levels in human kidney proximal tubule epithelial cells: contrasting responses to hypoxia and reoxygenation. J. Cell. Physiol. **160:** 61–68.
34. Lorsbach, R. B., W. J. Murphy, C. J. Lowenstein, S. H. Snyder & S. W. Russell. 1993. Expression of the nitric oxide synthase gene in mouse macrophages activated for tumor cell killing. J. Biol. Chem. **268:** 1908–1913.
35. Khokha, R. & D. T. Denhardt. 1987. On the use of antisense-RNA: down-regulation of mRNA encoding a metalloproteinase inhibitor. Anticancer Res. **7:** 653–660.
36. Hwang, S.-M. 1994. Regulation of nitric oxide synthesis by osteopontin in primary kidney tubule epithelial cells. PhD Thesis. Rutgers University, New Brunswick, NJ.
37. Denhardt, D. T. & A. F. Chambers. 1994. Overcoming obstacles to metastasis—de-

fenses against host defenses: osteopontin (OPN) as a shield against attack by cytotoxic host cells. J. Cell. Biochem. **56:** 48–51.

38. HOFFMAN, B. & D. A. LIEBERMANN. 1994. Molecular controls of apoptosis; differentiation/growth arrest primary response genes, proto-oncogenes, and tumor suppressor genes as positive & negative modulators. Oncogene **9:** 1807–1812.
39. SCHALLER, M. D. & J. T. PARSONS. 1993. Focal adhesion kinase: an integrin-linked protein tyrosine kinase. Trends Cell Biol. **3:** 258–262.
40. BURRIDGE, K., C. E. TURNER & L. H. ROMER. 1992. Tyrosine phosphorylation of paxillin and pp125^{fak} accompanies cell adhesion to extracellular matrix. J. Cell Biol. **119:** 893–903.
41. BOCKHOLT, S. M. & K. BURRIDGE. 1993. Cell spreading on extracellular matrix proteins induces tyrosine phosphorylation of tensin. J. Biol. Chem. **268:** 14565–14567.
42. PALLER, M. S. & T. V. NEUMANN. 1991. Reactive oxygen species and rat renal epithelial cells during hypoxia and reoxygenation. Kidney Int. **40:** 1041–1049.

The Role of Eta-1/Osteopontin in the Pathogenesis of Immunological Disorders

HARVEY CANTOR

Department of Pathology
Harvard Medical School
Laboratory of Immunopathology
Dana-Farber Cancer Institute
44 Binney Street
Boston, Massachusetts 02115

INTRODUCTION

Studies of soluble mediators elaborated by cells of the immune system which interact with nonimmunological cells have provided valuable insight into the role of the immune system in regulating cell growth and differentiation. One such soluble mediator is the early T lymphocyte activation-1 (Eta-1) gene, which may play an important role in immune resistance to certain bacterial and viral pathogens[1] and may link the immune system with several somatic organ systems. Since analysis of the structure of this protein has revealed that it is almost identical to several other proteins, discussed below, which are associated with neoplasia and bone structure/function, we shall use the designation Eta-1/Op for the purpose of this discussion.

Initial Description of Eta-1/Op

Our efforts to define genes associated with immunological resistance led to analysis of gene expression in activated CD4$^+$ T cells. To this end, we screened cDNA libraries from CD4$^+$ T cell clones for inserts selectively expressed in T cells after activation by antigen associated with class II major histocompatibility complex (MHC) products. One insert defined a gene that, unlike interleukin (IL)-2, IL-3 or γ-interferon (γ-IFN), increased markedly within 48 hours after infection of mice with bacteria or extracts of *Tuberculin bacillus,* suggesting that it might define an inducible T cell gene that differed in its activation requirements from known cytokine genes. The full-length cDNA insert (1569 base pairs (bp)) was characterized and found to contain an open-reading frame (ORF) of 882 nucleotides which encoded a polypeptide having a predicted molecular mass of 32,462 daltons. The first methionine in this ORF fit the consensus for eukaryotic translation initiation signals, and the 3′ noncoding sequence in the cDNA contained two potential polyadenylation signals. The ORF product of the sequenced Eta-1/Op cDNA is extremely hydrophilic and displays several features of a secreted protein. It also has a hydrophobic leader sequence similar to human and chicken transferrin, but lacks an obvious membrane anchoring region.

Eta-1/Op encodes a secreted 60-kDa protein having an RGD-containing subsequence. The RGD motif, although not unique, represents a major binding ligand for integrins on cellular adhesion receptors[2] and is present within a longer subsequence that is highly similar to that surrounding the RGD motif in fibronectin and other cell surface adhesion molecules. Biochemical analysis of both the recombinant[1] and natural[3] T cell product showed that the gene product found in supernatant fluids has the expected pI of 4.3 and displays a larger size than expected from its cDNA, most likely due to glycosylation at multiple sites.

The gene maps to a locus on murine chromosome 5 that more than a decade ago was found to confer unigenic susceptibility to infection by certain intracellular bacterial pathogens.[4] Studies of Eta-1/Op expression *in vivo* indicate that genetic resistance to the lethal effects of infection by intracellular bacterial pathogens is associated with a rapid and vigorous response of the Eta-1[a] allele and early production of antibacterial antibody.[1] Conversely, susceptibility to the lethal effects of infection is associated with a defective response of the Eta-1[b] allele and reduced antibody responses.

Eta-1 and Its Homologs

For over a decade, the activities of this protein in the context of neoplasia and bone formation have been studied by a number of groups.[5] The transformation-related phosphoprotein 2ar,[6] now designated secreted phosphoprotein-1 (Spp-1),[7] represents an intriguing marker of neoplastic transformation.[7–9] 2ar/Spp-1 and Eta-1 both map to the *Rickettsia tsutsugamushi* (RT) locus.[1,10] Independent studies of bone matrix proteins led to isolation of a phosphoprotein with variable molecular weight, designated 44 kDa-phosphoprotein, pp69, bone sialoprotein I and, most recently, osteopontin.[11,12] Although the physiologic function of this bone protein has not been established, it may be involved in cell attachment within the bone matrix, perhaps as an anchor element for osteoclasts.[13,14] Sequential analysis of these cDNAs from different species has shown that mouse osteopontin (2ar) and Eta-1 are virtually identical.[1,5,15]

Potential Functional Relationship of Dysregulated Eta-1/Op Production to Clinical Disorders of the Immune System: Murine Lupus

Studies of systemic autoimmune disease have led to the view that initiation and progression of this disease process reflects chronic and sustained B cell activation by unidentified polyclonal activating agents.[16,17] This abnormal B cell response is thought to lead to hypergammaglobulinemia and, eventually, to production of autoantibodies that mediate immunological tissue destruction.[18] This hypothesis is supported by numerous studies demonstrating high levels of polyclonal B cell activation in systemic lupus erythematosus (SLE) patients[19–21] as well as in animal models of autoimmune disease.[18,22] However, the identity of endogenous agent(s) responsible for chronic B cell activation has not been established.

Extensive studies have been undertaken using the MRL/lpr mouse model, which spontaneously develops high levels of autoantibodies and a syndrome with histopathological features of both rheumatoid arthritis and SLE.[23] Chronic activation of B cells in this murine model of autoimmune disease may reflect abnormal

T cell induction, since thymectomy or treatment with CD4 antibody prevents development of this disorder.[24–26] Prud'homme *et al.* have suggested that MRL/l T cells constitutively produce a soluble factor, termed MRL/L B cell differentiation factor (L-BCDF), which causes B cell activation and immunoglobulin production *in vitro*.[27]

We examined levels of cytokine gene expression in T cells during the course of murine autoimmune disease.[28] We found that the proliferating T cells in MRL/l, but not the congenic nonautoimmune MRL/+ strain, constitutively express very high levels of Eta-1/Op.[28] Selective dysregulation of Eta-1/Op expression, but not other T cell cytokines including IL-2, IL-3 and IL-4, were noted at the onset of this autoimmune disease and continued throughout the course of this disorder, resulting in 50–100-fold greater levels of gene expression than normal MRL/+ congenic mice. Moreover, the level of endogenous Eta-1/Op expression correlated closely with the degree of autoimmune disease expressed by mouse strains congenic at the lpr locus.[28] These findings suggested that Eta-1/Op might contribute directly to the pathogenesis of the autoimmune syndrome, thus prompting us to test the effects of this T cell cytokine on immunoglobulin production by B cells.[29]

We found that both recombinant and biochemically purified Eta-1/Op stimulate immunoglobulin (Ig)M and IgG production by mixtures of B cells and macrophages from the autoimmune MRL/l strain, and this activity was removed from T cell supernatants after absorption with immobilized anti-Eta-1/Op antibody.[29] Possibly, the effects of Eta-1/Op depend on the interaction between this ligand and macrophages resulting in expression of macrophage products (?cytokines) that act on B cells. Alternatively, some of the effects of Eta-1/Op may reflect a direct interaction between this cytokine and a receptor on B cells.

The genetic and cellular events leading to dysregulated Eta-1/Op expression by MRL/l T cells are not yet understood. If this cytokine directly contributes to the pathogenesis of this disorder, mice expressing high levels of transgenic Eta-1/Op might be expected to develop the histopathologic features of murine lupus and rheumatoid arthritis. Moreover, a subset of SLE patients may also be expected to display dysregulated levels of Eta-1/Op and thus represent a discrete nosologic disease entity within the autoimmune disease spectrum. If so, treatment of this biochemical disorder might be directed at correction of Eta-1/Op overexpression rather than current approaches that employ nonspecific immunosuppressive agents.

Murine Acquired Immunodeficiency Syndrome (MAIDS)

The immunological disorder caused by the murine retrovirus LP-BM5 is a useful animal model for studies of the acquired immunodeficiency syndrome (AIDS). Infection by the LP-BM5 retrovirus, like human immunodeficiency virus (HIV) in humans, causes early signs of B cell activation including lymphadenopathy and hypergammaglobulinemia followed by a severe T cell immunodeficiency and impaired resistance to bacterial pathogens.[30,31] This syndrome, termed MAIDS (for murine acquired immunodeficiency syndrome), is thought to reflect two major immunological defects in $CD4^+$ T cells. The early phase is associated with chronic T cell activation, leading to abnormal and excessive expression of TH_2-type cytokines and B cell expansion. The latter stages are associated with increased T cell unresponsiveness and immunosuppression.

The nature of dysregulated T cell cytokine expression associated with progression of MAIDS remains poorly understood. Although there is an early increase in TH_1 cytokines including IL-2, IL-5 and γ-interferon, by 3–4 weeks after LP-BM5 infection these cytokines are not expressed over background levels, even after stimulation of T cells *in vitro* with mitogens.[31] Moreover, with the exception of IL-10, cytokines typically synthesized by TH_2 cells, including IL-4 and IL-6, have not been shown to be increased *in vivo* during the course of this disorder.[31] The relevance of the observation that these and other cytokines can be secreted only after mitogen-dependent activation *in vitro* to their physiological expression *in vivo* is unclear.[31] The finding that *in vivo* expression of IL-10 is increased in MAIDS has led to the hypothesis that excessive production of this cytokine might be associated with impaired cell-mediated immunity. However, increased levels of this cytokine cannot account for chronic and excessive B cell activation.

We recently obtained evidence that dysregulated expression of Eta-1/Op may play a central role in the pathogenesis of MAIDS. We found that the level of expression of this cytokine is dramatically increased within four weeks after viral infection and reaches maximal levels by eight weeks, *i.e.*, increased expression is synchronous with the development of the major phenotypic markers of disease, including hypergammaglobulinemia and lymphadenopathy, while expression of IL-2, IL-3, IL-4 and γ-interferon were not significantly elevated compared with age-matched C57BL/6 controls (TABLE 1).

Elevated levels of Eta-1/Op RNA were detected in the lymph node and spleen from mice 6–12 weeks after infection, but not in lymphoid tissues of age-matched C57BL/6 controls. Eta-1/Op gene expression per 10^6 cells showed an approximate 6–8-fold increase by 8 weeks of infection. However, the absolute increase in Eta-1/Op expression in total lymphoid tissues of each animal is substantially higher because of the development of lymphadenopathy during this time: total levels of Eta-1/Op RNA in peripheral lymph node (LN) of infected mice by 8 weeks after viral infection increased by approximatley 10^3 (TABLE 2).

These findings do not directly implicate Eta-1/Op in the pathogenesis of MAIDS. However, the observations that (a) Eta-1/Op is persistently and dramatically elevated during the course of this disorder and (b) Eta-1/Op acts as a polyclonal B cell stimulant as well as an inhibitor of the macrophage response to intracellular bacterial pathogens makes this an attractive hypothesis, since both chronic B cell activation and impaired responses to intracellular bacterial pathogens are hallmarks of this disorder.

TABLE 1. RNA Expression during MAIDS[a]

	Cytokine Gene Expression (RDU)				
Week	Eta-1/Op	IL-2	IL-3	IL-4	γ-IFN
0	<0.5	<0.5	0.75	<0.5	<0.5
8	2.88	<0.5	1.35	<0.5	<0.5

[a] RNA was extracted from spleen and lymph node of C57BL/6 mice at 2- or 4-week intervals following LP-BM5 infection for a period of twelve weeks. Levels of Eta-1/Op, IL-2, IL-3, IL-4 and γ-interferon were quantitated by densitometric measurement of slot blots as described.[28,41] Gene expression at weeks 0 and 8 is shown as relative densitometric units (RDU) which represents the ratio of cytokine RNA levels to actin RNA levels in each cellular sample according to densitometric analysis. Data shown corresponds to the mean of at least three different samples.

TABLE 2. Total Eta-1/Op Activity in Lymph Nodes (LN) from C57BL/6 Mice after LP-BM5 Infection[a]

Week	LN Eta-1/Op Activity
0	<0.5
8	>2000

[a] The levels of Eta-1/Op RNA expressed as relative densitometric units (Eta-1/actin) (as described in the legend to TABLE 1) are multiplied by the ratio of LN weight at the age indicated to LN weight before viral infection. Beginning at week 4, animals displayed initial signs of disease and by week 8 showed a 50% mortality rate from the disease process. Two representative time points are shown.

MAIDS and AIDS: Evidence for an Imbalance of TH_2 Cytokine Production

Dominance of the TH_2 response is a central feature of MAIDS, as judged by chronic B cell activation, hypergammaglobulinemia and suppression of both cell-mediated immunity and resistance to infection by intracellular pathogens such as *Listeria monocytogenes* (LM), ectromelia, *cryptococcus, Leishmania* and *Toxoplasma gondi*.[32–34] The hypothesis that an imbalance between TH_2- and TH_1-type cytokines may also play an important role in progression of HIV infection in man is based on several recent observations. First, a prominent early component of HIV infection is polyclonal B cell activation and hypergammaglobulinemia.[35] Second, $CD4^+$ cells from the peripheral blood lymphocytes (PBL) of AIDS patients produce reduced amounts of TH_1 cytokines such as IL-2 and γ-interferon, but elevated levels of B cell growth and differentiation factors.[36–38] Third, recent studies suggest that TH_1 cytokine responses, which become downregulated during the progression of HIV infections, may be associated with protective immune responses to HIV. For example, peripheral blood mononuclear cells (PBMC) from a fraction of HIV sero-negative polymerase chain reaction (PCR)-negative high-risk homosexuals respond to gp160-derived HIV antigens according to proliferation and production of the TH_1 cytokine IL-2.[39,40] Possibly, HIV seroconversion reflects increased viral replication and associated expansion of the TH_2 cytokine response, resulting in polyclonal B cell activation, HIV-specific antibody and inhibition of protective TH_1-dependent cell-mediated immune responses to the virus and to bacterial pathogens.

It is important to avoid oversimplifying the relationship between the TH_2 cytokine response and subsets of TH cells in AIDS. For example, recent studies of *in situ* cytokine expression in lymph node by Fauci and colleagues documented a persistent elevation of the TH_2 cytokine IL-10 during the disease in lymph nodes, along with little or no expression of the TH_2 cytokine IL-4, according to semiquantitative PCR analysis (A. Fauci, personal communication). Moreover, almost all increased IL-10 expression was associated with either CD8 cells or non-T cells. Studies are currently underway to further define the relative contribution of different lymphoid and monocyte subpopulations responsible for elevated Eta-1/Op production in MAIDS, as well as the potential role of this cytokine in the pathogenesis of this disorder.

REFERENCES

1. PATARCA, R., G. J. FREEMAN, R. P. SINGH, F.-Y. WIE, T. DURFEE, F. BLATTNER, D. C. REGNIER, C. A. KOZACK, B. A. MOCK, H. C. MORSE, III, T. R. JERRELLS &

H. CANTOR. 1989. Structural and functional studies of the early T lymphocyte activation 1 (Eta-1) gene. Definition of a novel T cell-dependent response associated with genetic resistance to bacterial infection. J. Exp. Med. **170:** 145–161.
2. RUOHSLAHTI, E. & M. D. PIERSCHBACHER. 1986. Arg-Gly-Asp: a versatile cell recognition signal. Cell **44:** 517.
3. SINGH, R. P., R. PATARCA, J. SCHWARTZ, P. SINGH & H. CANTOR. 1990. Definition of a specific interaction between Eta-1 protein and murine macrophages *in vitro* and *in vivo*. J. Exp. Med. **171:** 1931–1942.
4. GROVES, M. G. & J. V. OSTERMAN. 1978. Host defenses in experimental scrub typhus: genetics of natural resistance to infection. Infect. Immun. **19:** 583.
5. PATARCA, R., R. A. SAAVEDRA & H. CANTOR. 1993. Molecular and cellular basis of genetic resistance to bacterial infection: the role of the early T-lymphocyte activation-1/osteopontin gene. Crit. Rev. Immunol. **13:** 225–246.
6. SMITH, J. H. & D. T. DENHARDT. 1987. Molecular cloning of a tumor promoter-inducible mRNA found in JB6 mouse epidermal cells: induction is stable at high, but not at low cell densities. J. Cell. Biochem. **34:** 13.
7. SENGER, D. R., C. A. PERRUZZI & A. PAPADOPOULOS. 1989. Elevated expression of secreted phosphoprotein I (osteopontin, 2ar) as a consequence of neoplastic transformation. Anticancer Res. **9:** 1291.
8. SENGER, D. R., B. A. ASCH, B. D. SMITH, C. A. PERRUZZI & H. F. DVORAK. 1983. A secreted phosphoprotein marker for neoplastic transformation of both epithelial and fibroblastic cells. Nature **302:** 714.
9. CRAIG, A. M., M. NEMIR & B. B. MUKHERJEE. 1988. Identification of the major phosphoprotein secreted by many rodent cell lines as 2ar/osteopontin: enhanced expression in H-ras-transformed 3T3 cells. Biochem. Biophys. Res. Commun. **157:** 166.
10. FET, V., M. E. DICKINSON & B. L. HOGAN. 1989. Localization of the mouse gene for secreted phosphoprotein-1 (Spp-1) (2ar, osteopontin, bone sialoprotein 1, 44-kDa bone phosphoprotein, tumor-secreted phosphoprotein) to chromosome 5, closely linked to Ric (*Rickettsia* resistance). Genomics **5:** 375.
11. BUTLER, W. T. 1989. The nature and significance of osteopontin. Connect. Tissue Res. **23:** 123.
12. GLIMCHER, M. J. 1989. Mechanism of calcification: role of collagen fibrils and collagen-phosphoprotein complexes *in vitro* and *in vivo*. Anat. Rec. **224:** 139.
13. OLDBERG, A., A. FRANZEN & D. HEINEGÅRD. 1986. Cloning and sequence analysis of rat bone sialoprotein (osteopontin) cDNA reveals an Arg-Gly-Asp cell-binding sequence. Proc. Natl. Acad. Sci. USA **83:** 8819.
14. REINHOLT, F. P., K. HULTENBY, A. OLDBERG & D. HEINEGÅRD. 1990. Osteopontin, a possible anchor of osteoclasts to bone. Proc. Natl. Acad. Sci. USA **87:** 4473.
15. CRAIG, A. M., J. H. SMITH & D. T. DENHARDT. 1989. Osteopontin, a transformation-associated cell adhesion phosphoprotein, is induced by 12-O-tetradecanoylphorbol 13-acetate in mouse epidermis. J. Biol. Chem. **264:** 9682.
16. HANG, L. M., J. H. SLACK, C. AMUNDSON, S. IZUI & A. N. THEOFILOPOULOS. 1983. Induction of murine autoimmune disease by chronic polyclonal B cell activation. J. Exp. Med. **157:** 874.
17. KLINMAN, D. M. & A. D. STEINBERG. 1987. Systemic autoimmune disease arises from polyclonal B-cell activation. J. Exp. Med. **165:** 1755–1760.
18. KLINMAN, D. M., Y. ISHIGATSUBO & A. D. STEINBERG. 1988. Acquisition and maturation of expressed B cell repertoires in normal and autoimmune mice. J. Immunol. **141:** 801.
19. BUDMAN, D. R., E. B. MERCHANT, A. D. STEINBERG, B. DOFT, M. E. GERSHWIN, E. LIZZIO & J. P. REEVES. 1977. Increased spontaneous activity of antibody-forming cells in the peripheral blood of patients with active SLE. Arthritis Rheum. **20:** 829.
20. BLAESE, R. M., J. GRAYSON & A. D. STEINBERG. 1980. Elevated immunoglobulin secreting cells in the blood of patients with active SLE: correlation of laboratory and clinical assessment of disease activity. Am. J. Med. **69:** 345.
21. UEDA, Y., T. SAKANE & T. TSUNEMATSU. 1989. Hyperreactivity of activated B cells to B cell growth factors in patients with SLE. J. Immunol. **143:** 3988.

22. IZUI, S. P., P. J. MCCONAHEY & F. J. DIXON. 1978. Increased spontaneous polyclonal activation of B lymphocytes in mice with spontaneous autoimmune disease. J. Immunol. **121:** 2213–2219.
23. THEOFILOPOULOS, A. N. & F. J. DIXON. 1985. Murine models of systemic lupus erythematosus. Adv. Immunol. **37:** 269.
24. STEINBERG, A. D., J. B. ROTHS, E. D. MURPHY, R. T. STEINBERG & E. S. RAVECHE. 1980. Effects of thymectomy or androgen administration upon the autoimmune disease of MRL/Mp-lpr/lpr mice. Cell **62:** 871–873.
25. THEOFILOPOULOS, A. N., R. S. BALDERAS, D. L. SHAWLER, S. LEE & F. J. DIXON. 1981. The influence of thymic genotype of the SLE-like disease and T-cell proliferation of MRL/Mp-lpr/lpr mice. J. Exp. Med. **153:** 1405–1414.
26. SANTORO, T. J., J. P. PORTANOVA & B. L. KOTZIN. 1988. The contribution of L3T4+ T cells to lymphoproliferation and autoantibody production in MRL-lpr/lpr mice. J. Exp. Med. **167:** 1713–1718.
27. PRUD'HOMME, G. J., C. L. PARK, T. M. FIESER, R. KOFLER, F. J. DIXON & A. N. THEOFILOPOULOS. 1983. Identification of a B cell differentiation factor(s) spontaneously produced by proliferating T cells in murine lupus strains of lpr/lpr genotype. J. Exp. Med. **157:** 730.
28. PATARCA, R., F.-Y. WEI, P. SINGH, M. I. MORASSO & H. CANTOR. 1990. Dysregulated expression of the T-cell cytokine Eta-1 in CD4-8-lymphocytes during the development of murine autoimmune disease. J. Exp. Med. **172:** 1177–1183.
29. LAMPE, M. A., R. PATARCA, M. V. IREGUI & H. CANTOR. 1991. Polyclonal B-cell activation by the Eta-1 cytokine and the development of autoimmune disease. J. Immunol. **147:** 2902–2906.
30. HITOSHI, Y., Y. OKADA, E. SONODA, A. TOMINAGA, M. MAKINO, K. SUZUKI, J. KINOSHITA, K. KOMURO, T. MIZUOCHI & K. TAKATSU. 1993. Delayed progression of a murine retrovirus-induced acquired immunodeficiency syndrome in X-linked immunodeficient mice. J. Exp. Med. **177:** 621–626.
31. GAZZINELLI, R. T., M. MAKINO, S. K. CHATTOPADHYAY, C. M. SNAPPER, A. SHER, A. W. HÜGIN & H. C. MORSE, III. 1992. CD4+ subset regulation in viral infection. Preferential activation of TH_2 cells during progression of retrovirus-induced immunodeficiency in mice. J. Immunol. **148:** 182–188.
32. DARBAN, H., J. ENRIQUEZ, C. R. STERLING, M. C. LOPEZ, G. CHEN, M. ABBASZADEGAN & R. R. WATSON. 1991. Cryptosporidiosis facilitated by murine retroviral infection with LP-BM5. J. Infect. Dis. **164:** 741.
33. SHER, A. & R. L. COFFMAN. 1992. Regulation of immunity to parasites by T cells and T cell-derived cytokines. Annu. Rev. Immunol. **10:** 385.
34. BULLER, R. M., R. A. YETTER, T. N. FREDERICKSON & H. C. MORSE, III. 1987. Abrogation of resistance to severe mousepox in C57BL/6 mice infected with LP-BM5 murine leukemia viruses. J. Virol. **61:** 383.
35. LUCEY, D. R., R. A. ZAJAC, G. P. MELCHER, C. A. BUTZIN & R. N. BOSWELL. 1990. Serum IgE levels in 622 persons with human immunodeficiency virus infection: IgE elevation with marked depletion of CD4+ T-cells from patients with AIDS. AIDS Res. Hum. Retroviruses **6:** 427.
36. MAGGI, E., D. MACCHIA, P. PARRONCHI, M. MAZZETTI, A. RAVINA, D. MILO & S. ROMAGNANI. 1987. Reduced production of interleukin 2 and interferon-gamma and enchanced helper activity of IgG synthesis by cloned CD4+ T cells from patients with AIDS. Eur. J. Immunol. **17:** 1685.
37. CLERICI, M., N. I. STOCKS, R. A. ZAJAC, R. N. BOSWELL, D. C. BERNSTEIN, D. L. MANN, G. M. SHEARER & J. A. BERZOFSKY. 1989. Interleukin-2 production used to detect antigenic peptide recognition by T-helper lymphocytes from asymptomatic HIV-seropositive individuals. Nature **339:** 383.
38. CLERICI, M., N. I. STOCKS, R. A. ZAJAC, R. N. BOSWELL, D. R. LUCEY, C. S. VIA & G. M. SHEARER. 1989. Detection of three distinct patterns of T helper cell dysfunction in asymptomatic, human immunodeficiency virus-seropositive patients. Independence of CD4+ cell numbers and clinical staging. J. Clin. Invest. **84:** 1892.
39. CLERICI, M., J. A. BERZOFSKY, G. M. SHEARER & C. O. TACKET. 1991. Exposure to

human immunodeficiency virus (HIV) type 1 indicated by HIV-specific T helper cell responses before detection of infection by polymerase chain reaction and serum antibodies. J. Infect. Dis. **164:** 178.

40. Clerici, M., J. V. Giorgi, C. C. Chou, V. K. Gudeman, J. A. Zack, P. Gupta, H.-N. Ho, P. G. Nishanian, J. A. Berzofsky & G. M. Shearer. 1992. Cell-mediated immune response to human immunodeficiency virus type 1 (HIV-1) in seronegative homosexuals with recent sexual exposure to HIV-1. J. Infect. Dis. **165:** 1012–1019.

41. Patarca, R., F.-Y. Wei, M. V. Iregui & H. Cantor. 1991. Differential induction of interferon-gamma gene expression after activation of CD4+ T-cells by conventional antigen and Mls superantigen. Proc. Natl. Acad. Sci. USA **88:** 2736–2739.

Engagement of the Osteoclast Integrin $\alpha_v\beta_3$ by Osteopontin Stimulates Phosphatidylinositol 3-Hydroxyl Kinase Activity[a,b]

KEITH A. HRUSKA, FELICE ROLNICK, MARGARET HUSKEY, ULISES ALVAREZ, AND DAVID CHERESH[c]

Division of Nephrology
Jewish Hospital/Washington University Medical Center
St. Louis, Missouri 63110
and
[c]Department of Immunology
Scripps Research Institute
La Jolla, California 92037

INTRODUCTION

Bone remodeling depends on the activity of osteoclasts which attach to and resorb bone. Initially, osteoclasts bind bone matrix and then develop a polarized plasma membrane resulting in local secretion of protons and collagenolytic enzymes by the ruffled border membrane serving to degrade the bone matrix.[1-3] Recently, efforts have focused on the signaling events activated by adhesion between osteoclasts and bone matrix leading to cell polarization. To this end, it has been shown that integrin $\alpha_v\beta_3$ mediates osteoclast binding to osteopontin in bone, resulting in the ultimate resorption of bone immediately below the osteoclast surface.[4-7] While integrin $\alpha_v\beta_3$ is required for this process, little is known of the signaling events that take place upon integrin occupancy. We have demonstrated that osteopontin or RGD-containing peptides bind to integrin $\alpha_v\beta_3$ on osteoclasts and stimulate an immediate reduction in cytosolic calcium.[8] Earlier studies had

[a] This work was supported by grants from the NIH: AR41677, DKO9976, AR39561 and T32-DK07679. DAC was supported by Grants 45726 and 50287 from the NCI. DAC is a recipient of an American Cancer Society Faculty Research Award. The data in this paper have been published elsewhere.[57]

[b] Abbreviations used in this paper: PtdInsP, phosphatidylinositol phosphate; PtdIns 3-OH kinase, phosphatidylinositol 3-hydroxyl kinase; RGD, Arg-Gly-Asp-cell adhesion sequence; ^{3}H, tritium; ^{32}Pi, radioactive phosphorus; OPN, osteopontin; OPN-A, osteopontin peptide (20 mer); HPLC, high pressure liquid chromatography; TLC, thin layer chromatography; LM609, anti-$\alpha_v\beta_3$ integrin antibody; CHAPS, 3-[(3-cholamidopropyl)dimelhylammonio]-1-propanesulfonate; EDTA, ethylene diamine tetraacetic acid; PMSF, phenyl methyl sulfonyl fluoride; ECL, enhanced chemiluminescence; CHAV, anti-chicken α_v integrin subunit antibody; PS, phosphatidylserine; PtdIns, phosphatidylinositol; PtdInsP$_2$, phosphatidylinositol bisphosphate; PtdIns P$_3$, phosphatidylinositol trisphosphate; Ins 1,3,4,5P$_4$, inositol tetrakisphosphate; SH2, *src* homology 2; SH3, *src* homology three domain; pp^{125FAK}, focal adhesion kinase; mab, monoclonal antibody; CSF-1, colony stimulating factor 1; c-*fms*, colony stimulating factor receptor; IP, immunoprecipitates.

shown that a reduction in cytosolic calcium is itself a stimulatory signal for bone resorption, participating in the cytoskeletal reorganization associated with the process.[9] Likewise, elevations in cytosolic Ca^{2+} produced by hormones (calcitonin) or the putative osteoclast Ca^{2+} sensor result in an inhibition of bone resorption.[10,11] The reduction in cytosolic Ca^{2+} stimulated by protons or matrix proteins is produced by an increase in Ca^{2+} efflux through activation of the plasma membrane Ca^{2+} ATPase.[12] The mechanism of Ca^{2+}-ATPase activation was unique since it was not through Ca^{2+}/calmodulin or protein kinase A.[8,9,12] Thus, we analyzed the effects of osteopontin on other regulators of Ca^{2+} ATPase. The inositol-containing phospholipids are potential regulators of the plasmalemmal Ca^{2+} ATPase pump activity.[13–16] In the studies reported herein, we analyzed ^{3}H-myoinositol and ^{32}Pi incorporation into phosphoinositides during osteoclast precursor stimulation. Evidence is presented that by binding to integrin $\alpha_v\beta_3$, osteopontin stimulates production of phosphatidylinositol phosphate (PtdInsP) and PtdIns 3,4,5P_3 by osteoclasts. The mechanism of osteoclast stimulation for phosphoinositide production may be activation of phosphatidylinositol 3-hydroxyl kinase (PtdIns 3-OH kinase), which was found to be immunoprecipitated by antibodies to the $\alpha_v\beta_3$ integrin. Also associated with the $\alpha_v\beta_3$ integrin, we found the c-*src* protoncogene product which might possibly be a link protein between the integrin and PtdIns 3-OH kinase. These findings demonstrate an integrin-associated signal-generating complex, which, when activated, may participate in changes in osteoclast function.

METHODS

Cell Preparation and Stimulation

Osteoclast precursors, prepared as previously reported,[17] were preincubated with ^{3}H-myoinositol (20–35 μCi/ml for 18 hr) or ^{32}Pi (15–50 μCi/ml for 4 hr) and stimulated by osteopontin (OPN) or an osteopontin peptide containing the Arg-Gly-Asp cell adhesion motif, OPN-A (TVDTYDGRGDSVVYGLRSKS). The preincubation media was inositol-free for the experiments using ^{3}H-myoinositol. OPN was a kind gift of Meena Chellaiah and OPN-A, a gift of Pam Robey.

Lipid Extraction and Analysis

Following stimulation, the osteoclast culture media were withdrawn, the cells were washed, and then treated with methanol at 4°C for 15 min. Lipids were extracted as previously described[18–19] and dried under N_2. Lipid extracts were reconstituted in chloroform : methanol (1 : 1), deacylated and injected into an HPLC system for separation of ^{3}H-inositol or ^{32}Pi containing glyceryl phosphoinositols as previously described.[18–19] The HPLC system was equipped with an ion exchange column (Dionex Ion Pak A55A-5 micron, Sunnyvale, CA), and a three-step gradient from 92/8 to 25/75 H_2O/150 mM NaOH, and a flow rate of 1 ml/min was used to elute the glyceryl phosphoinositols. The elution position of each metabolite was determined using standards. Recovery from extraction through HPLC ranged between 75 and 80%. The results were reported as cpm/mg/protein and were not corrected for recovery.

As an alternative to HPLC, lipid extracts were spotted on silica gel 60 thin layer chromatography (TLC) plates pretreated with 1.2% potassium oxalate in methanol : water (2 : 3). Chromatography was performed in chloroform : acetone : methanol : acetic acid : water (40 : 15 : 13 : 12 : 7), and autoradiograms were developed to detect the radiolabeled phospholipids. Unlabeled phospholipid standards were identified by exposure of the TLC plates to iodine vapor and phosphorylated phospholipids were identified by comigration with standards. Identification of ^{32}Pi-PtdIns 3,4,5P_3 was accomplished by removal of the putative phospholipid spots from TLC plates, deacylation, collection of the deacylated phospholipid, deglyceration, and HPLC. The product of glycerylphosphoinositol deglyceration comigrated with inositol tetrakisphosphate (Ins 1,3,4,5P_4).

Immunoprecipitations and Western Immunoblot Analyses

Cell lysates were prepared as described by Eiseman and Bolen[20] from osteoclast precursors in serum-free media for six hours prior to stimulation with OPN, OPN-A or vehicle. Immunoprecipitates were prepared using a monoclonal antibody (mab) to the $\alpha_v\beta_3$ integrin (LM609),[22] a polyclonal antibody to β_1 integrin subunits,[21] or a polyclonal antibody to the 85-kd regulatory subunit of PtdIns 3-OH kinase[23–25] (UBI Inc., Lake Placid, NY). Lysates were normalized for protein concentration and incubated with appropriate antibodies for 18 hr at 4°C. The lysis buffer was: 10 mM CHAPS, 50 mm Tris HCl (pH 8.0), 150 mM NaCl, 2 mM EDTA, 5 mM NaF, 10 μg/ml aprotinin, 10 μg/ml leupeptin, 1 mM vanadate, and 1 mM PMSF. The antigen-antibody complexes were collected with protein A or G sepharose beads (Sigma, St. Louis, MO) for 1 hr at 4°C on a shaker. The bead antigen antibody complexes were then consecutively washed with lysis buffer, 0.5 M lithium chloride, 50 mM tris (pH 7.4), and then 10 mM tris (pH 7.4) three times. After polyacrylamide gel electrophoresis, Western analyses of immunoprecipitates[26] were performed using a polyclonal antibody to PtdIns 3-OH kinase which cross-reacts with the chicken,[23–25] a monoclonal antibody to c-*src* (UBI Inc.), a monoclonal antibody to focal adhesion kinase (UBI, Inc.), or an antiphosphotyrosine antibody (UBI, Inc.). Detection of the primary antibody was by a second antibody and the chemiluminescence detection system (ECL).[26] Additional aliquots of the lysates were probed for the respective integrin subunit, α_v (using a monoclonal antibody to α_v (CHAV)[27] or β_1. The anti-integrin antibodies were equally effective in immunoprecipitating the $\alpha_v\beta_3$ and β_1 integrins. Osteopontin treatment had no effect on the amount of integrin in the immunoprecipitates.

In Vitro *Phospholipid Kinase and Protein Tyrosine Kinase Assays*

Osteoclast precursors were treated with vehicle, osteopontin or the osteopontin peptide prior to lysis. Immunoprecipitates were formed using the polyclonal antibody to PtdIns 3-OH kinase or the monoclonal antibody to $\alpha_v\beta_3$. Lysates were normalized for protein concentration and incubated with antibody for 18 hr at 4°C. The antigen-antibody complexes were collected with protein A sepharose beads (Sigma, St. Louis, MO) for 1 hr at 4°C on a shaker. The bead-antigen-antibody complexes were then consecutively washed with lysis buffer, then 0.5 M LiCl, 50 mM tris (pH 7.4), and then 10 mM tris (pH 7.4) three times. Immunoprecipitates were reacted with liposomes containing PS, PtdIns, and PtdIns(4,5)P_2 (142 μg/0.693 mg of protein) for 20 min at 4°C.[28] The liposome-immunoprecipitate

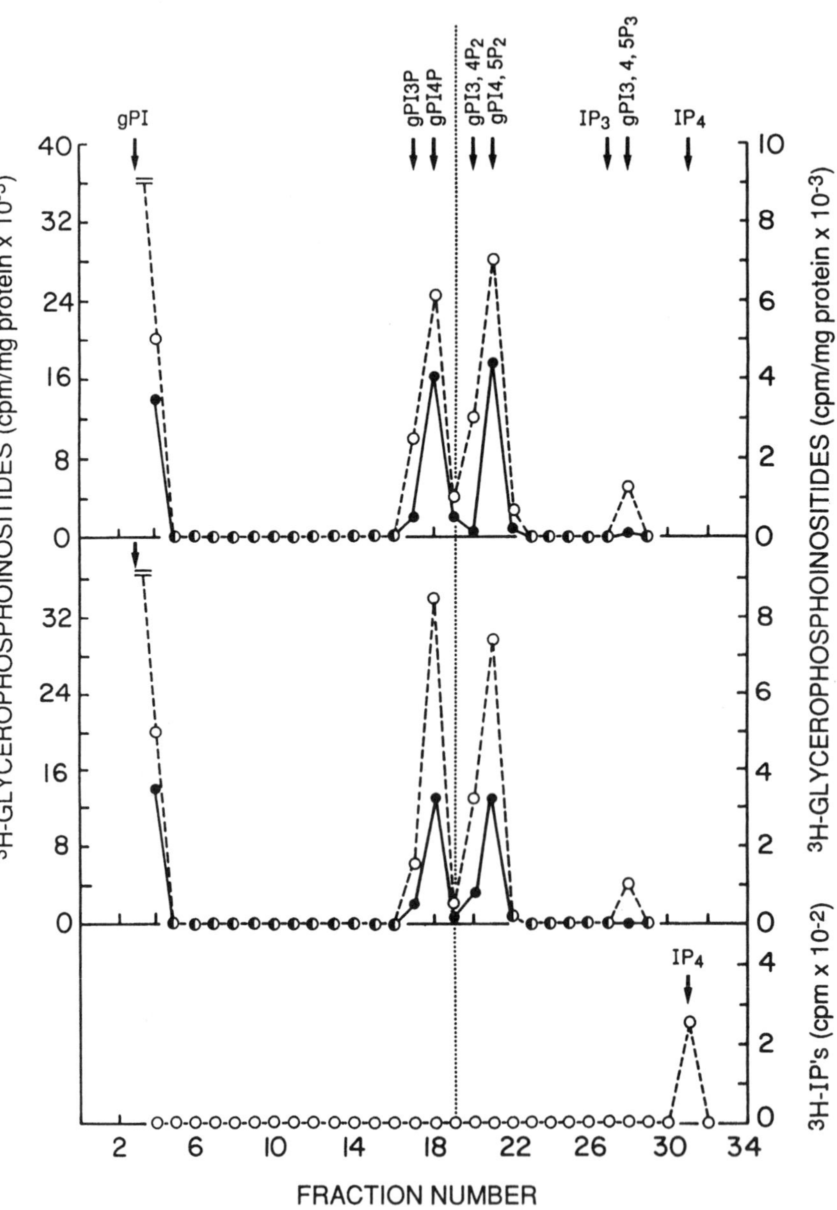

FIGURE 1. Stimulation of ^{3}H-myoinositol incorporation into phosphoinositides of osteoclast precursors by osteopontin. Lipid extracts from cells treated with osteopontin (25 μg/ml) or vehicle for 1–5 min were deacylated and separated by HPLC. OPN (○) increased ^{3}H-PtdInsP, ^{3}H-PtdInsP$_2$ and ^{3}H-PtdInsP$_3$ levels at 1 min (*top panel*) and 5 min (*middle panel*) compared to vehicle treated (*top panel*) or GRGES-treated (*middle panel*) cells (●). The scale changes to that on the *right side* of the figure after the gPIP peak and before the

TABLE 1. Time Course of OPN (25 μg/ml) Stimulation of ^{3}H-Myoinositol Incorporation into PtdInsP$_3$[a]

	^{3}H-PtdIns P$_3$ cpm/mg Cell Protein		
	15 sec	1 min	5 min
Vehicle	1900	2500	2000
OPN	4400	5300	2800

[a] The data are the means of two experiments performed in duplicate. (From Hruska *et al.*[57] Reprinted by permission from *Endocrinology*.)

mixture was incubated with γ^{32}P-ATP (30 μCi) for 20 min at 20°C. The reactions were stopped with 1 N HCl; the lipids were extracted; and then they were subjected to thin-layer chromatography or HPLC. Immune complex protein tyrosine kinase assays for determination of c-*src* activity were performed as described by Veillette *et al.*[29] using enolase as a kinase substrate. Autoradiograms were developed to detect ^{32}Pi incorporation into proteins from γ^{32}Pi-ATP.

RESULTS

Osteopontin Stimulation of Phosphoinositide Labeling

To identify the signaling events associated with recognition of osteopontin, osteoclast precursors were incubated with osteopontin or osteopontin peptides, and cell lysates were examined for the presence of inositol-containing phospholipids. Osteopontin and RGD-containing peptides from the osteopontin sequence stimulated ^{3}H-myoinositol incorporation into PtdInsP, PtdInsP$_2$, as well as PtdInsP$_3$ (FIG. 1 and TABLE 1). PtdInsP$_3$ was identified by its elution position between Ins 1,4,5P$_3$ and Ins 1,3,4,5P$_4$, and by deglyceration of deacylated lipids collected from HPLC elutions (FIG. 1, bottom panel) and repeat HPLC confirming coelution with authentic ^{3}H-inositol 1,3,4,5P$_4$.[30–33] The effect of OPN on PtdIns 3,4,5P$_3$ was rapid, occurring within 15 sec after exposure of cells to OPN (5×10^{-7} M) (TABLE 1). The peak effect was observed between 1 and 5 min and it was sustained for at least 15 min. The dose range of OPN used to study ^{3}H-myoinositol incorporation was 2×10^{-8}–10^{-6} M; all doses were stimulatory, but the threshold dose was not defined.

A peptide from the OPN sequence containing the RGD cell adhesion domain (OPN-A),[8] and the immediate flanking sequences also stimulated ^{3}H-myoinositol incorporation into phosphoinositides, albeit at higher doses than OPN (TABLE 2). An effect of the osteopontin peptide (OPN-A) on ^{3}H-PtdInsP$_2$ formation was not observed in these experiments. OPN-A also stimulated ^{32}Pi incorporation into the phosphoinositides when osteoclast precursors were prelabeled for one hr (FIG. 2). At this duration of prelabeling, ^{32}Pi incorporation reflects mainly new synthesis of phosphoinositides.[31–33] To confirm the identity of the ^{32}Pi-phosphoinositide

gPIP$_2$ peak. The ^{3}H-glycerol-phosphoinositols were normalized per mg cell protein. The putative gPIP$_3$ peaks of each injection which eluted between inositol 1,4,5P$_3$ and inositol 1,3,4,5P$_4$ were collected, deglycerated as described in Methods, and reinjected. Upon re-HPLC, they coeluted with the inositol 1,3,4,5P$_4$ standard (*bottom panel*). (From Hruska *et al.*[57] Reprinted by permission from *Endocrinology*.)

TABLE 2. Effect on an Osteopontin Peptide (OPN-A, 200 μg/ml) on the Incorporation of ^{3}H-Myoinositol into $PtdInsP_3$ in Osteoclast Precursors[a]

^{3}H-phosphoinositides, cpm/mg Cell Protein	OPN-A 200 μg/ml 0	1 min	5 min	15 min
$PtdInsP_3$	1,799	4,026	5,267	6,991
$PtdInsP_2$	21,914	20,727	15,473	15,213
PtdInsP	43,207	78,469	83,962	69,821
PtdIns	695,953	792,655	581,695	548,332

[a] The data are means of two experiments performed in duplicate. (From Hruska *et al.*[57] Reprinted by permission from *Endocrinology*.)

isomers, they were eluted from the TLC plates and deacylated, and HPLC was performed. The PtdIns 3,4,5P_3 from the TLC plate eluted as in FIGURE 1, except that other unidentified ^{32}Pi-containing compounds were also present in the HPLC elution profile, so that quantitation of $PtdInsP_3$ required HPLC as shown in FIGURE 1 and could not be performed by TLC alone. Peptides not containing the RGD sequence, GRGES, did not stimulate ^{3}H-myoinositol (FIG. 1) or ^{32}Pi incorporation. The GRGES control was selected on the basis of previous studies demonstrating that GRGDS, but not GRGES, stimulated efflux of $[Ca^{2+}]_i$.[8] However, the absence of the sequences flanking the RGD motif do influence the signaling properties of OPN (see Table 1 in Ref. 8) and, in this respect, GRGES was not the ideal control for OPN or the OPN-A peptide. The effect of OPN-A on phosphoinositide formation was not mimicked by the antibody to $\alpha_v\beta_3$ (LM609, 50 μg/ml), and the antibody blocked any additional stimulation by OPN-A (FIG. 2). This antibody markedly inhibits bone resorption by chicken osteoclasts.[7] These findings are in agreement with this antibody's ability to block OPN-stimulated Ca^{2+} efflux. Interestingly, $PtdInsP_3$ levels were detectable in unstimulated cells (TABLES 1, 2), possibly due to basal activation of $\alpha_v\beta_3$ by ligation with osteopontin produced by osteoclasts as an autocrine factor[34] or vitronectin from the culture media serum.

Role of PtdIns 3-OH Kinase in $\alpha_v\beta_3$-Dependent Signaling

The production of phosphoinositides phosphorylated in the 3 position such as $PtdInsP_3$ is typically due to the activity of PtdIns 3-OH kinase.[31,32] Therefore, we determined whether this enzyme was associated with the osteoclast $\alpha_v\beta_3$ integrin and was activated upon ligand binding. Immunoprecipitates formed by antibodies to integrin $\alpha_v\beta_3$ were analyzed for PtdIns 3-OH kinase. As shown in FIGURE 3, similar amounts of the 85-kd (regulatory subunit) component of PtdIns 3-OH kinase were detected by Western analysis in $\alpha_v\beta_3$ immunoprecipitates from control and osteopontin peptide-treated cells (FIG. 3, top panel, lanes 3 and 4). The antibody employed does not detect the P110 catalytic subunit of PtdIns 3-OH kinase. Since activity of PtdIns 3-OH kinase may be accomplished through tyrosine phosphorylation of the *src* homology 2 domain (SH2) of the P85 component,[23-25] immunoprecipitates of osteoclast cell lysates obtained using the antibody to the $\alpha_v\beta_3$ integrin were examined for PtdIns 3-OH kinase activity. As shown in TABLE 3, osteopontin treatment of osteoclast precursors increased PtdIns 3-OH kinase activity in immunoprecipitates formed by the $\alpha_v\beta_3$ integrin antibody. There was more than a twofold induction of immune complex PtdIns 3-OH kinase activity

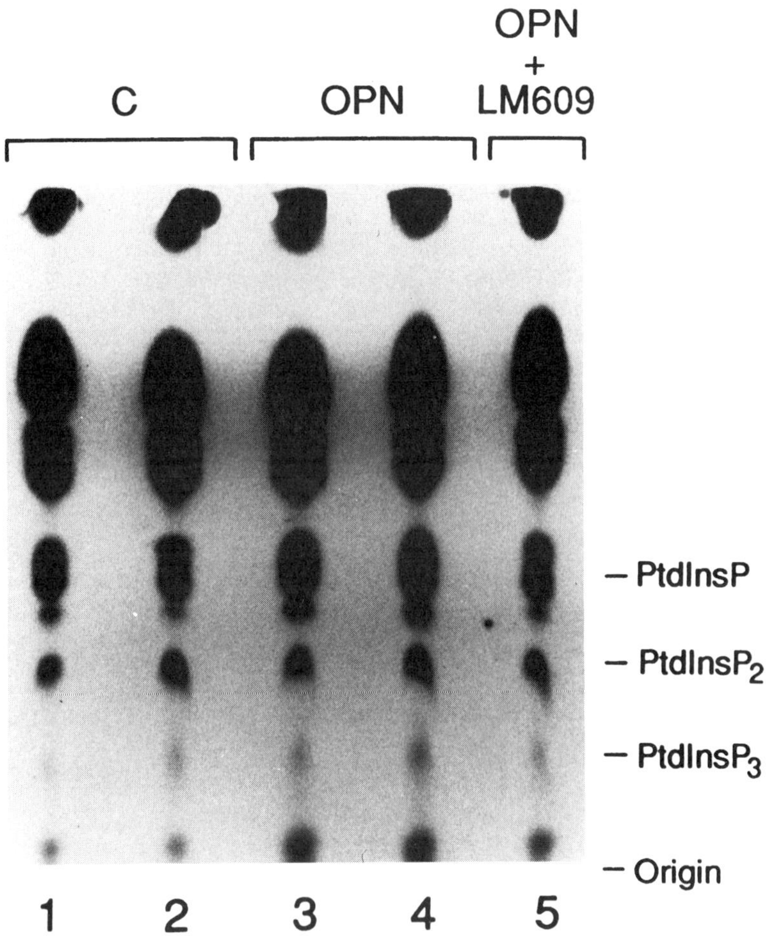

FIGURE 2. Effect of OPN-A and the mab to $\alpha_v\beta_3$, LM609, on ^{32}Pi incorporation into osteoclast phosphoinositides. OPN-A (100 μg/ml), vehicle or LM609 was added to osteoclast precursors for 1 min before reactions were stopped, lipids extracted and TLC performed. In lipid extracts from the OPN-A treated cells, the ^{32}Pi-PtdInsP was 140,220 cpm/mg protein compared to 78,540 in control, and the PtdInsP$_2$ was 85,690 cpm/mg protein compared to control of 62,240. Radioactivity in the putative ^{32}Pi PtdInsP$_3$ spots was 13,256 cpm/mg protein in the treated compared to 4,164 in control (see text for description of HPLC of the PtdInsP$_3$ spots). The counts were the mean of the duplicates. (From Hruska *et al.*[57] Reprinted by permission from *Endocrinology*.)

after 1 min of treatment with osteopontin (10 μg/ml). We also attempted to correlate the increase in PtdIns 3-OH kinase activity to tyrosine phosphorylation of the 85-kd regulatory subunit. Detection of phosphotyrosine by immunoblotting osteoclast lysates formed by anti-$\alpha_v\beta_3$ after electrophoresis revealed increased phosphorylation of an 85-kd protein (FIG. 4). There was substantial tyrosine phosphorylation of this protein detected in lysates from untreated control osteoclasts, but this and

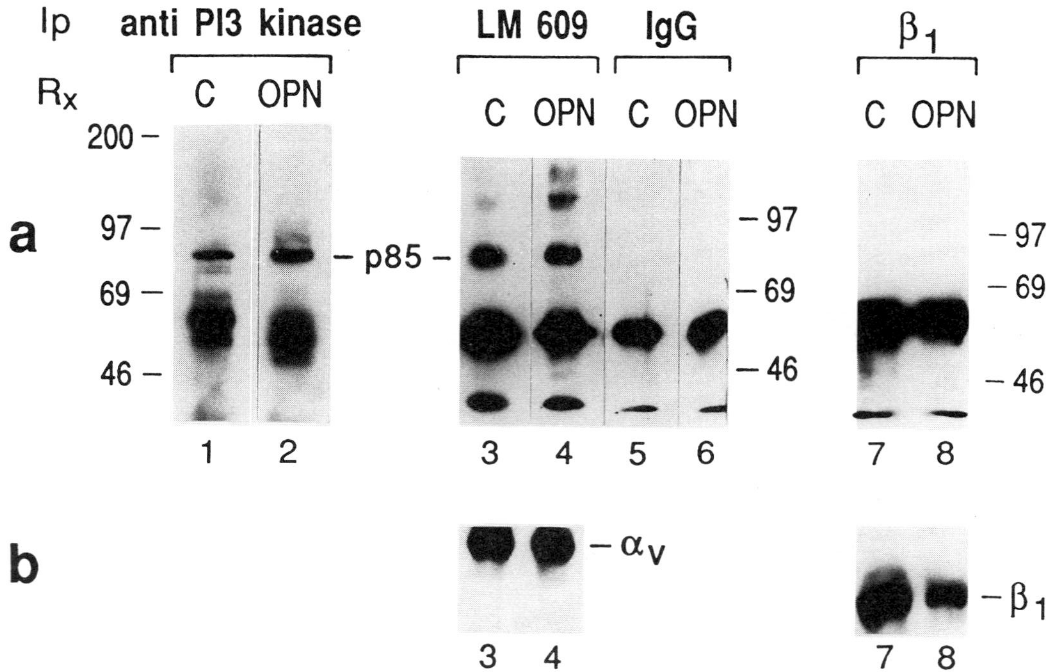

FIGURE 3. Immunoprecipitation of PtdIns 3-OH kinase by anti-$\alpha_v\beta_3$ and β_1 integrin antibodies. Immunoprecipitates (IP) formed by antibodies to PI-3 kinase (positive control), $\alpha_v\beta_3$ (LM609), β_1 integrin subunit, and nonimmune IgG were electrophoresed and immunoblotted with antibodies **(a)** to PI-3 kinase or **(b)** to α_v or β_1 (controls for adequacy of immunoprecipitation). *Lanes 1 and 2* are the immunoblots identifying the 85-kd regulatory component of PI-3 kinase in immunoprecipitates formed by antibodies to PI-3 kinase. *Lanes 3 and 4* are immunoblots of immunoprecipitates formed by anti-$\alpha_v\beta_3$. *Lanes 5 and 6* are immunoblots of immunoprecipitates formed by nonimmune serum. *Lanes 7 and 8* are immunoblots of immunoprecipitates formed by anti-β_1. PI-3 kinase was detected in immunoprecipitates formed by the anti-PI-3 kinase antibody and anti-$\alpha_v\beta_3$, but not anti-β_1. (From Hruska *et al.*[57] Reprinted by permission from *Endocrinology*.)

TABLE 3. Osteopontin Stimulation of PtdIns 3-OH Kinase Activity in Immunoprecipitates of Osteoclast Lysates Formed Using a Monoclonal Antibody to the $\alpha_v\beta_3$ Integrin[a]

	^{32}Pi-PtdInsP$_3$, cpm/mg Lysate Protein			
	1 min	5 min	10 min	30 min
Vehicle	700	770	—	800
OPN 25 μg/ml	1710	1210	1080	710

[a] The reactions of immunoprecipitates with liposomes and γ^{32}P-ATP were performed as described in Methods. The data are means of two experiments performed in duplicate. (From Hruska *et al.*[57] Reprinted by permission from *Endocrinology*.)

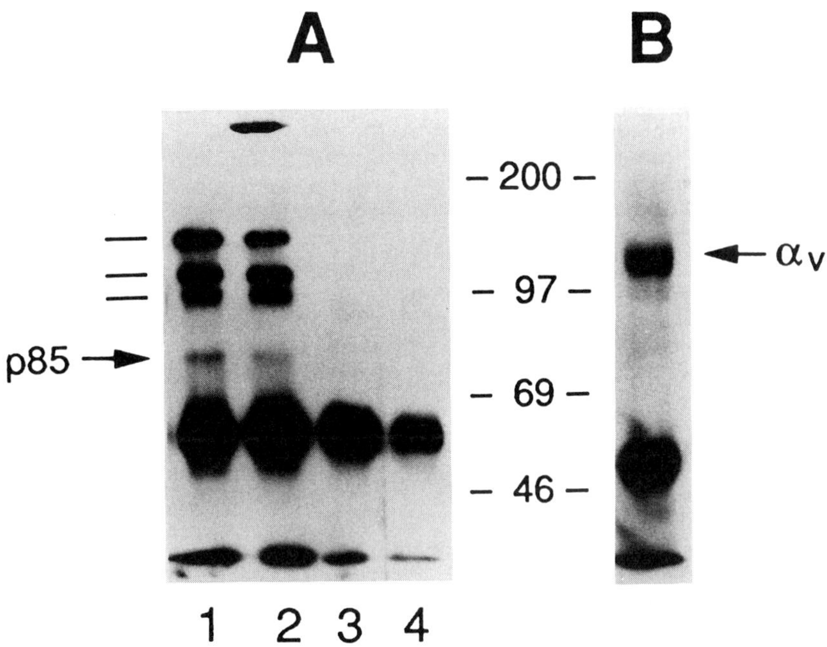

FIGURE 4. Detection of phosphotyrosine-containing proteins by the antibody to $\alpha_v\beta_3$. Immunoprecipitates of lysates from OPN-treated, *lanes 1 and 3,* or control cells, *lanes 2 and 4,* formed by anti-$\alpha_v\beta_3$ were electrophoresed and immunoblotted using an antibody to phosphotyrosine. **(A)** Osteopontin stimulation resulted in an increase in the phosphotyrosine content of an 85-kd protein (*lane 1* compared to *lane 2*). Several proteins of electrophoretic mobility 105, 110, and 125 (marked by *dashes*) were also detected by the anti-phosphotyrosine antibody in the immunoprecipitates formed by anti-$\alpha_v\beta_3$. *Lanes 1 and 2* are the immunoprecipitates from osteopontin-treated and control cells, respectively. *Lanes 3 and 4* are immunoprecipitates formed by a nonimmune IgG. Densitometry (not shown) of the 85-kd band revealed more than a twofold increase in phosphotyrosine in the OPN-treated lane. The treatment conditions were 25 μg/ml of OPN for 1 min. **(B)** An immunoprecipitate formed by anti-$\alpha_v\beta_3$ was immunoblotted using an antibody to α_v. α_v migrated as a 125-kd protein. (From Hruska *et al.*[57] Reprinted by permission from *Endocrinology*.)

the increase in phosphorylation stimulated by OPN, correlated with the PtdIns 3-OH kinase enzyme activity (TABLE 3). As can be seen in FIGURE 4, several phosphotyrosine-containing proteins with electrophoretic mobilities of approximately 105, 110, and 125 kd were present in the lysates formed by anti-$\alpha_v\beta_3$. The identity of these proteins is unknown. Tyrosine phosphorylation of α_v or β_3 has not been reported.[54] While the electrophoretic mobility of α_v is 125–130 kd under reducing conditions, and that of β_3 is 105–110 kd, it is unlikely that they represent the proteins observed in FIGURE 4. In addition, a tyrosine kinase, pp^{125FAK}, has been reported[38–40] to associate with several integrins in focal contacts. We have preliminary evidence that pp^{125FAK} is associated with $\alpha_v\beta_3$ (data not shown). In short, an extensive investigation of the nature of the tyrosine-phosphorylated proteins associated with $\alpha_v\beta_3$ is required which is beyond the scope of the present paper, and it is unknown whether these proteins participate in the association of PtdIns 3-OH kinase with the integrin.

Association of c-src with $\alpha_v\beta_3$

Since PtdIns 3-OH kinase is known to associate in signal-generating complexes with SH2 domain containing members of the c-*src* family,[35,37] and the fact that c-*src*-deficient transgenic mice express osteoclast dysfunction (osteopetrosis),[36] immunoprecipitates of osteoclast lysates obtained using a monoclonal antibody to $\alpha_v\beta_3$ (LM609) were probed with a monoclonal antibody to c-*src*. We found, in preliminary studies, equal amounts of the c-*src* protein associated with $\alpha_v\beta_3$ (data not shown) in immunoprecipitates from osteopontin-treated or control cells. In contrast, the antibody to the β_1 integrin subunit, despite coprecipitating several proteins including β_1 and two α subunits (not shown), failed to immunoprecipitate c-*src*. Thus, c-*src* may be associated with the $\alpha_v\beta_3$ integrin and account for the presence of PtdIns 3-OH kinase, and osteopontin treatment increases the activity of both kinases.

DISCUSSION

Our results demonstrate that osteopontin stimulated incorporation of ^{3}H-myo-inositol and ^{32}Pi into PtdInsP, PtdInsP$_2$ and PtdIns 3,4,5P$_3$ in osteoclast precursors. The stimulation of PtdIns 3,4,5P$_3$ production and the increase in PtdIns 3-OH kinase activity in $\alpha_v\beta_3$ integrin immunoprecipitates demonstrate that OPN binding to $\alpha_v\beta_3$ stimulates production of the 3-phosphorylated polyphosphoinositides which are thought to function as messengers in growth factor-stimulated cells.[30–33] Activation of PtdIns 3-OH kinase by OPN binding to $\alpha_v\beta_3$ is the first report of integrin activation of this enzyme. PtdIns 3-OH kinase is actin associated upon growth factor stimulation,[36] and it may play a role in cytoskeletal based cell shape changes associated with osteoclast organization for bone resorption.[6] The roles of PtdIns 3-phosphorylated phospholipids in osteoclast function require elucidation, but they may serve in the regulation of proteolipids which exhibit high-affinity binding to phosphoinositides. Included in the list of such substances are the plasma membrane Ca-ATPase, as indicated in the Introduction, and gelsolin, an actin filament regulatory protein.

A feature of our previous studies,[8] and those presented herein, is that soluble matrix peptides stimulated immediate cell signal transduction events. Integrin-based cell signaling in other reports[38–40,42] has required integrin clustering in focal adhesions. The ability of osteopontin to signal through $\alpha_v\beta_3$ when stimulated by a soluble protein suggests $\alpha_v\beta_3$ monomer function, and it may be related to the role of osteopontin as an autocrine factor in osteoclast function.[34] In addition, $\alpha_v\beta_3$ has been found in the osteoclast plasma membrane not opposed to bone matrix.[3,43] The function of osteopontin as an autocrine/paracrine factor in the bone is unknown, but recent reports in renal tubular cells indicate it has inhibitory actions on transcription of inducible nitric oxide synthase.[44] Osteoclasts are a rich source of this enzyme (Phil Osdoby, personal communication), and NO decreases bone resorption.[45] Such an action would be compatible with osteopontin's role of stimulating resorption.

While the mechanism of PtdIns 3-OH kinase association with the $\alpha_v\beta_3$ integrin is unknown, it is a heterodimer with *src* homology domains 2 (SH2) and 3 (SH3) in the 85-kd regulatory subunit.[23–25] SH2 domains specifically interact with phosphotyrosine-containing sequences. Formation of oligomeric complexes containing PtdIns 3-OH kinase are most often due to binding of the P85 component SH2 domain to a tyrosine-phosphorylated target including c-*src* itself, but activation of PtdIns 3-OH kinase by other mechanisms has been reported.[46] We present evidence that both PtdIns 3-OH kinase and c-*src* are associated with the osteoclast $\alpha_v\beta_3$ integrin, and that osteopontin binding stimulates the activity of both kinases. Thus, it is possible that association of c-*src* with $\alpha_v\beta_3$ is the means of bringing other enzymes with *src* homology domains into association with the integrin. The mechanism of c-*src* association with $\alpha_v\beta_3$ is also unknown, since the cytoplasmic tails of the heterodimer $\alpha_v\beta_3$ integrin are not known to have *src* homology domains or undergo tyrosine phosphorylation.[55] However, other receptor proteins not expressing *src* homology domains were recently shown to activate *src* kinase activity.[48,49] Thus, the mechanism of *src* activation or association with the integrin requires determination, but the activation of *src* kinase by osteopontin is a potential explanation for the finding of PtdIns 3-OH kinase association and activation by osteopontin.

Osteopontin stimulation of 3-phosphorylated phosphoinositide production and targeted disruption of the c-*src* gene[36] are compatible with a role for both PtdIns 3-OH kinase and c-*src* in integrin-based signal transduction. Osteoclast functional studies indicating blockage of osteopontin-stimulated reduction in cytosolic Ca^{2+} [10] and bone resorption by inhibitors of tyrosine kinase[50] and PtdIns 3-OH kinase[51] lend strong support to this suggestion. The finding of c-*src* associated and activated by the integrin, which is critical for bone resorptive activity of the osteoclast, provides a potential site of *src* function which produces the osteoclast defect in the targeted disruption studies. Cooper and Howell[52] recently speculated that *src* association with c-*fms,* the receptor protein tyrosine kinase for colony stimulating factor 1 (CSF-1), may be the site of *src* function knocked out in the gene disruption studies.[36] c-*src* is widely distributed on osteoclast intracellular membranes at high levels.[53] The component associated with the integrin $\alpha_v\beta_3$ or with c-*fms* is a small fraction of the total that Horne *et al.*[53] did not specifically detect. The critical role of *src* activity in the development of the osteoclast bone resorbing unit was demonstrated by Boyce *et al.*,[50] with the observation that the src-deficient mice[36] fail to develop ruffled borders (the osteoclast bone-resorbing plasma membrane) in their osteoclasts. Thus, our observations, that osteopontin which stimulates cell organization for bone resorption, signals through c-*src* kinase activity, are compatible with the observations of Boyce *et al.* and Soriano *et al.*[36,50]

Recent reports[42,54,55] demonstrated integrin-based stimulation of tyrosine kinases. Specifically, β_1 and β_3 integrin activation of pp^{125FAK} found in focal adhesions of fibroblasts and platelets were reported following adhesion to collagen or fibronectin.[38–40] We also found pp[125FAK] to be associated with $\alpha_v\beta_3$ in osteoclast precursors (data not shown). Since c-*src* has been shown to associate with pp[125FAK] in focal adhesions,[38–40] a complex of signal-generating enzymes similar to the focal adhesion is another possible explanation for the new finding of integrin activation of PtdIns 3-OH kinases by stimulation of $\alpha_v\beta_3$. This possibility is strengthened by the recent report of paxillin, a pp[125FAK]-associated protein, as a *src* homology 3-binding domain protein.[55,56]

ACKNOWLEDGMENTS

We thank Ed Skolnik for the gift of the antibody to PtdIns 3-OH kinase; Pamela Gehron Robey and M. C. Farach-Carson for peptides and osteopontin; Lou Reichardt for the antibody to α_v (CHAV); Richard Hynes for the antibody to β_1; and F. P. Ross and Steve Teitelbaum for valuable discussions. We thank Helen Odle and Debbie Windle for preparation of the manuscript.

REFERENCES

1. Gothlin, G. & J. L. E. Ericson. 1976. The osteoclast: review of ultrastructure, origin, and structure-function relationship. Clin. Orthop. Relat. Res. **120:** 201–231.
2. Holtrop, M. E. & G. J. King. 1977. The ultrastructure of the osteoclast and its functional implications. Clin. Orthop. Relat. Res. **123:** 177–196.
3. Lakkakorpi, P. T., M. A. Horton, M. H. Helfrich, E.-K. Karhukorpi & H. K. Väänänen. 1991. Vitronectin receptor has a role in bone resorption but does not mediate tight sealing zone attachment of osteoclasts to the bone surface. J. Cell Biol. **115:** 1179–1186.
4. Reinholt, F. P., K. Hultenby, A. Oldberg & D. Heinegård. 1990. Osteopontin—a possible anchor of osteoclasts to bone. Proc. Natl. Acad. Sci. USA **87:** 4473–4475.
5. Zambonin-Zallone, A., A. Teti, M. Grano, A. Rubinacci, M. Abbadini, M. Gaboli & P. C. Marchisio. 1989. Immunocytochemical distribution of extracellular matrix receptors in human osteoclasts: a β_1 integrin is colocalized with vinculin and talin in the podosomes of osteoclastoma giant cells. Exp. Cell Res. **182:** 645–652.
6. Horton, M. A., M. L. Taylor, T. R. Arnett & M. H. Helfrich. 1991. Arg-Gly-Asp (RGD) peptides and the anti-vitronectin receptor antibody 23C6 inhibit dentine resorption and cell spreading by osteoclasts. Exp. Cell Res. **195:** 368–375.
7. Ross, F. P., J. Chappel, J. I. Alvarez, D. Sander, W. T. Butler, M. C. Farach-Carson, K. A. Mintz, P. G. Robey, S. L. Teitelbaum & D. A. Cheresh. 1993. Interactions between the bone matrix proteins osteopontin and bone sialoprotein and the osteoclast integrin $\alpha_v\beta_3$ potentiate bone resorption. J. Biol. Chem. **268:** 9901–9907.
8. Miyauchi, A., J. Alvarez, E. M. Greenfield, A. Teti, M. Grano, S. Colucci, A. Zambonin-Zallone, F. P. Ross, S. L. Teitelbaum, D. Cheresh & K. A. Hruska. 1991. Recognition of osteopontin and related peptides by an $\alpha_v\beta_3$ integrin stimulates immediate cell signals in osteoclasts. J. Biol. Chem. **266:** 20369–20374.
9. Teti, A., H. C. Blair, P. Schlesinger, M. Grano, A. Zambonin-Zallone, A. J. Kahn, S. L. Teitelbaum & K. A. Hruska. 1989. Extracellular protons acidify osteoclasts, reduce cytosolic calcium, and promote expression of cell-matrix attachment structures. J. Clin. Invest. **84:** 773–780.
10. Malgaroli, A., J. Meldolesi, A. Zambonin-Zallone & A. Teti. 1989. Control of cytosolic free calcium in rat and chicken osteoclasts. The role of extracellular calcium and calcitonin. J. Biol. Chem. **264:** 14342–14347.
11. Miyauchi, A., K. A. Hruska, E. M. Greenfield, R. Duncan, J. Alvarez, R.

BARATTOLO, S. COLUCCI, A. ZAMBONIN-ZALLONE, S. L. TEITELBAUM & A. TETI. 1990. Osteoclast cytosolic calcium, regulated by voltage-gated calcium channels and extracellular calcium, controls, podosome assembly and bone resorption. J. Cell Biol. **111:** 2543–2552.

12. YAMAKAWA, K., R. DUNCAN & K. A. HRUSKA. 1994. An Arg-Gly-Asp peptide stimulates Ca^{2+} efflux from osteoclast precursors through a novel mechanism. Am. J. Physiol. **266:** F651–F657.
13. FILOTEO, A. G., A. ENYEDI & J. T. PENNISTON. 1992. The lipid-binding peptide from the plasma membrane Ca^{2+} pump binds calmodulin, and the primary calmodulin-binding domain interacts with lipid. J. Biol. Chem. **267:** 11800–11805.
14. ENYEDI, A., M. FLURA, B. SARKADI, G. GARDOS & E. CARAFOLI. 1987. The maximal velocity and the calcium affinity of the red cell calcium pump may be regulated independently. J. Biol. Chem. **262:** 6425–6430.
15. ZVARITCH, E., P. JAMES, T. VORHERR, R. FALCHETTO, N. MODYANOV & E. CARAFOLI. 1990. Mapping of functional domains in the plasma membrane Ca^{2+} pump using trypsin proteolysis. Biochemistry **29:** 8070–8076.
16. VROLIX, M., L. RAEYMAEKERS, F. WUYTACK, F. HOFMANN & R. CASTEELS. 1988. Cyclic GMP-dependent protein kinase stimulates the plasmalemmal Ca^{2+} pump of smooth muscle via phosphorylation of phosphatidylinositol. Biochem. J. **255:** 855–863.
17. ALVAREZ, J. I., S. L. TEITELBAUM, H. C. BLAIR, E. M. GREENFIELD, N. A. ATHANASOU & F. P. ROSS. 1991. Generation of avian cells resembling osteoclasts from mononuclear phagocytes. Endocrinology **128:** 2324–2335.
18. CIVITELLI, R., I. R. REID, S. WESTBROOK, L. V. AVIOLI & K. A. HRUSKA. 1988. PTH elevates inositol polyphosphates and diacylglycerol in a rat osteoblast-like cell line. Am. J. Physiol. **255:** E660–E667.
19. CIVITELLI, R., Y. S. KIM, S. L. GUNSTEN, A. FUJIMORI, M. HUSKEY, L. V. AVIOLI & K. A. HRUSKA. 1990. Nongenomic activation of the calcium message system by vitamin D metabolites in osteoblast-like cells. Endocrinology **127:** 2253–2262.
20. EISEMAN, E. & J. B. BOLEN. 1992. Engagement of the high-affinity IgE receptor activates *src* protein-related tyrosine kinases. Nature **355:** 78–80.
21. HYNES, R. O., E. E. MARCANTONIO, M. A. STEPP, L. A. URRY & G. H. YEE. 1989. Integrin heterodimer and receptor complexity in avian and mammalian cells. J. Cell Biol. **109:** 409–420.
22. CHERESH, D. A. & R. C. SPIRO. 1987. Biosynthetic and functional properties of an Arg-Gly-Asp-directed receptor involved in human melanoma cell attachment to vitronectin, fibrinogen, and von Willebrand factor. J. Biol. Chem. **262:** 17703–17711.
23. SKOLNIK, E. Y., B. MARGOLIS, M. MOHAMMADI, E. LOWENSTEIN, R. FISCHER, A. DREPPS, A. ULLRICH & J. SCHLESSINGER. 1991. Cloning of P13 kinase-associated p85 utilizing a novel method for expression/cloning of target proteins for receptor tyrosine kinases. Cell **65:** 83–90.
24. ESCOBEDO, J. A., S. NAVANKASATTUSAS, W. M. KAVANAUGH, D. MILFAY, V. A. FRIED & L. T. WILLIAMS. 1991. cDNA cloning of a novel 85 kd protein that has SH2 domains and regulates binding of PI-3 kinase to the PDGF β-receptor. Cell **65:** 75–82.
25. OTSU, M., I. HILES, I. GOUT, M. J. FRY, F. RUIZ-LARREA, G. PANAYOTOU, A. THOMPSON, R. DHAND, J. HSUAN, N. TOTTY, A. D. SMITH, S. J. MORGAN, S. A. COURTNEIDGE, P. J. PARKER & M. D. WATERFIELD. 1991. Characterization of two 85 kd proteins that associate with receptor tyrosine kinases, middle-T/pp60$^{c\text{-}src}$ complexes, and PI-3 kinase. Cell **65:** 91–104.
26. BRONSTEIN, I., J. C. VOYTA, O. J. MURPHY, L. BRESNICK & L. J. KRICKA. 1992. Improved chemiluminescent Western blotting procedure. Biotechniques **12:** 748–753.
27. NEUGEBAUER, K. M., C. J. EMMETT, K. A. VENSTROM & L. F. REICHARDT. 1991. Vitronectin and thrombospondin promote retinal neurite outgrowth: developmental regulation and role of integrins. Neuron **6:** 345–358.
28. ŠUŠA, M., M. KEELER & L. VARTICOVSKI. 1992. Platelet-derived growth factor activates membrane-associated phosphatidylinositol 3-kinase and mediates its translocation from the cytosol. J. Biol. Chem. **267:** 22951–22956.

29. VEILLETTE, A., M. A. BOOKMAN, E. M. HORAK & J. B. BOLEN. 1988. The CD4 and CD8 T cell surface antigens are associated with the internal membrane tyrosine-protein kinase $p56^{lck}$. Cell **55:** 301–308.
30. TRAYNOR-KAPLAN, A. E., A. L. HARRIS, B. L. THOMPSON, P. TAYLOR & L. A. SKLAR. 1988. An inositol tetrakisphosphate-containing phospholipid in activated neutrophils. Nature **334:** 353–356.
31. HAWKINS, P. T., T. R. JACKSON & L. R. STEPHENS. 1992. Platelet-derived growth factor stimulates synthesis of PtdIns(3,4,5)P_3 by activating a PtdIns(4,5)P_3 3-OH kinase. Nature **358:** 157–159.
32. AUGER, K. R., L. A. SERUNIAN, S. P. SOLTOFF, P. LIBBY & L. C. CANTLEY. 1989. PDGF-dependent tyrosine phosphorylation stimulates production of novel polyphosphoinositides in intact cells. Cell **57:** 167–175.
33. STEPHENS, L. R., K. T. HUGHES & R. F. IRVINE. 1991. Pathway of phosphatidylinositol(3,4,5)-trisphosphate synthesis in activated neutrophils. Nature **351:** 33–39.
34. IKEDA, T., S. NOMURA, A. YAMAGUCHI, T. SUDA & S. YOSHIKI. 1992. In situ hybridization of bone matrix proteins in undecalcified adult rat bone sections. J. Histochem. Cytochem. **40:** 1079–1088.
35. GUTKIND, J. S., P. M. LACAL & K. C. ROBBINS. 1990. Thrombin-dependent association of phosphatidylinositol-3 kinase with $p60^{c\text{-}src}$ and $p59^{fyn}$ in human platelets. Mol. Cell. Biol. **10:** 3806–3809.
36. SORIANO, P., C. MONTGOMERY, R. GESKE & A. BRADLEY. 1990. Targeted disruption of the c-*src* proton-oncogene leads to osteopetrosis in mice. Cell **64:** 693–702.
37. MITCHELL, C. A., A. B. JEFFERSON, B. E. BEJECK, J. S BRUGGE, T. F. DEUEL & P. W. MAJERUS. 1990. Thrombin-stimulated immunoprecipitation of phosphatidylinositol 3-kinase from human platelets. Proc. Natl. Acad. Sci. USA **87:** 9396–9400.
38. BURRIDGE, K., C. E. TURNER & L. H. ROMER. 1992. Tyrosine phosphorylation of paxillin and pp^{125FAK} accompanies cell adhesion to extracellular matrix: a role in cytoskeletal assembly. J. Cell Biol. **119:** 893–903.
39. LIPFERT, L., B. HAIMOVICH, M. D. SCHALLER, B. S. COBB, J. T. PARSONS & J. S. BRUGGE. 1992. Integrin-dependent phosphorylation and activation of the protein tyrosine kinase pp^{125FAK} in platelets. J. Cell Biol. **119:** 905–912.
40. SCHALLER, M. D., C. A. BORGMAN, B. S. COBB, R. R. VINES, A. B. REYNOLDS & J. T. PARSONS. 1992. pp^{125FAK}, a structurally distinctive protein-tyrosine kinase associated with focal adhesions. Proc. Natl. Acad. Sci. USA **89:** 5192–5196.
41. ZHANG, J., M. J. FRY, M. D. WATERFIELD, S. JAKEN, L. LIAO, J. E. B. FOX & S. E. RITTENHOUSE. 1992. Activated phosphoinositide 3-kinase associates with membrane skeleton in thrombin-exposed platelets. J. Biol. Chem. **267:** 4682–4692.
42. KORNBERG, L. J., H. S. EARP, C. E. TURNER, C. PROCKOP & R. L. JULIANO. 1991. Signal transduction by integrins: increased protein tyrosine phosphorylation caused by clustering of β_1 integrins. Proc. Natl. Acad. Sci. USA **88:** 8392–8396.
43. LAKKAKORPI, P. T., M. H. HELFRICH, M. A. HORTON & H. K. VÄÄNÄNEN. 1993. Spatial organization of microfilaments and vitronectin receptor, $\alpha_v\beta_3$, in osteoclasts. J. Cell Sci. **104:** 663–670.
44. HWANG, S., C. A. LOPEZ, D. E. HECK, C. R. GARDNER, D. L. LASKIN, J. D. LASKIN & D. T. DENHARDT. 1994. Osteopontin inhibits induction of nitric oxide synthase gene expression by inflammatory mediators in mouse kidney epithelial cells. J. Biol. Chem. **269:** 711–715.
45. MACINTYRE, I., M. ZADI, A. S. M. TOWHIDUL ALAM, H. K. DATTA, B. S. MOONGA, P. S. LIDBURY, M. HECKER & J. R. VANE. 1991. Osteoclastic inhibition: an action of nitric oxide not mediated by cyclic GMP. Proc. Natl. Acad. Sci. USA **88:** 2936–2940.
46. VLAHOS, C. J. & W. F. MATTER. 1992. Signal transduction in neutrophil activation. Phosphatidylinositol 3-kinase is stimulated without tyrosine phosphorylation. FEBS Lett. **309:** 242–248.
47. BARTFELD, N. S., E. B. PASQUALE, J. E. GELTOSKY & L. R. LANGUINO. 1993. The $\alpha_v\beta_3$ integrin associates with a 190-kDa protein that is phosphorylated on tyrosine in response to platelet-derived growth factor. J. Biol. Chem. **268:** 17270–17276.
48. SIMONSON, M. S. & W. H. HERMAN. 1993. Protein kinase C and protein tyrosine

kinase activity contribute to mitogenic signaling by endothelin-1. J. Biol. Chem. **268:** 9347–9357.

49. THURSTON, JR., A. W., S. G. RHEE & S. D. SHUKLA. 1993. Role of guanine nucleotide-binding protein and tyrosine kinase in platelet-activating factor activation of phospholipase C in A431 cells: proposal for dual mechanisms. J. Pharmacol. Exp. Ther. **266:** 1106–1112.
50. BOYCE, B. F., T. YONEDA, C. LOWE, P. SORIANO & G. R. MUNDY. 1992. Requirement of $pp60^{c\text{-}src}$ expression for osteoclasts to form ruffled borders and resorb bone in mice. J. Clin. Invest. **90:** 1622–1627.
51. NAKAMURA, I. *et al.* 1994. Wortmannin, an inhibitor of phosphatidylinositol 3-kinase, blocks the formation of ruffled border membrane in osteoclast. J. Bone Miner. Res. **9**(Suppl.). In press.
52. COOPER, J. A. & B. HOWELL. 1993. The when and how of *src* regulation. Cell **73:** 1051–1054.
53. HORNE, W. C., L. NEFF, D. CHATTERJEE, A. LOMRI, J. B. LEVY & R. BARON. 1992. Osteoclasts express high levels of $pp60^{c\text{-}src}$ in association with intracellular membranes. J. Cell Biol. **119:** 1003–1013.
54. KANNER, S. B., A. B. REYNOLDS, R. R. VINES & J. T. PARSONS. 1990. Monoclonal antibodies to individual tyrosine-phosphorylated protein substrates of oncogene-encoded tyrosine kinases. Proc. Natl. Acad. Sci. USA **87:** 3328–3332.
55. GUAN, J. L., J. E. TREVITHICK & R. O. HYNES. 1991. Fibronectin/integrin interaction induces tyrosine phosphorylation of a 120-kDa protein. Cell Regul. **2:** 951–964.
56. WENG, Z., J. A. TAYLOR, C. E. TURNER, J. S. BRUGGE & C. SEIDEL-DUGAN. 1993. Detection of *src* homology 3-binding proteins, including paxillin, in normal and v-*src* transformed Balb/c cells. J. Biol. Chem. **268:** 14956–14963.
57. HRUSKA, K. A., F. ROLNICK, M. HUSKEY, U. ALVAREZ & D. CHERESH. 1995. Engagement of the osteoclast integrin $\alpha_v\beta_3$ by osteopontin stimulates phosphatidylinositol 3-hydroxyl kinase activity. Endocrinology. In press.

Localization and Expression of Osteopontin in Mineralized and Nonmineralized Tissues of the Periodontium[a]

R. L. MacNEIL,[b,d] J. BERRY,[b] J. D'ERRICO,[b]
C. STRAYHORN,[b] AND M. J. SOMERMAN[b,c]

[b]*Department of Periodontics/Prevention and Geriatrics*
School of Dentistry
University of Michigan
Ann Arbor, Michigan 48109-1078

[c]*Department of Pharmacology*
School of Medicine
University of Michigan
Ann Arbor, Michigan 48109-1078

INTRODUCTION

A major consequence of periodontal disease is loss of periodontal support, *i.e.*, destruction of supporting periodontal ligament, cementum and bone and subsequently tooth loss if left untreated. The recognition that periodontal regeneration can be achieved, *i.e.*, formation of new bone, new cementum and supportive periodontal ligament, has resulted in increased attempts to understand the mechanisms and factors promoting regeneration.[1–4] The ultimate goal of these efforts is the development of improved therapies for the treatment of periodontal diseases.

Substantial progress has been made towards understanding the events, factors/proteins and mechanisms regulating periodontal regeneration.[3] These advances, coupled with a better appreciation of variables inherent in the host response (*e.g.*, health status, genetics history, behavioral patterns and composite of oral flora), have resulted in improved strategies and, consequently, better clinical outcomes using regenerative therapies. However, current treatment modalities are not predictable and clearly there is a need for increased studies in many areas in order to develop more reliable and efficacious clinical therapies. In order to further our understanding of both the potential and limitation of periodontal regeneration, we established a list of questions to address, with a specific focus on adhesion molecules.

- Can we achieve regeneration predictably?
- What are the timed and spatial events required?
- Does generation recapitulate development?
- What precursor cells/differentiated cells are required?
- What protein/factors are required?

[a] Source of funding for this study: NIH Grant DE09532.
[d] Corresponding author.

- Is cementum different from bone?
- Is acellular cementum different from cellular cementum?

Our laboratory has focused on determining the role of adhesion molecules during development, maintenance and regeneration of the periodontium with a major emphasis on adhesion molecules associated with cementum. The rationale for our approach is discussed below and depicted in FIGURE 1.

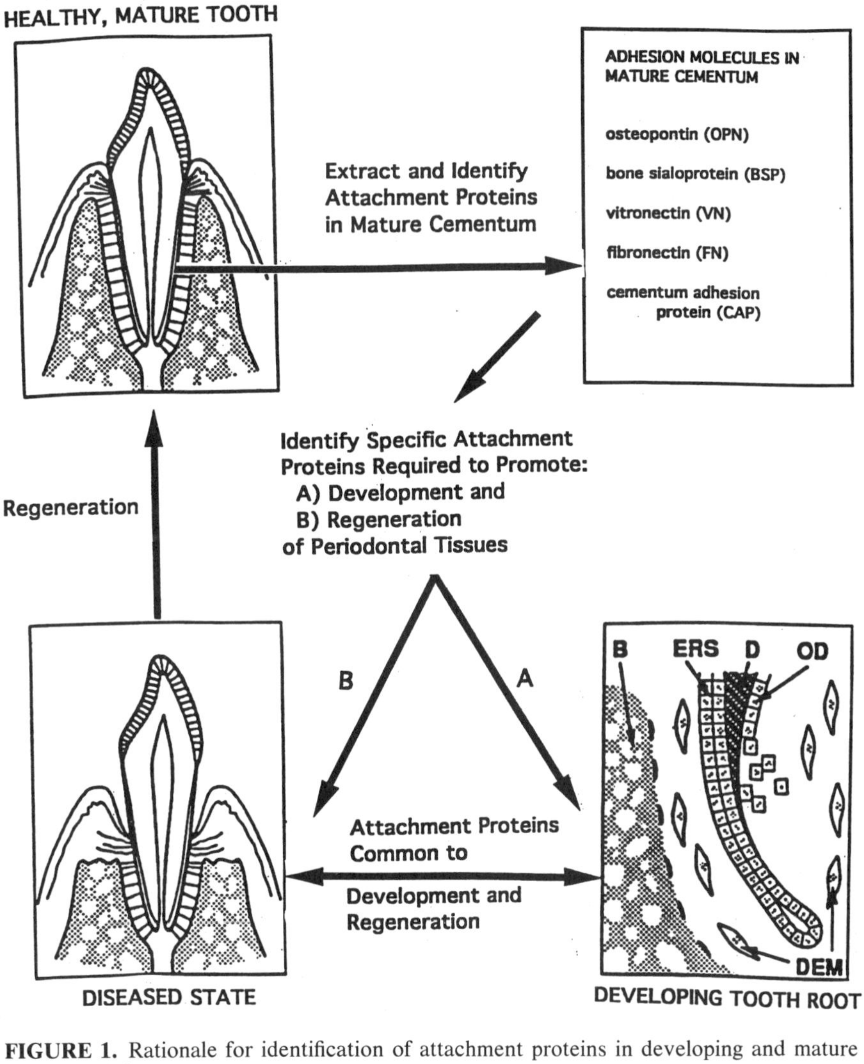

FIGURE 1. Rationale for identification of attachment proteins in developing and mature periodontal tissues.

RATIONALE

Convincing data, generated from numerous studies, exists supporting the possibility that cell attachment proteins and their corresponding cell surface receptors play a major role in regulating cell differentiation and tissue regeneration in various organ and tissue systems, including the periodontium.[5–7] Periodontal disease, *i.e.*, periodontitis, results in the lysis and destruction of the fibrous connective tissue attachment which connects tooth root cementum and surrounding alveolar bone. As illustrated in FIGURE 1, this inflammatory process results in a denuded root surface devoid of its supporting tissue, the periodontal ligament (PDL). The concept of periodontal regeneration, *i.e.*, reformation of a new PDL, is based on the premise that surrounding tissues (the remaining PDL, alveolar bone, and other tissues) possess progenitor stem cell populations wth regenerative wound healing potential. To this end, specific surgical debridement techniques have been developed to allow specific cell groups to migrate to and repopulate the root.[4,8] It is believed that a critical and potentially limiting event in this process is the adhesion of periodontal cells to the root surface. Consequently, clinical attempts have been made to promote cell adhesion to the tooth root using methods such as acid demineralization of the root surface,[9–12] application of adhesive proteins (*e.g.*, fibronectin) to the root,[11] and placement of surgical barriers to protect and stabilize the early root-fibrin clot interface.[4,13] While these approaches have met with some success, the clinical goal of predictable periodontal regeneration has not yet been achieved. A better understanding of the role of adhesion proteins in regenerative wound healing would greatly assist in the development of these therapeutic strategies.

Our investigational approach revolves around two basic concepts: i) the process of cementum formation, *i.e.*, cementogenesis, is the pivotal biologic event in periodontal ligament development; and ii) the molecular processes occurring during developmental and regenerative periodontal tissue formation may be similar (see FIG. 1). In order to determine the role of adhesion molecules in these processes, we first attempted to identify cell attachment proteins in mature cementum. With this information, we then determined the expression of these molecules during cementogenesis. We hypothesize that a precisely timed and spatial expression of specific adhesion molecules is critical to the process of cementum formation and that the identification of the proteins responsible for the promotion of developmental cementogenesis will provide important information regarding regenerative cementum formation.

In order to appreciate the importance of attachment proteins to the process of cementogenesis, a brief review of histological events occurring during cementogenesis is presented below and in FIGURE 1. Depicted in the diagram are the major events occurring in the process of cementogenesis, during molar root formation. In brief, the disruption and migration of cells of the epithelial root sheath (ERS) away from the root surface is timed with the migration and subsequent attachment of cementoblast progenitor cells derived from the ectomesenchyme (*i.e.*, the dental follicle and dental papillae).[14–16] The attachment of these cells is followed by their differentiation to a cementoblastic phenotype; henceforth, cementoblast cells secrete a collagenous matrix onto the root surface which is then mineralized to form acellular cementum in the coronal half of the root and cellular cementum in the apical half of the root. The specific cellular and extracellular factors which determine the formation of acellular and cellular cementum types remain undefined. Cells of the ectomesenchyme, *e.g.*, follicle and papilla, are thought to be capable, when appropriately triggered, of forming not only cementum, but also periodontal ligament (PDL) and bone.[14,17,18] As previously discussed, we proposed that adhesion molecules are critical for differentiation of these cells and play important regulatory roles in the development of the mature periodontium.

Mature Cementum

To demonstrate the importance of adhesion molecules during cell differentiation of periodontal tissues we first identified specific adhesion molecules in mature human and bovine cementum. The adhesion molecules identified in mature cementum by our laboratory are listed in FIGURE 1 and include osteopontin (OPN), bone sialoprotein (BSP), vitronectin, and fibronectin. Also included is a partially characterized cementum adhesion molecule (CAP).[19–21] As a result of these studies, we have determined that cementum may contain unique adhesion molecules, as yet to be fully characterized.[22]

Cementum proteins were extracted from human and bovine teeth by sequential chaotropic extraction using guanidine (Gu) followed by Gu/EDTA as reported previously.[22] The Gu/EDTA extracts of cementum promoted attachment of periodontal cells, *e.g.*, human gingival fibroblasts, human periodontal ligament cells and mouse ectomesenchymal cells. Furthermore, this activity was blocked by arginine-glycine-aspartic acid (RGD) peptides suggesting that RGD-containing ligands must be present in the extract and are responsible, at least in part, for mediating cell attachment. Fractionation of the Gu/EDTA extract by anion exchange, mono Q column (Pharmacia) attached to a Waters 650E advanced purification system, resulted in several bands exhibiting attachment activity for mature human periodontal cells. Further studies on these extracts have shown that both bone sialoprotein and fibronectin are found in high concentrations.[22] Additional known attachment proteins were identified in the Gu/EDTA extract using slot and Western blot analyses. Specific adhesion molecules evaluated included osteopontin, thrombospondin, tenascin, BAG-75 and vitronectin. These studies confirmed the presence of both vitronectin and OPN in cementum. We were unable to identify thrombospondin or BAG-75 in cementum, although these proteins have been identified in bone.[23,24] However, these studies are preliminary, and additional studies are needed to further clarify the spectrum of adhesion molecules present in mature cementum and to compare their role in cementum versus bone.

With the increasing evidence that two particular bone-associated sialoproteins, OPN and BSP, are involved in the mineralization process, we then focused our attention on the spatial and temporal expression of these two molecules during cementogenesis, using immunological and nucleic acid probes. (For comprehensive reviews on OPN see this volume and Denhardt and Gao (1993),[25] Patarca *et al.* (1993);[26] for reviews on BSP see Fisher *et al.* (1992),[27] Sodek *et al.* (1992).[28]) Briefly, OPN, an RGD-containing glycoprotein originally identified in bone is now recognized to be present in many tissues. In bone tissues, OPN has been identified prior to mineralization suggesting a role in regulation of crystal growth.[29,30] In fact, studies are accumulating to suggest OPN may inhibit crystal formation.[30–32] OPN may also serve a role as i) a chemotactic and attachment factor where OPN has been shown to recruit and stimulate macrophages perhaps as a mechanism for protection against infection,[26] ii) a modulator of osteoclast function through Ca^{+2} regulation and control of nitric oxide production,[25] and iii) an adhesion promoter to regulate matrix-cell and matrix-matrix interactions.[33] Its possible role in cementum formation is discussed in the next section. In contrast BSP is selective to mineralized tissues and evidence is accumulating that it may have critical role as a nucleator of mineralization.[29] BSP expression is restricted to bone, cementum, hypertrophic cartilage, with a lesser amount found in dentin, cartilage and decidua.

Cementogenesis

Materials and Methods

First molars obtained from CD-1 mice, gestation day 19–60, were initially fixed with several solutions including: 80% ethanol, 4% paraform aldehyde and Bouin's solution (Polysciences, Inc. Warrington, PA). We found that Bouin's fixation maintained tissue integrity and was suitable for both immunological and nucleic acid probes. For demineralization of tissues, which was necessary at day 27, a solution was used containing 10% acetic acid, 0.85 g NaCl/100 ml and 4% formaldehyde.[34] Tissues were dehydrated and embedded in paraffin. Cut sections, 7 μm, were placed on 3-aminopropyltriethoxy-silane(TES)-coated slides.

To ensure similar results and reproducible findings, several antibodies for OPN and BSP were used, as well as two staining procedures, fluorescin isothiocyanate conjugate (FITC), and streptavidin alkaline phosphatase conjugate (SAP). Antibodies for BSP provided by Dr. Larry Fisher (NIDR/NIH) included antibodies against human, bovine and rat BSP, all made in rabbit, and all gave similar results in mouse tissues.[27] OPN antibodies include 2arN[25] (mouse fusion protein antibody) and control cro-β-galactosidase[35] provided by Dr. David Denhardt, Rutgers University, bovine OPN antibody prepared by our laboratory and human milk OPN antibody[36] provided by Dr. Don Senger, Harvard University. OPN antibodies all gave similar results, with 2arN and milk OPN antibodies having stronger intensities than the bovine antibody. For *in situ* hybridization, probes used included a mouse BSP cDNA[37–39] from Dr. Marion Young, NIDR/NIH and two OPN cDNAs—mouse 2arN cDNA[25,35] and human OP-10 cDNA,[39] with similar results. The *in situ* method used is a modification of Zeller and Rogers (1987).[40] Both sense (control) and antisense probes were linearized and then ^{35}S-UTP labeled, using the Riboprobe® Gemini System (Promega Biotec, Madison, WI).

Results and Discussion

The results of these studies are summarized in TABLE 1. Outlined in TABLE 2 are events occurring at different stages of root development corresponding to the time periods described in TABLE 1. Results from both *in situ* hybridization and immunohistochemical studies indicate that BSP localized to and was expressed by alveolar bone and cementum at time points associated with initial mineralization of these tissues. In contrast, OPN was identified in several tissues.

Note: Days reported for rats and mice by other investigators in the following discussion were converted to our numbering system to facilitate comparison of data among different laboratories.

Day 21

Pre root formation; OPN localized to the follicle region (stratum reticulum, stratum intermedium) with intense staining in the region of the gubernaculum.[41,42] This latter area is considered to be the pathway for tooth eruption. *In situ* hybridiza-

TABLE 1. Timed and Spatial Expression/Localization of BSP and OPN during Tooth Root Development[a]

Day	Cells/ Region	Immunocytochemistry		*In situ* Hybridization		
		BSP	OPN	BSP	OPN	COL1
Day 21	OD	+/−	+/−	−	+/−	+++
	OST	++	+	++	++	++
	SR/SI	−	+++	−	−	−
Day 27	OD	−	++	−	+/−	+++
	OST	++	+/−	++	++	++
	RS	++	−	−	−	−
Day 42	OD	−	+/−	−	+/−	+++
	OST	++	+/−	++	++	+
	RS	++	+	++	++	++
	PDL	−	+++	−	+++	+++

[a] *Abbreviations:* COL1, collagen type I; OD, odontoblasts; OST, osteoblasts; SR/SI, stratum reticulum/stratum intermediate; RS, root surface; PDL, periodontal ligament.

tion revealed intense positive bone cells, in the area of eruption. It is noteworthy that OPN was localized in the pathway of tooth eruption where bone resorptive activity must occur to allow for the eruptive movement of the tooth. OPN is thought to be an adhesion molecule for osteoclasts,[43] although its absence from the sealing zone of osteoclasts as reported by Lakkakorpi *et al.* (1991)[44] has led to some question as to the exact function of OPN in regulation of osteoclast behavior. OPN did not stain (or weakly stained) odontoblasts at this stage, and this is in agreement with studies on day 20–22-rats by Helder *et al.* (1993)[45] and Mark *et al.* (1988).[46] Chen *et al.* (1993) reported OPN staining for odontoblasts and odontoblast processes, but not peritubular dentin in day 35/50 porcine fetuses.[47] Our *in situ* hybridization studies indicated weak/no expression for OPN by odontoblasts with a positive response in osteoblasts (TABLE 1).

At day 21 BSP localization/expression was limited to surrounding alveolar bone. A very weak stain for BSP was noted in the odontoblast region, but *in situ* hybridization indicated that odontoblasts at this developmental stage do not express BSP. In contrast, Chen *et al.* (1992) reported BSP expression in odontoblasts of day-21 rats.[48]

TABLE 2. Tooth Development in Mice[a]

Age	Stage of Tooth Development
10	tooth development/bone initiated
11	dental lamina—odontogenesis
11–25	enamel/dentin crown formation
25–27	disruption of Hertwig's root sheath
27–40	acellular cementum formation, physiologic tooth eruption
40–60	cellular cementum formation, root partially formed
110	fully formed roots

[a] This aging is based on vaginal plug date = day 0. Note: mice are usually born on day 19. By day 35 eruption of molars into the oral cavity can be seen.

Day 27

At this stage, dentinogenesis at the root region is noted, the ERS begins to disintegrate and cementogenesis is initiated.

With initiation of root formation OPN was localized to ERS at the area of newly forming root and also noted in the odontoblast region. *In situ* hybridization studies indicated a weak response for odontoblasts. Surrounding bone cells and matrix were positive for OPN. This is in agreement with Bronckers *et al.,* who report that OPN was localized to cellular, acellular cementum, PDL, bone, and odontoblasts in day-34/59 rat molars.[49]

At this same time point, BSP was expressed and localized to the region of primary deposition of extracellular matrix onto root dentin. Staining was also noted in surrounding bone. Immunostaining in the dental papilla was close to background. In porcine molars, day-35/50 fetuses, Chen *et al.* (1993) reported positive staining for odontoblasts, but weak expression by *in situ* hybridization.[47]

Day 42

At this stage in mice, first molars are in early phases of eruption into the oral cavity and root formation is well advanced but incomplete.

At day 42, strong staining for OPN was noted at the cementum surface and throughout the PDL. Dental papillae did not stain strongly. *In situ* hybridization suggested a similar pattern of expression, strong in cells throughout the PDL and the root surface with weak or no expression noted in odontoblasts.

For the most part these findings are in agreement with other studies. The description here of strong localization/expression of OPN in the PDL contrasted to weak expression in odontoblasts is both supported and challenged by previous studies. OPN has been identified in the PDL by Takano-Yamamoto *et al.* (1994)[50] using approximate day-68 rat molars and *in situ* hybridization (digoxigenin RNA probes) and by Bronckers *et al.* (1994)[49] using rat molars at 34 and 59 days and immunolocalization. Conversely, McKee and Nanci (1994) noted only weak staining for OPN in acellular cementum and cellular cementum and an absence of OPN in the PDL proper using a rat model.[33] Even more uncertainty exists whether OPN is expressed by odontoblasts at latter stages of tooth development. Takano-Yamamoto *et al.* (1994), using *in situ* hybridization, did not observe expression of OPN by rat odontoblasts at day 68;[50] this finding contrasts with immunohistochemical data reported here and by others (Bronckers *et al.,* 1994) which suggests that odontoblasts express OPN.[49] The fact that OPN has been biochemically identified in mature dentin,[51] albeit to a lesser degree than in bone, and immunolocalized in mantle dentin,[33,49] indirectly suggests that odontoblasts synthesize OPN. Hence the inability to confirm the localization/expression of OPN in odontoblasts may be largely due to the limits of the specific techniques employed. Future studies, using a variety of probes, different species and freshly isolated mRNA cell preparations may help to resolve this ongoing controversy.

BSP localization and expression was distinct at this stage of root development. BSP localized to the root surface starting at the CEJ and continued along the root surface to its apical terminus. Surrounding bone also stained for BSP. Cells lining the root surface associated with both cellular and acellular cementum expressed BSP. This expression was restricted to cells lining the root surface and also to cells lining the surrounding bone. Cells within the PDL were not positive for BSP.

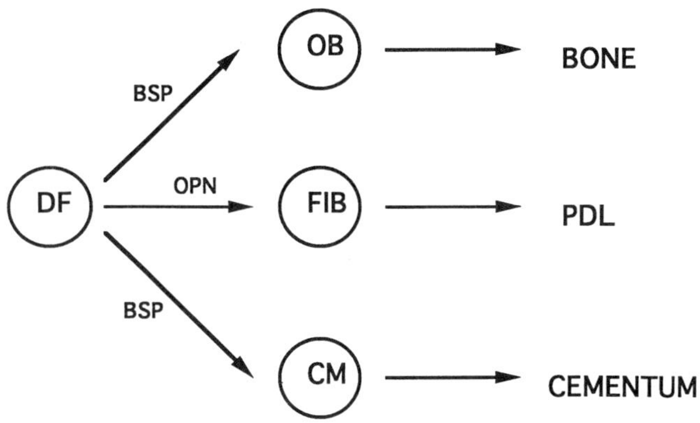

FIGURE 2. Proposed pathways for differentiation of periodontal tissues. *Abbreviations:* DF, dental follicle; OB, osteoblast; FIB, fibroblast; CM, cementoblast; BSP, bone sialoprotein; OPN, osteopontin.

These findings are in agreement with Chen *et al.* (1992)[48] where they reported BSP expression by rat cementoblasts, day 80.

SUMMARY

To summarize results from various studies focusing on determining the expression/localization of BSP and OPN during tooth root development, there is general agreement that OPN is expressed/localized to the root surface during cementogenesis and is also seen throughout the PDL region. The expression/localization of OPN to odontoblasts and its role in dentinogenesis is less apparent. Recent studies directed at establishing odontoblast cell lines should help to resolve this conflict.

Studies on BSP expression during tooth root formation indicate a very precise expression and localization of this molecule during cementogenesis indicating that this molecule may play an important role in the formation of this mineralized tissue.[52] However, as with OPN, the expression of BSP and its role in dentin formation is not clearly defined.

FUTURE DIRECTIONS

The studies outlined here demonstrate clearly that BSP is expressed in a highly specific fashion during root molar development consistent with the spatial and temporal initiation of cementogenesis. Most recently, we reported that RNA isolated from day-27 follicle cells express BSP mRNA. Thus we predict that BSP secretion onto the root surface is required for normal development of cementum and that follicle cells in the local environment, when triggered appropriately, are the cells which express and secrete BSP onto the root surface. Alternatively OPN may allow normal periodontal ligament alignment, by controlling the rate of root mineralization, thus preventing ankylosis (FIG. 2). Furthermore, the timed and spatial expression of specific surface receptors on cells during molar development is required for controlling migration and attachment of cementoblast progenitor cells onto the root surface. We consider these events/factors, critical for initiating cementogenesis and for maintaining a functional periodontium. Future studies are directed at establishing that BSP and follicle cells are responsible for inducing cementum formation and that OPN is important as a modulator of periodontal ligament function.

REFERENCES

1. AUKHIL, I., K. NISHIMURA & W. FERNYHOUGH. 1990. Experimental regeneration of the periodontium. Crit. Rev. Oral Biol. Med. **1:** 101–105.
2. CAFFESSE, R. G. & C. R. QUINONES. 1993. Polypeptide growth factors and attachment proteins in periodontal wound healing and regeneration. J. Periodontol. 2000 **1:** 69–79.
3. MCCULLOCH, C. A. G. 1993. Basic considerations in periodontal wound healing to achieve regeneration. J. Periodontol. 2000 **1:** 16–25.
4. O'NEAL, R., H-L. WANG, R. L. MACNEIL & M. J. SOMERMAN. 1994. Cells and materials involved in guided tissue regeneration. Cur. Opin. Periodontol. **1994:** 141–156.
5. ALBELDA, S. M. & C. A. BUCK. 1990. Integrins and other cell adhesion molecules. FASEB J. **4:** 2868–2880.
6. HYNES, R. D. 1992. Integrins: versatility, modulation, and signaling of cell adhesion. Cell **69:** 11–25.
7. MARIOTTE, A. 1993. The extracellular matrix of the periodontium: dynamic and interactive tissues. J. Periodontol. 2000 **3:** 39–63.
8. BECKER, W., B. BECKER, L. BERG, J. PRICHARD, R. CAFFESSE & E. ROSENBERG. 1988. New attachment after treatment with root isolation procedures: class III and class II furcations and vertical osseous defects. Int. J. Periodontal Res. Dent. **8:** 9–23.
9. CAFFESSE, R. G., C. E. NASJLETI, G. B. ANDERSON, D. E. LOPATIN, B. A. SMITH & E. C. MORRISON. 1991. Periodontal healing following guided tissue regeneration with citric acid and fibronectin application. J. Periodontol. **62:** 21–29.
10. HANDELSMAN, M., M. DAVARPANAH & R. CELLETTI. 1991. Guided tissue regeneration with and without citric acid treatment in vertical osseous defects. Int. J. Periodontal. Res. Dent. **11:** 350–363.
11. KERSTEN, B. G., A. D. CHAMBERLAIN, S. KHORSANDI, U. M. WIKESJO, K. A. SELVIG & R. E. NILVEUS. 1992. Healing of the intrabony periodontal lesion following root conditioning with citric acid and wound closure including an expanded PTFE membrane. J. Periodontol. **63:** 876–882.
12. PARASHIS, A. O. & F. J. MITSIS. 1993. Clinical evaluation of the effect of tetracycline root preparation on guided tissue regeneration in the treatment of class II furcation defects. J. Periodontol. **64:** 133–136.
13. WIKESJO, U. M. E. & R. NILVEUS. 1990. Periodontol repair in dogs: effect of wound stabilization on healing. J. Periodontol. **61:** 719–724.
14. TEN CATE, A. R., C. MILLS & G. SOLOMON. 1971. The development of the periodontium: a transplantation and autoradiographic study. Anat. Rec. **170:** 365–380.
15. CHO, M-I. & P. R. GARANT. 1988. Ultrastructural evidence of directed cell migration during initial cementoblast differentiation. J. Periodontal. Res. **23:** 268–276.
16. THOMAS, H. F. & E. J. KOLLAR. 1988. Tissue interactions in normal murine root development. *In* The Biological Mechanisms of Tooth Eruption and Root Resorption. Z. Davidovitch, Ed. 107–116. EBSCO Media. Birmingham, AL.
17. PALMER, R. M. & A. S. LUMSDEN. 1987. Development of periodontal ligament and alveolar bone in homografted recombinations of enamel organs and papillary, pulpal and follicular mesenchyme in the mouse. Arch. Oral Biol. **32:** 281–289.
18. OSBORN, J. W. & D. G. PRICE. 1988. An autoradiographic study of periodontal development in the mouse. J. Dent. Res. **67:** 455–461.

19. McAllister, B., A. S. Narayanan, Y. Miki & R. C. Page. 1990. Isolation of a fibroblast attachment factor from cementum. J. Periodontal Res. **25:** 99–105.
20. Arzate, H., S. W. Olson, R. C. Page, A. M. Gown & A. S. Narayanan. 1992. Production of a monoclonal antibody to an attachment protein derived from human cementum. FASEB J. **6:** 2990–2995.
21. Pitaru, S., N. Savion, H. Hekmati, S. Olson & S. A. Narayanan. 1993. Molecular and cellular interactions of a cementum attachment protein with periodontal cells and cementum matrix components. J. Periodont. Res. **28:** 560–572.
22. Somerman, M. J., J. J. Sauk, R. A. Foster, K. Dickerson, K. Norris & W. S. Argraves. 1991. Cell attachment activity of cementum: bone sialoprotein II identified in cementum. J. Periodontal. Res. **26:** 10–16.
23. Robey, P. G., M. F. Young, L. W. Fisher & T. D. McClain. 1989. Thrombospondin is an osteoblast-derived component of mineralized extracellular matrix. J. Cell Biol. **108:** 719–727.
24. Sato, M., W. Grasser, S. Harm, C. Fullenkamp & J. P. Gorski. 1992. Bone acidic glycoprotein 75 inhibits resorption activity of isolated rat and chicken osteoclasts. FASEB J. **6:** 2966–2976.
25. Denhardt, D. T. & X. Gao. 1993. Osteopontin: a protein with diverse functions. FASEB J. **7:** 1475–1482.
26. Patarca, R., R. A. Saavedra & H. Cantor. 1993. Molecular and cellular basis of genetic resistance to bacterial infection: the role of the early T-lymphocyte activation-1/osteopontin gene. Crit. Rev. Immunol. **13:** 225–246.
27. Fisher, L. W. 1992. Structure/function studies of the sialoglycoproteins and proteoglycans of bone. It is still the early days. *In* Chemistry and Biology of Mineralized Tissues. H. Slavkin & P. Price, Eds. 177–187. Excerpta Medica. Amsterdam-New York.
28. Sodek, J., J. Chen, S. Kasugai, T. Nagata, Q. Zhang, M. D. McKee & A. Nanci. 1992. Elucidating the functions of bone sialoprotein and osteopontin in bone formation. *In* Chemistry and Biology of Mineralized Tissues. H. Slavkin & P. Price, Eds. 297–307. Excerpta Medica. Amsterdam-New York.
29. Hunter, G. K. & H. A. Goldberg. 1994. Modulation of crystal formation by bone phosphoproteins: role of glutamic acid-rich sequences in the nucleation of hydroxyapatite by bone sialoprotein. Biochem. J. **302.** In press.
30. Hunter, G. K., C. L. Kyle & H. A. Goldberg. 1994b. Modulation of crystal formation by bone phosphoproteins: structural specificity of the osteopontin-mediated inhibition of hydroxyapatite formation. Biochem. J. **300:** 723–728.
31. Shiraga, H., W. Min, W. J. VanDusen, M. D. Clayman, D. Miner, C. H. Terrell, J. R. Sherbotie, J. W. Foreman, C. Przysiecki, E. G. Neilson & J. R. Hoyer. 1992. Inhibition of calcium oxalate crystal growth *in vitro* by uropontin: another member of the aspartic acid-rich protein superfamily. Proc. Natl. Acad. Sci. USA **89:** 426–430.
32. Boskey, A. L., M. Maresca, W. Ullrich, S. B. Doty, W. T. Butler & C. W. Prince. 1993. Osteopontin-hydroxyapatite interactions *in vitro:* inhibition of hydroxyapatite formation and growth in a gelatin-gel. Bone Miner. **22:** 147–159.
33. McKee, M. D. & A. Nanci. 1994. Osteopontin at mineralized tissue interfaces in bone, teeth and osseointegrated implants: ultrastructural distribution and implications for mineralized tissue formation, turnover and repair. Microsc. Res. Tech. In press.
34. D'Souza, R. N., R. P. Happonen, N. M. Ritter & W. T. Butler. 1990. Temporal and spatial patterns of transforming growth factor-β_1 expression in developing rat molars. Arch. Oral. Biol. **35:** 957–965.
35. Craig, A. M., M. Nemir, B. B. Mjkherjee, A. F. Chambers & D. T. Denhardt. 1988. Identification of the major phosphoprotein secreted by many rodent cell lines as 2ar/osteopontin: enhanced expression in H-Ras transformed 3T3 cells. Biochem. Biophys. Res. Commun. **157:** 166–173.
36. Senger, D. R., C. A. Perruzzi, A. Papadopoulos & D. G. Tenen. 1989. Purification of a human milk protein closely similar to tumor-secreted phosphoproteins and osteopontin. Biochem. Biophys. Acta **996:** 43–48.

37. FISHER, L. W., O. W. MCBRIDE, J. D. TERMINE & M. F. YOUNG. 1990. Human bone sialoprotein: deduced protein sequence and chromosomal localization. J. Biol. Chem. **265:** 2347–2351.
38. YOUNG, M. F., J. M. KERR, K. IBARAKI, P. DOMINGUEZ, L. W. FISHER & P. G. ROBEY. 1992a. Unusual DNA structures within the bone sialoprotein and osteonectin genes are potential candidates for transcriptional control. *In* Chemistry and Biology of Mineralized Tissues. H. Slavkin & P. Price, Eds. 243–252. Excerpta Medica. Amsterdam-New York.
39. YOUNG, M. F., J. M. KERR, J. D. TERMINE, U. M. WEWER, M. G. WANG, O. W. MCBRIDE & L. W. FISHER. 1992b. cDNA cloning, mRNA distribution and heterogeneity, chromosomal location, and RFLP analysis of human osteopontin (OPN). Genomics **7:** 491–502.
40. ZELLER, R. & M. ROGER. 1987. *In situ* hybridization and immunocytochemistry. *In* Current Protocols in Molecular Biology. F. M. Ausubel, R. Brent, R. E. Kingston, D. D. Moore, J. G. Seldman, J. A. Smith & K. Struhlk, Eds. 14.0.1–14.6.13. Greene Publishing Associates & John Wiley and Sons. New York.
41. SOMERMAN, M. J., B. SHROFF, W. S. ARGRAVES, G. MORRISON, A. M. CRAIG, D. T. DENHARDT, R. A. FOSTER & J. J. SAUK. 1990. Expression of attachment proteins during cementogenesis. J. Biol. Buccale. **18:** 207–214.
42. SOMERMAN, M. J., B. SHROFF, R. A. FOSTER, W. T. BUTLER & J. J. SAUK. 1992. Mineral-associated adhesion proteins are linked to root formation. Proc. Fin. Dent. Soc. **88**(Suppl. I): 451–462.
43. REINHOLT, F. P., K. HULTENBY, A. OLDBERG & D. HEINEGÅRD. 1990. Osteopontin—a possible anchor of osteoclasts to bone. Proc. Natl. Acad. Sci. USA **87:** 4473–4475.
44. LAKKAKORPI, P. T., M. A. HORTON, M. H. HELFRICH, E-K. KARHUKORPI & H. K. VÄÄNÄNEN. 1991. Vitronectin receptor has a role in bone resorption but does not mediate tight sealing zone attachment of osteoclasts to the bone surface. J. Cell Biol. **115:** 1179–1186.
45. HELDER, M. N., A. L. J. J. BRONCKERS & J. H. M. WOLTGENS. 1993. Dissimilar expression patterns for the extracellular matrix proteins osteopontin (OPN) and collagen type 1 in dental tissues and alveolar bone of the neonatal rat. Matrix **13:** 415–425.
46. MARK, M. P., W. T. BUTLER, C. W. PRINCE, R. D. FINDELMAN & J-V. RUCH. 1988. Developmental expression of 44-kDa bone phosphoprotein (osteopontin) and bone γ-carboxyglutamic acid (Gla)-containing protein (osteocalcin) in calcifying tissues of rat. Differentiation **37:** 123–136.
47. CHEN, J., C. A. G. MCCULLOCH & J. SODEK. 1993. Bone sialoprotein in developing porcine dental tissues: cellular expression and comparison of tissue localization with osteopontin and osteonectin. Arch. Oral Biol. **38:** 241–249.
48. CHEN, J., H. S. SHAPIRO & J. SODEK. 1992. Developmental expression of bone sialoprotein mRNA in rat mineralized connective tissues. J. Bone Miner. Res. **7:** 987–997.
49. BRONCKERS, A. L. J. J., M. C. FARACH-CARSON, E. V. WAVEREN & W. T. BUTLER. 1994. Immunolocalization of osteopontin, osteocalcin, and dentin sialoprotein during dental root formation and early cementogenesis in the rat. J. Bone Miner. Res. **9:** 833–841.
50. TAKANO-YAMAMOTO, T., T. TAKEMURA, Y. KITAMURA & S. NOMURA. 1994. Site-specific expression of mRNAs for osteonectin, osteocalcin, and osteopontin revealed by *in situ* hybridization in rat periodontal ligament during physiological tooth movement. J. Histochem. Cytochem. **42:** 885–896.
51. FUJISAWA, R., W. T. BUTLER, J. L. BRUNN, H. Y. ZHOV & Y. KUBOKI. 1993. Differences in composition of cell-attachment sialoproteins between dentin and bone. J. Dent. Res. **72:** 1222–1226.
52. MACNEIL, R. L., N. SHENG, C. STRAYHORN, L. W. FISHER & M. J. SOMERMAN. 1994. Bone sialoprotein is localized to the root surface during cementogenesis. J. Bone Miner. Res. **9:** 1597–1606.

Osteopontin and the Bone Remodeling Sequence

Colloidal-Gold Immunocytochemistry of an Interfacial Extracellular Matrix Protein[a]

MARC D. McKEE AND ANTONIO NANCI

Faculty of Dentistry
Department of Stomatology
University of Montreal
P.O. Box 6128, Station Centre-Ville
Montreal, QC, Canada H3C 3J7

INTRODUCTION

Within the past decade or so, a variety of extracellular matrix proteins have been isolated and characterized from bone, cartilage, dentin, cementum and enamel. When comparing the organic constituents found within the extracellular milieu of these mineralized versus other nonmineralized tissues, it becomes immediately apparent that in mammalian hard tissues such as bones and teeth, organic components must be designed and conserved by evolution to routinely interface with, or function in the immediate vicinity of, a local environment rich in mineral ions and calcium phosphate salts.[1] Furthermore, although there is continual turnover of extracellular organic matrix in bone as there is for many other hard and soft tissues, this particular phenomenon in bone requires an additional process involving extracellular dissolution of mineral. Moreover, the very nature of the bone remodeling sequence in normal mammalian tissues dictates that following resorption, new bone be deposited onto a rigid substrate of older bone (or calcified cartilage) while simultaneously maintaining the overall integrity and functional status of this weight-bearing tissue.[2] In the context of discussing the participation of extracellular matrix in regulating certain osteogenic processes, the present article is designed to review the ultrastructural distribution and potential role of osteopontin (OPN), one of the major noncollagenous sialoproteins found in this and other collagen-based mineralized tissues.[3,4] Hypotheses are presented in which it is proposed that OPN in laminae limitantes at bone surfaces may first act as a substrate for osteoclast adhesion, and then for initial sealing zone attachment, during osteoclast migration and bone matrix resorption, respectively. During the reversal phase of the remodeling sequence, it is proposed that OPN secreted early in the life cycle of differentiating (pre)osteoblasts accumulates at the resorbed bone surface to form a cement line ('plane'). Although the deposition of this planar arrangement of OPN initially may serve to influence early matrix organization and mineralization, and possibly (pre)osteoblast adhesion at these sites, ultimately it may function in a broader sense as a matrix-matrix/mineral 'biological glue' to

[a] This work was supported by the MRC of Canada and the FRSQ of Quebec.

attach newly-formed bone to older bone in order to maintain overall tissue integrity and biomechanical strength during bone remodeling.

Overview of Cell-Matrix and Matrix-Matrix/Mineral Relationships in Bone

This brief overview is intended to reinforce the notion that cell dynamics and extracellular matrix/mineral interactions in bone are best understood when knowledge of the protein chemistry of a molecule is melded with an appreciation of its ultrastructural distribution. For mineralized tissues in general, electron microscopic techniques have been invaluable in allowing us to understand the details of tissue structure, particularly matrix/mineral relationships. Indeed, modern immunocytochemical approaches used to identify the presence of individual proteins at specific sites in bone have validated earlier histochemical reports describing more generalized associations of organic matrix with mineral.[5] Despite the inherent difficulties in preserving mineral in its native state during routine tissue processing for electron microscopy,[6] substantial progress has been made with regard to understanding the way in which mineral interfaces with matrix by carefully integrating and interpreting a variety of morphological, cytochemical and immunocytochemical data.[5] To this end, initial calcification occurs at discrete sites among the collagen fibrils of the osteoid, with the mineral subsequently becoming more confluent at the mineralization front.[7] Calcification at both these sites is associated with an accretion of numerous, mineral-binding, noncollagenous organic moieties,[8] a major one of which is OPN. The structural relationships between collagen, on the other hand, and the various noncollagenous proteins of bone have yet to be elucidated.

Cell-matrix relationships in bone are of paramount importance in bone cell biology when one considers the highly dynamic nature of bone modeling and remodeling and the observed coupling between bone deposition and resorption.[9–12] Indeed, the whole process of bone formation and turnover appears to be tightly regulated by a multitude of local and systemic cell signalling mechanisms, many of which seem to reside in the way in which bone cells respond to stimuli generated by the substrates with which they come into contact. In bone, whose primary weight-bearing and protective roles are defined by its rigid, collagenous and mineral-impregnated extracellular matrix, exposed bony surfaces are abundant, yet these are populated by a relatively small set of osteogenic cells that includes primarily only four cell types—osteoblasts, osteocytes, bone-lining cells and osteoclasts.[13] Each of these cells interacts in a well-defined way with its subjacent matrix to ensure that the various metabolic processes occurring in bone do not occur haphazardly, but rather proceed in a coordinated, appropriate manner that is capable of responding to the biomechanical demands,[14] and homeostatic activity,[15,16] intrinsic to this tissue. In light of these considerations, the recent identification of OPN as a major, *interfacial* extracellular matrix protein in bone and a variety of other mineralized tissues was an exciting development in this field and led to a number of proposals for the role of this particular protein in hard tissue formation, turnover and repair.[17]

General Considerations for the Role of OPN in Bone

Among the various mineralized tissues, OPN was originally identified as a major secretory product of osteoblasts that accumulates in bone and for the most part associates with the solid mineral phase contained therein.[18–20] OPN is certainly not unique in this regard, as numerous other extracellular matrix proteins are produced by bone cells and include osteocalcin, bone sialoprotein, osteonectin, bone acidic glycoprotein-75, matrix Gla protein, thrombospondin and fibronectin.[21–23] Initial proposals for the function of OPN in bone originated from its biochemical characterization and the identification of several conserved peptide sequences shared among species and known to be present in other well-characterized extracellular matrix molecules with demonstrated function.[24] Based on considerations interpreting the significance of the highly acidic nature of this sialylated phosphoprotein, its polyaspartic acid domain and its Arg-Gly-Asp (RGD) motif, original hypotheses regarding the potential ability of OPN to bind to mineral and to participate in bone cell adhesion[25] remain intact and are currently under intensive further investigation. In fact, similar hypotheses have been extended to other normal and pathological tissues, and OPN is now implicated in a variety of scenarios involving extracellular matrix assembly, calcification and turnover in teeth, urinary calculi, atherosclerotic plaque and some bone diseases.[17] The ability of OPN to bind to mineral and inhibit crystal growth has now been well documented *in vitro*,[26,27] data for which correlates closely with ultrastructural immunocytochemical observations *in vivo* describing the presence of OPN at sites of tissue mineralization.[28–30] Abundant evidence is also available describing the cell adhesion and spreading properties of this molecule for a number of cell types,[25,31–34] thus lending substantial credence to the hypothesis that RGD/integrin-mediated attachment of cells to OPN plays a significant role not only in regulating cell dynamics in mineralized tissues, but in mediating similar processes in soft tissues and tumors as well.[24] Unquestionably, our understanding of these processes remains to be more clearly defined, yet substantial progress has been made in understanding the protein chemistry, molecular biology and function of this protein, and how its precise distribution at various tissue interfaces in bone[17] (and as discussed below) has allowed us to speculate on structure/function relations in this tissue during endochondral and intramembranous ossification.

OPN as an Interfacial Extracellular Matrix Protein in Bone: Terminology, Distinctions and Commonalities of Cement Lines and Laminae Limitantes

Interfaces in bone are numerous and can be broadly classified as being either cell-matrix or matrix-matrix interfaces. As such, the former represents predominantly the interaction of the various bone cell lineages with surfaces of bone matrix, and the latter represents the situation where newer bone is deposited onto older bone (or calcified cartilage in the case of endochondral ossification) to create an interface between two spatially- and temporally-distinct matrices. As visualized by transmission electron microscopy, these interfaces are usually characterized by the presence of either a lamina limitans or a cement line.[17,35] Cement lines are found at matrix-matrix interfaces where bone deposition has been preceded by a resorptive event (also called reversal lines). Embedded within the bone, cement lines (or better yet, 'cement planes') may vary in profile, often being rather irregular and thus termed crenated. Cement lines are also observed at bony sites where there has been a temporary cessation of bone matrix production, not predicated by a resorptive event, and these structures generally have a smoother outline and have thus been considered as being noncrenated (also called resting lines). Laminae limitantes,[36] on the other hand, are generally somewhat thinner structures found at bone surfaces interfacing with cells (*i.e.*, cell-matrix interfaces)—these sites include the periphery of osteocyte lacunae and bone canaliculae, and quiescent bone surfaces covered by bone-lining cells. Ultimately, in the instance where resting lines are created, a preexisting lamina limitans may actually become part of a cement line as new bone is deposited onto this older bone surface after additional cement line components are secreted by newly-arrived preosteoblasts (see below). It should be noted that on occasion, cement lines are observed in continuity with laminae limitantes, and thus the distinction between these two structures is somewhat operational. Osteoclast-mediated resorption events impinge on preexisting cement lines and laminae limitantes within bone, and thus, during the reversal phase of the bone remodeling sequence, newly-formed cement lines can be observed that are continuous with these preexisting structures.[17]

Particularly striking with regard to the distribution of noncollagenous proteins in mineralized tissues in general is the unique accumulation of OPN at various tissue interfaces.[17] These interfaces include not only the distinctive matrix-matrix interfaces traditionally known as cement lines in bone, but also a wide variety of other natural and artificially-created interfaces where newly-formed organic matrix containing OPN is directly apposed to one type of solid substrate or another. Such additional substrates coated with OPN include the spicules of calcified cartilage formed during endochondral ossification, the dentin of the forming root of the tooth during dentinogenesis and cementogenesis in the rat, the peripheral edges of bone, dentin and cementum defects (root resorption) during healing after wounding, particulate bone debris generated by surgical manipulation, bone surfaces rendered nonvital by heat treatment (*e.g.*, laser irradiation or high-speed drilling), cell culture dishes, and ceramic (hydroxyapatite) and metal (titanium) implants used as biomaterials for tissue reconstruction.[17,37] Clearly, each of these scenarios represents an occasion where the primary tissue response is to establish (as during normal tissue modeling), or reestablish (as during tissue remodeling or repair), mineralized tissue integrity and continuity at planar interfaces where natural or experimentally-created matrix, mineral or biomaterial surfaces are discontinuous from surrounding mineralized tissue. In each case, whether involving the modeling or remodeling of mineralized tissues, or the integration of new endogenous or exogenous solid-phase substrates, a series of cellular and extracellular matrix events occurs in which a population of cells of a particular 'blast' lineage assembles at the substrate surface and there produces a 'coating' of OPN to delineate the future mineralized tissue interface. The implications of this process are important in that they require site-directed recruitment, organization and stimulation of mineralized tissue cells and possibly vectorial secretion of noncollagenous protein. Subsequent extracellular matrix assembly and mineralization may indeed be dependent on this initial secretory event, or may occur secondarily and independent of the existence of the surface-modified substrate. Regardless of the precise role of this OPN-containing 'coating' in guiding subsequent extracellular matrix assembly, organization and mineralization, its consistent presence and accumulation at bone interfaces may be defined as a hallmark of the cellular phenotype involved in its production and may represent one of the earliest (if indeed not the initial) secretory products of *de novo* mineralized tissue formation by osteoblasts.[17,29,30,37–39] The possibility that OPN may also be a late or final secretory product of a given cell lineage—to form the cell-matrix interface termed the lamina limitans[13,17]—and a discussion of the potential significance of these early and late events during mineralized tissue formation is given below for osteoblasts.

Immunocytochemical Observations Related to OPN and the Bone Remodeling Sequence

The finding that OPN is concentrated in cement lines and laminae limitantes has lead to exciting new proposals for the role of this protein in osteogenesis and hard tissue development in general. Ultrastructural immunocytochemistry has provided a means by which to examine, with the high resolution necessary to appreciate the details of these fine structures and their relations, the way in which OPN molecules might interact with other elements of the extracellular matrix and with cells. Structure/function considerations of the active domains of OPN[40] are consistent with immunocytochemical observations indicating that this protein is present and potentially capable of acting in multiple ways at bone surfaces,[17] while serving in a similar or different fashion within the bulk of the bone matrix where it accumulates in small, irregular patches of organic material among the collagen fibrils.[28,41] Outlined below is a summary regarding the distribution and potential role of OPN specifically related to the bone remodeling sequence.

The initial osteogenic mechanisms that guide bone modeling, and the subsequent remodeling events that couple bone formation with resorption, undoubtedly rely in part on a series of intricate cell-cell and cell-matrix interactions. At sites where cellular activity is extensive and bone remodeling has taken place, cement lines[42] are a prominent feature of the extracellular matrix and can be observed by both light and electron microscopy. FIGURE 1A presents an overview light micrograph of remodeling, rat alveolar bone and FIGURES 1B–D represent a series of electron micrographs taken after application of immunocytochemistry for OPN using the postembedding protein A-gold technique. These latter images were selected to illustrate only certain successive stages of the bone remodeling sequence and cement line formation, a process depicted schematically in its entirety in FIGURE 2 and presented elsewhere by immunocytochemistry.[17] Initially, quiescent bone surfaces are covered by an OPN-rich lamina limitans and bone-lining cells (FIG. 2A). Upon local activation of the bone remodeling cycle, bone-lining cells at this site disperse and lamellipodia of motile osteoclasts, including podosome-like protrusions, appear directly against the OPN-containing lamina limitans (FIGS. 1B and 2B). As the osteoclast enters the resorptive phase and develops a ruffled border, the lamina limitans is degraded (FIGS. 1C and 2C), and the osteoclast proceeds to resorb into the mineralized bone matrix proper (FIG. 2D). When the osteoclast terminates its resorptive activity, it loses its specialized ruffled and sealing zones and detaches from the bone (FIG. 2E). At this point, osteoblast precursors arrive at the resorption site (FIG. 2E) and differentiate into preosteoblasts, exhibiting intense intracellular labeling for OPN (FIG. 2F). These cells appear against the variable, and generally irregular, outline of the resorbed matrix and 'coat' this substrate with an organic layer rich in OPN to begin the formation of a cement line (FIG. 2F). With time, and with further differentiation of these cells to become typical, collagen-secreting osteoblasts, a collagenous osteoid matrix is deposited onto the cement line. This matrix subsequently mineralizes and incorporates OPN within a network of 'patches' of organic matrix among the collagen fibrils of the mineralized bone compartment, while at the same time leaving behind an OPN-containing cement (reversal) line embedded in the bone (FIGS. 1D and 2G). As one of the final activities of the osteoblast phenotype, and just before it becomes a bone-lining cell, the osteoblast secretes another 'coating' of OPN to form a lamina limitans at what becomes a relatively-smooth, quiescent bone surface, thus completing the bone remodeling cycle (FIG. 2H). If not removed at a later time by resorptive osteoclasts following local activation of the bone remodeling

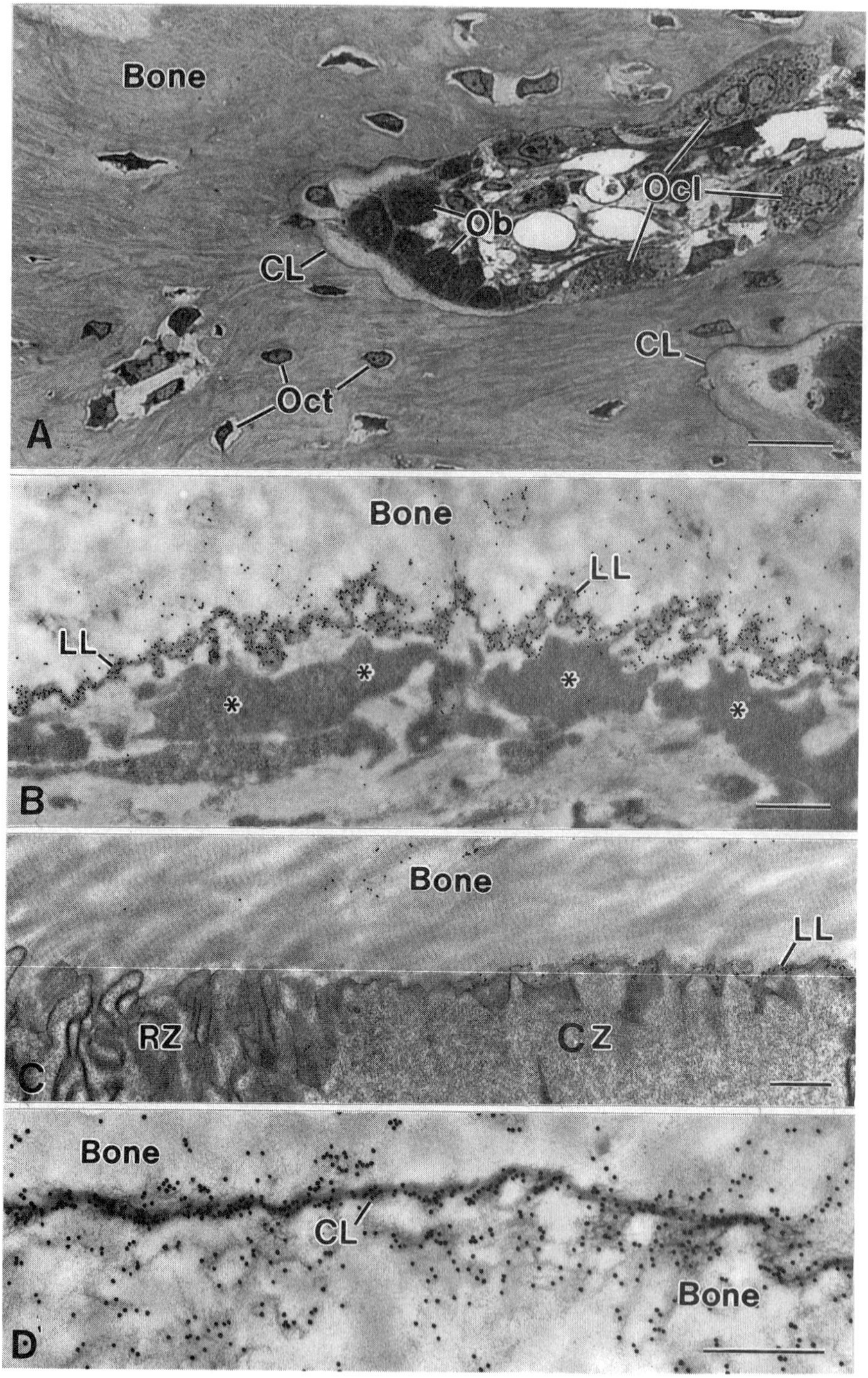
Bone
Ocl
Ob
CL
CL
Oct
A
Bone
LL
LL
*
B
Bone
LL
RZ
CZ
C
Bone
CL
Bone
D

sequence, the lamina limitans has the potential to be incorporated into the bone as a resting (cement) line following osteoblast differentiation and secretory activity at that site.

A final comment regarding OPN at bone surfaces, yet unrelated to the bone remodeling cycle, concerns the observation that OPN is also occasionally found as a major component of laminae limitantes at various other cell-matrix interfaces in bone including those of canaliculae surrounding osteoblast, bone-lining cell and osteocyte cell processes, and the lamina limitans surrounding osteocyte lacunae.[17,30] The cells immediately associated with these particular laminae limitantes appear to be responsible for secreting the OPN contained therein as evidenced by labeling of the Golgi apparatus and secretory granules in these cells, and on some occasions, multiple, concentric laminae limitantes may line a single osteocyte lacunae.[17]

OPN and Bone Remodeling: Discussion and Conclusions

The implications of a concentration of OPN at cell- and matrix-matrix interfaces involved in bone remodeling is intriguing from the point of view of the composition and primary and secondary structure of OPN, and its functional properties.[43] This well-characterized molecule contains a number of amino acid motifs which make it a strong candidate for regulating mineralization and/or mediating local cell dynamics and function.[3,23] Its overall acidic nature and its polyaspartic acid domain appear to be responsible for the strong binding of OPN to mineral.[27,44] Furthermore, the demonstrated ability of OPN to form complexes with other extracellular matrix proteins,[45] and the fact that OPN is a substrate for the protein cross-linking activity of transglutaminase,[46,47] all indicate that OPN may play an important role in extracellular matrix organization and adhesion.

During initial bone deposition onto 'older' bone after the reversal phase of the bone remodeling sequence, and onto spicules of calcified cartilage during endochondral ossification (or apparently against most other substrates, for that matter), committed osteoprogenitor cells[48,49] are recruited to proliferate and differentiate into preosteoblasts which then begin to secrete bone matrix constituents at these sites (see TABLE 1). Although the source of these precursors is not completely resolved, it is clear that soon after they arrive at the surface of a solid-phase substrate, they differentiate into preosteoblasts and are stimulated to

FIGURE 1. Light micrograph (A) and electron micrographs after colloidal-gold immunocytochemistry for OPN (B–D) of remodeling rat bone. **(A)** As part of the remodeling sequence, resorption of bone by osteoclasts (Ocl) is followed by matrix deposition by osteoblasts (Ob). Cement lines (CL) are present at the interface between the newer bone and the older bone. Oct; osteocyte. **(B)** At a potential resorption site, ultrastructural immunocytochemistry for OPN on thin sections of decalcified rat bone initially reveals the presence of thin lamellipodia (*asterisks*), originating from motile osteoclasts, against the OPN-labeled lamina limitans (LL) normally present at bone surfaces. **(C)** Following further differentiation of osteoclasts to form ruffled (RZ) and clear (CZ) zones closely apposed to the OPN-containing bone surface, bone resorption begins and the lamina limitans (LL) is initially degraded under the ruffled border of the osteoclast. **(D)** After resorption of a given volume of bone proper by osteoclasts, these cells detach, and osteoblast lineage cells arrive and produce a 'coating' of OPN to form the cement line at the interface between the older bone and newer bone. *Bars* equal 50 μm (A) and 0.5 μm (B–D).

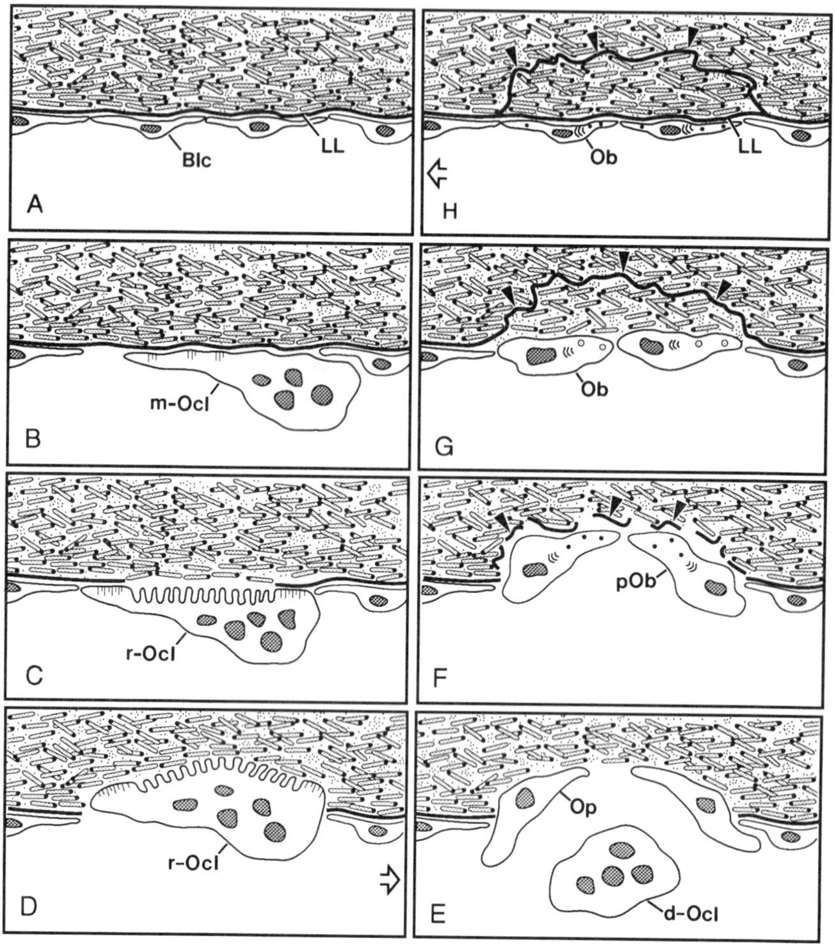

FIGURE 2. Schematic diagram illustrating various cellular and extracellular matrix events related to the secretion and distribution of OPN in cement lines and laminae limitantes during the bone remodeling cycle (see text for complete description). Blc, bone-lining cell; m-Ocl, motile osteoclast; r-Ocl, resorptive osteoclast; d-Ocl, detached osteoclast; Op, osteoprogenitor cell; pOb, preosteoblast; Ob, osteoblast; LL, lamina limitans; *arrowheads,* cement line; *cylindrical rods,* collagen fibrils in bone; *stippling,* OPN in bone matrix proper; *clusters of fine lines,* filaments of osteoclast podosomes and clear zones.

produce nascent bone matrix components such as OPN.[17] Indeed, the frequency of intracellular labeling for OPN observed in this 'young' cell type, and the observation by transmission electron microscopy of the simultaneous appearance of the OPN coating (cement line) at the resorbed bone surface, indicate that predominantly *local* cellular secretory activity (versus plasma-derived accretion of OPN) is required for establishing this new interface. Ultimately, knowing the importance of the exact composition and physical properties of this interface cannot be overemphasized. Indeed, existing and new hypotheses concerning the molecular mechanisms involved in maintaining overall bone integrity, efficacious bone repair and the 'bonding' of biomaterials with bone must take into account the fact that the network of collagen fibrils and other components residing on one side of the interface are not in direct contact with those fibrils (or biomaterials) on the other side. Thus, barring that spatiotemporally distinct packets of bone (basic structural units, BSUs[11,50]) adhere to each other purely by means of mechanical interlock, it is certainly reasonable to predict that the molecular interactions of OPN in cement lines provide a matrix-matrix/mineral adhesion ('bonding') mechanism for maintaining the biomechanical integrity of bone during remodeling.[17] In this context, a contiguous, coherent, macromolecular assembly including OPN as a major component may serve not only in the capacity just described, but initially, also as a means to retain those very cells involved in its production to the substrate for the subsequent secretion and assembly of bone matrix proper during the formative phase of the bone remodeling sequence. Indeed, the acidic nature of OPN, together with its constituent polyaspartic acid domain, phosphoamino acid residues, and cell (integrin)-binding RGD tripeptide sequence, are certainly consistent with these concepts. Here, it should be emphasized that if indeed involved in adhering two similar or dissimilar matrices or materials together, then it is the combination of the organic components present at this junction, together with their biochemical interactions with the contacting materials on either side, that will determine the nature and/or bonding strength of a particular biological interface. Alternatively, it may be possible that OPN at these interfaces acts to influence early mineralization events and/or participates in the assembly and organization (orientation) of both collagenous and noncollagenous components of the bone matrix proper that subsequently accumulates at these sites. Regarding matrix and mineral organization in, and properties of, cement lines in bone, these structures may participate in controlling microscopic fatigue damage and prevent microcrack propagation across neighboring osteons.[51]

TABLE 1. Stages of the Remodeling Sequence[a]

Stage	Events
I. Activation	a. Recruitment of precursor cells (osteoclast lineage)
	b. Cell fusion and differentiation to form osteoclasts
	c. Migration and attachment of osteoclasts to bone surfaces
II. Resorption	a. Initiation of resorption by osteoclasts
	b. Progression of resorption by osteoclasts
	c. Termination of resorption by osteoclasts
III. Reversal	a. Detachment of osteoclasts from bone surface
	b. Recruitment of precursor cells (osteoblast lineage)
	c. Proliferation, migration and differentiation of osteoprogenitor cells to form preosteoblasts
IV. Formation	a. Initiation of matrix production by preosteoblasts (cement line-OPN)
	b. Secretion of collagenous and noncollagenous matrix by osteoblasts
	c. Mineralization of matrix
	d. Termination of matrix secretion by osteoblasts (lamina limitans-OPN)
	e. Maturation of mineral and matrix

[a] Adapted from Martin and Burr.[62]

The presence of the cell adhesion RGD peptide in OPN points to a role for this protein in promoting cell attachment and spreading. Although clearly not involved in providing biomechanical support and stability to bones, the OPN-containing lamina limitans present at bone surfaces subjacent to bone-lining cells may serve to accurately position and tightly adhere these cells to the adjacent extracellular matrix via an integrin/RGD-mediated attachment mechanism, while at the same time possibly regulating mineral dynamics (*i.e.*, crystal growth and dissolution) that are operative in such close proximity to the cell.[17] Although not yet substantiated, another attractive hypothesis is that the OPN in this lamina limitans may act to 'prime' the surface for future osteoclastic activity at these sites. Indeed, our immunocytochemical data are consistent with the notion that the RGD sequence of OPN exposed at bone surfaces may be important initially for integrin-mediated podosome attachment[52] of motile osteoclasts to the extracellular matrix of bone, and then subsequently for initial sealing zone fixation to the bone surface.[29,53]

Despite the recent advances in characterizing cement line composition, more complete elucidation of their role awaits a better understanding of the biochemical interactions between matrix molecules (including proteoglycans and lipids), mineral and the alloplasts currently used in medicine and dentistry. Furthermore, despite a tendency to attribute specific functions to particular domains within a molecule, it may be useful to emphasize that it is probably the coordinate activities and interactions between molecules that elicit particular cellular responses or extracellular matrix events at a given site within bone. Surely other, as yet unidentified, interactions must exist among the molecules in bone, and it is clearly the nature of these interactions that allow osteogenic processes to proceed as they do.

In conclusion, available evidence from a variety of reports investigating different aspects of mineralized tissue formation, turnover and repair is predominantly consistent with the notions of OPN functioning as an inhibitor of mineralization and/or as a mediator of cell-matrix and matrix-matrix/mineral adhesion at tissue interfaces. More specifically, its function in the cement lines of bone may be to act as a biological adhesion promotor at the interface between older bone and newer bone, as created during bone remodeling. The coating of OPN comprising laminae limitantes at various bone interfaces might be considered as a means by which a mineralized tissue compartment is temporarily 'sealed off' and/or 'primed' for immediate or future cellular events, one of which presumably involves osteoclast integrin ($\alpha_v\beta_3$) binding to the RGD motif of OPN.[54–61]

SUMMARY

Relative to other noncollagenous, extracellular matrix proteins in mineralized tissues, colloidal-gold immunocytochemistry has demonstrated that the ultrastructural distribution of osteopontin (OPN) is unique in that this protein preferentially accumulates at mineralized tissue interfaces. In bone, this protein is present as a major component of cell-matrix and matrix-matrix interfacial structures called cement lines and laminae limitantes. In the present article, the implications of this distinct tissue distribution are discussed in terms of the bone remodeling sequence, and a detailed account of the secretion, accumulation and potential role of OPN is presented and related to current theory on the cellular and extracellular matrix events associated with basic multicellular unit (BMU)-based bone remodeling. In this context, a proposal is made describing the production of this protein as one of the earliest, and latest, secretory activities of the osteoblastic lineage, and that this activity manifests itself morphologically as a cement line ('plane') and a lamina limitans, respectively, at bone matrix interfaces. When integrated with other, known functional characteristics of this protein, the present morphological and compositional data indicate that OPN in cement lines and laminae limitantes may participate in initial and late extracellular matrix organization and mineralization, matrix-matrix/mineral adhesion and/or cell adhesion at bone interfaces.

ACKNOWLEDGMENTS

The authors are grateful to S. Boudrias, M. Fortin, I. Turgeon and S. Zalzal for participating in tissue processing and immunocytochemical aspects of this work, J. Krishnamoorthy and G. Lambert for preparing photographic reproductions and D. Clément for the artwork. We also thank Drs. W. T. Butler and M. C. Farach-Carson (University of Texas at Houston) for their generous donation of anti-OPN antibody.

REFERENCES

1. Weiner, S. 1986. CRC Crit. Rev. Biochem. **20:** 365–408.
2. Frost, H. M. 1990. Anat. Rec. **226:** 414–422.
3. Butler, W. T. 1989. Connect. Tissue Res. **23:** 123–136.
4. Butler, W. T. 1991. J. Biol. Buccale. **19:** 83–89.
5. McKee, M. D. & A. Nanci. 1994. Post-embedding colloidal-gold immunocytochemistry of non-collagenous extracellular matrix proteins in mineralized tissues. Microsc. Res. Tech. In press.
6. Landis, W. J., M. C. Paine & M. J. Glimcher. 1977. J. Ultrastruct. Res. **59:** 1–30.
7. Christoffersen, J. & W. J. Landis. 1991. Anat. Rec. **230:** 435–450.
8. Grynpas, M. D., J. H. Tupy & J. Sodek. 1994. Bone **15:** 505–513.
9. Baylink, D. J., R. D. Finkelman & S. Mohan. 1993. J. Bone Miner. Res. **8**(Suppl. 2): S565–S572.
10. Chow, J. W. M., S. Badve & T. J. Chambers. 1993. Bone **14:** 355–360.
11. Parfitt, A. M. 1990. *In* Bone: The Osteoblast and Osteocyte. B. K. Hall, Ed. Vol. 1: 351–429. The Telford Press. Caldwell.
12. Puzas, J. E. & M. Ishibe. 1992. *In* Biology and Physiology of the Osteoclast. B. R. Rifkin & C. V. Gay, Eds. 337–356. CRC Press. Boca Raton.
13. McKee, M. D. & A. Nanci. 1993. Cells Mater. **3:** 219–243.
14. Lanyon, L. E. 1992. J. Bone Miner. Res. **7**(Suppl. 2): S369–S375.
15. De Bernard, B. 1992. *In* Bone: Bone Metabolism and Mineralization. B. K. Hall, Ed. Vol. 4: 73–98. CRC Press. Boca Raton.
16. Parfitt, A. M. 1987. Bone **8:** 51–58.
17. McKee, M. D. & A. Nanci. 1994. Osteopontin at mineralized tissue interfaces in bone, teeth and osseointegrated implants: ultrastructural distribution and implications for mineralized tissue formation, turnover and repair. Microsc. Res. Tech. In press.
18. Franzén, A. & D. Heinegård. 1985. Biochem. J. **232:** 715–724.
19. Prince, C. W., T. Oosawa, W. T. Butler, T. Tomana, A. S. Bhown, M. Bhown & R. E. Schrohenloher. 1987. J. Biol. Chem. **262:** 2900–2907.
20. Fisher, L. W., G. R. Hawkins, N. Tuross & J. D. Termine. 1987. J. Biol. Chem. **262:** 9702–9708.
21. Young, M. F., K. Ibaraki, J. M. Kerr & A.-M. Heegard. 1993. *In* Cellular and Molecular Biology of Bone. M. Noda, Ed. 191–234. Academic Press. San Diego.
22. Gehron Robey, P., N. S. Fedarko, T. E. Hefferan, P. Bianco, U. K. Vetter, W. Grzesik, A. Friedenstein, G. van der Pluijm, K. P. Mintz, M. F. Young,

J. M. Kerr, K. Ibaraki & A.-M. Heegaard. 1993. J. Bone Miner. Res. **8**(Suppl. 2): S483–S487.
23. Gorski, J. P. 1992. Calcif. Tissue Int. **50:** 391–396.
24. Denhardt, D. T. & X. Guo. 1993. FASEB J. **7:** 1475–1482.
25. Oldberg, Å., A. Franzén & D. Heinegård. 1986. Proc. Natl. Acad. Sci. USA **83:** 8819–8823.
26. Boskey, A. L., M. Maresca, W. Ullrich, S. B. Doty, W. T. Butler & C. W. Prince. 1993. Bone Miner. **22:** 147–159.
27. Hunter, G. K., C. L. Kyle & H. A. Goldberg. 1994. Biochem. J. **300:** 723–728.
28. McKee, M. D., A. Nanci, W. J. Landis, Y. Gotoh, L. C. Gerstenfeld & M. J. Glimcher. 1990. Anat. Rec. **228:** 77–92.
29. McKee, M. D., M. J. Glimcher & A. Nanci. 1992. Anat. Rec. **234:** 479–492.
30. McKee, M. D., M. C. Farach-Carson, W. T. Butler, P. V. Hauschka & A. Nanci. 1993. J. Bone Miner. Res. **8:** 485–496.
31. Somerman, M. J., C. W. Prince, J. J. Sauk, R. A. Foster & W. T. Butler. 1987. J. Bone Miner. Res. **2:** 259–265.
32. Somerman, M. J., L. W. Fisher, R. A. Foster & J. J. Sauk. 1988. Calcif. Tissue Int. **43:** 50–53.
33. Somerman, M. J., C. W. Prince, W. T. Butler, R. A. Foster, J. M. Moehring & J. J. Sauk. 1989. Matrix **9:** 49–54.
34. Gotoh, Y., M. D. Pierschbacher, J. J. Grzesiak, L. C. Gerstenfeld & M. J. Glimcher. 1990. Biochem. Biophys. Res. Commun. **173:** 471–479.
35. Sauren, Y. M. H. T., R. H. P. Mieremet, C. G. Groot & J. P. Scherft. 1992. Anat. Rec. **232:** 36–44.
36. Scherft, J. P. 1972. J. Ultrastruct. Res. **38:** 318–331.
37. Nanci, A., G. F. McCarthy, C. Clokie, H. Warshawsky, S. Zalzal & M. D. McKee. 1994. Cells Mater. **4:** 1–30.
38. Davies, J. E., P. Ottensmeyer, X. Shen, M. Hashimoto & S. A. F. Peel. 1991. *In* The Bone-Biomaterial Interface. J. E. Davies, Ed. 214–228. University of Toronto Press. Toronto.
39. Shen, X., E. Roberts, S. A. F. Peel & J. E. Davies. 1993. Cells Mater. **3:** 257–272.
40. Sodek, J., J. Chen, S. Kasugai, T. Nagata, Q. Zhang, M. D. McKee & A. Nanci. 1992. *In* Chemistry and Biology of Mineralized Tissues. H. Slavkin & P. Price, Eds. 297–306. Elsevier Science Publishers. Amsterdam.
41. McKee, M. D., A. Nanci, W. J. Landis, Y. Gotoh, L. C. Gerstenfeld & M. J. Glimcher. 1991. J. Bone Miner. Res. **6:** 937–945.
42. De Ricqles, A., F. J. Meunier, J. Castanet & H. Francillon-Vieillot. 1991. *In* Bone: Bone Matrix and Bone Specific Products. B. K. Hall, Ed. Vol. 3: 1–78. CRC Press. Boca Raton.
43. Prince, C. W. 1989. Connect. Tissue Res. **21:** 15–20.
44. Hunter, G. K. & H. A. Goldberg. 1994. Biochem. J. **302:** 175–179.
45. Ritter, N. M., M. C. Farach-Carson & W. T. Butler. 1992. J. Bone Miner. Res. **7:** 877–885.
46. Prince, C. W., D. Dickie & C. L. Krumdieck. 1991. Biochem. Biophys. Res. Commun. **177:** 1205–1210.
47. Beninati, S., D. R. Senger, E. Cordella-Miele, A. B. Mukherjee, I. Chackalaparampil, V. Shanmugam, K. Singh & B. B. Mukherjee. 1994. J. Biochem. **115:** 675–682.
48. Aubin, J. E., K. Turksen & J. N. M. Heersche. 1993. *In* Cellular and Molecular Biology of Bone. M. Noda, Ed. 1–45. Academic Press. San Diego.
49. Tenenbaum, H. C. 1990. *In* Bone: The Osteoblast and Osteocyte. B. K. Hall, Ed. Vol. 1: 41–69. The Telford Press. Caldwell.
50. Frost, H. M. 1990. Anat. Rec. **226:** 403–413.
51. Burr, D. B., M. B. Schaffler & R. G. Frederickson. 1988. J. Biomech. **21:** 939–945.
52. Aubin, J. E. 1992. J. Bone Miner. Res. **7:** 365–368.
53. Reinholt, F. P., K. Hultenby, Å. Oldberg & D. Heinegård. 1990. Proc. Natl. Acad. Sci. USA **87:** 4473–4475.

54. Helfrich, M. H., S. A. Nesbitt, E. L. Dorey & M. A. Horton. 1992. J. Bone Miner. Res. **7:** 335–343.
55. Shinar, D. M., A. Schmidt, D. Halperin, G. A. Rodan & M. Weinreb. 1993. J. Bone Miner. Res. **8:** 403–414.
56. Chuntharapai, A., S. Bodary, M. Horton & K. Jin Kim. 1993. Exp. Cell Res. **205:** 345–352.
57. Ross, F. P., J. Chappel, J. I. Alvarez, D. Sander, W. T. Butler, M. C. Farach-Carson, K. A. Mintz, P. G. Robey, S. L. Teitelbaum & D. A. Cheresh. 1993. J. Biol. Chem. **268:** 9901–9907.
58. Nesbitt, S., A. Nesbit, M. Helfrich & M. Horton. 1993. J. Biol. Chem. **268:** 16737–16745.
59. Grano, M., P. Zigrino, S. Colucci, G. Zambonin, L. Trusolino, M. Serra, N. Baldini, A. Teti, P. C. Marchisio & A. Z. Zallone. 1994. Exp. Cell Res. **212:** 209–218.
60. van der Pluijm, G., H. Mouthaan, C. Baas, H. de Groot, S. Papapoulos & C. Löwik. 1994. J. Bone Miner. Res. **9:** 1021–1028.
61. Mimura, H., X. Cao, F. P. Ross, M. Chiba & S. L. Teitelbaum. 1994. Endocrinology **134:** 1061–1066.
62. Burr, D. B. & R. B. Martin. 1989. Function and Adaptation of Compact Bone. Raven Press. New York.

Interaction of Osteopontin with Osteoclast Integrins[a]

MICHAEL A. HORTON,[b] M. ANDREW NESBIT,[c] AND MIEP H. HELFRICH[d]

[b]*Department of Medicine*
University College London
Thorn Institute
The Middlesex Hospital
London W1N 8AA, England

[c]*Neuromuscular Unit*
Department of Paediatrics and Neonatal Medicine
Hammersmith Hospital
London, W12 0NN, England

[d]*Department of Medicine and Therapeutics*
Aberdeen University Medical School
Aberdeen, AB2 2ZD, Scotland

INTRODUCTION

Osteoclasts are the cells responsible for the breakdown of the extracellular matrix of bone.[1] Resorption of bone involves a series of developmental and regulatory steps which include: the proliferation and homing to bone of hemopoietic progenitors; the differentiation and fusion of osteoclast precursors; and migration of osteoclasts to the area of bone to be remodelled where they attach to the bone surface and polarize to create a functional end cell. During resorption, the tight sealing zone (or 'clear zone') forms as a membrane domain closely opposed to the bone matrix.[2,3] This forms a diffusion barrier and so permits the localized accumulation of high concentrations of protons and proteases secreted via the ruffled border into an extracellular resorption compartment underneath the cell.[1] Many of these steps could involve adhesion events between developing or mature osteoclasts with components of the extracellular matrix. Such interactions are mediated in part by integrin cell adhesion molecules[4] and encompass events not only of a purely adhesive nature but also signal transduction.[5,6]

In this paper we provide evidence for the synthesis of osteopontin, but not bone sialoprotein, by rat and human osteoclasts, and for the mechanism of interaction of rat osteoclasts with osteopontin compared with the other major sialoprotein of skeletal tissue, bone sialoprotein. The possible functional consequences of integrin-mediated recognition of sialoproteins in bone is discussed.

MATERIALS AND METHODS

Assays of osteoclast adhesion to matrix proteins osteopontin (rat, human) and bone sialoprotein (human) (gifts of L. W. Fisher, NIDR, MA, USA and W. T.

[a] We acknowledge the financial support of the Imperial Cancer Research Fund.

Butler, University of Texas at Houston, TX, USA) compared to fetal calf serum were carried out as previously detailed.[7] Likewise, the effects of anti-receptor antibodies to rat β_1 (gift of D. Gullberg[8]), α_v (gift of K. J. Kim, Genentech, Inc., CA, USA) and β_3 (monoclonal F11[9]) integrins and inhibitory cyclic RGD (50 μM, G4120, gift of T. Gadek, Genentech, Inc., CA, USA) and fibrinogen peptides[10] (500 μM, H12 [His-His-Leu-Gly-Gly-Ala-Lys-Gln-Ala-Gly-Asp-Val; HHLGGAK-QAGDV], KQAGD and AGDV) (Peninsula Labs., Belmont, CA, USA) upon osteoclast adhesion were performed as described.[11]

The expression of β_1 and β_3 integrins in osteoclasts was analyzed by indirect immunofluorescence,[9] and the distribution of F-actin by staining with rhodamine-phalloidin (Molecular Probes, Eugene, OR, USA).[12–14] The presence of osteopontin and bone sialoprotein protein in osteoclasts was assessed by indirect immunofluorescence in monensin-treated[15] (1 μM, 3 hr, 37°C) rat limb bone osteoclasts adhering to fetal calf serum (using monoclonal antibodies to rat osteopontin and bone sialoprotein MPIIIB10 and WV1D1, respectively, both from M. Solursh, University of Iowa, IA, USA) and in formalin-fixed wax sections of human breast cancer with osteoclastic infiltrate (using rabbit polyclonal antibodies to osteopontin and bone sialoprotein, LF7 and LF6, respectively, gift of L. W. Fisher, NIDR, MA, USA). The effect of preincubating the monoclonal to osteopontin (neat supernatant) with osteopontin and bone sialoprotein (50 μg/ml, 12 hr, 4°C) upon staining of rat osteoclasts was used to assess the specificity of the reaction.

The expression of mRNA for osteopontin and bone sialoprotein was analyzed by *in situ* hybridization as detailed previously.[4] cDNAs for osteopontin (OP10) and bone sialoprotein (B65G) were subcloned from plasmids obtained from L. Fisher (NIDR, MA, USA) into pGEM-1 vector (Promega, Madison, WI, USA) and [α-35S]UTP (Amersham, UK) labelled sense and antisense riboprobes generated by transcription with SP6 or T7 polymerases (Stratagene, UK), respectively. Expression was assayed in formalin-fixed undecalcified sections of rat long bones and human breast cancer with osteoclastic infiltrate, controls consisting of hybridization with sense RNAs.

RESULTS

Adhesion of Osteoclasts to Osteopontin: Morphological Comparison with Cells Adhering to Serum Proteins

Rat (and mouse or human, not shown) osteoclasts when plated upon glass coated with either osteopontin (rat or human isolated from bone matrix; or recombinant, not shown) (FIG. 1a,b) or bone sialoprotein (not shown) show a characteristically irregular and extended morphology (as initially described in Helfrich *et al.*[7]). In contrast, osteoclasts adhering to serum proteins (mainly vitronectin) show a much more regular and spread morphology (FIG. 1d,e).

Osteoclasts adhering to osteopontin when stained for F-actin with rhodamine-phalloidin show numerous punctate accumulations (or podosomes[16–18]) concentrated at the cellular projections (FIG. 1c, arrowheads). In contrast, cells plated on serum show some peripherally placed F-actin dots or, infrequently, accumulations into ring-like structures (FIG. 1f) reminiscent of those found in osteoclasts resorbing bone.[12–14]

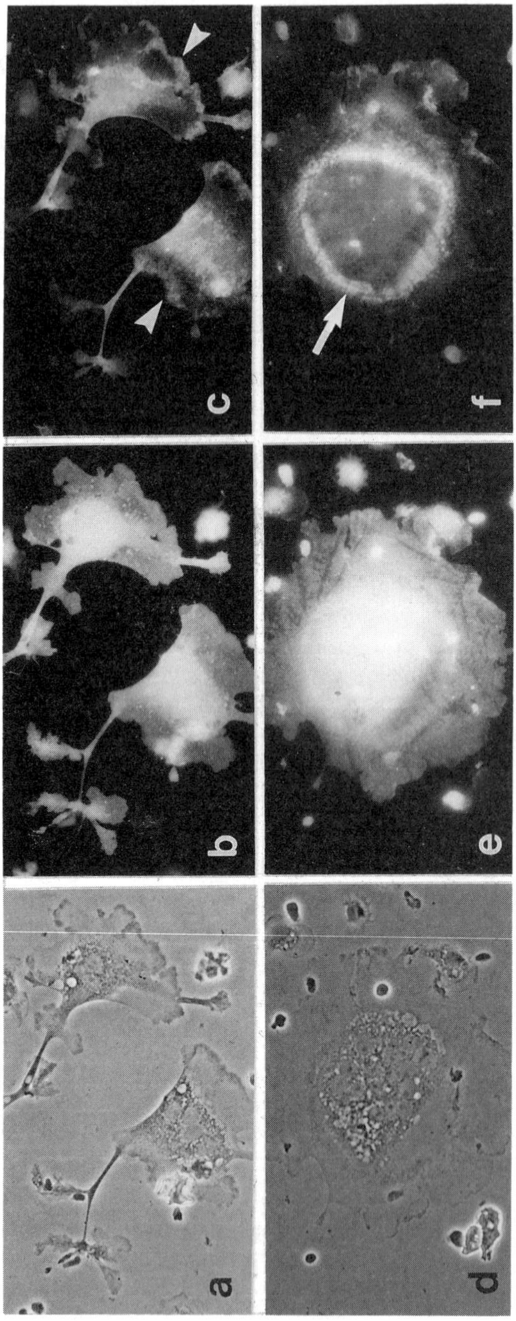

FIGURE 1. Photomicrographs of rat osteoclasts plated on osteopontin **(a–c)** and fetal calf serum **(d–f)**: phase **(a,d)**, immunofluorescence **(b,** anti-rat β_3 antibody, F11; **e,** anti-rat β_1 serum) and rhodamine-phalloidin stained to show F-actin **(c,f)**. *Arrowheads:* F-actin dots; *arrow:* F-actin ring. Magnification × 200.

Role of Integrin Receptors in Rat Osteoclast Adhesion to Osteopontin and Bone Sialoprotein

Osteoclast binding to osteopontin (and serum, FIG. 2a,iii and Helfrich *et al.*[7]) was sensitive to blocking antibodies to rat β_3 (F11) (90% inhibition, FIG. 2a,i) and α_v with no additive effects from exposure to both antibodies. Adhesion to osteopontin was not affected by an inhibitory polyclonal antibody to rat β_1; there was no significant enhancement of α_v inhibition when osteoclasts were exposed to a mixture of α_v and β_1 antibodies (FIG. 2a,i). This suggests that the recognition of osteopontin by rat osteoclasts was dependent upon $\alpha_v\beta_3$ only.

Adhesion of rat osteoclasts to bone sialoprotein was less sensitive to β_3 or α_v monoclonals, but was abolished by antibody coaddition (FIG. 2a,ii). In contrast to the case of osteopontin, adhesion to bone sialoprotein was inhibited somewhat by the antibody to β_1 (FIG. 2a,ii) and adhesion was significantly enhanced by a mixture of α_v and β_1 antibodies. These data imply that bone sialoprotein recognition involves $\alpha_v\beta_1$ (or other α_v or β_1 integrins) in addition to $\alpha_v\beta_3$.

Peptide Sequences Involved in Osteoclast Adhesion to Osteopontin

As previously reported,[7,19] the adhesion of osteoclasts to osteopontin (and bone sialoprotein, see Ref. 7) is inhibited by RGD-containing peptides (>90%, FIG. 2b,i). Adhesion was not inhibited by Lys-Gly-Asp (KGD)- or Arg-Gly-Glu (RGE)-containing peptides (not shown).

An alternative adhesive peptide sequence (H12 sequence) has been localized to the N-terminus of the γ-chain of fibrinogen[10] and is recognized by the platelet integrin, gpIIbIIIa (αIIbβ3). It has been suggested that osteoclast adhesion to sialoproteins might be influenced by such peptides.[20] Adhesion of osteoclasts to both osteopontin and serum proteins is unaffected by the full-length H12 peptide, but the shorter peptides KQAGD and AGDV induce cell 'rounding' on both substrates. Adhesion of osteoclasts to osteopontin is unaffected (FIG. 2b,i) but AGDV significantly inhibited osteoclast adhesion to serum (FIG. 2b,ii). Adhesion to bone sialoprotein has not been examined.

Osteoclasts Express Osteopontin, but Not Bone Sialoprotein, mRNA by in situ *Hybridization*

In situ mRNA hybridization was used to examine whether osteoclasts express sialoproteins. In contrast to a previous report,[21] we could not detect bone sialoprotein message in rat (FIG. 3c) or human (not shown) osteoclasts. However, and confirming previous publications, both rat limb bone and human breast cancer associated osteoclasts express high levels of osteopontin mRNA (FIGS. 3b and e).

Osteopontin Protein Expression by Osteoclasts

Rat osteoclasts settled on serum-coated glass coverslips stain weakly with a monoclonal antibody to rat osteopontin (not shown). The signal was greatly amplified by culturing the settled osteoclasts in the presence of monensin[15] and localized in the Golgi region (monoclonal MPIIIB10, FIG. 4b). No staining was seen with a monoclonal to rat bone sialoprotein (monoclonal WV1D1, FIG. 4e). The specific-

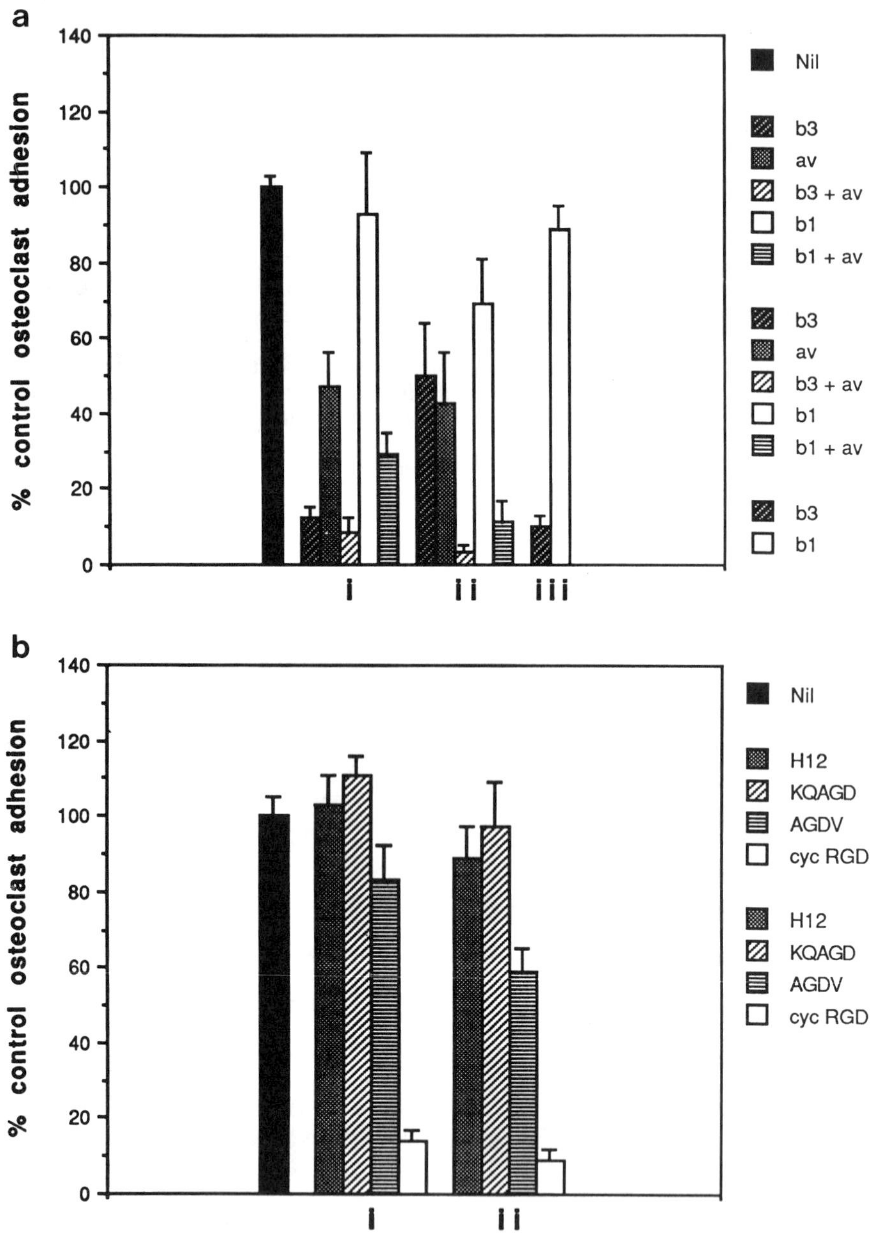

FIGURE 2. Adhesion of rat osteoclasts to osteopontin **(a,i; b,i)**, bone sialoprotein **(a,ii)** and fetal calf serum **(a,iii; b,ii)** coated glass coverslips showing the effects of anti-integrin antibodies **(a)** and RGD-containing and fibrinogen peptides **(b)** expressed as percentage of control adhesion in the absence of inhibitor.

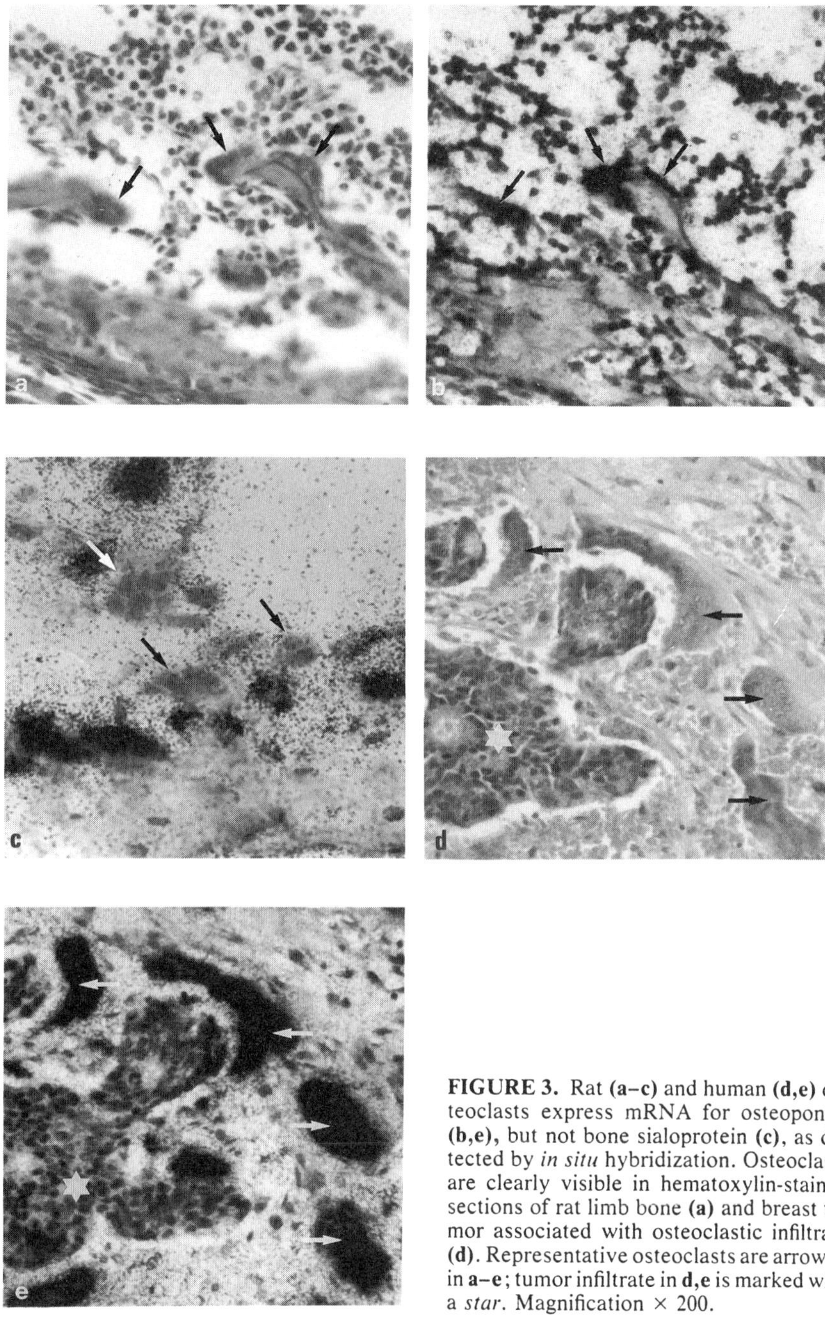

FIGURE 3. Rat **(a–c)** and human **(d,e)** osteoclasts express mRNA for osteopontin **(b,e)**, but not bone sialoprotein **(c)**, as detected by *in situ* hybridization. Osteoclasts are clearly visible in hematoxylin-stained sections of rat limb bone **(a)** and breast tumor associated with osteoclastic infiltrate **(d)**. Representative osteoclasts are arrowed in **a–e**; tumor infiltrate in **d,e** is marked with a *star*. Magnification × 200.

ity of osteopontin staining was validated by demonstrating that the reaction with antibody was inhibited by absorption with osteopontin protein but not bone sialoprotein (FIG. 4c and d).

Similar results were found with human osteoclasts in a varient of breast cancer associated with an osteoclastic infiltrate. Here also osteoclasts express osteopontin (antibody LF7, FIG. 4g) but not bone sialoprotein (not shown).

DISCUSSION

Adhesion of osteoclasts to the bone surface involves the interaction of osteoclast integrins with extracellular matrix proteins within the bone matrix. Osteoclasts use the $\alpha_v\beta_3$ vitronectin receptor to adhere to a wide variety of proteins containing the Arg-Gly-Asp (RGD) sequence.[7,19] In addition, mammalian osteoclasts adhere to type I collagen via a vitronectin receptor-independent, RGD-sensitive mechanism[7] probably involving β_1 integrins (Helfrich *et al.*, in preparation). In addition, it has been shown that bone resorption is inhibited by blocking osteoclast adhesion with RGD-containing peptides and anti-receptor antibodies,[11,13,22–24] implying that integrin-mediated mechanisms are operative during matrix recognition in bone resorption.

The results of our analysis of the mechanism by which rat osteoclasts adhere to osteopontin bear out the data which we previously published[7] showing that it is predominantly an RGD and β_3-dependent process. Additional experiments using blocking antibodies to α_v and β_1 show that adhesion is $\alpha_v\beta_3$ dependent and does not involve β_1. In contrast, adhesion to bone sialoprotein appears to involve β_1 (probably $\alpha_v\beta_1$) in addition to $\alpha_v\beta_3$. Flores *et al.*[20] have reported that a peptide (KQAGD) from an alternative adhesion sequence, the H12 region from the fibrinogen γ-chain,[10] is involved in osteoclast adhesion to bone sialoprotein but not osteopontin. To date, our experiments have shown that the short H12-derived peptides KQAGD and ADGV, but not the parental dodecapeptide, induce osteoclast 'rounding' and only minimal cell detachment (unlike RGD-peptides which cause stellate retraction and detachment[11]). This effect is not sialoprotein-specific as similar results were seen with osteoclasts adhering to serum proteins.

There is controversy as to whether any bone proteins or receptors are 'specifically' localized under resorbing osteoclasts, thus indicating a role in initial cell attachment or tight seal formation and maintenance. It has been suggested that the vitronectin receptor ($\alpha_v\beta_3$) is localized in the sealing zone of osteoclasts and interacts with osteopontin in the bone matrix underneath.[19,25–27] Others, however, have been unable to confirm this and, indeed, have reported that vitronectin receptor is undetectable in the sealing zone of resorbing osteoclasts[12–14] and that osteopontin, although enriched in lamina limitans-like structures of bone and thus encountered by resorbing osteoclasts, is not specifically localized under sealing zones.[28] An added complexity is that there is increasing evidence that osteoclasts synthesize osteopontin (originally suggested in data from Mark *et al.*[29] and confirmed by others [30–35]) and that osteopontin might be involved in the regulation of bone resorption, following osteoclast adhesion to bone matrix.[36]

Our *in situ* and immunohistological data on the distribution of osteopontin mRNA and protein in rat and human osteoclasts illustrated herein is entirely in line with the majority of published work. That is, osteoclasts have high levels of osteopontin mRNA and protein,[29–36] but have undetectable bone sialoprotein. This latter finding contrasts with the immunohistological results of Bianco *et al.*,[21] but it agrees with the recent *in situ* hybridization data of Chen *et al.*[37]

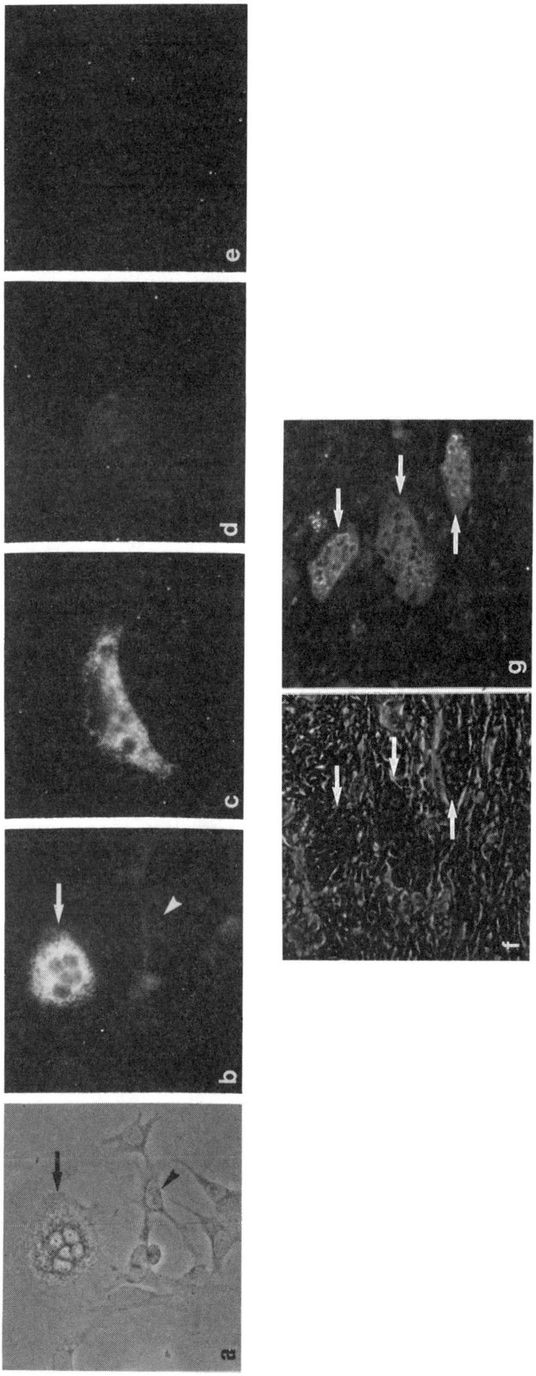

FIGURE 4. Osteoclasts express osteopontin (rat, **b**; human, **g**; phase contrast to demonstrate osteoclasts which are *arrowed*, **a,f**) but not bone sialoprotein **(e)** protein as detected by immunofluorescence. *Arrowheads* **(a,b)** indicate osteoblasts which also express osteopontin. Immunostaining of osteopontin in monensin-treated osteoclasts disaggregated from rat limb bone **(b)** is inhibited by incubation with osteopontin **(d)** but not bone sialoprotein **(c)**. Magnification × 200.

The observation that osteoclast attachment to matrix-coated glass or bone is interrupted by integrin inhibitors led to the view that tight osteoclast attachment may be mediated by integrins. Indeed, studies from Heinegård and co-workers[25,26] suggested that vitronectin receptor may be enriched in clear zones and podosomes.[16,17,38] There is no really convincing evidence so far that integrins form tight junctions in other systems (for example, integrin-rich focal adhesions are 'dark' by interference microscopy and thus a minimum of depth of 20 nm and integrins have not been detected in zona occludens tight junctions). Moreover, the dimensions of the integrin molecule (20 nm long by rotary EM shadowing) versus the membrane to bone gap (of much less than 1 nm [2]) would seem to suggest that maintenance of the tight seal during resorption could not be mediated by integrins. On the other hand, the physical dimensions of the initial attachment site, and during dynamic receptor turnover associated with cell migration or formation of complex resorption lacunae,[12] would not be so (Refs. 13, 14 and unpublished data). Thus in our view, the molecular mechanism of the attachment process in the established clear zone of a resorbing, nonmigratory osteoclast remains to be established.

Our finding that osteoclasts, when adhering to an osteopontin (or bone sialoprotein) substrate, have a characteristic morphology which is quite distinct from that observed when osteoclasts are plated upon other matrix proteins (herein and Ref. 7) does, though, raise the possibility that sialoproteins induce some specific function in osteoclasts that is not solely adhesive; this could be the stimulation of a nonresorptive phenotype, for instance, as podosome-rich F-actin rings do not form when osteoclasts are on osteopontin (FIG. 1) when compared to serum and their shape is irregular. Both these characteristics would mitigate against formation of the required structures for the generation of the specific cytoskeletal structures in the tight seal during bone resorption. In support of these ideas is the finding that other functional domains appear to be detectable in both osteopontin and bone sialoprotein which are distinct from the RGD sequences and can mediate adhesion[39,40] (though osteoclasts have not been tested). This hypothesis is testable using, for example, recombinant expressed sialoprotein fragments.

In conclusion, we present data on the mechanism of osteoclast adhesion to sialoproteins, demonstrating a predominant, though not exclusive, utilization of vitronectin receptor, and the synthesis of osteopontin, but not bone sialoprotein, by osteoclasts. These findings are discussed in the context of the disposition of integrins in osteoclasts and their role in osteoclast adhesion and during bone resorption.

SUMMARY

Mammalian osteoclasts express three integrin receptors—$\alpha_v\beta_3$ (vitronectin receptor), $\alpha_2\beta_1$ and $\alpha_v\beta_1$. The vitronectin receptor recognizes bone matrix proteins, including bone sialoproteins, in an RGD-dependent manner, whereas adhesion to collagen involves β_1 integrins. Interference with integrin function, by anti-receptor antibodies or RGD-peptides, blocks bone resorption. Data on the mechanism of osteoclast adhesion to sialoproteins and the differential synthesis of osteopontin and bone sialoprotein by osteoclasts is presented. Thus, osteoclasts adhere to both osteopontin and bone sialoprotein with a characteristic irregular morphology with numerous, peripherally placed, actin-rich podosomes. Adhesion is predominantly RGD and β_3 dependent, though $\alpha_v\beta_1$ may also be involved in adhesion to bone sialoprotein. KQAGD and AGDV, but not H12, fibrinogen peptides induce osteoclast 'rounding' on osteopontin suggesting there is an alternative anti-adhesive signal to 'RGD.' However, adhesion is not completely inhibited and is not specific for osteopontin as equivalent effects are seen with adhesion to serum. The role of sialoproteins in osteoclast adhesion *in situ* in the skeleton is complicated by the finding of endogenous synthesis of osteopontin, but not bone sialoprotein, by osteoclasts. The disposition of osteoclast integrins during resorption and the role of integrins and sialoprotein-derived peptides in osteoclast adhesion and function is also reviewed.

ACKNOWLEDGMENTS

We thank the donors of various reagents for their kind gifts.

REFERENCES

1. Rifkin, B. R. & C. V. Gay. 1992. Biology and Physiology of the Osteoclast. CRC Press. Boca Raton, FL.
2. Holtrop, M. E. & G. J. King. 1977. Clin. Orthop. **123:** 177–196.
3. Zhou, H., R. Chernecky & J. E. Davies. 1993. Cells Mater. **3:** 141–150.
4. Nesbitt, S., A. Nesbit, M. H. Helfrich & M. Horton. 1993. J. Biol. Chem. **268:** 16737–16745.
5. Hynes, R. O. 1992. Cell **69:** 11–25.
6. Juliano, R. L. & S. J. Haskill. 1993. J. Cell Biol. **120:** 577–585.
7. Helfrich, M. H., S. A. Nesbitt & M. A. Horton. 1992. J. Bone Miner. Res. **7:** 345–351.
8. Gullberg, D., D. C. Turner, T. K. Borg, L. Terracio & K. Rubin. 1990. Exp. Cell Res. **190:** 254–264.
9. Helfrich, M. H., S. A. Nesbitt, E. L. Dorey & M. A. Horton. 1992. J. Bone Miner. Res. **7:** 335–344.
10. Kloczewiak, M., S. Timmons, T. J. Lukas & J. Hawiger. 1984. Biochemistry **23:** 1767–1774.
11. Horton, M. A., E. L. Dorey, S. A. Nesbitt, J. Samanen, F. E. Ali, J. M. Stadel, A. Nichols, R. Greig & M. H. Helfrich. 1993. J. Bone Miner. Res. **8:** 239–247.
12. Lakkakorpi, P. T. & H. K. Väänänen. 1991. J. Bone Miner. Res. **6:** 817–826.
13. Lakkakorpi, P. T., M. A. Horton, M. H. Helfrich, E-K. Karhukorpi & H. K. Väänänen. 1991. J. Cell Biol. **115:** 1179–1186.
14. Lakkakorpi, P. T., M. H. Helfrich, M. A. Horton & H. K. Väänänen. 1993. J. Cell Sci. **104:** 663–670.
15. Hembry, R. M., G. Murphy & J. J. Reynolds. 1985. J. Cell Sci. **73:** 105–119.
16. Zambonin-Zallone, A., A. Teti, M. Grano, A. Rubinacci, M. Abbadini, M. Gaboli & P. C. Marchisio. 1989. Exp. Cell Res. **182:** 645–652.
17. Teti, A., P. C. Marchisio & A. Zambonin-Zallone. 1991. Am. J. Physiol. **261:** C1–C7.
18. Aubin, J. E. 1992. J. Bone Miner. Res. **7:** 365–368.
19. Flores, M. E., M. Norgard, D. Heinegård, F. P. Reinholt & G. Andersson. 1992. Exp. Cell Res. **201:** 526–530.
20. Flores, M. E., D. Heinegård, F. P. Reinholt & G. Andersson. 1993. J. Bone Miner. **8**(Suppl. 1): S397.
21. Bianco, P., L. W. Fisher, M. F. Young, J. D. Termine & P. G. Robey. 1991. Calcif. Tissue Int. **49:** 421–426.
22. Chambers, T. J., K. Fuller, J. A. Darby, J. A. S. Pringle & M. A. Horton. 1986. Bone Miner. **1:** 127–133.

23. Sato, M., M. K. Sardana, W. A. Grasser, V. Garsky, J. M. Murray & R. J. J. Gould. 1990. J. Cell Biol. **111,** 1713–1723.
24. Horton, M. A., M. L. Taylor, T. R. Arnett & M. H. Helfrich. 1991. Exp. Cell Res. **195:** 368–375.
25. Reinholt, F. P., K. Hultenby, A. Oldberg & D. Heinegård. 1990. Proc. Natl. Acad. Sci. USA **87:** 4473–4475.
26. Hultenby, K., F. P. Reinholt & D. Heinegård. 1993. Eur. J. Cell Biol. **62:** 86–93.
27. Carlson, C. S., H. M. Tulli, M. J. Jayo, R. F. Loeser, R. P. Tracy, K. G. Mann & M. R. Adams. 1993. J. Bone Miner. Res. **8:** 71–81.
28. McKee, M. D. & A. Nanci. 1993. Cells Mater. **3:** 219–243.
29. Mark, M. P., C. W. Prince, T. Oosawa, S. Gay, A. L. Bronckers & W. T. Butler. 1987. J. Histochem. Cytochem. **35:** 707–715.
30. Ikeda, T., S. Nomura, A. Yamguchi, T. Suda & S. Yoshiki. 1992. J. Histochem. Cytochem. **40:** 1079–1088.
31. Tezuka, K., T. Sato, H. Kamioka, P. J. Nijweide, K. Tanaka, T. Matsuo, M. Ohta, N. Kurihara, Y. Hakeda & M. Kumegawa. 1992. Biochem. Biophys. Res. Commun. **186:** 911–917.
32. Frank, J. D., B. Raffaella, P. Masarachia, J. G. Seedor & M. E. Cartwright. 1993. Histochemistry **99:** 295–301.
33. Merry, K., R. Dodds, A. Littlewood & M. Gowen. J. Cell Sci. **104:** 1013–1020.
34. Nakase, T., K. Takaoka, K. Hirakawa, S. Hirota, T. Takemura, H. Onoue, K. Takebayashi, Y. Kitamura & S. Nomura. 1994. Bone Miner. **26:** 109–122.
35. Takano-Yamamoto, T., T. Takemura, Y. Kitamura & S. Nomura. 1994. J. Histochem. Cytochem. **42:** 885–896.
36. Dodds, R. A., J. R. Connor, E. I. James, E. Applebaum, E. Dul & M. Gowen. 1994. Bone Miner. **25**(Suppl. 1): S41–42.
37. Chen, J., M. D. McKee, A. Nanci & J. Sodek. 1992. Histochem. J. **26:** 67–78.
38. Turksen, K., J. Kanehisa, M. Opas, J. N. M. Heersche & J. E. Aubin. 1988. J. Bone Miner. Res. **3:** 389–400.
39. Mintz, K. P., W. J. Grzesik, R. J. Midura, P. Gehron Robey, J. D. Termine & L. W. Fisher. 1993. J. Bone Miner. Res. **8:** 985–995.
40. van Dijk, S., J. A. D'Errico, M. J. Somerman, M. C. Farach-Carson & W. T. Butler. 1993. J. Bone Miner. Res. **8:** 1499–1506.

Interaction of Osteopontin with Fibronectin and Other Extracellular Matrix Molecules[a]

BARID B. MUKHERJEE,[b,e] MOHAMED NEMIR,[b]
SIMONE BENINATI,[c] ELEONORA CORDELLA-MIELE,[c]
KRISHNA SINGH,[b,d] ISAAC CHACKALAPARAMPIL,[b]
VIJAYALAKSHMI SHANMUGAM,[b]
MICHAEL W. DeVOUGE,[b] AND ANIL B. MUKHERJEE[c]

[b]*Departments of Biology and Human Genetics*
McGill University
1205 Doctor Penfield Avenue
Montreal, Quebec, Canada H3A 1B1

[c]*Section on Developmental Genetics*
Human Genetics Branch
NICHD, NIH
Bethesda, Maryland 20892

INTRODUCTION

Senger *et al.*[1] in 1979 reported that virally and spontaneously transformed cell lines derived from several mammalian species secrete a major phosphoprotein with an approximate molecular mass of 60 kDa (transformation-specific phosphoprotein). Later, in 1985, Franzen and Heinegard[2] isolated a major 57-kDa sialoprotein from calcified matrix of rat bone,[3] and Chackalaparampil *et al.*[4] reported that normal rat kidney (NRK) cells in culture secrete a 69-kDa major phosphoprotein (pp69). Oldberg *et al.*[5] in 1986 isolated the cDNA clone of the sialoprotein identified by Franzen and Heinegard[2] from a rat osteosarcoma (RO7 17/2.8) phage λgt11 library. Analysis of the amino acid sequence deduced from the cDNA revealed a GRGDS cell-binding sequence and likely sites for N- and O-glycosidically linked oligosaccharides. Purified protein also showed cell attachment activity. They named this protein "osteopontin." In 1987 Prince *et al.*[5] independently purifed and characterized a 44-kDa phosphorylated glycoprotein from rat bone which is also synthesized and secreted by ROS17/2.8 cells. Denhardt *et al.* in 1988 showed that mouse 2ar, a tumor promoter-inducible protein secreted by mouse JB6 epidermal cells is homologous to rat OPN. Craig *et al.*[8] in the same year noted that the sequence of nine N-terminal amino acids of the rat transformation-specific phosphoprotein reported by Senger *et al.*[9] is identical to the N-terminal 17 to 25 amino acid residues of the polypeptide deduced from the cDNA sequence of rat

[a] This work was supported by grants from the Medical Research Council, Canada (MT-11628 to B.B.M.). S.B. was the recipient of a professional services contract from the National Institute of Child Health and Human Development, NIH.

[d] Present address: Cardiovascular Division, Brigham Woman's Hospital, Boston, MA.
[e] To whom correspondence should be addressed.

osteopontin, and reported that 2ar,[10,11] pp69 secreted by NRK cells,[4] and 44-kDa bone phosphoprotein,[6] all are in fact osteopontin (OPN).[5]

OPN is a major phosphoprotein of the extracellular bone matrix (osteoid). It is the most abundant product of osteoblasts, secreted by these cells into the extracellular matrix (ECM), and becomes localized in osteoid.[12] OPN interacts with the $\alpha_v\beta_3$ integrin on the cell surface,[13] and OPN-$\alpha_v\beta_3$ interaction appears to be a transient event prior to its translocation into the ECM.[14] Although OPN is an important constituent of the ECM, the identity of the ECM proteins that interact with OPN and the molecular mechanisms of stable incorporation of OPN into the ECM remain largely unknown.

Normal Rat Kidney (NRK) Cells Secrete Both Phosphorylated and Nonphosphorylated Forms of OPN

In 1989 Nemir *et al.*[15] demonstrated for the first time that NRK cells secrete OPN in both phosphorylated and nonphosphorylated forms. NRK cells were labeled with [^{35}S]methionine or [^{32}P]orthophosphate, and labeled proteins in the conditioned media were immunoprecipitated, using an anti-OPN serum, and analyzed by one- and two-dimensional SDS-PAGE. FIGURE 1A shows that immunoprecipitated [^{32}P]-labeled OPN migrates as a single 69-kDa band in one-dimensional gels (lane 1), but the same protein labeled with [^{35}S]methionine forms a doublet (lane 2). In addition, SDS-PAGE analysis of immunoprecipitated [^{35}S]-labeled proteins from the conditioned media showed additional high molecular weight bands, besides OPN (lane 2). On analysis of [^{35}S]-labeled secreted proteins by IEF/SDS-PAGE (FIG. 1B, panel 1) two 69-kDa proteins were identified, one with an estimated pI of 3.8 and the other having an approximate pI of 4.5. However, [^{32}P]-labeled proteins, upon identical analysis showed only one 69-kDa protein with a pI of 3.8 (FIG. 1B, panel 2). When [^{35}S]-labeled secreted proteins were digested with alkaline phosphatase, which efficiently dephosphorylated [^{32}P]-labeled OPN (FIG. 2, panel A), only one 69-kDa protein with an estimated pI of 4.5 was detected (FIG. 2B). This observation demonstrated that immunoprecipitated [^{35}S]-labeled 69-kDa OPN which migrated as a doublet in one-dimensional SDS-PAGE gels (FIG. 1A) is composed of phosphorylated (pp69) and nonphosphorylated (np69) forms of OPN. It became obvious, therefore, that NRK cells secrete both pp69 and np69 forms of OPN.

Nonphosphorylated OPN Complexes with Soluble Fibronectin and Phosphorylated OPN Interacts with Cell-Surface Fibronectin

When [^{35}S]-labeled proteins in the conditioned media from NRK cell cultures were immunoprecipitated with anti-OPN serum, several other bands, in addition to both np69 and pp69, were observed (FIG. 3A, lane 1). However, anti-fibronectin (Fn) serum immunoprecipitated, in addition to Fn, a 69-kDa protein which is the np69 form (lower band of the doublet) of OPN (lane 3). When the protein samples were heated to 95°C in the presence of 0.2% SDS prior to immunoprecipitation, anti-OPN serum immunoprecipitated both pp69 OPN and np69 OPN but not Fn (lane 2), and anti-Fn serum immunoprecipitated only Fn (lane 4). These observations demonstrated complex formation between np69 OPN and Fn, and that such complexes can be dissociated by heat denaturation.[15] Western blot analysis of unlabeled proteins from serum-free conditioned media using either anti-OPN serum

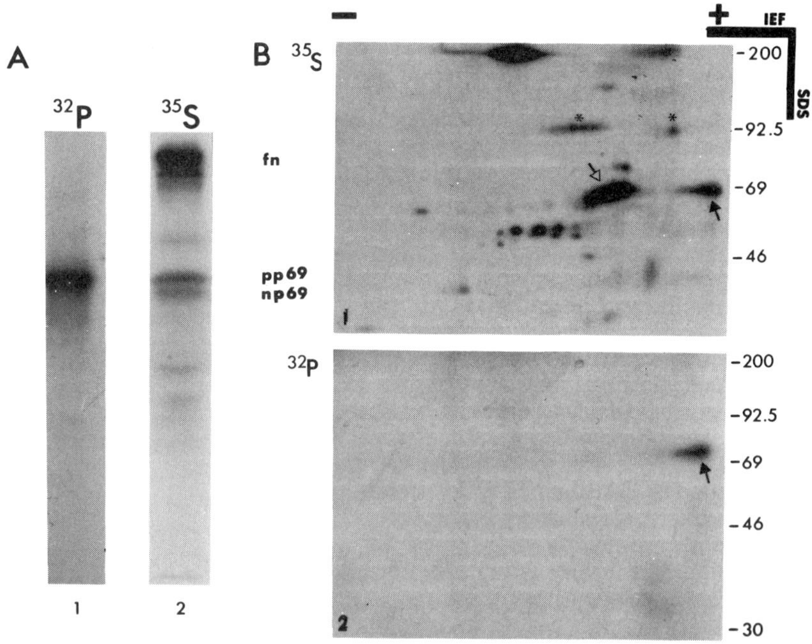

FIGURE 1. One- and two-dimensional gel electrophoresis of secreted proteins from cells. Proteins were metabolically labeled with [^{35}S]methionine or [^{32}P]orthophosphate for 4 h, and aliquots containing equal amounts of trichloroacetic acid-precipitable radioactivity were analyzed by one- and two-dimensional SDS-PAGE. **(A)** ^{32}P-labeled (*lane 1*) and ^{35}S-labeled (*lane 2*) secreted proteins immunoprecipitated with anti-69-kDa OPN. Positions of the np69, pp69, and fibronectin (*fn*) are indicated. **(B)** Total secreted proteins labeled with [^{35}S]methionine (*panel 1*) and [^{32}P]orthophosphate (*panel 2*). The position of the np69 in panel 1 is marked by an *open arrow* and the positions of the pp69 in panels 1 and 2 are marked by *solid arrows*. *Asterisks* mark the positions of reference proteins. The positions of the prestained molecular weight markers are shown in the *right margin*. The approximate isoelectric points of pp69 (3.8) and np69 (4.5) were determined by estimating pH values in four separate blank IEF gels. (From Nemir *et al.*[15] Reprinted by permission from the *Journal of Biological Chemistry*.)

(FIG. 3B, lane 1), or anti-Fn serum (lane 2) showed that anti-OPN serum stains only OPN, and anti-Fn serum detects only Fn. These observations confirmed complex formation between np69 OPN and soluble Fn. FIGURE 3A, lane 1, shows that at least two other proteins, in addition to np69 OPN, pp69 OPN and Fn, can be detected after immunoprecipitation of [^{35}S]-labeled secreted proteins with anti-OPN serum. The identities of these proteins still remain to be determined.

When monolayer cultures of NRK cells were subjected to cell-surface radioiodination and the lysates were immunoprecipitated with anti-OPN serum, only a 69-kDa band was observed (FIG. 4, lane 2), demonstrating that OPN is localized on the cell surface. Immunoprecipitated [^{125}I]-labeled 69-kDa OPN analyzed by two-dimensional gel electrophoresis (FIG. 4B) showed only one spot with an identical isoelectric point to that of pp69 OPN (panel 2). No [^{125}I]-labeled protein

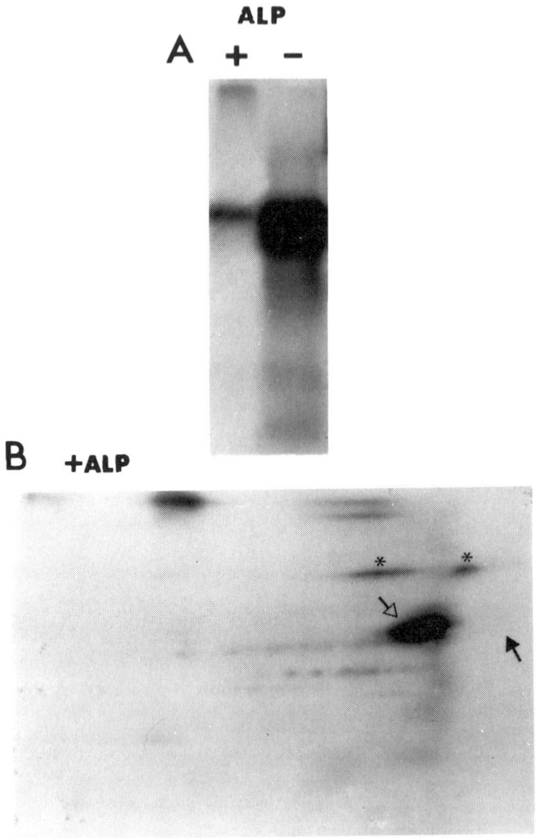

FIGURE 2. Relationship between two 69-kDa secreted proteins. **(A)** SDS-PAGE analysis of ^{32}P-labeled total secreted proteins incubated with (+ *lane*) or without (− *lane*) alkaline phosphatase (ALP). **(B)** Two-dimensional analysis of ^{35}S-labeled total secreted proteins incubated with alkaline phosphatase. *Open and solid arrows* show the positions of np69 and pp69, respectively. *Asterisks* mark the positions of reference proteins. (From Nemir *et al.*[15] Reprinted by permission from the *Journal of Biological Chemistry*.)

migrated to the pI range of np69 OPN. Comigration of [^{125}I]-labeled cell-surface-associated OPN with [^{32}P]-labeled pp69 OPN on two-dimensional gels (panel 2) demonstrated that pp69 OPN, not its nonphosphorylated form, interacts with the cell surface.[15]

Nonphosphorylated OPN but Not Phosphorylated OPN Contains N-Linked Oligosaccharides Which Are Necessary for Complex Formation between npOPN and Fn

SDS-PAGE analyses of immunoprecipitated [^{35}S]-labeled proteins in the conditioned media of tunicamycin-treated NRK cells with anti-OPN serum, and immunoprecipitated [^{35}S]methionine and [^{32}P]orthophosphate-labeled OPN digested with peptide : N-glycosidase F showed that np69 OPN but not pp69 OPN contains N-linked oligosaccharides.[16] FIGURE 5A shows that immunoprecipitation of [^{35}S]-labeled proteins from the conditioned media of NRK cells with anti-Fn serum detects 69-kDa OPN and Fn plus two unidentified proteins (lane 1), but when the protein sample was heated at 95°C before immunoprecipitation the 69-kDa OPN and 44-kDa proteins could no longer be detected (lane 2). Immunoprecipitation of proteins from the conditioned media of tunicamycin-treated cells with anti-Fn serum could detect only Fn but not 69-kDa OPN (lane 3). Anti-OPN serum which immunoprecipitates both 69-kDa OPN and Fn from the conditioned media of untreated NRK cells (lane 4), could immunoprecipitate only 69-kDa OPN and a 44-kDa protein but not from the conditioned media of tunicamycin-treated cells (lane 5). These observations suggest that np69 OPN from tunicamycin-treated cells which does not contain N-linked oligosaccharides fails to form complexes

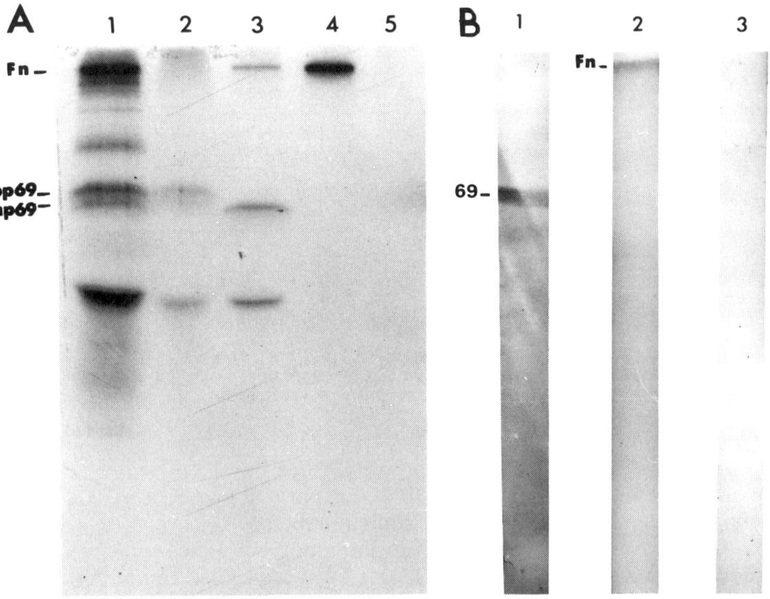

FIGURE 3. Coprecipitation of np69 with fibronectin (*Fn*). **(A)** ^{35}S-labeled secreted proteins from NRK cells were immunoprecipitated with either anti-69-kDa OPN (*lanes 1 and 2*) or with anti-Fn serum (*lanes 3 and 4*), prior to (*lanes 1 and 3*) or after (*lanes 2 and 4*) heat denaturation. *Lane 5*, with nonimmune rabbit serum. Aliquots of conditioned medium from NRK cell cultures containing approximately 20,000 cpm were immunoprecipitated using either 30 μl of anti-69-kDa OPN or 20 μl of anti-Fn serum. For denaturation, the samples were heated at 95°C for 5 min in the presence of 0.2% SDS and then diluted with RIPA buffer lacking SDS and immunoprecipitated. **(B)** Western blotting. Unlabeled secreted proteins from serum-free conditioned medium of NRK cells were separated on a 10% polyacrylamide gel and transferred to a nitrocellulose filter, and individual strips of the filter were probed using either a 1 : 250 dilution of anti-69-kDa osteopontin (*lane 1*) or a 1 : 500 dilution of anti-Fn serum (*lane 2*). *Lane 3* was probed with nonimmune rabbit serum (1 : 250 dilution). (From Nemir *et al.*[15] Reprinted by permission from the *Journal of Biological Chemistry*.)

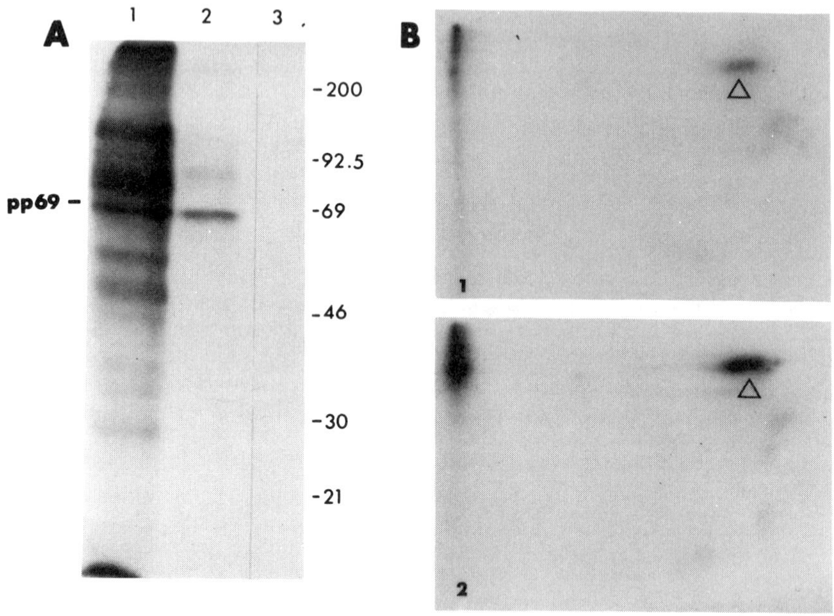

FIGURE 4. Cell surface localization of pp69 OPN. Subconfluent NRK cell cultures were surface-labeled with ^{125}I for 30 min at room temperature and examined under the microscope, and intact cultures were lysed with RIPA buffer and immunoprecipitated. **(A)** SDS-PAGE analysis of total ^{125}I-labeled 69-kDa cell-surface-associated OPN (*lane 1*) and immunoprecipitated ^{125}I-labeled cell-surface proteins with anti-OPN serum (*lane 2*). **(B)** Two-dimensional analysis of immunoprecipitated ^{32}P-labeled pp69 OPN (*panel 1*) and a mixture of ^{32}P-labeled and ^{125}I-labeled pp69 OPN showing comigration (*panel 2*). (From Nemir *et al.*[15] Reprinted by permission from the *Journal of Biological Chemistry*.)

with Fn, suggesting that glycosylation of OPN or Fn or both plays a role in the complex formation between np69 OPN and plasma Fn. FIGURE 5B shows the immunoprecipitation of [^{32}P]orthophosphate-labeled 69-kDa OPN with anti-Fn serum. Although anti-69-kDa OPN serum immunoprecipitates pp69 from the conditioned medium of NRK cells (lane 1), anti-Fn serum failed to immunoprecipitate the protein (lane 2), confirming our previous observation that only np69 and not pp69 forms complexes with plasma Fn.[15]

Formation of Stable Complex between pp69 OPN and Cell-Surface-Associated Fn

By cell-surface iodination, immunoprecipitation, and SDS-PAGE it was shown previously that only the pp69 OPN is localized to the NRK cell surface (FIG. 4). FIGURE 6 shows that the cell surface of NRK cells contains a 69-kDa protein which can be immunoprecipitated with either anti-69-kDa OPN or anti-Fn serum (lanes 1 and 2). In addition, anti-69-kDa OPN serum immunoprecipitates Fn, a 44-kDa protein, and three other proteins (lane 1). The anti-Fn serum immunoprecipitates a protein complex containing Fn, the 69-kDa OPN, and two other proteins (lane 2). Since only pp69-kDa OPN can be found on the cell surface, the above observations show that pp69-kDa OPN complexes with the cell-surface-associated Fn, and because of this complex formation, both proteins can be immunoprecipitated by either anti-69-kDa OPN or anti-Fn serum. Immunoprecipitation of ^{125}I-labeled cell-surface proteins from tunicamycin-treated NRK cells with anti-69-kDa OPN or anti-Fn serum also detected the same proteins present on the cell surface of untreated cells (lanes 4 and 6). This suggests that tunicamycin treatment does not have any effect on the cell-surface association of pp69-kDa OPN. However, when ^{125}I-labeled protein samples were heated at 95°C before immunoprecipitation, such treatment failed to dissociate pp69-kDa OPN from the cell surface Fn (lanes 3 and 5). These data, therefore, demonstrate that pp69-kDa OPN and the cell-surface-associated Fn form a heat-stable complex, and imply that pp69-kDa OPN becomes incorporated into the fibrillar network of the cell-surface Fn.[16]

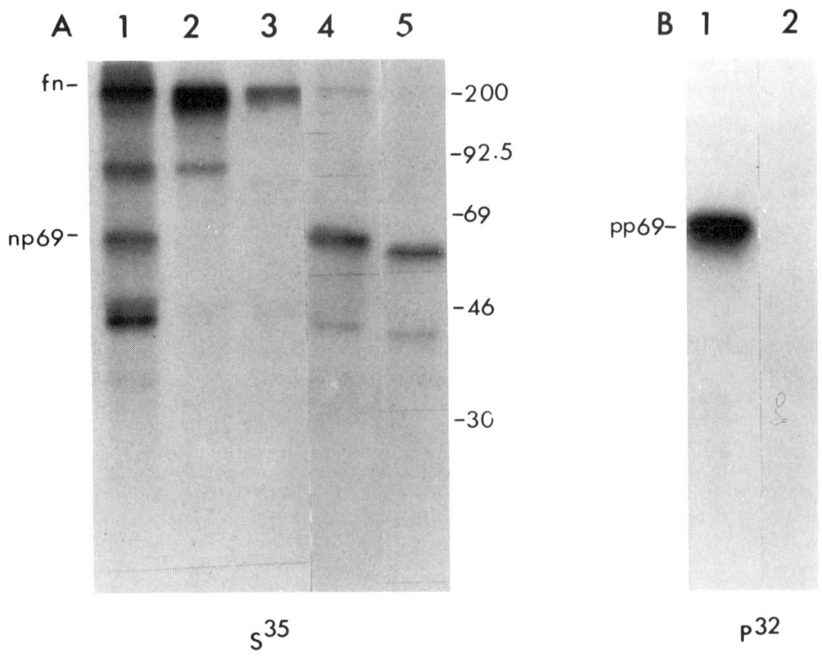

FIGURE 5. Interaction of ^{35}S- and ^{32}P-labeled 69-kDa OPN with plasma fibronectin. **(A)** ^{35}S-labeled secreted proteins from tunicamycin treated (*lanes 3 and 5*) and untreated NRK cells (*lanes 1,2,4*) were immunoprecipitated with either anti-fibronectin (*lanes 1,2 and 3*) or anti-69 kDa osteopontin serum (*lanes 4 and 5*). The sample of the conditioned medium from untreated NRK cells was heated at 95°C for 5 min in the presence of 0.2% SDS and then diluted with RIPA buffer lacking SDS and immunoprecipitated (*lane 2*). **(B)** ^{32}P-labeled secreted proteins from NRK cells were immunoprecipitated with either anti-69-kDa osteopontin (*lane 1*) or with anti-fibronectin serum (*lane 2*). Positions of the prestained molecular weight markers are as indicated. (From Singh *et al.*[16] Reprinted by permission from the *Journal of Biological Chemistry*.)

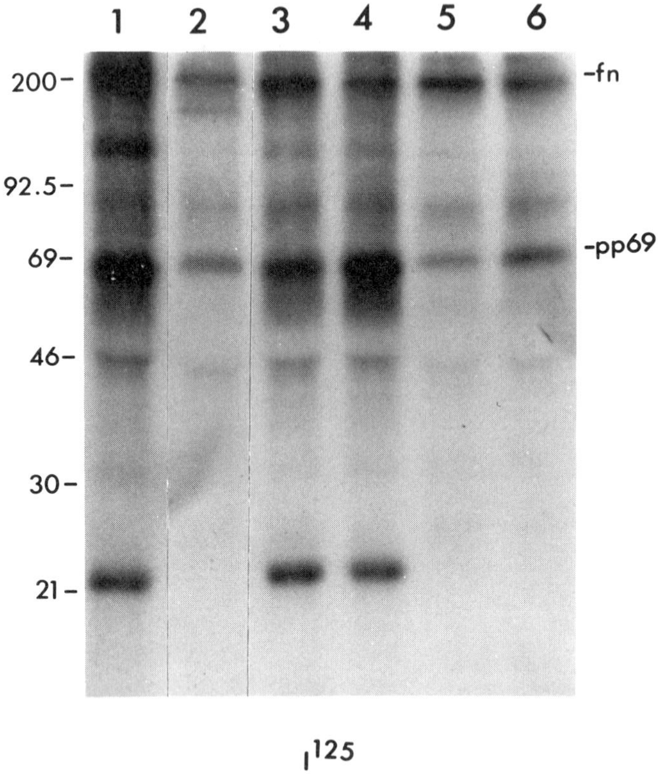

FIGURE 6. Cell-surface association of pp69 osteopontin and fibronectin. Tunicamycin-treated (8 h) and untreated NRK cells were surface labeled with ^{125}I for 30 min at room temperature and examined under the microscope. Intact cultures were lysed with RIPA buffer, immunoprecipitated and analyzed by SDS-PAGE as described in the text. ^{125}I-labeled cell-surface proteins from the tunicamycin-treated (*lanes 4 and 6*) or control (*lanes 1,2,3 and 5*) cultures were immunoprecipitated with either anti-69-kDa osteopontin (*lanes 1,3 and 4*) or anti-fibronectin serum (*lanes 2,5 and 6*) prior to (*lanes 1 and 2*) or after (*lanes 3,4,5 and 6*) heat denaturation. Positions of pp69 and fibronectin (fn) are indicated. (From Singh *et al.*[16] Reprinted by permission from the *Journal of Biological Chemistry*.)

OPN-Fn Cross-Linking on the Cell Surface

Since it has been shown that pp69 OPN secreted by NRK cells forms a heat-stable complex with cell-surface Fn,[15,16] the question arises how ppOPN becomes stably incorporated into the ECM of which Fn is a major constituent.[17] It has been shown that both OPN and Fn are known substrates of transglutaminase (TG).[18–21] TGs are a class of enzymes found both intra- and extracellularly, and are widely distributed in body fluids.[22] These enzymes catalyze a Ca^{2+}-dependent acyl transfer reaction, causing the formation of an ε-(γ-glutamyl)lysine isopeptide bond and promoting intra- and/or intermolecular cross-linking in proteins.[23] Therefore, purified OPN, fibronectin and TG were incubated in a reaction mixture at room temperature for 1 h and the products of the reaction were subjected to SDS-PAGE analysis. Two identical gels were run and the proteins were electrotransferred to PVDF membranes. Proteins in one membrane were interacted with anti-OPN serum, and the other with an anti-Fn serum. Both membranes were then incubated with [^{125}I]protein-A, and autoradiographed. Both lanes 2 and 4 in FIGURE 7A showed three identical high molecular weight bands. The fact that all three bands could be visualized after immunoblotting with either anti-OPN or anti-Fn serum suggests that they are the cross-linked products of OPN and Fn.[27]

To examine whether OPN-Fn cross-linking occurs *in vivo*, NRK cells were grown in medium containing TG, monolayers were cell-surface iodinated, and the lysates were immunoprecipitated with either anti-OPN or anti-Fn serum, resolved by SDS-PAGE, and autoradiographed. FIGURE 7B shows that both antibodies immunoprecipitated a very high molecular weight protein which migrated at the same position into the resolving gel. This suggests that this high-molecular-weight

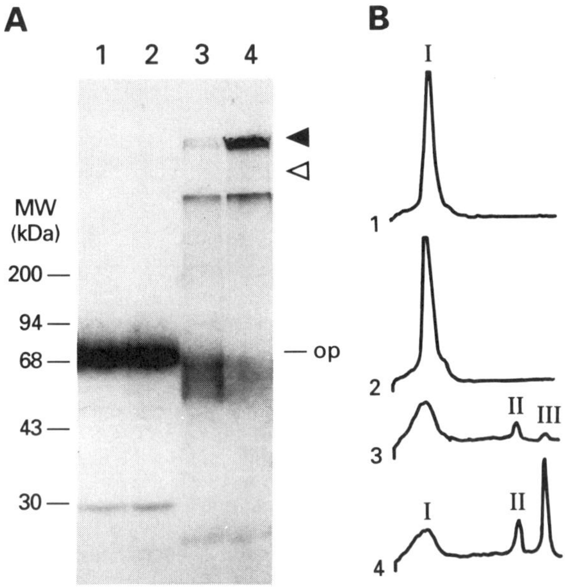

Lane #	Peak #	Area %
1.	I	100.0
2.	I	100.0
3.	I	86.0
	II	9.0
	III	5.0
4.	I	44.0
	II	14.0
	III	42.0

FIGURE 7. Autoradiographic and densitometric analysis of TG-catalyzed cross-linking of [^{125}I] OPN and Fn. TG-catalyzed reactions using [^{125}I]OPN and nonradioactive Fn were carried out as described in the text. **(A)** Autoradiograph of OPN-Fn interaction: *lane 1,* [^{125}I]OPN; *lane 2,* [^{125}I]OPN + Fn; *lane 3,* [^{125}I]OPN + TG; and *lane 4,* [^{125}I]OP + Fn + TG. Note the appearance of two very-high-molecular-weight bands in *lanes 3 and 4,* and the significant increase in labeling of the *top band* in *lane 4. Open and solid arrow heads* point to the tops of stacking and resolving gels, respectively. **(B)** Densitometric analysis of radioactive protein bands shown in (A). Relative densities of the bands represented as % area under the curve shown in the *inset.* 1, peak I; 2, peak II; and 3, peak III. Note that the apparent density of the putative intramolecularly cross-linked OPN band (peak I, *lanes 3 and 4*) is reduced, with a simultaneous increase in the density of two high-molecular-weight bands (peaks II and III, *lanes 3 and 4*). Also note that the 30-kDa band in *lanes 3 and 4* has a higher electrophoretic mobility than that in *lanes 1 and 2*. (From Beninati *et al.*[27] Reprinted by permission from the *Journal of Biochemistry*.)

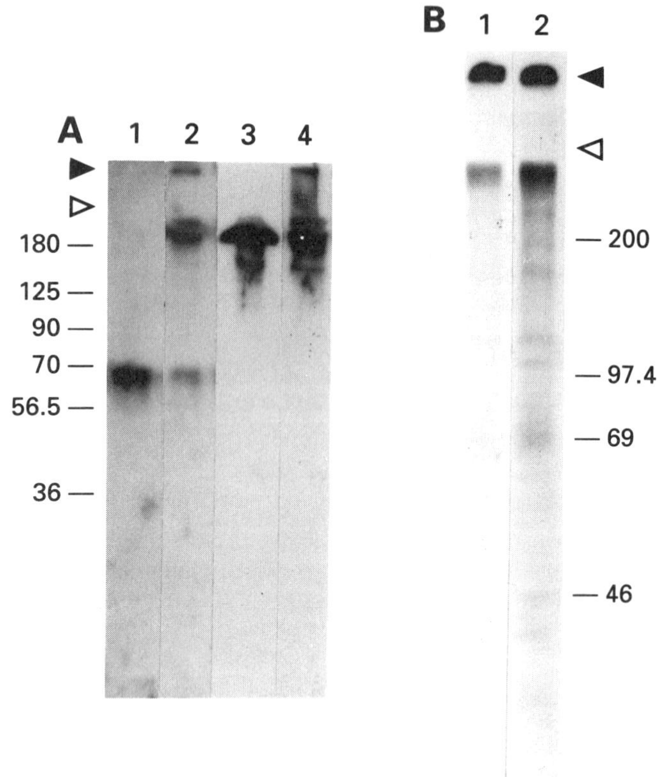

FIGURE 8. OP-Fn cross-linking demonstrated by Western blotting and immunoprecipitation of ^{125}I-labeled cell-surface proteins. **(A)** Proteins in *lanes 1 and 2* were interacted with anti-OPN antibody, and *lanes 3 and 4* with anti-Fn antiserum. *Lane 1,* OPN; *lane 2,* OPN + Fn + TG; *lane 3,* Fn; *lane 4,* OPN + Fn + TG. The amount of OPN and Fn used per reaction was 5 μg each. A catalytic amount of TG (1.3 nM) was used and the Ca^{2+} concentration was 1 mM. *Lanes 2 and 4* show three very-high-molecular-weight bands which interacted with both Fn and OPN antibodies. *Lanes 1 and 3* show monomeric forms of OPN and Fn, respectively. **(B)** ^{125}I-labeled NRK cell-surface proteins immunoprecipitated with either anti-OPN (*lane 1*) or anti-Fn (*lane 2*) antibody. Both showed a high-molecular-weight band which migrated into the resolving gel, and these showed an identical migration pattern. *Open and solid arrowheads* point to the top of stacking and resolving gels, respectively. Note a very-high-molecular-weight protein product reacting with both OPN and Fn antibodies which does not enter the resolving gel. (From Beninati *et al.*[27] Reprinted by permission from the *Journal of Biochemistry.*)

band contains both OPN and Fn, and since this band cannot be dissociated even after boiling, OPN and Fn in this band is probably cross-linked covalently.[27]

Analysis of OPN-Fn Cross-Linking Using [^{125}I]OPN

[^{125}I]OPN and unlabeled Fn in equimolar concentrations were incubated in the presence of a catalytic amount of transglutaminase (TG), and the reaction products were then subjected to SDS-PAGE and autoradiography. FIGURE 8 shows that when OPN was reacted with TG (lane 3), two very-high-molecular-weight bands appeared which must have resulted from TG-mediated intermolecular cross-linking of OPN. Furthermore, the electrophoretic mobility of the monomeric form of OPN in lane 3 increased as compared to those observed in lanes 1 and 2. This may be the result of intramolecular cross-linking of OPN by TG. When [^{125}I]OPN was incubated with unlabeled Fn in the presence of TG, the density of the radioactive band which did not enter the stacking gel increased significantly. Since the only variable between lanes 3 and 4 was the presence of Fn in lane 4, and only OPN was radiolabeled, it was concluded that OPN-Fn cross-linking was responsible for the increased labeling intensity of this high-molecular-weight band. These results demonstrate that TG catalyzes covalent cross-linking of these proteins.[27]

OPN is a component of the extracellular, bone, and urinary stone matrices[3,6,14,24] and immunohistochemical studies of osteoid suggest that interaction of OPN with collagen may involve Fn, since OPN does not possess a collagen attachment domain,[25] but Fn binds to type-I collagen.[26] Taken together, the above observations strongly suggest the possibility that one of the molecular mechanisms by which OPN becomes firmly incorporated into the extracellular matrices is TG-mediated OPN-Fn covalent cross-linking.[27]

In summary, we first demonstrated that normal rat kidney (NRK) cells secrete osteopontin (OPN) with a molecular weight of 69-kDa, and in two forms, phosphorylated (ppOPN) and nonphosphorylated (npOPN). Phosphorylated OPN interacts with the cell surface, and npOPN complexes with soluble fibronectin (Fn).[15] Nonphosphorylated OPN, but not ppOPN, contains N-linked carbohydrates, and npOPN lacking N-linked glycosylation fails to complex with soluble Fn. Phosphorylated OPN, however, forms a heat-stable complex with cell-surface (insoluble) Fn, suggesting that ppOPN is a component of the extracellular matrix (ECM) of NRK cells.[16] We have also demonstrated that OPN is a substrate of transglutaminase (TG),[20,27] and TG catalyzes covalent cross-linking of OPN with cell-surface Fn, both *in vitro* and *in vivo*. These observations clearly suggest that TG-mediated cross-linking between OPN and Fn represents one of the most likely mechanisms by which OPN becomes stably incorporated into the extracellular, bone, and urinary stone matrices.[27]

REFERENCES

1. SENGER, D. R., D. F. WIRTH & R. O. HYNES. 1979. Cell **16:** 885–893.
2. FRANZEN, A. & D. HEINEGÅRD. 1985. Biochem. J. **232:** 715–724.
3. FRANZEN, A. & D. HEINEGÅRD. 1985. *In* Chemistry and Biology of Mineralized Tissues. W. T. Butler, Ed. 132–141. EBSCO Media. Birmingham, AL.
4. CHACKALAPARAMPIL, I., D. BANERJEE, Y. POIRIER & B. B. MUKHERJEE. 1985. J. Virol. **53:** 841–850.
5. OLDBERG, A., A. FRANZEN & D. HEINEGÅRD. 1986. Proc. Natl. Acad. Sci. USA **83:** 8819–8823.
6. PRINCE, C. W., T. OOSAWA, W. T. BUTLER, M. TOMANA, A. BHOWN, M. BHOWN & R. E. SCHROHENLOHER. 1987. J. Biol. Chem. **262:** 2900–2907.
7. DENHARDT, D. T., A. M. CRAIG & J. H. SMITH. 1988. *In* Genes and Signal Transduction in Multistage Carcinogenesis. N. Colburn, Ed. 167–189. Alan R. Liss Inc. New York.
8. CRAIG, A. M., M. NEMIR, B. B. MUKHERJEE, A. F. CHAMBERS & D. T. DENHARDT. 1988. Biochem. Biophys. Res. Commun. **157:** 166–173.
9. SENGER, D. R., C. A. PERRUZZI, C. A. GRACEY, C. F. PAPADOPOULOS & D. G. TENEN. 1988. Cancer Res. **48:** 5770–5774.
10. SMITH, J. H. & D. T. DENHARDT. 1987. J. Cell Biochem. **34:** 13–22.

11. Craig, A. M., J. H. Smith & D. T. Denhardt. 1989. J. Biol. Chem. **264:** 9682–9689.
12. Sodek, J., Q. Zhang, H. A. Goldberg, C. Dominicucci, S. Kasugai, J. L. Warna, H. S. Shapiro & J. Chen. 1991. *In* The Bone Biomaterials Interface. J. E. Davies, Ed. 97–110. University of Toronto Press. Toronto.
13. Reinholt, F. P., K. Hultenby, A. Oldberg & D. Heinegård. 1990. Proc. Natl. Acad. Sci. USA **87:** 4473–4475.
14. Gerstenfeld, L. C., Y. Gotoh, M. D. McKee, A. Nanci, W. J. Landis & M. J. Glimcher. 1990. Anat. Record **228:** 93–103.
15. Nemir, M., M. W. DeVouge & B. B. Mukherjee. 1989. J. Biol. Chem. **264:** 18202–18208.
16. Singh, K., M. W. DeVouge & B. B. Mukherjee. 1990. J. Biol. Chem. **265:** 18696–18701.
17. Hynes, R. O. 1992. Cell **69:** 11–25.
18. Mosher, D. F., P. E. Schad & J. M. Vann. 1980. J. Biol. Chem. **255:** 1181–1188.
19. Keski-Oja, J., D. F. Mosher & A. Vahen. 1976. Cell **9:** 29–35.
20. Beninati, S., D. R. Senger, A. B. Mukherjee, K. Singh & B. B. Mukherjee. 1991. J. Cell Biol. **115:** 135a (abstract).
21. Prince, C. W., D. Dickie & C. L. Krumdieck. 1991. Biochem. Biophys. Res. Commun. **177:** 1205–1210.
22. Folk, J. E. 1980. Annu. Rev. Biochem. **49:** 517–531.
23. Folk, J. E. & S. I. Chung. 1973. Adv. Enzymol. Relat. Areas Mol. Biol. **38:** 109–191.
24. Kohri, K., Y. Suzuki, K. Yoshida, K. Yamamoto, N. Amasaki, T. Yamate, T. Umekawa, M. Iguchi, H. Sinohara & T. Kurita. 1992. Biochem. Biophys. Res. Commun. **184:** 859–864.
25. Butler, W. T. 1989. Connect. Tissue Res. **23:** 123–136.
26. Kleinman, H. K., E. B. McGoodwin, G. R. Martin, R. J. Klebe, P. P. Fietzek & D. E. Woolley. 1978. J. Biol. Chem. **253:** 5642–5646.
27. Beninati, S., D. R. Senger, E. Cordella-Miele, A. B. Mukherjee, I. Chackalaparampil, V. Shanmugam, K. Singh & B. B. Mukherjee. 1994. J. Biochem. **115:** 675–682.

Roles of Osteopontin in Bone Remodeling[a]

DICK HEINEGÅRD,[b] GÖRAN ANDERSSON,[c] AND FINN P. REINHOLT[c]

[b]*Department of Medical and Physiological Chemistry*
Lund University
P.O. Box 94
S-221 00 Lund, Sweden
and
[c]*Division of Pathology*
The Karolinska Institute
Huddinge University Hospital
S-141 86 Huddinge, Sweden

INTRODUCTION

Bone contains several proteins with a very high abundance of sialic acid-containing oligosaccharides, *e.g.*, BSP, BAG-75 and osteopontin. The one initially studied by Herring as early as the 1960s was bone sialoprotein (BSP).[1] It was later found, after the introduction by Oegema *et al.*[2] of proteinase inhibitors in connective tissue extraction, that these early studies described a fragment of a much larger protein.[3,4] The introduction of this approach in the field of bone research[5] made it possible to isolate and characterize an increasing number of intact proteins from bone.

One of these proteins is osteopontin, which was first identified and isolated in 1985.[4] It is somewhat less acidic than BSP, but from its primary sequence[6] it is clear that it contains stretches of aspartic acid residues, one stretch being nine residues long. The protein also contains O-glycosidically linked oligosaccharides[4] and serine phosphate groups.[7] It is distinct from the other bone proteins and its sequence has been determined from several sources following the initial sequencing of cDNA representing rat osteopontin.[6]

Interestingly, the purification showed that osteopontin can bind quite strongly to hydroxyapatite, which may be of relevance in considering its function in bone.[4] It is of particular interest that upon sequencing, it was found that osteopontin, similarly to what was later found for BSP,[8] contains the classical RGD-sequence interacting with integrin receptors at the cell surface, which was first identified in the cell-binding domain of fibronectin.[9] As discussed below, although over the years a number of bone matrix proteins have been isolated and characterized, we only have some ideas on the functional roles of a few of these, *i.e.,* collagen I, osteopontin and osteonectin. None of these proteins, however, is uniquely found in bone, but they are present in other tissues, with osteopontin exhibiting a somewhat more restricted distribution than the others. Thus, osteopontin mRNA levels

[a] This study was supported by grants from the Swedish Medical Research Council, the Bank of Sweden Tercentenary Foundation, Faculty of Odontology, Karolinska Institutet and the Axel and Margaret Ax:son Johnson Foundation.

are highest in bone, but clearly also significant in the kidney.[6,10] The protein has been shown to be a component of kidney stones, and it most likely influences kidney stone formation.[11] Furthermore, osteopontin is found at locations in the brain and in the placenta.[10] It is also expressed by transformed cells and by cells of the immune and inflammatory system.[12] Here we will give an account of some of the properties and putative functions of osteopontin in bone.

Osteopontin Function

Sequence determination of osteopontin indicates that the protein may bind to a cell in the matrix and furthermore it is known from the purification procedure that it binds very tightly to hydroxyapatite. To learn more about its role in bone, we first studied the effects on osteopontin mRNA-levels in ROS 17/2.8 osteoblastic osteosarcoma cells by factors known to be involved in the regulation of bone turnover. We found that 1,25-dihydroxyvitamin-D_3 (kindly provided by Sandoz Pharmaceuticals) greatly increased the mRNA-levels as well as the secretion of the protein from the cells. Interestingly, stimulation of osteopontin secretion was much more pronounced than for any other protein labeled in cultures of ROS 17/2.8 osteoblast cells.[13] Meanwhile, our results were corroborated in a study from Rodan's group.[14] We were also able to show that BSP exhibited a distinctly different regulation and was actually downregulated by vitamin-D_3, while upregulated by glucocorticoids.[15] Taken together, these data thus indicated that osteopontin synthesis is upregulated by 1,25-vitamin-D_3, a factor inducing calcium mobilization from bone, apparently by inducing osteoclast activity.

In view of the protein having an RGD putative cell binding sequence and also having hydroxyapatite binding capabilities, we hypothesized that osteopontin is involved in osteoclast binding to the mineral surface. Indeed, we could show that osteoclasts bind to osteopontin coated either to plastic or to glass surfaces[16] (FIG. 1). Notably, BSP bound osteoclasts less efficiently.[16] In subsequent experiments we demonstrated that the binding of isolated rat osteoclasts to osteopontin could be inhibited by synthetic RGD-peptides, but not by a control RGE-peptide. This prompted us to try to determine the subunit composition of the integrin involved. Only antibodies to the $\alpha_v\beta_3$ integrin inhibited binding to OPN (FIG. 1), while antibodies against a number of other integrins and their individual α- or β-chains were ineffective. The data show that binding of isolated osteoclasts to osteopontin is apparently mediated by the $\alpha_v\beta_3$ integrin, *i.e.*, a vitronectin receptor. This integrin had actually previously been demonstrated on isolated human osteoclast-like giant cells, by the use of a monoclonal antibody to an antigen referred to as the osteoclast-specified antigen, which later turned out to represent the β_3-subunit of the integrin.[17] Although the receptor is the commonly occurring $\alpha_v\beta_3$, it appears that the specificity of the cell binding can be made even more restricted since the microenvironment at the osteoclast surface may induce a specific conformation of the integrin only allowing binding of a specific subset of binding polypeptides.[18]

Studies on isolated cells can only serve as a model. Proof of the biological relevance of an interaction is actually that the prerequisites are found in the tissue by, for instance, demonstrating colocalization of the components involved. Therefore, we proceeded to probe the localization in bone of osteopontin as well as other proteins with a potential for interaction with cells at the light microscopic and ultrastructural levels. The primary focus of the study was on osteopontin, and the technology used was low-temperature embedding, which allows quantitative

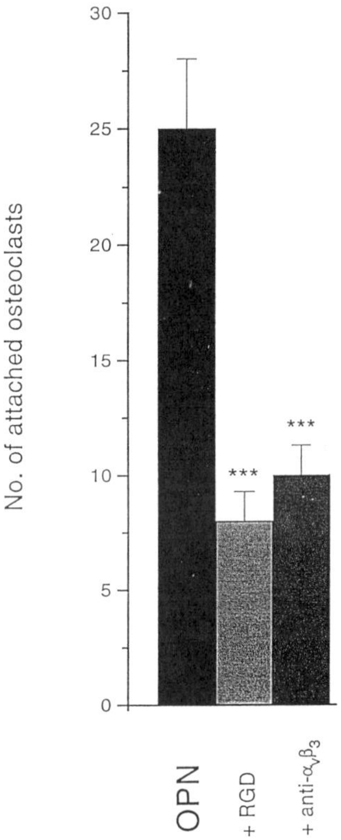

FIGURE 1. Binding of isolated rat osteoclasts to glass surfaces coated with osteopontin (*left-hand column*). Cells were identified by TRAP-staining. It is also shown that preincubation of the cells with an RGD-containing hexapeptide (GRGDSL) and anti-$\alpha_v\beta_3$, respectively, inhibits osteoclast attachment.

recovery of the proteins present in the tissue. It should be noted that most other techniques are prone to variability and irreproducibility due to extraction of components during processing. After low-temperature embedding, ultrathin sections were prepared and used to localize osteopontin by immunogold technology. We showed that osteopontin was much enriched at the bone surface facing the clear zone of osteoclasts,[19] shown in FIG. 2 and schematically in FIG. 3. At this site, osteopontin was enriched more than tenfold compared to most other sites in the tissue, with the exception of some surfaces facing osteoblasts. This latter observation is expected in view of the fact that osteopontin is secreted primarily by osteoblasts.[20] In contrast, no enriched localization of osteopontin was observed in bone matrix at the ruffled border of the osteoclast,[19] where the resorption takes place. Thus, it appears that osteopontin is involved in binding of the osteoclast since it is very selectively localized to the attachment zones. Only low levels of osteopontin were noted in central bone.[19]

In subsequent studies we gained further support for the contention that osteopontin is the primary attachment factor for osteoclasts, since other proteins with cell-binding capacity present in bone were not enriched at the adequate site. These proteins include fibronectin and BSP.[16]

In support of the role of osteopontin as an osteoclast anchor in bone, we could localize the receptor, $\alpha_v\beta_3$, much enriched at the clear zone area of the plasma

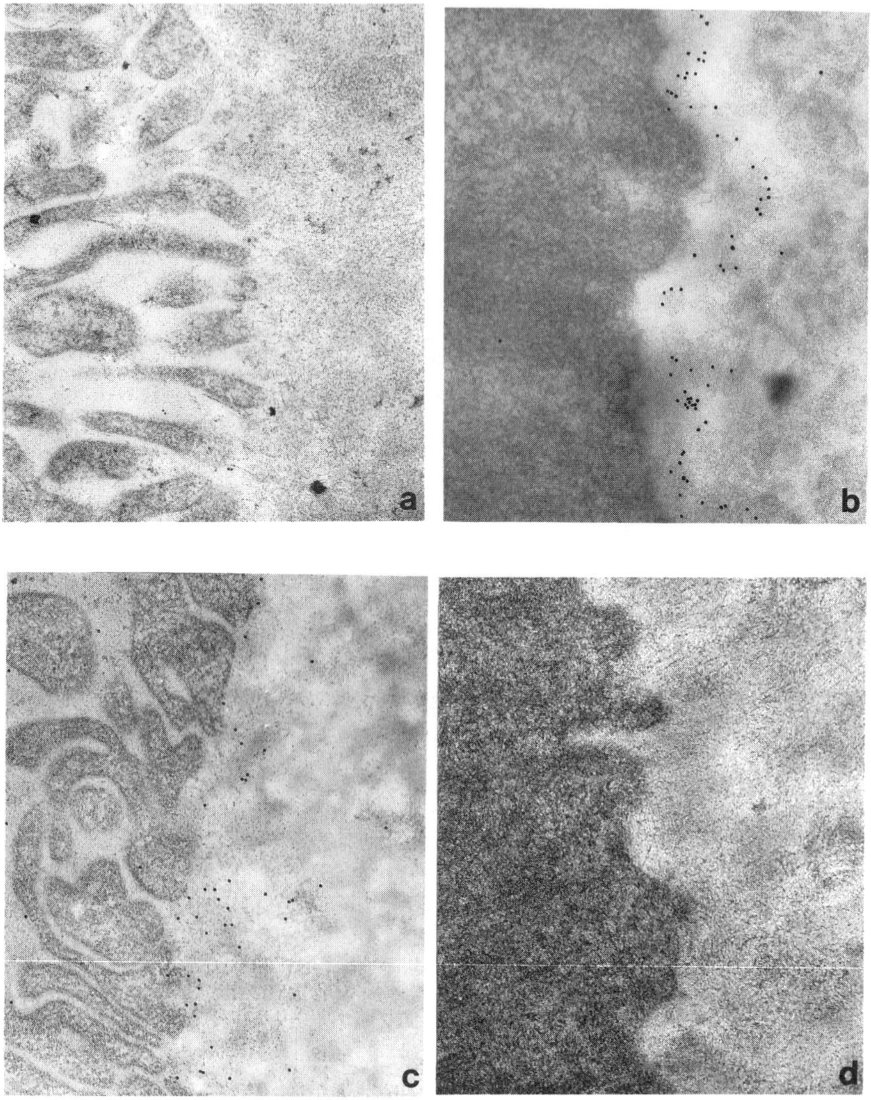

FIGURE 2. Immunolocalization of osteopontin **(a, b)** and tartrate-resistant and purple acid phosphatase (TRAP) **(c, d)** in metaphyseal rat bone facing osteoclasts. Osteopontin is virtually absent in bone facing the ruffled border (a), while accumulated at surfaces facing the clear zone (b). TRAP is enriched in the ruffled border area (c) but almost absent in the clear zone (d). Immunogold labeling of low-temperature-embedded bone. Magnification 48,000×.

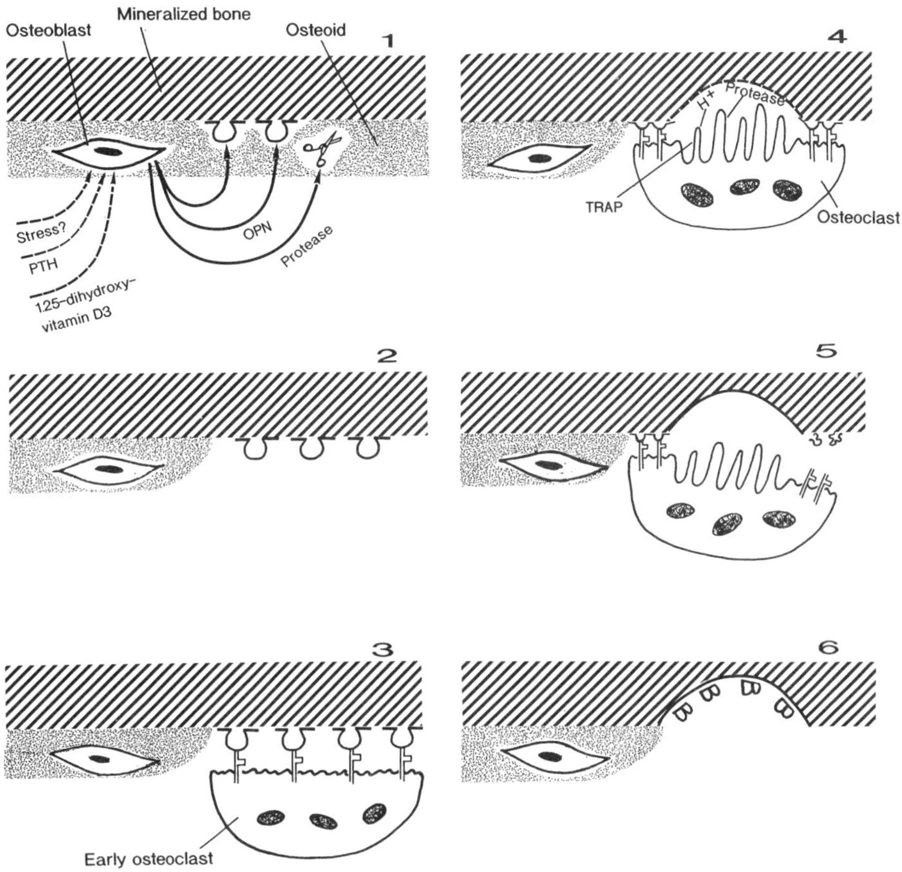

FIGURE 3. Schematic illustration of key steps in bone resorption. Initially, osteoblasts (lining cells) in nonmineralized matrix at the bone surface are induced to secrete osteopontin that binds to the bone mineralized surface. At the same time secreted proteinases are activated and remove the nonmineralized tissue covering the bone surface, thereby exposing osteopontin. Osteoclast precursor cells are recruited and differentiate/polarize to become the resorbing osteoclast. It should be noted that presently available data allow no conclusions to be drawn on the initial binding of the osteoclast precursor, since technology to unambiguously identify this cell is not available. As resorption continues the osteopontin is dephosphorylated, which may provide the mechanism for osteoclast detachment.

membrane opposing the site of osteopontin localization.[19,21] This provides strong corroborating evidence, since it is clear that osteopontin can bind to this particular receptor. Whether osteopontin represents the primary ligand for binding the osteoclasts to the mineralized bone surface could not be unequivocally answered from these studies, since the technology selects for well differentiated, characteristic osteoclasts which may represent a later differentiation stage than the primary attaching, nonresorbing osteoclast. However, in studies of the osteopetrotic *ia*/*ia*-rat, where osteoclast activity is limited, we found little development of the ruffled border region, while attachment of osteoclasts was adequate.[22] Also in this case we could show normal levels of osteopontin as well as the $\alpha_v\beta_3$ receptor at the predominating clear zone area.

In other studies using sections prepared by cryoultrotomy we found that no other representative integrins were demonstrable at the osteoclast plasma membrane corroborating the role of only osteopontin and the $\alpha_v\beta_3$ integrin in osteoclast development and attachment. Thus, we detected neither β_1- nor β_5-integrins. Interestingly, the $\alpha_v\beta_3$ integrin appears rather specific for the osteoclast since our studies revealed no or only low expression of this integrin on the osteoblast.[19,21]

In more recently published work by Lakkakorpi *et al.*[23] it was suggested that the $\alpha_v\beta_3$ integrin actually localized to the ruffled border region of the osteoclast and not at the clear zone area. The reason for these contradictory results is not clear, but the observation has, however, compelled us to go back and repeat the study with the use of similar techniques that these authors used. Our extended study offered little explanation for the difference since we still observed selective localization at the clear zone area and definitively no labeling at the ruffled border region. Lakkakorpi *et al.*[23] did not offer any explanation for the contradictory results and one should bear in mind that in their study no attempt was made to localize a potential ligand. Actually, we detected no candidate ligand in our studies at sites corresponding to the ruffled border region.[16,19] Since it is unlikely that the receptor would localize to a particular domain of the plasma membrane without having a ligand to tie it there, the support for an enrichment of $\alpha_v\beta_3$ to the ruffled border area is rather weak, particularly in view of the continuous breakdown of matrix at this site. Lakkakorpi *et al.*[23] observed highest labeling at the basolateral membrane, where we observed higher labeling than in the plasma membrane at the ruffled border region, but still severalfold less than at the clear zone.[19] One possible explanation is that the technology and antibody used in the Lakkakorpi study does not readily detect the integrin at the clear zone area, where it should be involved in interactions with its ligand. The several different antibodies to the complete $\alpha_v\beta_3$-integrin and its subunits used in our subsequent study[21] confirmed consistent enrichment of the integrin to the clear zone. It is unlikely that these different antibodies would all fail to detect the integrin at the ruffled border or the basolateral side.

In summary, compelling evidence demonstrates that the osteoclasts, at least the fully differentiated cells, attach to bone in their clear zone area and this involves binding, via an $\alpha_v\beta_3$ integrin, to osteopontin localized at the mineralized bone surface.

In support of a mechanism including an RGD-dependent osteoclast attachment, it has been shown that RGD-peptides as well as the snake venom RGD-containing polypeptide echistatin inhibit osteoclast binding. In fact echistatin can actually also inhibit osteoclastic bone resorption.[18] This information provides further support for the role of osteopontin in attaching osteoclasts via binding to $\alpha_v\beta_3$ integrins.

It is tempting to speculate on a role of integrin binding in the early phases of osteoclast differentiation. It is possible that signals are elicited to the mononuclear precursor, when binding via the integrins to the mineralized bone surface, to differentiate and polarize into the differentiated osteoclast. It is likely that at some stage, the osteoclast is attached via a continuous clear zone with no ruffled border region at the attachment area and that as polarization continues, the ruffled border region and resorption area develop. This is schematically illustrated in FIGURE 3, where a model for a sequence of events in bone resorption and remodeling is depicted. Some support for this hypothesis is the finding of a much larger clear zone area with normal levels of $\alpha_v\beta_3$ integrins and osteopontin in the *ia/ia*-rat.[22]

At the same time the ruffled border region appears not to develop properly as might be expected from the limited capacity for bone resorption with ensuing inability of the osteoclasts to detach in this osteopetrotic rat.

In support of the role of osteopontin in bone resorption, we and others have observed[20] high levels of osteopontin mRNA in osteoblasts in the metaphyseal region where osteoclastic activities are the most pronounced in growing animals (FIG. 4). Furthermore, intense osteopontin message is often noted in osteoblasts around osteoclasts.[20] We did not observe OPN message in the osteoclasts themselves, identified by presence of TRAP-mRNA (FIG. 4). Further conclusive evidence will have to await better identification of the osteoclasts, only adequately done by *in situ* hybridization at the electron microscopic level.

Interestingly, immunolabeling for osteopontin was not detected in the secretory vesicles accumulated on the cytosolic side of the clear zone of the *ia*/*ia*-osteoclast,[22] having a secretory dysfunction. This observation provides further indication that the osteoclast, at least at this stage of differentiation, is not a significant source of the osteopontin detected at the surface of the bone.

The normally localized differentiated osteoclasts secrete tartrate-resistant acid phosphatase (TRAP), which can be demonstrated also at the ultrastructural level by using immunogold technology.[24] It was found that the TRAP is much enriched in the ruffled border region and not detected in the clear zone area (FIG. 2),

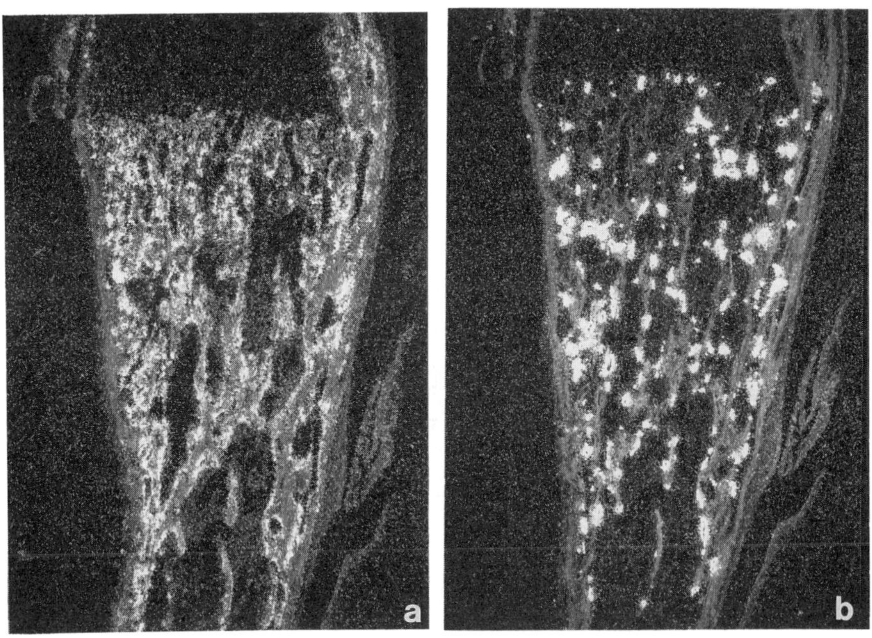

FIGURE 4. *In situ* hybridization of rat metaphyseal bone using RNA-probes for osteopontin **(a)** and tartrate-resistant and purple acid phosphatase **(b)**. Note the extensive message for osteopontin throughout the bone trabeculae with a somewhat less intensive signal at the epiphyseal/metaphyseal border and the distinct message for the enzyme in large cells—the osteoclasts. Magnification 40×.

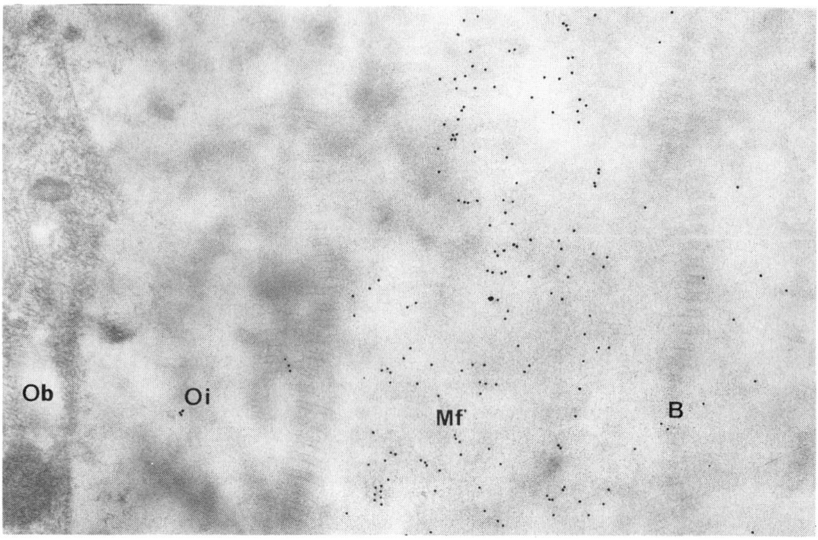

FIGURE 5. Immunolocalization of osteopontin in metaphyseal bone. Note the accumulation of immunoreactivity at the mineralizing front. Ob, osteoblast; Oi, osteoid; Mf, mineralizing front; B, bone. Immunogold labeling of low-temperature-embedded bone. Magnification 48,000×.

consistent with the secretion of the enzyme by the osteoclasts. The question then is whether this phosphatase has a role extracellularly. We therefore searched for potential substrates among the bone matrix phosphoproteins we had access to. We demonstrated that TRAP, *in vitro,* could dephosphorylate the three bone proteins tried, *i.e.*, osteopontin, BSP[25] as well as osteonectin (Ek-Rylander and Andersson, unpublished data). In search of consequences of the dephosphorylation of osteopontin we found that dephosphorylated osteopontin no longer supported binding of osteoclasts.[25] This is an interesting observation and it may show a potential mechanism for detachment of the cells when the resorption has proceeded to a certain stage. At this stage, the number of osteopontin molecules that have become dephosphorylated and therefore no longer support binding, may actually represent such a large proportion of the total number of osteopontin molecules available, that the total binding strength for the cell is too low for it to remain bound, schematically shown in FIGURE 3.

Evidence has thus been accumulating over recent years providing a strong indication that one essential protein involved in bone resorption is osteopontin.

It appears that osteopontin has more than one role in bone, like many other proteins that have several roles. Thus we could early show, by quantitative immunohistochemical studies on low-temperature embedded tissue explored at the ultrastructural level, that osteopontin is enriched at the mineralization front in the bone[26] (FIG. 5). Although BSP and osteopontin have similar overall physical properties, the pattern with BSP was strikingly different in that this protein showed no such localization.[20] It is possible that the osteopontin localized at this site serves to regulate the mineralization process. In support, work from the laboratory of Goldberg[27] has corroborated this finding and furthermore shown that osteopontin can actually inhibit nucleation and crystal growth of hydroxyapatite in model systems.[27] Interestingly, osteopontin dephosphorylated by alkaline phosphatase did not inhibit hydroxyapatite formation in these *in vitro* systems. These observations infer potential different roles for dephosphorylated compared to intact osteopontin during bone turnover. It is thus possible that osteopontin actually has several functions, both in regulating mineralization of the skeleton and in the resorption of already formed mineralized matrix.

CONCLUDING REMARKS

Studies of osteopontin structure, localization and metabolism have thus provided new insights into one facet of bone remodeling. The information may actually help us target the event in this process by making use of specific competitors and/or modulating the expression of osteopontin.

ACKNOWLEDGMENTS

This overview represents much work by Barbro Ek-Rylander, Myra Flores, Kjell Hultenby, Maria Norgård and Mikael Wendel.

REFERENCES

1. HERRING, G. M. 1972. The organic matrix of bone. *In* The Biochemistry and Physiology of Bone. G. H. Bourne, Ed. Vol. 1: 127–189. Academic Press. New York.
2. OEGEMA, T., V. HASCALL & D. DZIEWIATKOWSKI. 1975. Isolation and characterization of proteoglycans from the Swarm rat chondrosarcoma. J. Biol. Chem. **250:** 6151–6159.
3. FISHER, L. W., S. W. WHITSON, L. V. AVIOLI & J. D. TERMINE. 1983. Matrix sialoprotein of developing bone. J. Biol. Chem. **258:** 12723–12727.
4. FRANZÉN, A. & D. HEINEGÅRD. 1985. Isolation and characterization of two sialoproteins present only in bone calcified matrix. Biochem. J. **232:** 715–724.
5. FISHER, L. W., J. D. TERMINE, S. W. DEIJTER, S. W. WHITSON, M. YANAGISHITA, J. H. KIMURA, V. C. HASCALL, H. K. KLEINMAN, J. R. HASSEL & B. NILSSON. 1983. Proteoglycans of developing bone. J. Biol. Chem. **258:** 6588–6594.
6. OLDBERG, Å., A. FRANZÉN & D. HEINEGÅRD. 1986. Cloning and sequence analysis of rat bone sialoprotein (osteopontin) cDNA reveals an Arg-Gly-Asp cell-binding sequence. Proc. Natl. Acad. Sci. USA **83:** 8819–8823.
7. HEINEGÅRD, D., K. HULTENBY, Å. OLDBERG, F. REINHOLT & M. WENDEL. 1989. Macromolecules in bone matrix. Connect. Tissue Res. **21:** 3–14.
8. OLDBERG, Å., A. FRANZÉN & D. HEINEGÅRD. 1988. The primary structure of a cell binding bone sialoprotein. J. Biol. Chem. **263:** 19430–19432.
9. RUOSLAHTI, E. & M. D. PIERSCHBACHER. 1987. New perspectives in cell adhesion: RGD and integrins. Science **238:** 491–497.
10. NOMURA, S., A. J. WILLS, D. R. EDWARDS, J. K. HEATH & B. L. M. HOGAN. 1988. Developmental expression of 2ar (osteopontin) and SPARC (osteonectin) RNA as revealed by *in situ* hybridization. J. Cell Biol. **106:** 441–450.
11. NEMIR, M., M. W. DEVOUGE & B. B. MUKHERJEE. 1989. Normal rat kidney cells secrete both phosphorylated and nonphosphorylated forms of osteopontin showing different physiological properties. J. Biol. chem. **264**(30): 18202–18208.
12. CRAIG, A. M., J. H. SMITH & D. T. DENHARDT. 1989. Osteopontin, a transformation-

associated cell adhesion phosphoprotein, is induced by 12-*O*-tetradecanoylphorbol 13-acetate in mouse epidermis. J. Biol. Chem. **264**(16): 9682–9689.

13. HEINEGÅRD, D., B. JIRSKOG-HED, Å. OLDBERG, F. REINHOLT & M. WENDEL. 1990. Bone macromolecules. *In* Calcium Regulation and Bone Metabolism. Basic and Clinical Aspects. D. V. Cohn, F. H. Glorieux & T. J. Martin, Eds. Vol. 10: 181–187. Excerpta Medica. Elsevier. Amsterdam.
14. YOON, K., R. BUENAGA & G. A. RODAN. 1987. Tissue specificity and developmental expression of rat osteopontin. Biochem. Biophys. Res. Commun. **148**(3): 1129–1136.
15. OLDBERG, Å., B. JIRSKOG-HED, S. AXELSSON & D. HEINEGÅRD. 1989. Regulation of bone sialoprotein (BSP) mRNA by steroid hormones. J. Cell Biol. **109:** 3183–3186.
16. FLORES, M., M. NORGÅRD, D. HEINEGÅRD, F. REINHOLT & G. ANDERSSON. 1992. RGD-directed attachment of isolated rat osteoclasts to osteopontin, bone sialoprotein and fibronectin. Exp. Cell Res. **201:** 526–530.
17. HORTON, M. A. 1988. Osteoclast-specific antigens. ISI Atlas Sci. Immunol. **1:** 35–43.
18. SATO, M., V. GARSY, R. J. MAJESKA, T. A. EINHORN, J. MURRAY, A. H. TASHJIAN, JR. & R. J. GOULD. 1994. Structure-activity studies of the s-echistatin inhibition of bone resorption. J. Bone Miner. Res. **9**(9): 1441–1449.
19. REINHOLT, F. P., K. HULTENBY, Å. OLDBERG & D. HEINEGÅRD. 1990. Osteopontin—a possible anchor of osteoclasts to bone. Proc. Natl. Acad. Sci. USA **87:** 4473–4475.
20. HULTENBY, K., F. REINHOLT, M. NORGÅRD, Å. OLDBERG, M. WENDEL & D. HEINEGÅRD. 1994. Distribution and synthesis of bone sialoprotein in metaphyseal bone of young rats show a distinctly different pattern from that of osteopontin. Eur. J. Cell Biol. **63:** 230–239.
21. HULTENBY, K., F. REINHOLT & D. HEINEGÅRD. 1993. Distribution of integrin subunits on rat metaphyseal osteoclasts and osteoblasts. Eur. J. Cell Biol. **62:** 86–93.
22. HULTENBY, K., F. P. REINHOLT, D. HEINEGÅRD, G. ANDERSSON & S. C. MARKS, JR. 1994. Osteoclast attachment in osteopetrotic (*ia*/*ia*) rats—Ultrastructural distribution of osteopontin and integrin subunits α_v and β_3. Submitted for publication.
23. LAKKAKORPI, P. T., M. A. HORTON, M. H. HELFRICH, E.-K. KARHUKORPI & H. K. VÄÄNÄNEN. 1991. Vitronectin receptor has a role in bone resorption but does not mediate tight sealing zone attachment of osteoclasts to the bone surface. J. Cell Biol. **115**(4): 1179–1186.
24. REINHOLT, F. P., S. MENGARELLI WIDHOLM, B. EK-RYLANDER & G. ANDERSSON. 1990. Ultrastructural localization of a tartrate-resistant acid ATPase in bone. J. Bone Miner. Res. **5**(10): 1055–1061.
25. EK-RYLANDER, B., M. FLORES, M. WENDEL, D. HEINEGÅRD & G. ANDERSSON. 1994. Dephosphorylation of osteopontin and bone sialoprotein by osteoclastic tartrate-resistant acid phosphatase. Modulation of osteoclast adhesion *in vitro*. J. Biol. Chem. **269:** 14853–14856.
26. HULTENBY, K., F. REINHOLT, Å. OLDBERG & D. HEINEGÅRD. 1991. Ultrastructural immunolocalization of osteopontin in metaphyseal and cortical bone. Matrix **11:** 206–213.
27. HUNTER, G. K., C. L. KYLE & H. A. GOLDBERG. 1994. Modulation of crystal formation by bone phosphoproteins: structural specificity of the osteopontin-mediated inhibition of hydroxyapatite formation. Biochem. J. **300:** 723–728.

Regulation of Osteopontin Expression in Osteoblasts

JARO SODEK,[a,b] JINKUN CHEN,[c] TOSHIHIKO NAGATA,[d]
SHOHEI KASUGAI,[e] REYNALDO TODESCAN, JR.,[b]
IVAN W. S. LI,[b] AND RICHARD H. KIM[b]

[b]*Medical Research Council Group in Periodontal Physiology*
Faculty of Dentistry
and
Department of Biochemistry
Univeristy of Toronto
Toronto, Ontario, Canada M5S 1A8

[c]*Department of Pediatric Dentistry*
University of Texas Health Science Center at San Antonio
Dental School
San Antonio, Texas 78284-7888

[d]*Department of Periodontics and Endodontology*
Tokushima University School of Dentistry
Tokushima 770, Japan

[e]*Department of Pharmacology*
Faculty of Dentistry
Tokyo Medical and Dental University
Tokyo 113, Japan

Previous studies have identified osteopontin (OPN) as a prominent component of the mineralized bone extracellular matrix.[1–4] In most mammalian species only a single form of the protein has been recognized despite the knowledge that considerable variation could be generated through alternate splicing of the gene transcript,[5,6] and through post-translational modification of the nascent protein involving glycosylation, phosphorylation and sulphation.[6–8] Indeed OPNs produced by different tissues within a species are likely to show discrete differences in their structure that may be important to the specific function of OPN in those particular tissues.

The primary structures of OPNs from a number of mammalian[10–14] and avian[15,16] species have been determined. Since conserved structural features are likely to be important in the function of the protein, comparison of these sequences can be highly informative. Thus, while there is a fairly extensive conservation of sequence in the mammalian proteins (~50% sequence identity and a further ~10% of the sequence conserved), in which several extraskeletal functions are apparent,[17] when the avian sequence is included, relatively few common sequences are evident

[a] To whom correspondence should be addressed: Dr. J. Sodek, MRC Group in Periodontal Physiology, 4383, Medical Sciences Building, University of Toronto, Toronto, Ont. M5S 1A8, Canada.

(FIG. 1). These include the polyaspartic acid sequence, the RGD cell attachment sequence, the thrombin susceptibility site and several sites of serine phosphorylation.

Immunohistochemical staining of bone tissues with antibodies to OPN have shown the protein to be present in the mineralized bone matrix, in the osteoid seam[18,19] and beneath lining cells of older bone (FIG. 2), and in reversal lines of remodelling alveolar bone.[20] Examination of remodelling cortical bone has revealed concentric seams of OPN labelling that appear to represent the initial deposition of a cement layer and, following the subsequent formation of bone, an OPN-enriched seam that defines the outer limit of the bone that encircles the canniculi (FIG. 2). Electron microscopic examination of immunogold staining for OPN has shown a strong association of gold particles with electron-dense material described as "grey patches" within the bone and in cement lines, and as a lamina limitans lining the surface of mineralized matrix in bone and in cartilage.[21–23] The lamina limitans appears to provide the attachment structure for lining cells, for osteoblasts and for osteoclasts,[23] indicating a role for OPN in cell attachment through the RGD motif which recognizes the $\alpha_v\beta_3$ vitronectin integrin present on the surface of bone cells and osteoclasts.[24,25]

Studies on the developmental expression of OPN using *in situ* hybridization have shown that OPN is expressed by osteoblast precursors early in bone formation, and by osteoblastic cells during *de novo* bone formation.[26] However, in newly-forming calvarial bone the strongest signals are observed in cells on the lower, resorbing surface, while in growing tibiae the strongest signals are evident in cells lining the bone trabeculae in the secondary spongiosa, which is also undergoing rapid remodelling.[27] That OPN is associated with bone formation is, nevertheless, apparent from the lower, but significant expression in the osteoblastic cells in the epiphysis and in some of the hypertrophic cartilage cells.[27–29] The results of *in situ* hybridization analyses are consistent with results of Northern hybridizations demonstrating expression of OPN mRNA early in bone formation, with a dramatic increase in expression occurring following the initial formation of bone.[29] Notably, when embryonic tibiae are cultured in the presence of vitamin D_3 the expression of OPN mRNA is strongly stimulated as shown in FIGURE 3. In these embryonic bones the signal in the hypertrophic cartilage cells is especially enhanced and strong hybridization is also evident in the epiphysis and in the periosteal cells. These observations reflect the importance of vitamin D_3 regulation of OPN in promoting the formation and remodelling of bone during growth and development.

While the temporal expression and localization of OPN *in vivo,* in conjunction with its structural properties, have provided insights into potential functions of OPN, more detailed analysis of its involvement in bone remodelling has come from studies of its regulated expression and its tissue compartmentalization in bone formation *in vitro*. The regulation of OPN has been reported in a number of normal and transformed bone cell lines that can be readily maintained in culture. Earlier studies demonstrated a marked increase in OPN expression by vitamin D_3,[30,31] and TGF-β, which stimulates matrix formation by connective tissue cells, was also shown to increase OPN expression significantly.[32] However, it was subsequently recognized that, in contrast to larger mammals which appear to produce a single bone OPN, rodent bone cells produce multiple forms of OPN.[8] Moreover, since these early studies had used transformed (osteosarcoma) cells, which characteristically produce an OPN which may be associated with cellular transformation, the physiological significance of these earlier findings need to be re-evaluated.

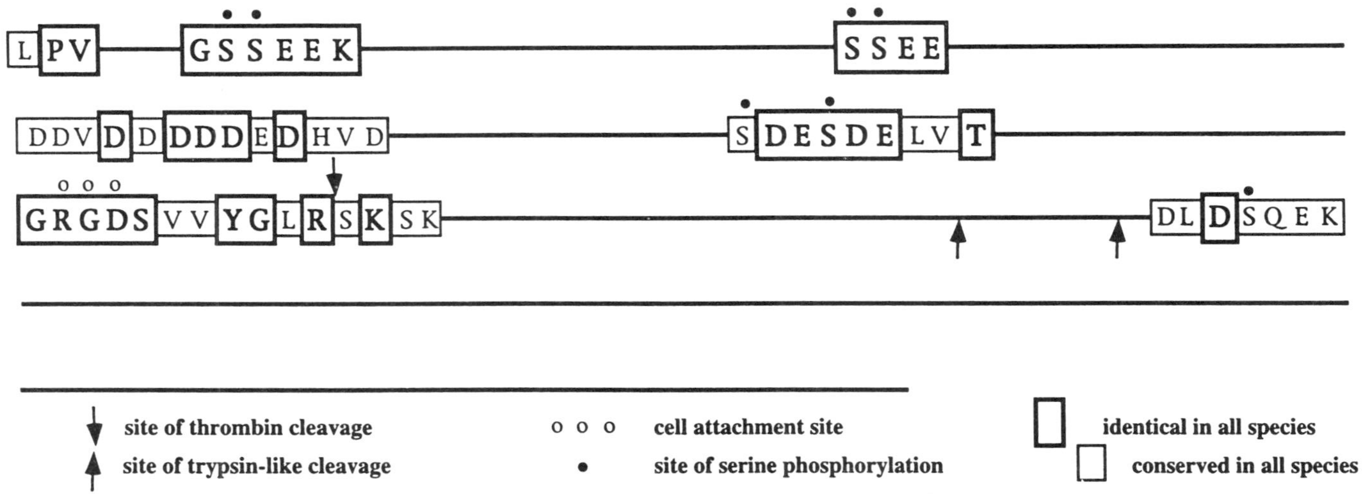

FIGURE 1. Comparison of osteopontin sequences. The primary sequences of OPNs derived from cDNA analysis of the human,[6] cow,[13,14] pig,[12] rat,[10] mouse,[11] and chicken[15,16] transcripts are compared to reveal conserved regions. Sequence identity shown represents ~11% of the total amino acids, while conservation of amino acids is ~20%. The conserved regions include the polyaspartic acid sequence, the RGD cell attachment site, the thrombin susceptibility site and several sites of serine phosphorylation.

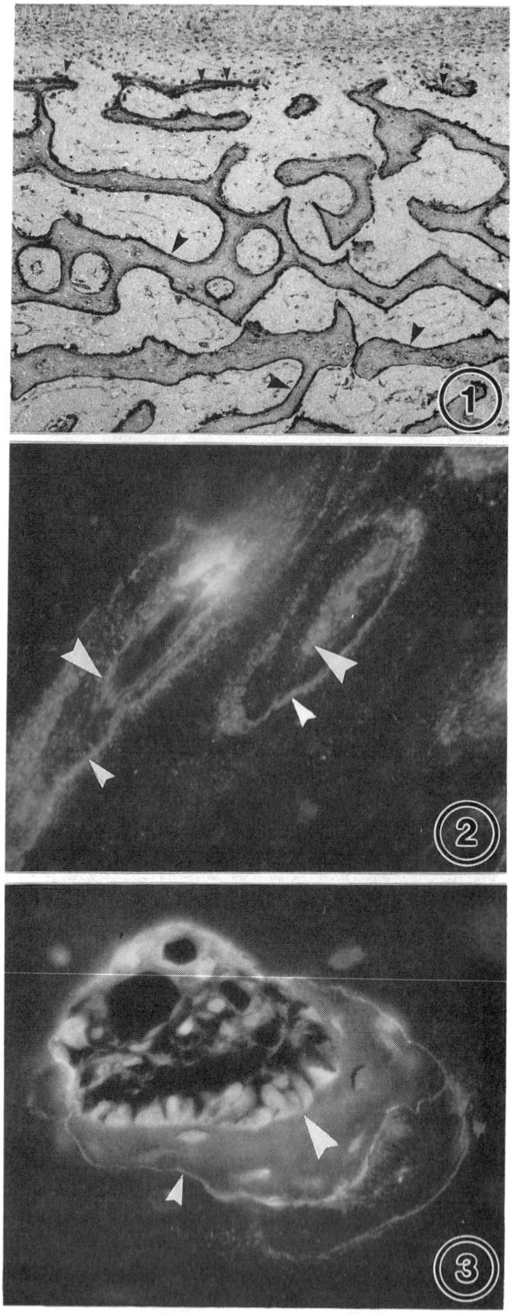
1
2
3

Although not identified as OPN in the original studies, two forms of phosphorylated proteins termed pp62 and pp69 were shown to be expressed by transformed cells.[11] The pp69 has since been shown to associate with fibronectin on the cell surface, whereas a nonphosphorylated, N-glycosylated form, np69, immunoprecipitates with soluble fibronectin.[7,9] Two principal forms of OPN, a 55-kDa and a 44-kDa species, have been demonstrated as products of rat bone cells.[8] By using [^{35}S]-methionine, [$^{35}SO_4$] and [$^{32}PO_4$] to label protein synthesized by bone cells, it has been shown that the 55-kDa OPN is less phosphorylated and less sulphated than its 44-kDa counterpart[35] and that it is expressed by less differentiated osteoblastic cells. This latter conclusion is based on the synthesis of this form of OPN by bone-derived cell lines which do not differentiate into osteoblasts[36] and from the temporal expression of these proteins by differentiating bone cells during bone formation *in vitro*.[37,38] In comparison, the more highly phosphorylated and sulphated 44-kDa OPN corresponds to the protein isolated from rat bone[2] and is expressed by differentiated osteoblasts. Together with osteocalcin and bone sialoprotein (BSP), it is considered a phenotypic marker of fully differentiated osteoblastic cells.[37] Thus, the 44-kDa OPN is expressed only after the differentiation of osteoblasts during the formation of bone *in vitro*.[37–39]

Notably, a highly phosphorylated 44-kDa OPN is also expressed together with a 55-kDa OPN by the well-characterized osteosarcoma cell line, ROS 17/2.8,[40] which are considered to have an osteoblastic phenotype.[41] However, since the expression of OPN as pp62 and pp69 by transformed cells has been shown in nonbone cells it is unclear whether the OPN forms expressed by ROS 17/2.8 cells represent the transformation-associated proteins or bone proteins. Although there is an apparent discrepancy in the size of these proteins, it is known that the M_r of OPN on SDS-PAGE varies according to the acrylamide concentration.[42] We sought, therefore, to compare the OPN forms from the different cells. When [$^{32}PO_4$]-labelled OPNs produced by ROS 17/2.8 cells, and NRK 49F cells which produce pp62 and pp69, were analyzed on SDS-PAGE the mobilities of the two forms of OPNs were indistinguishable from the mobilities of the 55-kDa OPN synthesized by the rat calvarial cell line A11 and confluent freshly isolated rat calvarial (RC) cells, and the 44-kDa forms produced by bone-forming RC cells (FIG. 4A). Moreover, digestions with trypsin and thrombin, which cleaves at a specific site in OPN, produced products of 33 and 27 kDa, and 14 and 10 kDa, respectively, from each OPN form (FIG. 4B). The results of a number of studies comparing radiolabelled OPNs produced by different bone cell populations are summarized in FIGURE 5. Although these studies indicate close similarity of the bone and transformed cell OPNs, the elution of the 44-kDa protein from anionic

FIGURE 2. Immunolocalization of OPN. **(1)** Immunoperoxidase labelling of fetal porcine calvaria showing strong staining for OPN at sites of bone formation on newly-forming trabeculae on the dorsal surface (*small black arrowheads*) and on bone surfaces of older trabeculae (*large black arrowheads*). The mineralized bone matrix shows a lower level of staining which may reflect the loss of OPN as mineralization proceeds. **(2)** Immunofluorescent staining of OPN in adult rat cortical bone (tibia) showing seams of OPN in remodelled bone lamellae. The outer seam (*small white arrowheads*) appears to represent a cement layer laid down following osteoclastic resorption while the inner seam (*large white arrowheads*) appears to represent OPN laid down after the cessation of bone formation. **(3)** Immunofluorescent staining of an osteoclast resorbing adult rat cortical bone. OPN staining is seen as an outer seam (*small white arrowhead*) and an inner seam (*large white arrowhead*) that enclose the newly-formed bone matrix. Bone-forming cells line the upper surface of the inner seam.

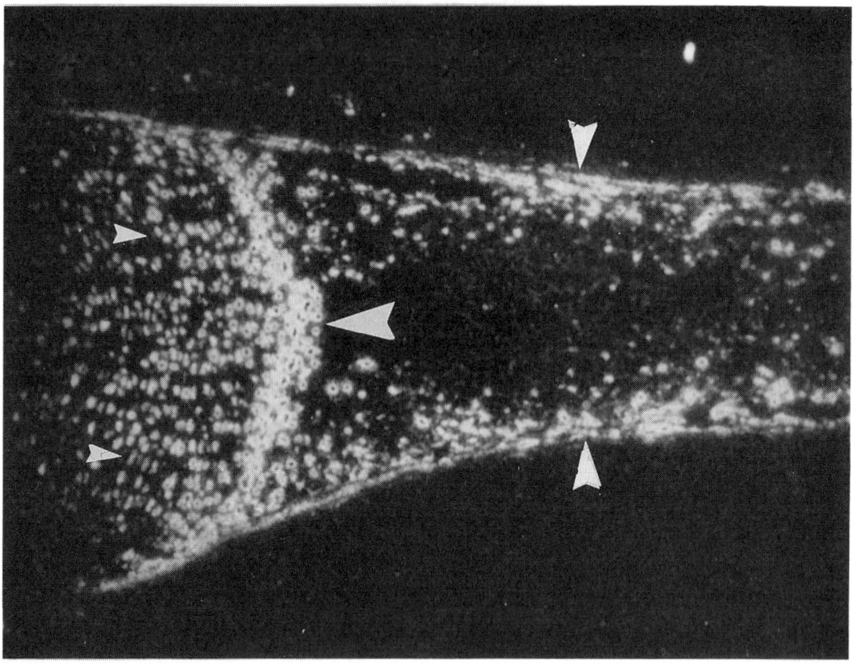

FIGURE 3. *In situ* hybridization analysis of OPN mRNA expression in rat tibial bone. The cellular expression of OPN mRNA is shown in embryonic tibia maintained in culture in the presence of 10^{-8} M vitamin D_3. *In situ* hybridization with an OPN cRNA probe was performed as described previously.[27] Strong hybridization, evident as white dots in the dark field micrograph, is seen in cells of the epiphysis (*large white arrowhead*), in cells associated with cortical bone (*medium white arrowheads*) and in the hypertrophic cartilage cells (*small white arrowheads*). At this stage the hybridization signal for the OPN mRNA in the control tibial bone was very low.

exchange resin 'Mono Q' (Pharmacia, FPLC) indicated a stronger binding of the ROS 17/2.8 44-kDa OPN compared to the corresponding RC cell protein and a slightly faster migrating thrombin fragment has been observed.[40] Also, while 2-D electrophoresis reveals essentially identical patterns for the OPNs produced by ROS 17/2.8 and NRK cells, the corresponding patterns for the bone cell OPNs do show some apparent differences in mobility (FIG. 6). Most notable, however, is that [^{35}S]-methionine incorporation into the 44-kDa protein expressed by transformed cells has been difficult to detect, as reported previously for the pp62 in transformed cells,[34,43,44] in comparison to the 44-kDa OPN produced by RC osteoblasts.[40] Since rat OPN has two methionines, which are present at positions 85 and 178 in the primary sequence of the nascent protein,[10] and since alternate splicing of OPN could potentially remove one or both of these methionines, we analyzed for differences in the size of mRNAs in the different cell populations. To increase the sensitivity of this analysis we performed RT-PCR on the mRNA preparations with primers designed to amplify each half of the OPN mRNA, one methionine being present in each half, and separated the amplified products on a

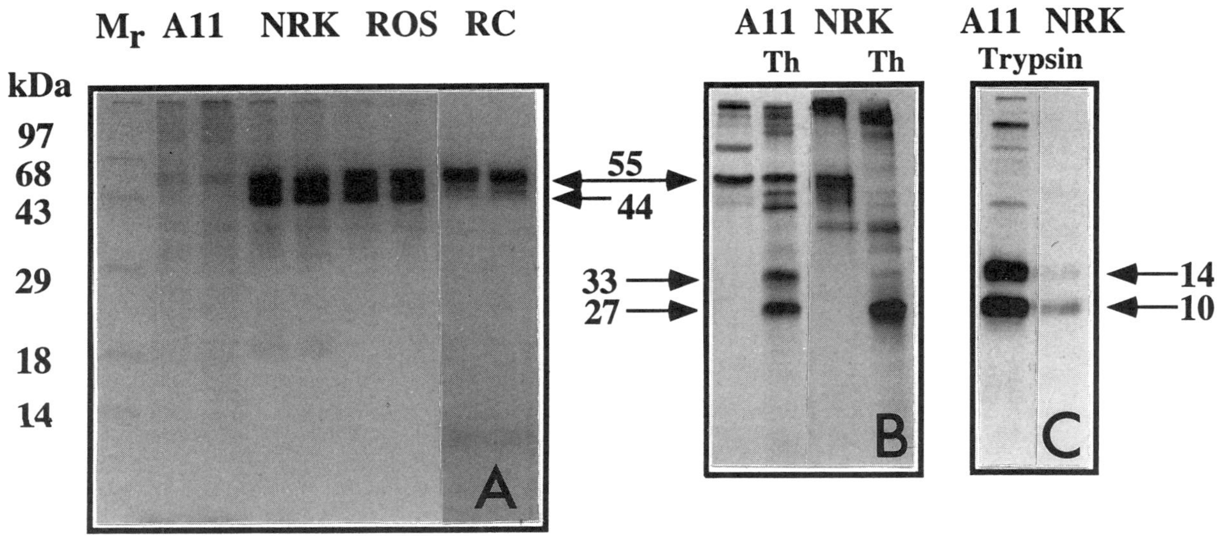

FIGURE 4. Comparison of rat OPNs on SDS-PAGE. **(A)** The electrophoretic mobility of [$^{32}PO_4$]-labelled OPNs secreted by A11, NRK 49F, ROS 17/2.8 and RC cells on 10% SDS-PAGE Tris/glycine gels are compared, showing the production of 44-kDa and 55-kDa forms. **(B)** Thrombin (Th) digest of [^{35}S]-methionine-labelled OPNs produced by A11 and NRK cells and analyzed on 15% SDS-PAGE gels. Two fragments of 33 kDa and 27 kDa were generated from the 55-kDa OPN synthesized by the A11 cells, whereas the 27-kDa fragment was the prominent fragment generated from the 44-kDa and 55-kDa OPNs synthesized by the NRK 49F cells. **(C)** Tryptic fragments of 14 kDa and 10 kDa obtained from the same OPNs as shown in (B) analyzed on 15% SDS-PAGE Tris/glycine gels.

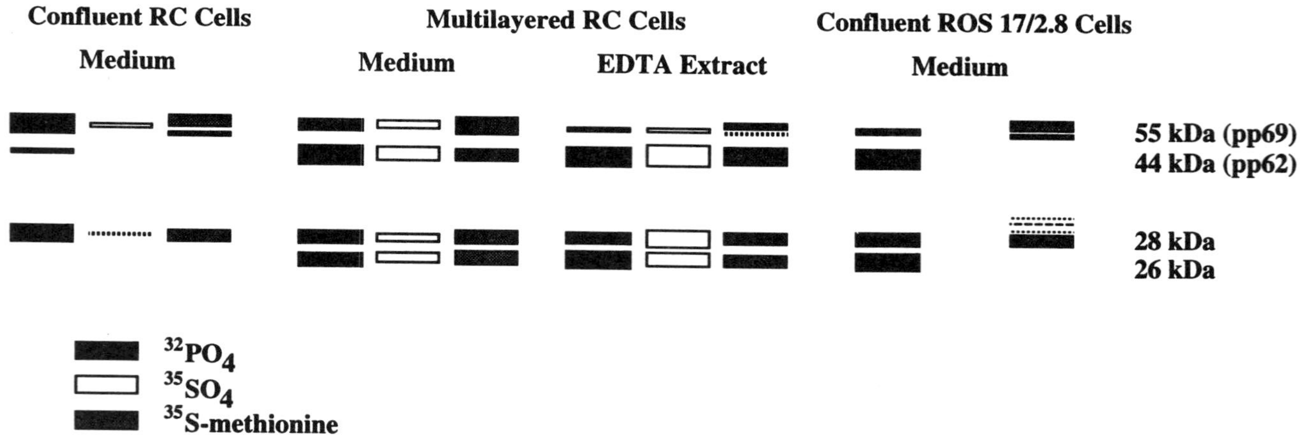

FIGURE 5. Multiple forms of rat OPNs. A summary of the electrophoretic characteristics of the various forms of radiolabelled OPN on SDS-PAGE are compared for normal rat calvarial (RC) and transformed (ROS 17/2.8 cells) bone cells. The results are from experiments in which secreted OPNs labelled with either [^{35}S]-methionine, [$^{35}SO_4$] or [$^{32}PO_4$] were immunoprecipitated and separated on 15% acrylamide Tris/glycine gels (sulphation of ROS 17/2.8 cells was not analyzed). In addition, the position of the thrombin fragments generated from these radiolabelled proteins is also shown. Note the absence of a band for the [^{35}S]-methionine-labelled 44-kDa OPN produced by the ROS 17/2.8 cells and the low sulphation of the OPN produced by the RC cells at confluence.

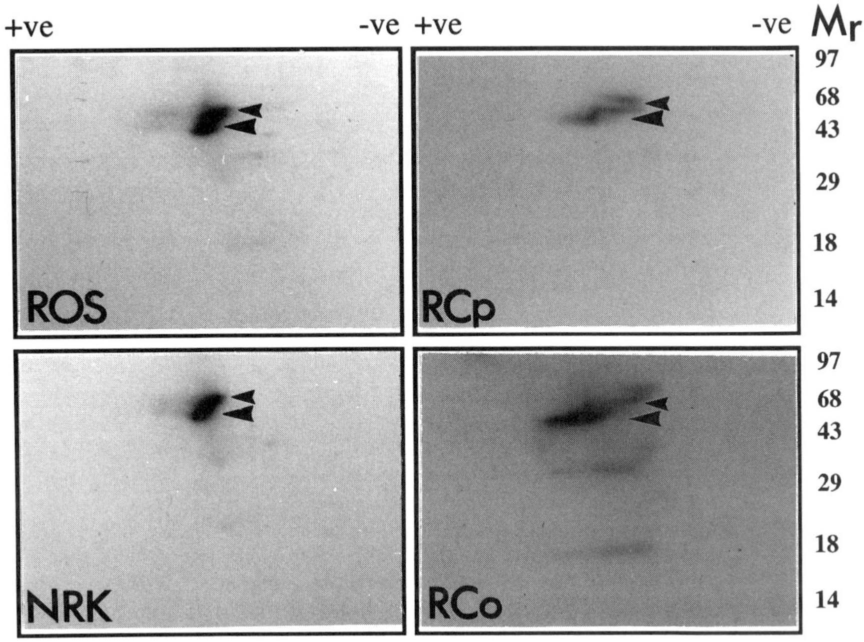

FIGURE 6. Two-dimensional electrophoretic analysis of rat OPNs. [$^{32}PO_4$]-labelled OPNs secreted by NRK 49F, ROS 17/2.8 and RC cells (RCp, preosteoblasts; RCo, osteoblasts) analyzed by 2-dimensional electrophoresis. The proteins were first separated by isoelectric focussing followed by SDS-PAGE on 15% Tris/glycine gels. Note the close similarity in the electrophoretic properties of the 44-kDa and 55-kDa forms of OPN produced by the NRK and ROS 17/2.8 cells whereas the 44-kDa and 55-kDa OPNs produced by the RC cells appear to be further resolved on the electro-focussing.

2% agarose gel (FIG. 7). In each case essentially identical patterns of amplification were obtained with prominent bands present at positions expected for the product's 'unspliced' mRNA. These results indicate, therefore, that the low incorporation of [^{35}S]-methionine in the transformed cell 44-kDa (pp62) OPN combined with its ease of [$^{32}PO_4$]-labelling must reflect an extremely high level of phosphorylation in the transformation-associated protein, compared to the corresponding bone protein. Importantly, these results also indicate that the 44-kDa product of the ROS 17/2.8 cells is the minor form of OPN compared to the 55-kDa form, and that these OPNs likely represent the transformation-associated OPNs.

Analysis of osteotropic hormones, vitamin D_3 and retinoic acid, and the cytokines PDGF, EGF, TGF-β, together with nonphysiological agents such as phorbol esters and concanavalin A have all shown stimulatory effects on OPN expression.[45] However, the magnitude of the stimulation and the effects on the different forms of OPN and their posttranslational modification is clearly dependent upon the target bone cell. Thus, while ROS 17/2.8 cells, the established bone cell line A11 which produces only the 55-kDa OPN, and osteoblastic RC cells (population IV) all respond strongly to vitamin D_3, the response of the ROS 17/2.8 cells to other agents was generally lower when compared to the A11 cells and particularly when

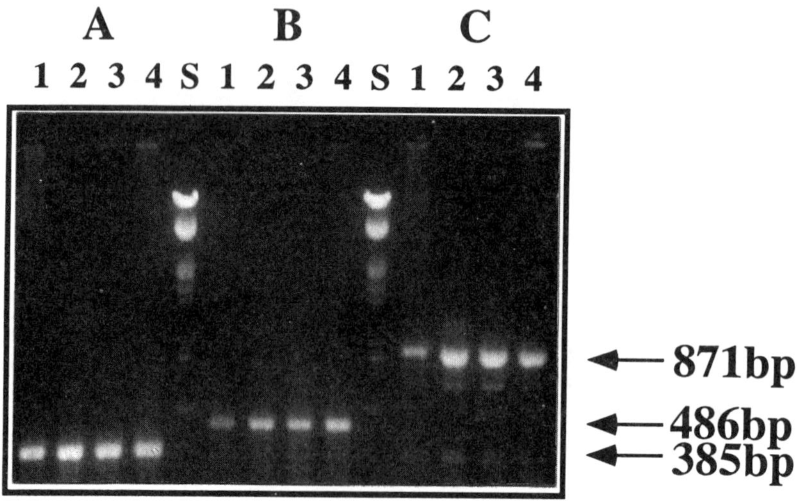

FIGURE 7. RT-PCR amplification of OPN mRNAs. To determine whether multiple copies of OPN mRNA were expressed in bone cells, total mRNA was isolated from UMR 106-06 (*lanes 2, 7, 12*), ROS 17/2.8 (*lanes 3, 8, 13*) RCo (osteoblast-enriched; *lanes 4, 9, 14*) and RCp (preosteoblast enriched; *lanes 5, 10, 15*) cells. Oligonucleotide primer pairs (20-mers) 5′ 137 to 156 and 3′ 522 to 503 (*lanes A1–4*) were used to amplify a 385-bp segment in the 5′ half of the rat OPN cDNA sequence; 5′ 522 to 541 and 3′ 1008 to 987 (*lanes B1–4*) to amplify a 486-bp segment in the 3′ half of the sequence; and 5′ 137 to 156 to 3′ 1008 to 987 (*lanes C1–4*) to amplify 871 bps of the total sequence. *Lanes S* contained a λ*Hind*III/*Eco*RI DNA marker digest. Following reverse transcriptase (RT) incubation to convert the mRNA to cDNA, 40 cycles of polymerase chain reaction (PCR) were completed and the amplified products analyzed on a 2% agarose gel. The single amplified species for each bone cell population indicates the absence of multiple mRNAs generated by alternate splicing.

compared to the RC cells. In particular, the A11 and RC cells respond strongly to TGF-β, and the RC cells also respond strongly to retinoic acid and concanavalin A, and to a lesser degree to the phorbol ester, 12-*O*-tetradecanoyl-phorbol-13-acetate (TPA), whereas the ROS 17/2.8 cells demonstrate a more modest response to these agents. In addition, the same studies also revealed that some agents had a differential effect on the phosphorylation of the 55-kDa and 44-kDa forms in the RC cell populations. Thus, the phosphorylation of the 44-kDa OPN was preferentially increased by EGF, PDGF and TPA, whereas the phosphorylation of the 55-kDa OPN was preferentially phosphorylated by TGF-β. Since the RC cell populations comprise preosteoblastic cells and osteoblastic cells, it is possible that these differential effects may reflect the different responses of these cells. The changes in the 44-kDa would reflect the response of osteoblasts, while the 55-kDa would reflect the response of the preosteoblasts.

While the stimulation of OPN expression by vitamin D_3 and retinoic acid reflects a response of bone cells to osteotropic hormones, it is notable that a third osteotropic hormone, parathyroid hormone (PTH), has only a marginal effect on RC cells[45] and slightly suppresses expression in ROS 17/2.8 cells.[46] In general the effects of osteotropic hormones are to promote bone resorption and remodelling, and vitamin D_3 can inhibit the differentiation of bone cells. Thus, it might be anticipated that the increase in OPN induced by these hormones is associated with the proposed function of OPN in promoting osteoclast attachment[47] and stimulation of osteoclastic activity.[48,49] This is supported by the recent demonstration (Mukherjee *et al.*, reported in this conference) that in transgenic mice that do not express the *src* gene the associated lack of OPN expression appears to be linked to the osteopetrosis that afflicts these animals. It is also notable that OPN expression is also increased in response to other mutations that abrogate osteoclast activity,[50] further supporting a role for OPN in promoting osteoclastic bone resorption.

The stimulation of OPN synthesis by growth factors, particularly TGF-β, and by physical stress,[51] which promote bone matrix formation, supports the involvement of OPN in the formation of bone. In recent studies we analyzed the molecular mechanisms of OPN regulation in bone-forming embryonic rat calvarial (RC) cells which first undergo differentiation *in vitro* before bone formation is initiated by differentiated osteoblasts.[52] The temporal expression of OPN mRNA during the differentiation of these cells[53] revealed a pattern that is consistent with that observed during the formation of the calvariae *in vivo*.[27,54] An early expression of OPN can be linked to the activity of preosteoblasts which produce the 55-kDa OPN. Notably, an especially high level of OPN mRNA is expressed during this same time period by adult rat bone marrow cells (RBMC), as shown in FIGURE 8, which are induced to differentiate in the presence of the synthetic glucocorticoid dexamethasone.[55,56] Examination of bone tissue formation by the RBMCs revealed the presence of microscopic mineral accretions that are associated with processes extending from osteoblastic cells.[57] These globules subsequently coalesce to form a 'cement layer' upon which bone tissue matrix is then deposited. The possibility that OPN might be involved in the formation of these globules is indicated by immunocytochemical analysis of the cell surface. In nonpermeabilized cells patches of immunostaining have been observed on the tips of cell processes.[57] That it is the 55-kDa form of OPN which is present at the cell surface is indicated by the demonstration that the synthesis of this OPN is prevalent at this stage of bone cell differentiation and the observation that the corresponding OPN in NRK cells is preferentially attached to the cell surface.[7,9]

Although previous studies argued against the involvement of OPN in the initial mineralization of newly-formed bone, based on the temporo-spatial expression of OPN[58] and the tendency of OPN to inhibit hydroxyapatite formation,[59,60] it is recognized that acidic bone proteins can act as nucleators or inhibitors of bone formation depending upon their physical state.[61] Thus, in an immobilized form OPN could initiate HA crystal formation, and OPN is clearly immobilized on the surface of cells with the appearance of preosteoblasts. Thus, it is conceivable that OPN may have a function in the initiation of hydroxyapatite crystal nucleation during cement formation. Consistent with this concept is the observation, discussed above, that OPN is enriched in cement lines that form between periods of bone formation and following osteoclastic resorption, and in the cutting cones of osteoclasts involved in remodelling bone lamellae (FIG. 2).

In recent studies (Li, Cheifetz, Sampath & Sodek, in preparation) we found that human recombinant BMP-7 (hOP-1), a potent bone inductive agent *in vivo*[62] which stimulates bone formation by RC cells[62] and RBMCs, selectively stimulates OPN expression <5-fold by preosteoblasts within several hours of administration (FIG. 9). Analysis of the secreted protein revealed that it is the 55-kDa OPN that is elevated, the maximal elevation of the protein being evident ~9 hr following the exposure of cells to 25 ng/ml hOP-1. These results, together with an induction

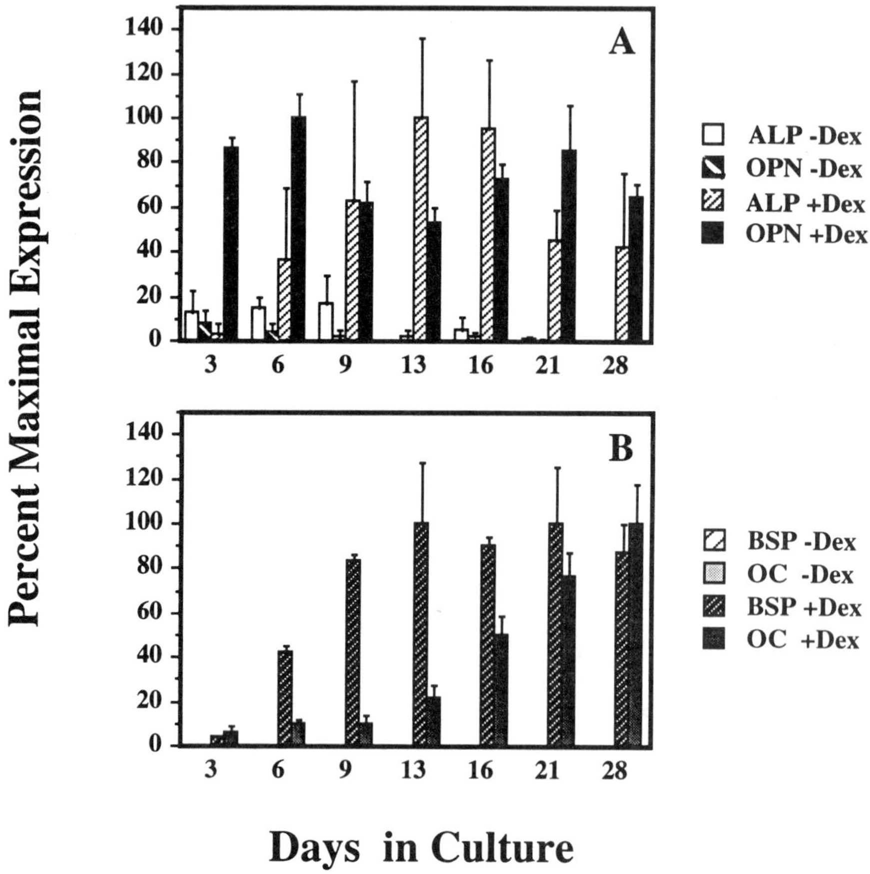

FIGURE 8. Temporal expression of rat OPN mRNA during bone formation *in vitro*. The expression of OPN mRNA is compared to the expression of alkaline phosphatase (ALP), bone sialoprotein (BSP) and osteocalcin (OC) in rat bone marrow cells cultured in the presence and absence of 10^{-8} M dexamethasone. The dexamethasone induces osteoblast differentiation and bone formation, as indicated by the appearance of BSP and OC, whereas OPN and ALP are strongly stimulated by dexamethasone. Note the strong stimulation of OPN early in the cultures when preosteoblastic cells are present.

of bone sialoprotein,[63] show that BMP-7 targets the preosteoblasts to promote the formation of a cement matrix and to stimulate their differentiation into osteoblasts. In contrast, however, exposure of differentiated, bone-forming osteoblasts to hOP-1 produces only a marginal stimulation of OPN mRNA (not shown). Thus, the stimulation of OPN by hOP-1 clearly reflects a role for OPN early in bone formation.

While almost all factors and conditions studied to date have produced a stimulation of OPN synthesis, a marked suppression of OPN expression is observed with bisphosphonates, which are used clinically to ameliorate bone resorption by

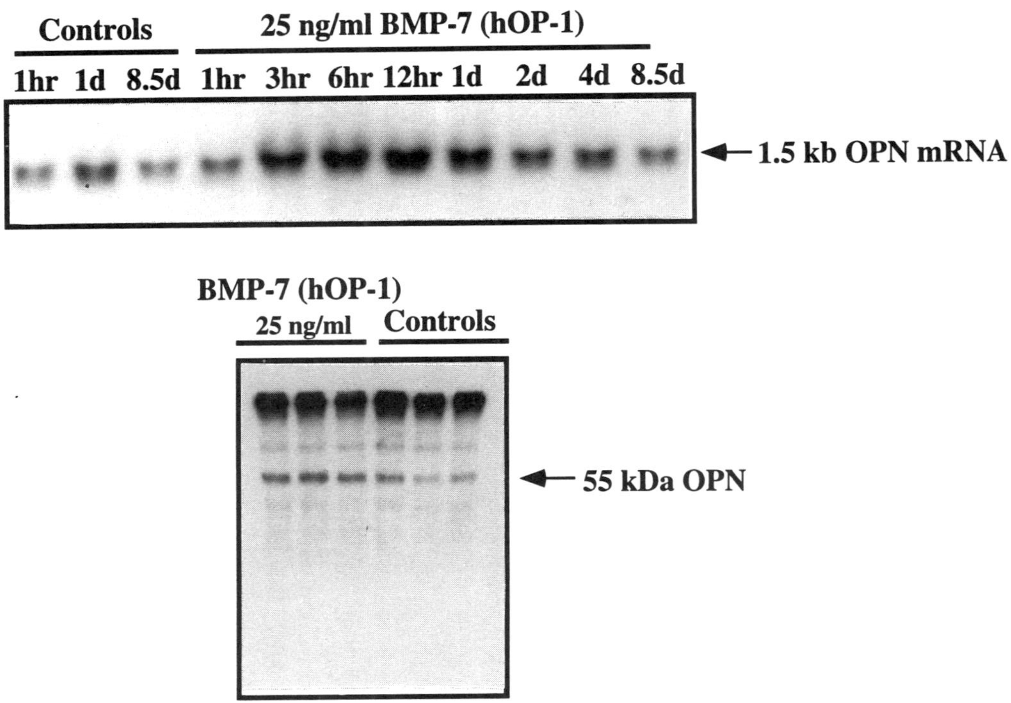

FIGURE 9. Regulation of rat bone OPN by bone morphogenic protein OP-1. The effects of human recombinant osteogenic protein hOP-1 on the expression of OPN were analyzed in cultures of rat calvarial (RC) cells. Confluent cells were incubated with 25 ng/ml of hOP-1 and analyzed for OPN mRNA (*upper panel*) and OPN protein (*lower panel*) at various time points. The OPN mRNA was increased <5-fold, beginning at 1–3 hr and reaching maximal levels at 6–12 hr before declining. The increase in mRNA was reflected in an increase in [^{35}S]-methionine-labelled 55-kDa OPN immunoprecipitated from the culture medium. Maximal levels of radiolabelled protein appeared, as shown, when cells were pulse-labelled 9 hr following the addition of the hOP-1.

osteoclasts. Although the precise mechanism whereby the bisphosphonates exert their effects is not entirely clear, it is known that they bind tightly to hydroxyapatite, and it is thought that they can stabilize the crystals in the acidic environment of the osteoclast lacunae and on release are toxic to the osteoclasts themselves. Here we show that the bisphosphonates, 1-hydroxyethylidene-1,1-bisphosphonate (EHDP) and the more potent 3-amino-1-hydroxypropylidene-1,1-bisphosphonate (APD), at pharmacological levels, selectively reduce OPN mRNA expression by osteoblasts. At concentrations of 10^{-5} M to 10^{-8} M, EHDP and APD suppress OPN mRNA levels ~50%. Although the expression of type I collagen mRNA is also suppressed by APD and to a lesser extent by EHDP, the expression of mRNAs coding for two other noncollagenous proteins, bone sialoprotein and osteocalcin, were not affected significantly (FIG. 10). These observations demonstrate a selective effect of the bisphosphonates on the expression of OPN which could impair the activity of OPN in stimulating bone resorption.

Based on current knowledge of the properties of OPN, its temporal expression and its regulation by hormones, cytokines and pharmacological agents, it is apparent at OPN has a multiplicity of functions in bone formation and remodelling as depicted in FIGURE 11. Thus, it would appear that the OPN synthesized by preosteoblastic cells, which is a less phosphorylated and sulphated form in rats (55-kDa form), associates with the $\alpha_v\beta_3$ integrin receptor on cell processes and directs the formation of a 'cement' layer upon which mineralized bone tissue is formed. Synthesis of this OPN is stimulated by bone morphogenic proteins which stimulate bone formation, and by growth factors, such as TGF-β which promote bone matrix formation. Notably, vitamin D_3 and retinoic acid, as well as glucocorticoids[56] and OP-1, which influence bone cell differentiation, also stimulate OPN expression by preosteoblastic cells. This synthesis of OPN by 'preosteoblastic' cells is associated with the formation of a cement layer, which is expected to form, in response to bone formation signals following osteoclastic resorption, on surfaces of mineralized cartilage, on damaged bone and on implanted biomaterials.

OPN synthesized by osteoblastic cells, which is highly phosphorylated and sulphated in rats (44-kDa form), is also upregulated by vitamin D_3 and retinoic acid, and especially EGF and PDGF, but only marginally by OP-1. This form of rat OPN associates rapidly with preexisting crystals of hydroxyapatite[37-39] possibly regulating their growth. In porcine bone the mineral-associated OPN has been shown to be fragmented in a process that may also reflect regulation of crystal growth.[4]

The expression of OPN (44-kDa form in rats) is apparently increased as osteoblasts mature[27,56] and become osteocytes and lining cells, the OPN sealing off the surface of the newly-formed bone and providing a substratum (lamina limitans) for the attachment of the cells.[22] Inflammatory cytokines, such as IL-1α and β, TNF-α and lipopolysaccharides (LPS), which promote bone remodelling, also stimulate OPN expression,[64] whereas bisphosphonates that suppress bone resorption also suppress OPN expression. These signals and the response of the osteoblasts are consistent with the proposed role of OPN in mediating the attachment and of osteoclasts and in their regulation of osteoclastic activity.

SUMMARY

Osteopontin (OPN) is a prominent bone matrix protein that is synthesized by osteoblastic cells. To elucidate the function of OPN in bone we studied the regu-

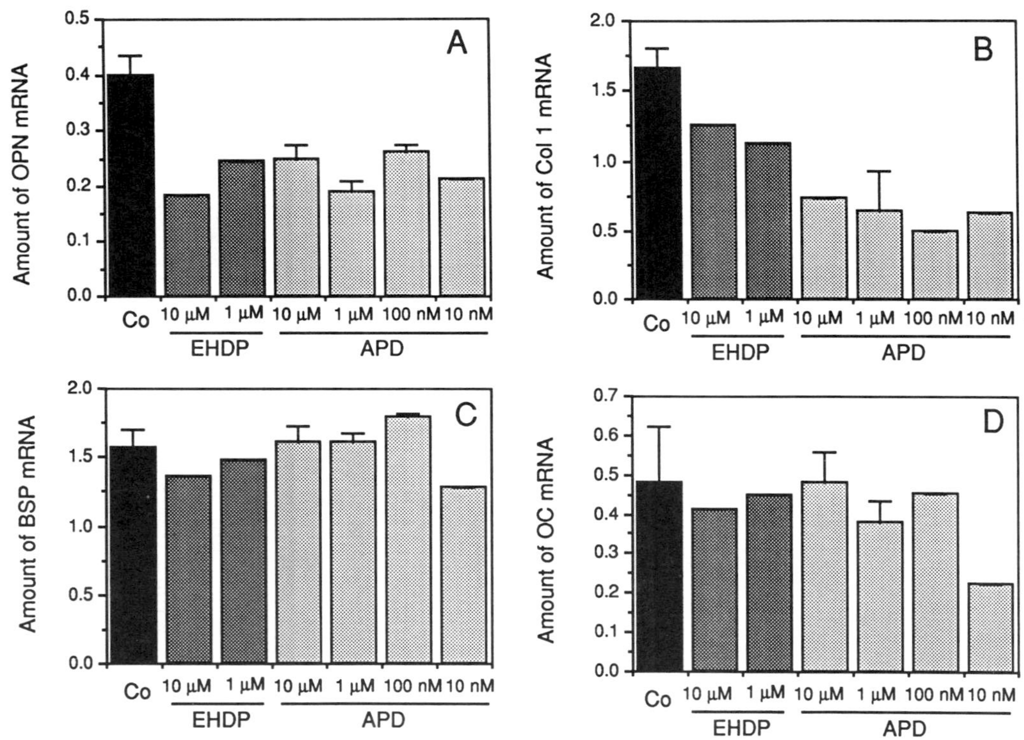

FIGURE 10. Suppression of rat bone OPN mRNA expression by bisphosphonates. The effects of various concentrations of the bisphosphonates EHDP and APD on OPN mRNA expression by cultures of rat calvarial cells were analyzed by Northern hybridization and compared to the effects on collagen type I, bone sialoprotein (BSP) and osteocalcin (OC) mRNA. A significant reduction of OPN mRNA and type I collagen mRNA was observed, while mRNAs for BSP and OC were generally unaffected.

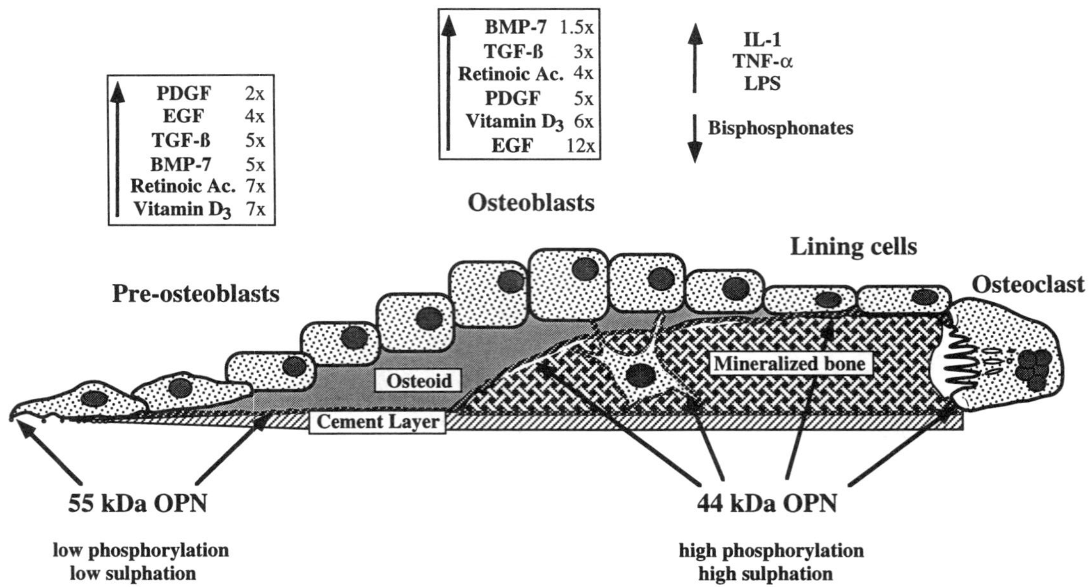

FIGURE 11. Summary of rat OPN synthesis during different phases of bone formation. The diagram depicts the differentiation of osteoblastic cells to form osteoblasts, osteocytes and lining cells. The early preosteoblasts are shown depositing a mineralized 'cement layer' which is initiated as microscopic accretions associated with cell processes that express the 55-kDa OPN. The cement layer represents the first material laid down onto preexisting bone or onto the surface of implanted biomaterials. As these cells differentiate into cuboidal osteoblasts they secrete an organic matrix (osteoid) over the cement layer. The osteoid subsequently mineralizes, entrapping some osteoblasts (osteocytes). After bone matrix formation ceases the osteoblasts become quiescent, flattening into lining cells which secrete increased levels of 44-kDa OPN to stop the mineralization beneath the cells and provide a substratum to which osteoclasts can attach following appropriate bone resorptive signals. Various hormones and cytokines that affect the expression of OPNs by the preosteoblasts, osteoblasts and lining cells are shown.

lated expression of the rat OPN protein during bone formation *in vivo* and *in vitro*. OPN mRNA is expressed by preosteoblastic cells early in bone formation, but the highest expression is observed in mature osteoblasts at sites of bone remodelling. A low-phosphorylated, 55-kDa form of OPN is produced by the preosteoblastic cells, whereas osteoblasts produce a highly phosphorylated, 44-kDa protein; the two forms of OPN corresponding to pp69 and pp62 in transformed rat cells. The synthesis of the 55-kDa OPN correlates with the formation of a 'cement' matrix that is synthesized prior to bone deposition, whereas the 44-kDa OPN synthesized by osteoblasts associates rapidly with hydroxyapatite, possibly regulating crystal growth, and may also provide a substratum for osteoclast attachment. Expression of OPN mRNA is upregulated by growth and differentiation factors (PDGF, EGF, TGF-β and BMP-7/OP-1) and by mechanical stress, which promote bone formation, as well as by osteotropic hormones (retinoic acid and vitamin D_3), which can promote bone resorption and remodelling. However, OPN mRNA is downregulated by bisphosphonates, which abrogate bone resorption. Regulation of OPN expression is, therefore, consistent with a multiplicity of functions for OPN that involve specific structural motifs in both the synthesis and resorption of bone.

ACKNOWLEDGMENTS

We thank Dr. Kuber Sampath of Creative Biomolecules (Hopkinton, MA) for providing the hOP-1 used in these studies, and Dr. Marc Grynpas (Mount Sinai Hospital, Toronto) for supplying the APD and EHDP. Expert technical assistance was provided in these studies by Carol Laschinger, Brigitte Hawrylyshyn, Yale Farache and Kam-Ling Yao.

REFERENCES

1. FRANZÉN, A. & D. HEINEGÅRD. 1985. Biochem. J. **232:** 715–724.
2. PRINCE, C. W., T. OOSAWA, W. T. BUTLER, M. TOMANA, A. S. BHOWN, M. BHOWN & R. E. SCHROHENLOHER. 1987. J. Biol. Chem. **262:** 2900–2907.
3. FISHER, L. W., G. R. HAWKINS, N. TUROSS & J. D. TERMINE. 1987. J. Biol. Chem. **262:** 9702–9708.
4. ZHANG, Q., C. DOMENICUCCI, H. A. GOLDBERG, J. L. WRANA & J. SODEK. 1990. J. Biol. Chem. **265:** 7583–7589.
5. YOUNG, M. F., J. M. KERR, J. D. TERMINE, U. M. WEWER, M. G. WANG, O. W. MCBRIDE & L. W. FISHER. 1990. Genomics **7:** 491–502.
6. SINGH, K., A. B. MUKHERJEE, M. W. DEVOUGE & B. B. MUKHERJEE. 1992. J. Biol. Chem. **267:** 23847–23851.
7. NEMIR, M., M. W. DEVOUGE & B. B. MUKHERJEE. 1989. J. Biol. Chem. **264:** 18202–18208.
8. KUBOTA, T., Q. ZHANG, J. L. WRANA, R. BER, J. E. AUBIN, W. T. BUTLER & J. SODEK. 1989. Biochem. Biophys. Res. Commun. **162:** 1453–1459.
9. SINGH, K., M. W. DEVOUGE & B. B. MUKHERJEE. 1990. J. Biol. Chem. **265:** 18696–18701.
10. OLDBERG, A., A. FRANZÉN & D. HEINEGÅRD. 1986. Proc. Natl. Acad. Sci. USA **83:** 8819–8823.
11. CRAIG, A. M., J. H. SMITH & D. T. DENHARDT. 1989. J. Biol. Chem. **264:** 9682–9689.
12. WRANA, J. L., Q. ZHANG & J. SODEK. 1989. Nucleic Acids Res. **17:** 10119.
13. KERR, J. M., L. W. FISHER, J. D. TERMINE & M. F. YOUNG. 1991. Gene **108:** 237–243.
14. CRIVELLO, J. F. & E. DELVIN. 1992. J. Bone Miner. Res. **7:** 639–699.
15. CASTAGNOLA, P., P. BET, R. QUARTO, M. GENNARI & R. CANCEDDA. 1991. J. Biol. Chem. **266:** 9944–9949.

16. MOORE, M. A., Y. GOTOH, K. RAFIDI & L. C. GERSTENFELD. 1991. Biochemistry. **30:** 2501–2508.
17. DENHART, D. T. & X. GUO. 1993. FASEB J. **7:** 1475–1482.
18. MARK, M. P., C. W. PRINCE, T. OOSAWA, S. GAY, A. L. J. J. BRONCKERS & W. T. BUTLER. 1987. **35:** 707–715.
19. CHEN, J., Q. ZHANG, C. A. G. MCCULLOCH & J. SODEK. 1991. Histochem. J. **23:** 281–289.
20. CHEN, J., C. A. G. MCCULLOCH & J. SODEK. 1993. Arch. Oral Biol. **38:** 241–249.
21. MCKEE, M. D., M. J. GLIMCHER & A. NANCI. 1992. Anat. Rec. **234:** 479–492.
22. MCKEE, M. D. & A. NANCI. 1993. Cells Mater. **3:** 219–243.
23. CHEN, J., M. D. MCKEE, A. NANCI & J. SODEK. 1994. Histochem. J. **26:** 67–78.
24. ROSS, F. P., J. CHAPPEL, J. I. ALVAREZ, D. SANDER, W. T. BUTLER, M. C. FARACH-CARSON, K. A. MINTZ, P. GEHRON-ROBEY, S. L. TEITELBAUM & D. A. CHERESH. 1993. J. Biol. Chem. **268:** 9901–9907.
25. HELFRICH, M. H., S. A. NESBITT, E. L. DOREY & M. A. HORTON. 1992. J. Bone Miner. Res. **7:** 335–343.
26. NOMURA, S., A. J. WILLS, D. R. EDWARDS, J. K. HEATH & B. L. M. HOGAN. 1988. J. Cell Biol. **106:** 441–450.
27. CHEN, J., K. SINGH, B. B. MUKHERJEE & J. SODEK. 1993. Matrix **13:** 113–123.
28. HULTENBY, K., F. P. REINHOLT, M. NORGARD, A. OLDBERG & D. HEINEGARD. 1994. Eur. J. Cell Biol. **63:** 230–239.
29. HULTENBY, K., F. P. REINHOLT, A. OLDBERG & D. HEINEGARD. 1991. Matrix **11:** 206–213.
30. PRINCE, C. W. & W. T. BUTLER. 1987. Collagen Relat. Res. **7:** 305–313.
31. NODA, M., R. L. VOGEL, A. M. CRAIG, J. PRAHL, H. F. DELUCA & D. T. DENHARDT. 1990. Proc. Natl. Acad. Sci. USA **87:** 9995–9999.
32. NODA, M., K. YOON, C. W. PRINCE, W. T. BUTLER & G. A. RODAN. 1988. J. Biol. Chem. **263:** 13916–13921.
33. CHACKALAPARAMPIL, I., D. BANNERJEE, Y. POIRIER & B. B. MUKHERJEE. 1985. J. Virol. **53:** 841–850.
34. LAVERDURE, G. R., D. BANERJEE, I. CHACKALAPARAMPIL & B. B. MUKHERJEE. 1987. FEBS Letts. **222:** 261–265.
35. NAGATA, T., R. TODESCAN, H. A. GOLDBERG, Q. ZHANG & J. SODEK. 1989. Biochem. Biophys. Res. Commun. **165:** 234–240.
36. BER, R., T. KUBOTA, J. SODEK & J. AUBIN. 1991. Biochem. Cell Biol. **69:** 132–140.
37. NAGATA, T., C. G. BELLOWS, S. KASUGAI, W. T. BUTLER & J. SODEK. 1991. Biochem. J. **274:** 513–520.
38. KASUGAI, S., R. TODESCAN, JR., T. NAGATA, K.-L. YAO, W. T. BUTLER & J. SODEK. 1991. J. Cell Physiol. **147:** 111–120.
39. KASUGAI, S., T. NAGATA & J. SODEK. 1992. J. Cell. Physiol. **152:** 467–477.
40. WRANA, J. L., T. KUBOTA, Q. ZHANG, C. M. OVERALL, J. E. AUBIN, W. T. BUTLER & J. SODEK. 1991. Biochem. J. **273:** 523–531.
41. MAJESKA, R. J., S. B. RODAN & G. A. RODAN. 1980. Endocrinology **107:** 1494–1503.
42. BUTLER, W. T. 1989. Connect. Tissue Res. **23:** 123–136.
43. CRAIG, A. M., M. NEMIR, B. B. MUKHERJEE, A. F. CHAMBERS & D. T. DENHARDT. 1988. Biochem. Biophys. Res. Commun. **157:** 166–173.
44. CHAN, J. C., M. SCANLON, D. T. DENHARDT, K. SINGH, B. B. MUKHERJEE, M. C. FARACH-CARSON & W. T. BUTLER. 1990. Int. J. Cancer **46:** 864–870.
45. KASUGAI, S., Q. ZHANG, C. M. OVERALL, J. L. WRANA, W. T. BUTLER & J. SODEK. 1991. Bone Miner. **13:** 235–250.
46. NODA, M. & G. A. RODAN. 1989. J. Cell Biol. **108:** 713–718.
47. REINHOLT, F. P., K. HULTENBY, A. OLDBERG & D. HEINEGÅRD. 1990. Proc. Natl. Acad. Sci. USA **87:** 4473–4475.
48. MIYAUCHI, A., J. ALVAREZ, E. M. GREENWOOD, A. TETI, M. GRANO, S. COLUCCI, A. ZAMBONIN-ZALLONE, F. P. ROSS, S. M. TEITELBAUM, D. CHERESH & K. A. HRUSHKA. 1991. J. Biol. Chem. **266:** 20369–20374.
49. ZIMOLO, Z., G. WESOLOWSKI, H. TANAKA, J. HYMAN, J. R. HOYER & G. A. RODAN. 1994. Am. J. Physiol. **266:** 376–381.

50. SHALHOUB, V., B. BETTENCOURT, M. E. JACKSON, C. A. MACKAY, M. J. GLIMCHER, S. C. MARKS, G. A. STEIN & J. B. LIAN. 1994. J. Cell. Physiol. **158:** 110–120.
51. KUBOTA, T., M. YAMAUCHI, J. ONAZAKI, S. SATO, Y. SUZUKI & J. SODEK. 1993. Arch. Oral Biol. **38:** 23–30.
52. LIAN, J. & G. S. STEIN. 1992. Crit. Rev. Oral Biol. Med. **3:** 269–305.
53. STEIN, G. S., J. B. LIAN & T. A. OWEN. 1990. FASEB J. **4:** 3111–3123.
54. CHEN, J., H. S. SHAPIRO & J. SODEK. 1992. J. Bone Miner. Res. **7:** 987–997.
55. MANIATOPOULOS, C., J. SODEK & A. H. MELCHER. 1988. Cell Tissue Res. **254:** 317–330.
56. YAO, K.-L., R. TODESCAN, JR. & J. SODEK. 1994. J. Bone Miner. Res. **9:** 231–240.
57. SHEN, X., E. ROBERTS, S. A. F. PEEL & J. E. DAVIES. 1993. Cells Mater. **3:** 257–272.
58. SODEK, J., J. CHEN, S. KASUGAI, T. NAGATA, Q. ZHANG, M. D. MCKEE & A. NANCI. 1992. Elucidating the functions of bone sialoprotein and osteopontin in bone formation. *In* The Chemistry and Biology of Mineralized Tissues. H. Slavkin & P. Price, Eds. 297–306. Elsevier B.V. Amsterdam.
59. BOSKEY, A. L., M. MARESCA, W. ULLRICH, S. B. DOTY, W. T. BUTLER & C. W. PRINCE. Bone Miner. Res. **22:** 147–159.
60. HUNTER, G. K., C. L. KYLE & H. A. GOLDBERG. 1994. Biochem. J. **300:** 723–728.
61. LINE, A. & A. LUSSI. 1989. Connect. Tissue Res. **21:** 197–203.
62. SAMPATH, T. K., J. C. MALIAKAL, P. V. HAUSCHKA, W. K. JONES, H. SASAK, R. F. TUCKER, K. H. WHITE, J. E. COUGHLIN, M. M. TUCKER, R. H. L. PANG, C. CORBETT, E. OZKAYNAK, H. OPPERMANN & D. C. RUEGER. 1992. J. Biol. Chem. **267:** 20352–20362.
63. SODEK, J., K.-L. YAO, C. G. BELLOWS & K. SAMPATH. 1994. J. Dent. Res. **73:** 205. Abs# 829.
64. JIN, C., C. MIYAURA, I. ISHIMA, M. HONG, T. SATO, E. ABE & T. SUDA. 1990. Mol. Cell. Endocrinol. **74:** 221–228.

Regulation of Osteopontin Gene Expression in Osteoblasts[a]

MASAKI NODA[b,c] AND DAVID T. DENHARDT[d]

[c]Department of Molecular Pharmacology
Division of Functional Disorder Research
Medical Research Institute
Tokyo Medical and Dental University
3-10, Kanda-Surugadai 2-Chrome, Chiyoda-Ku
Tokyo 101, Japan
and
[d]Department of Biological Sciences
Nelson Biological Laboratories
Rutgers University
P.O. Box 1059
Piscataway, New Jersey 08855-1059

Osteoblasts form bone by secreting many extracellular matrix proteins, including type I collagen, osteopontin, osteocalcin, osteonectin, biglycan, and decorin. Osteopontin was first purified from rat bone matrix and has been considered to play an important role in the cascade of events required for the formation of mature bone matrix.[1,2] Transforming growth factor β (TGFβ) is one of the most abundant cytokines stored in bone matrix. Osteoblasts secrete TGFβ and it is accumulated in the bone matrix in an inactive form. Osteoblasts possess receptors for TGFβ and this cytokine is considered to be an autocrine or paracrine hormone regulating bone development.[3]

To examine the role of TGFβ in the regulation of the osteoblastic function, ROS 17/2.8 cells and other osteoblast-like cells were treated with TGFβ and expression of genes encoding phenotype-related proteins in these cells was examined by Northern blot analysis. Noda and colleagues[4] reported that TGFβ treatment enhanced the expression of several genes encoding extracellular matrix proteins, including osteopontin, in these cells. The osteopontin protein level was increased after 72 h treatment with TGFβ as deduced from a Western blot analysis (FIG. 1). This increase was due to *de novo* synthesis of osteopontin, since a metabolic labeling study indicated that TGFβ enhanced the level of newly synthesized osteopontin protein, which was pulse-labeled and secreted into the medium by the osteoblast-like cells.

An approximately 1.8-kb OPN mRNA species was observed in the Northern blot analysis of the cells and the abundance of the message was increased severalfold by the treatment with TGFβ. This increase was time dependent, starting at 48 h after the initiation of the TGFβ treatment and lasting up to 72 h. The effect was observed in a dose-dependent manner with a maximum enhancement at approximately 2 ng/ml (FIG. 2). To examine the mode of TGFβ action, the cells

[a] Abbreviations: BMP, bone morphogenetic protein; OPN, osteopontin; PTH, parathyroid hormone; TGF, transforming growth factor.

[b] Corresponding author.

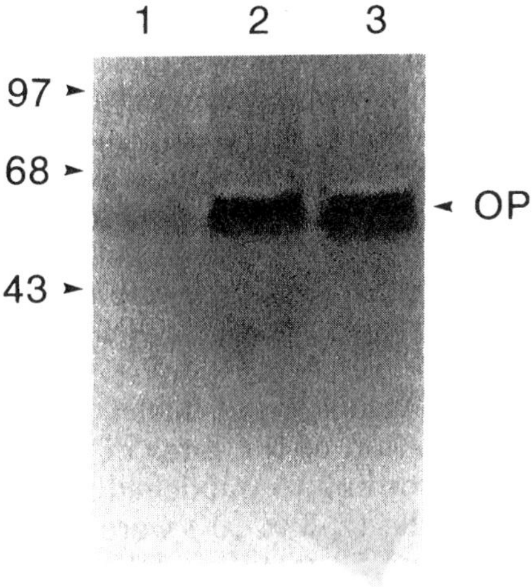

FIGURE 1. Effect of TGFβ on osteopontin production by ROS 17/2.8 cells. A confluent culture was not treated (*lane 1*) or treated with 4 ng/ml of TGFβ1 (*lane 2*) or TGFβ2 (*lane 3*) for 96 h in serum-free medium. Samples were concentrated 24-fold and analyzed on SDS-10% polyacrylamide gels. A Western immunoblot was performed as described.[4] Numbers on the left indicate the position of MW markers in kDa. (From Noda *et al.*[4] Reprinted by permission from the *Journal of Biological Chemistry*.)

were treated with TGFβ in the presence or absence of inhibitors of transcription or of protein synthesis. The results of these analyses suggested that TGFβ enhanced osteopontin message expression via transcriptional events that required new protein synthesis. Nuclear run-on studies confirmed that TGFβ treatment increased the level of the transcriptional activity of the osteopontin gene.[4]

Parathyroid hormone (PTH) is a calcitropic peptide hormone responsible for regulating the plasma calcium level in the body. Osteoblasts possess receptors for PTH and respond to this hormone by accumulating intracellular cyclic AMP, which subsequently activates protein kinase A and further downstream targets.[5] Another organ that responds to PTH is the kidney, which also expresses osteopontin (uropontin) abundantly. Noda and Rodan[6] examined the effects of PTH on osteopontin gene expression in osteoblasts by using the same osteoblast models as used in the studies on TGFβ.

Contrary to the actions of TGFβ, PTH treatment reduced the osteopontin protein level in the conditioned medium of these osteoblasts.[6] PTH treatment also suppressed *de novo* synthesis of osteopontin protein as revealed by pulse-labeling experiments. Northern blot experiments showed that PTH treatment lowered the osteopontin mRNA level in a dose-dependent manner, starting at approximately 10^{-10} M and saturating at 10^{-7} M with an ED50 at about 5×10^{-8} M (FIG. 3). Kinetic studies showed that the PTH effect started within 12 h and the maximum suppression was observed 24 h after the initiation of the treatment. Nuclear run-on experiments showed that the PTH suppression was basically via transcriptional events. The combinatory experiments indicated that PTH treatment neutralized the enhancing effects of TGFβ. These observations suggest that the osteopontin gene is one of the targets of the two major calcitropic effectors, and it appears that the osteopontin level is maintained by the positive and negative regulation of the osteopontin promoter activity by these two, and presumably other, modulators.

Osteopontin gene expression is enhanced by other calcitropic factors, including vitamin D. Noda *et al.*[7] investigated the promoter region of the murine osteopontin gene and found that the activity of the −777-bp portion was enhanced by vitamin D treatment[7] (FIG. 4). Further analysis identified a direct repeat sequence AGGTT-CACG to be the vitamin D response element capable of binding to nuclear proteins purified from osteoblastic nuclei containing the vitamin D receptor (FIG. 5).

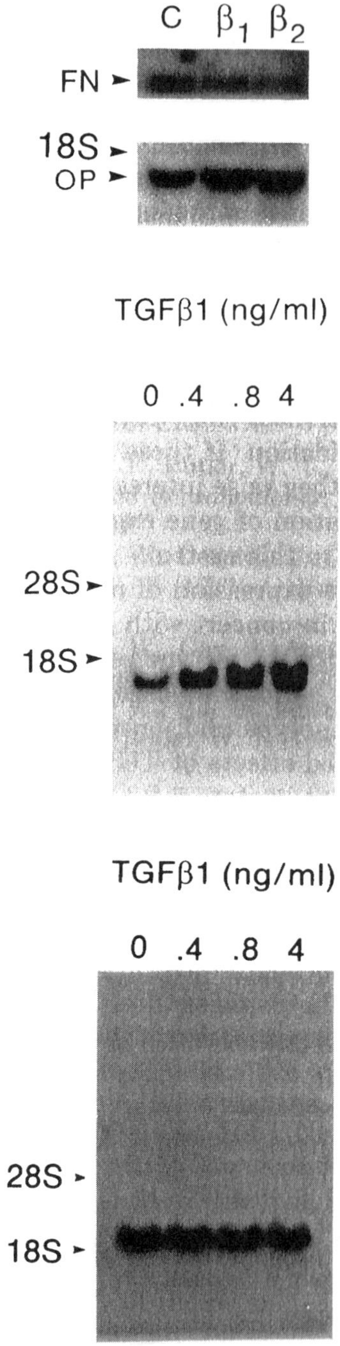

FIGURE 2. Effect of TGFβ on steady state OPN mRNA levels in ROS 17/2.8 cells. Cells were cultured in the absence or presence of the indicated concentration of TGFβ1 or 2 for 72 h. Northern blots of the RNA extracted from these cells were performed as described.[4] The *top panel* shows the signal for fibronectin (FN) and osteopontin (OP) in cells treated with 4 ng/ml TGFβ1 or TGFβ2. The *middle panel* shows the dose dependence of OPN mRNA levels on the concentration of TGFβ1. The *bottom panel* shows the same blot stripped and probed for β-actin. (From Noda *et al.*[4] Reprinted by permission from the *Journal of Biological Chemistry*.)

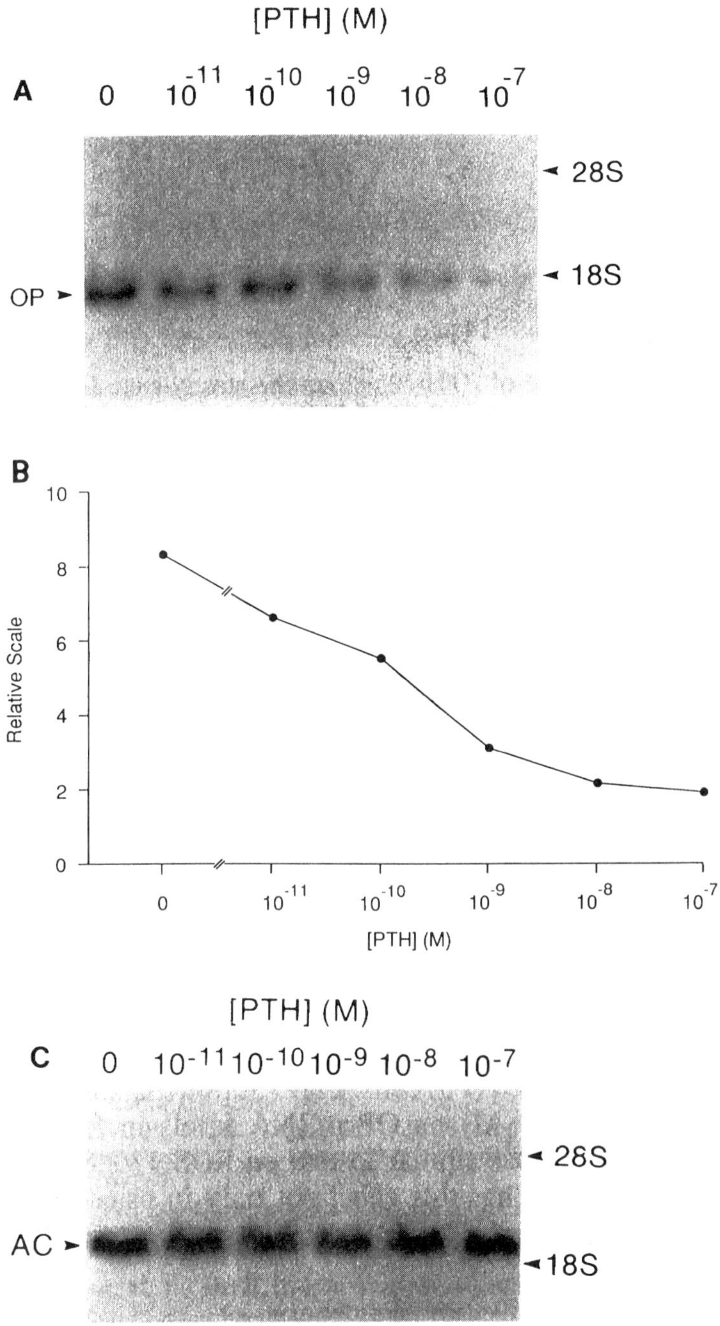

FIGURE 3. Effect of PTH on the steady-state level of OPN mRNA levels in ROS17/2.8 cells. Cells were treated with the indicated concentration of PTH for 72 h. RNA was extracted and a Northern blot analysis performed as described.[6] **(A)** shows the osteopontin (OP) mRNA levels while **(C)** shows the same blot probed for β-actin (AC) mRNA. **(B)** shows a quantitative analysis of the relative strengths of the signals seen in (A). (From Noda & Rodan.[6] Reprinted by permission from the *Journal of Cell Biology*.)

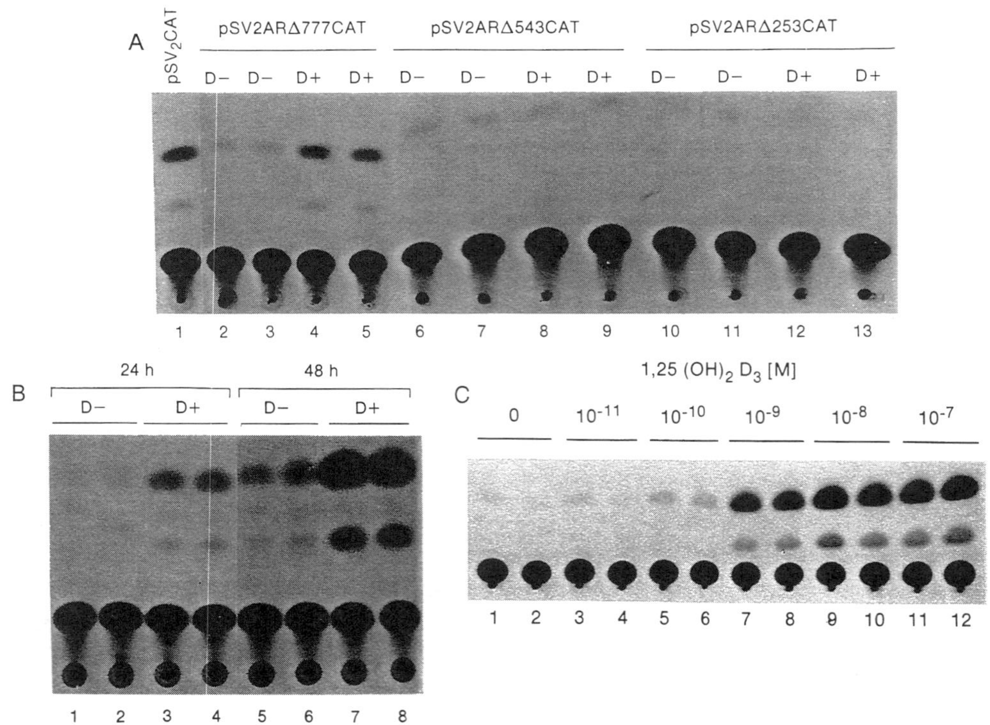

FIGURE 4. Effect of $1,25(OH)_2D_3$ on the activity of the OPN promoter as judged by expression of chloramphenicol acetyl transferase (CAT) transcribed by various portions of the OPN promoter and expressed in transfected ROS 17/2.8 cells. **(A)** Cells transfected with the indicated construct in which CAT expression was driven by an element of the OPN promoter extending to −777, −543 or −253 as indicated. D− and D+ indicate the absence or presence (10 nM) of $1,25(OH)_2D_3$. **(B)** Time course for the promoter extending to the −777 construct. **(C)** Dependence on the $1,25(OH)_2D_3$ concentration of expression from the −777 construct. Further details can be found in Noda *et al.*[7] (From Noda *et al.*[7] Reprinted by permission from the National Academy of Sciences.)

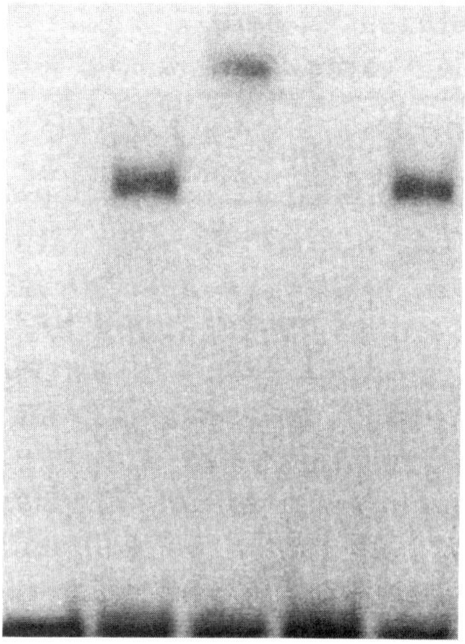

FIGURE 5. Localization of the vitmain D receptor in the complex of the OPN promoter with the vitamin D response element (VDRE). The ^{32}P-labeled 2.5S tandem repeat element ACAAGGTTCACGAGGTTCAC-GTCT was incubated with a nuclear extract in the absence (*lane 2*) or presence of the monoclonal antibodies to the vitamin D receptor (XVIE10C6 in *lane 3* and VIIID8C12 in *lane 4*—the latter but not the former recognizes the DNA-binding domain) and then subjected to a gel-shift analysis. Lanes 1 and 5 are controls with probe only or probe plus nuclear extract and nonimmune IgG. Further details can be found in Noda *et al.*[7] (From Noda *et al.*[7] Reprinted by permission from the National Academy of Sciences.)

Shioide and Noda[8] recently reported the ability of the vasoconstrictive peptide endothelin to increase modestly (twofold) OPN mRNA levels in ROS 17/2.8 cells. This is interesting in light of the ability of endothelin to upregulate nitric oxide (NO) production.[9] Because OPN has been shown to inhibit the induction of NO synthesis by inflammatory mediators,[10] it is attractive to consider OPN as an antagonist of the action of the endothelins, acting to regulate negatively their action.

Bone morphogenetic protein (BMP) is a member of the TGFβ super family and is known to be a potent inducer of ectopic bone formation by implantation assay.[11,12] BMP treatment can enhance the osteopontin message levels in the osteoblast-like cells (Noda *et al.*, in preparation). Since BMP regulates expression of certain helix-loop-helix genes in these osteoblasts,[13] the promoter region of the murine osteopontin gene was analyzed. E-box motifs, CANNTG, were found within the promoter, and proteins capable of binding to these motifs were observed in the nuclear extracts of the osteoblasts by using gel shift assays (Tamura and Noda, unpublished observations). An E-box in the osteoclacin gene appears to be important for specific expression in osteoclasts.[14] These observations reveal that osteotropic cytokines and hormones regulate osteopontin gene expression at least in part via the regulation of the promoter activity of the gene.

REFERENCES

1. Butler, W. T. 1989. The nature and significance of osteopontin. Connect. Tissue Res. **23:** 123–136.
2. Sodek, J., J. Chen, S. Kasugai, T. Nagata, Q. Zhang, M. D. McKee & A. Nanci. 1992. Elucidating the functions of bone sialoprotein and osteopontin in bone forma-

tion. *In* Chemistry and Biology of Mineralized Tissues. H. Slavkin & P. Price, Eds. 297–306. Elsevier Science Publishers B.V. Amsterdam.

3. TEN DIJKE, P., H. YAMASHITA, T. K. SAMPATH, A. H. REDDI, M. ESTEVEZ, D. L. RIDDLE, H. ICHIJO, C.-H. HELDIN & K. MIYAZONO. 1994. Identification of type I receptors for osteogenic protein-1 and bone morphogenetic protein-4. J. Biol. Chem. **269:** 16985–16988.
4. NODA, M., K. YOON, C. W. PRINCE, W. T. BUTLER & G. A. RODAN. 1988. Transcriptional regulation of osteopontin production in rat osteosarcoma cells by TGFβ transforming growth factor. J. Biol. Chem. **263:** 13916–13921.
5. FUKAYAMA, S., E. SCHIPANI, H. JUPPNER, B. LANSKI, H. M. KRONENBERG, A. B. ABOU-SAMRA & F. R. BRINGHURST. 1994. Role of protein kinase A in homologous down-regulation of parathyroid hormone (PTH)/PTH-related peptide receptor mRNA in human osteoblast-like SaOS-2 cells. Endocrinology **134:** 1851–1858.
6. NODA, M. & G. A. RODAN. 1989. Transcriptional regulation of osteopontin production in rat osteoblast-like cells by parathyroid hormone. J. Cell Biol. **108:** 713–718.
7. NODA, M., R. L. VOGEL, A. M. CRAIG, J. PRAHL, H. F. DELUCA & D. T. DENHARDT. 1990. Identification of a DNA sequence responsible for binding of the 1,25-dihydroxyvitamin D_3 receptor and 1,25-dihydroxyvitamin D_3 enhancement of mouse secreted phosphoprotein 1 (Spp-1 or osteopontin) gene expression. Proc. Natl. Acad. Sci. USA **87:** 9995–9999.
8. SHIOIDE, M. & M. NODA. 1993. Endothelin modulates osteopontin and osteocalcin messenger ribonucleic acid expression in rat osteoblastic osteosarcoma cells. J. Cell. Biochem. **53:** 176–180.
9. TSUKAHARA, H., H. ENDE, H. I. MAGAZINE, W. F. BAHOU & M. S. GOLIGORSKY. 1994. Molecular and functional characterization of the non-isopeptide-selective ET_B receptor in endothelial cells. J. Biol. Chem. **269:** 21778–21785.
10. HWANG, S.-M., C. A. LOPEZ, D. E. HECK, C. R. GARDNER, D. L. LASKIN, J. D. LASKIN & D. T. DENHARDT. 1994. Osteopontin inhibits induction of nitric oxide synthase gene expression by inflammatory mediators in mouse kidney epithelial cells. J. Biol. Chem. **269:** 711–715.
11. SAMPATH, T. K., J. C. MALIAKAL, P. V. HAUSCHKA, W. K. JONES, H. SASAK, R. F. TUCKER, K. H. WHITE, J. E. COUGHLIN, M. M. TUCKER, R. H. L. PANG, C. CORBETT, E. OZKAYNAK, H. OPPERMANN & D. C. RUEGER. 1992. Recombinant human osteogenic protein-1 (hOP-1) induces new bone formation *in vivo* and with a specific activity comparable with natural bovine osteogenic protein and stimulates osteoblast proliferation and differentiation *in vitro*. J. Biol. Chem. **267:** 20352–20362.
12. KINGSLEY, D. M. 1994. The TGF-β superfamily: new members, new receptors, and new genetic tests of function in different organisms. Genes Dev. **8:** 133–146.
13. OGATA, T. & M. NODA. 1991. Expression of Id, a member of the HLH protein family, is down-regulated at confluence and enhanced by dexamethasone in a mouse osteoblastic cell line, MC3T3E1. Biochem. Biophys. Res. Commun. **180:** 1194–1199.
14. TAMURA, M. & M. NODA. 1994. Identification of a DNA sequence involved in osteoblast-specific gene expression via interaction with helix-loop-helix (HLH)-type transcription factors. J. Cell Biol. **126:** 773–782.

Osteopontin and Related Phosphorylated Sialoproteins: Effects on Mineralization[a]

ADELE L. BOSKEY

The Hospital for Special Surgery
535 East 70th Street
New York, New York 10021

There are at least four phosphorylated sialoproteins in the matrices of the mineralized connective tissues (*i.e.*, bone, calcified cartilage and dentin). Osteopontin, first isolated from bovine bone by Franzen and Heinegard[1] and rat bone by Prince *et al.*[2] is more abundant than the bone sialoprotein II (BSP)[2] and bone acidic glycoprotein (BAG-75).[3] These proteins account for about 15% of the noncollagenous proteins in most species' bones.[4] Dentin sialoprotein, recently isolated and cloned by Butler's group, shares extensive sequence homology with osteopontin.[5] Each of these has been associated with the mineralization process by both direct and indirect evidence.

Phosphoproteins have long been linked to the mineralization process based on their accumulation at the mineralization front,[6-8] and more recently and significantly, on the inability of dephosphorylated bone matrices to support mineralization in metastable calcium phosphate solutions.[9,10] There is also indirect evidence from cell cultures deprived of ATP that matrix protein phosphorylation is a mandatory step in the formation of a mineralized matrix.[11,12]

Our initial studies with a crude preparation of rat bone osteopontin suggested that osteopontin was an inhibitor of both *de novo* apatite (bone-like mineral) formation and apatite growth.[13] The studies were done in a system designed to utilize minimal amounts of matrix proteins.[14] In this system, the osteopontin (or other protein) is placed in a 50-μl band in a gelatin gel at the point at which mineral is expected to form based on control, protein-free gels. Buffered Ca and P solutions flow from "infinite" reservoirs entering opposite ends of the gel. Accumulation of Ca and P in 0.3-mm-thick slices throughout the gel is compared to similar double diffusion protein-free gels and to single diffusion gels to assess Ca and/or P chelation by the protein. Although Ca (or P) chelated by the protein may actually take part in mineral nucleation, this contribution is ignored in these experiments, and single diffusion results are subtracted from double diffusion data, prior to comparison between experimental and control gels. This system is similar in many aspects to that used by Hunter and Goldberg for the study of BSP,[15] except their matrix was agar rather than gelatin, the protein was contained throughout the matrix, and the ions did not recirculate; thus the BSP was exposed to a "steady-state" concentration of ions.

In the studies of osteopontin in the gelatin-gel system, (FIG. 1) a highly purified bovine bone osteopontin was found to inhibit both apatite formation and growth in a dose-dependent manner.[13] In contrast to studies with the dentin phosphoprotein,

[a] Supported by NIH Grant DE04141.

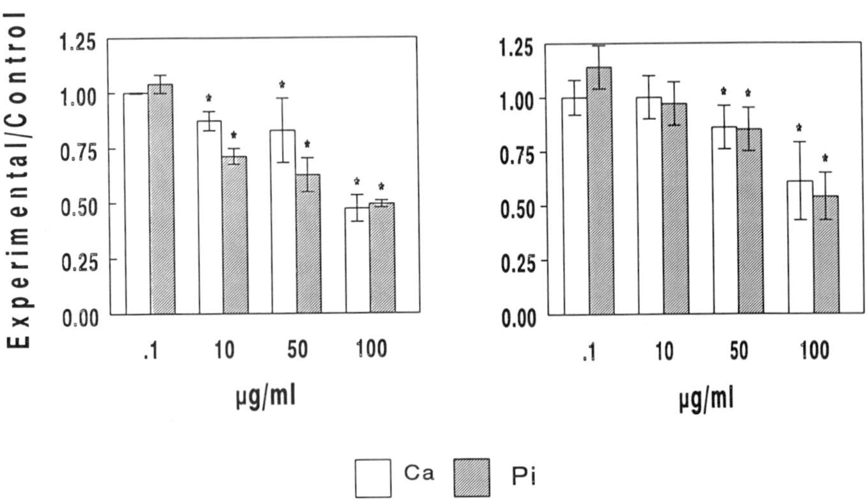

FIGURE 1. Dose-dependent inhibition of hydroxyapatite formation (*left*) and seeded-growth (*right*) by bovine osteopontin. Data show 5-day ion accumulation in experimental relative to control gels whose 50-μl precipitant band contained the indicated concentration of osteopontin (formation) or osteopontin + 0.5 mg/ml apatite crystals (growth). Data (published in different form in Ref. 13) shows mean ± SD for n ≥ 3 experiments, $*p \leq 0.05$ relative to control.

phosphophoryn,[16] which inhibited apatite growth at high concentrations and facilitated growth at lower concentrations, low concentrations of osteopontin did not affect either process. In each of these studies (with osteopontin,[13] BSP,[15] and phosphophoryn[16]), the mineral formed in the experimental and control systems was shown by X-ray diffraction to be a poorly crystalline hydroxyapatite, analogous in crystal size and perfection to bone mineral.

The observation that osteopontin was an effective inhibitor of apatite formation and growth was again demonstrated in studies by Hunter *et al.*[17] In metastable solutions where pH was maintained constant by use of an autotritator, they found significant inhibition of apatite formation by as little as 0.1 μg/ml osteopontin, and almost complete inhibition by 1.0 μg/ml. In their agarose gel system, 100 μg/ml osteopontin reduced Ca + P accumulation by 24 percent.

In the gelatin-gel, dephosphorylated osteopontin[13] had no ability to inhibit hydroxyapatite formation and growth, again indicating the importance of the phosphate residues, and explaining, in part why different osteopontins from different tissues, which have varying degrees of phosphorylation,[18] may have diverse effects on mineral formation and growth. Hunter *et al.*[17] also reported greater than a 40% decrease in inhibition in the presence of dephosphorylated osteopontin, while cross-linking carboxylic acid groups did not block osteopontin's inhibitory ability. Which phosphorylated residues and which protein domains are required for this inhibition remain to be determined, both in solution, and when the protein is bound to a matrix in the conformation it would have *in situ*.

Based on the EM appearances of apatite grown in the presence of 0–100 μg/ml osteopontin, it appeared that this protein blocked crystal elongation,[13] rather than secondary nucleation as was the case for phosphophoryn.[16] This implied that osteopontin bound with high affinity to one or more crystal faces. The binding of osteopontin, and several related macromolecules, based on Langmuir adsorption isotherms have been determined, and the data are summarized in TABLE 1. From this table, it is apparent that the extent of phosphorylation as well as the number and arrangement of other anionic residues which determine the proteins conformation, also determine the abilities of these macromolecules to bind to hydroxyapatite.

Proteins that bind to the hydroxyapatite can affect mineralization in several, not mutually exclusive, ways.[19] They can stabilize crystal nuclei, facilitating growth; they can act as epitaxial/foreign nucleators negating the energy demand of the process of nucleation (forming the first stable apatite crystal); they can bind to a particular face of the crystal preventing growth upon that face and thereby changing the shape of the crystal; or they can engulf the entire crystal, preventing its proliferation and/or growth. Proteins that bind with high specificity (low N) are more likely to be bound to well defined faces than those which bind nonspecifically (higher N). Models of binding can only be developed when knowledge of the protein's conformation is available.

Although the conformation of osteopontin, BSP, and BAG-75 have been predicted from sequence analysis,[20] little is known about their actual conformation *in situ*. Evans's elegant NMR studies of phosphophoryn's conformation[21] at different pHs illustrate the power of one of the techniques available to address this question, showing that phosphophoryn consists of structurally organized repeating charged clusters separated by neutral amino acids, which may serve as flexible hinges. The charge clusters are involved in calcium chelation, and most likely in interactions with the mineral. Similar studies of osteopontin and its analogues with varying extents of phosphorylation and varying sialic acid contents are anxiously awaited. The availability of expression systems for the native protein[18] should soon make these studies possible.

Osteopontin is not the only phosphorylated sialoprotein that inhibits apatite formation and growth. Preliminary studies (FIG. 2) indicate that the dentin sialoprotein (DSP) has a comparable effect. This is not surprising in view of the high structural homology with osteopontin. One would predict that phosphorylated DSP, which like osteopontin has repeating aspartate residues, might have an affinity for apatite comparable to phosphorylated osteopontin.

Bone sialoprotein with its polyglutamic acid sequences, in contrast, is apt to bind in a different way. Hunter and Goldberg[15,17] were the first to demonstrate that this specific bone cell product is an apatite "nucleator" rather than an inhibitor of apatite formation and growth. In the agar gel system, BSP was shown to induce apatite formation at Ca + P products significantly below the threshold for

TABLE 1. Comparative Affinities of Osteopontin and Related Molecules for Hydroxyapatite (HA) or Calcium Oxalate (OX)[a]

	N (μmole/m^2)	K (ml/μmol)	Ref.
Osteopontin, HA	0.026	1087	13
Osteopontin, OX	0.026	($1/3.7 \times 10^{58}$)	28
Aspartic acid, HA	0.44	0.30	24
Dentin phosphophoryn, HA	0.0016	8370	33
Salivary phosphoprotein, HA	0.21–0.24	18000–26000	33

[a] All data based on Langmuir adsorption isotherms.

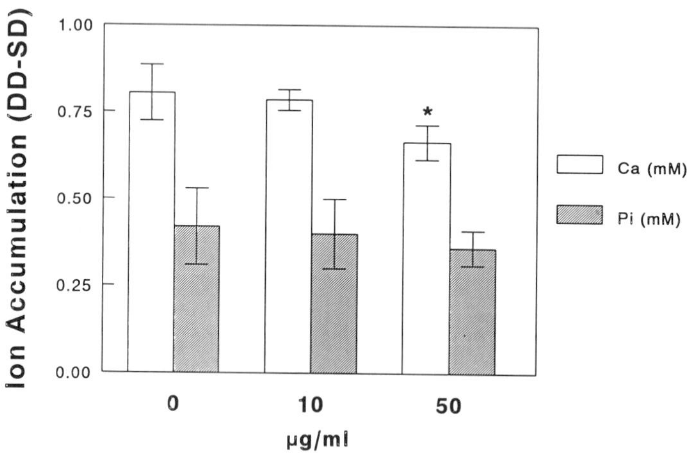

FIGURE 2. Dose-dependent inhibition of hydroxyapatite formation by dentin sialoprotein (DSP). Preliminary data shown represent ion accumulation in the double diffusion (DD) gel in the presence of 1, 10 and 50 μg/ml DSP, corrected for accumulation in single diffusion (SD) studies. Values are mean ± SD, n = 3. $*p \leq 0.05$ relative to 0 μg/ml control.

spontaneous precipitation.[17] These studies in the agar system have been duplicated in the gelatin system (FIG. 3), demonstrating that results are not system specific. Modification of BSP by cross-linking carboxylic acid residues markedly blocked their nucleation abilities, while dephosphorylation with alkaline phosphatase had less of an effect.[17] These data clearly suggest the nature of the interaction of BSP with apatite, although structural studies are still necessary.

TABLE 2 describes the known structural features of the connective tissue phosphorylated sialoproteins. The domains most likely to interact with apatite, based on the studies above, are indicated. Of those phosphorylated sialoproteins shown, only BAG-75 has not yet been studied in mineralizing systems, but it is likely that it too will have phosphorylated conformations that can interact with matrix and mineral.[20] Polyasparte is known to bind strongly to apatite and prevent/retard mineral formation and growth.[22,17] Thus aspartate groups in soteopontin and DSP may facilitate interaction with the mineral. Cross-linking glycine ethyl ester to the carboxyl residues of osteopontin did reduce its inhibitory potential suggesting this is a likely mechanism. Polyglutamate as found in BSP has less of an affinity for apatite, and little inhibitory potential.[17] Since the phosphorylated sialoproteins bind collagen, cells, and mineral, it will be important to know the conformation of the mineral-binding domain when the protein is bound to collagen, and vice versa. Further, it is possible that there are domains of these molecules that may not be required for these interactions which can be removed or modified without affecting mineralization. Using NMR, conformational changes have been detected when a small peptide containing the RGD-adhesion sequence binds to cells.[23] Whether these conformations remain the same when the intact protein binds also remains to be determined. It is of interest to note that the abnormally calcified elastin fibers contain BSP,[25] while the calcified aortic placque contains osteopontin,[23] revealing a tissue specificity for these interactions.

The kidney analogue of osteopontin, uropontin,[26] has been reported to inhibit

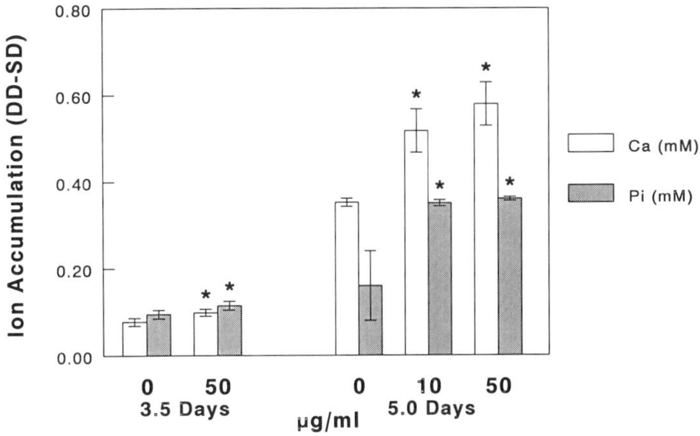

FIGURE 3. Dose-dependent proliferation of hydroxyapatite formation by bone sialoprotein (BSP) at 3.5 and 5 days in the gelatin gel. Data shown represent ion accumulation in the double diffusion (DD) gel in the presence of 0, 10, and 50 μg/ml BSP corrected for accumulation in single diffusion (SD) studies. Values are mean ± SD, n = 3, $*p \leq 0.05$ relative to 0 μg/ml control.

calcium oxalate formation *in vivo*[27] and *in vitro*.[28] Based on constant composition analyses and adsorption studies, it was suggested that in the oxalate system, the binding of the protein to one or more crystal surfaces is responsible for its potent inhibitory effect.[28] In urine, it thus may be suggested that osteopontin is present to prevent/reduce crystal formation. In bone, osteopontin, whose expression in bone cell culture parallels rather than precedes mineral deposition,[29] may play more of a regulatory role. (It should be noted that in a different culture system[30] while osteopontin expression parallelled that of alkaline phosphatase expression, BSP increased after mineralization.) Since osteopontin is localized on the bone

TABLE 2. Primary Structures of the Phosphorylated Sialoproteins of Bone and Dentin

Feature	Osteopontin	BSP	BAG-75	DSP
RGD sequences	throughout	C-terminal	+	+
Poly-Asp repeats[a]	+	−	+	+
Poly-Glu repeats[a]	−	+	+	−
Phosphorylated residues[a]	% variable	up to 30%	8%	?
Sulfation[a]	some species	variable	?	?
Carbohydrate (%)/ location	throughout	N-terminal half	7%	30% throughout
% sialate	10%	12–18%	7%	9%
NH_2-terminal	Leu-Pro-Val-Lys-Val-Ala-Glu-Phe	Phe-Ser-Met . . .	Leu-Pro-Val-Ala-Arg-Tyr-Gln-Asn	Ile-Pro-Val-Pro-Gln-Leu-Val-Pro

[a] Features possibly important for binding to HA.

surface where the osteoclasts are anchored,[31] and is expressed in osteoblasts and osteocytes at late stages of ectopic-bone induction when osteoclast recruitment is apparent[32] it and by analogy DSP may function to recruit matrix-forming cells, regulate the growth of hydroxyapatite crystals formed on that matrix, and assist in the recruitment of osteoclasts.

SUMMARY

Osteopontin is one of a family of phosphorylated sialoproteins found in the extracellular matrices of mineralized connective tissues. Solution studies from a variety of laboratories have shown that while some of these proteins (*e.g.*, osteopontin, dentin sialoprotein) inhibit calcium phosphate production, others, such as bone sialoprotein, can act as a nucleator of calcium phosphate formation. The differences in the behaviors of these proteins can be related to their interactions with mineral crystal nuclei and crystal growth sites. The specificity of these interactions, in turn, can be related to differences in the primary structures of the sialoproteins and to the extent to which they are phosphorylated. *In vitro* systems for the study of osteopontins and osteopontin-related protein effects on mineral deposition are described, along with mechanisms explaining the contrasting abilities of the polyglutamate-containing bone sialoprotein to act as a nucleator, while the polyaspartate-containing osteopontin and dentin sialoproteins inhibit calcium phosphate formation and growth.

ACKNOWLEDGMENTS

The BSP and DSP used for the studies described here were kindly provided by Dr. W. T. Butler.

REFERENCES

1. Franzen, A. & H. Heinegård. 1985. Isolation and characterization of two sialoproteins present only in bone calcified matrix. Biochem. J. **232:** 715–724.
2. Prince, C. W., T. Oosawa, W. T. Butler, M. Toman, A. S. Bhon & R. E. Schrohenloher. 1987. Isolation, characterization and biosynthesis of a phosphorylated glycoprotein formed at bone. J. Biol. Chem. **262:** 2900–2907.
3. Gorski, J. F. & K. Shimizu. 1988. Isolation of new phosphorylated glycoprotein from mineralized phase of bone that exhibits limited homology to adhesive protein osteopontin. J. Biol. Chem. **252:** 15938–15945.
4. Gehron-Robey, P. 1989. The biochemistry of bone. Endocrinol. Metab. Clin. North Am. **18:** 859–901.
5. Butler, W. T., M. Bhown, J. C. Brunn, R. N. D'Souza, M. C. Farach-Carson, R. P. Happonen, R. E. Schrohenloher, J. M. Seyer, M. J. Sommerman, R. A. Foster, M. Tomena & S. Van Dijk. 1992. Isolation, characterization and immunolocalization of a 53 kDa dentin sialoprotein (DSP). Matrix **12:** 343–351.
6. Weinstock, M. & C. P. Leblond. 1973. Radioautographic visualization of the deposition of a phosphoprotein at the mineralization front in the dentin of the rat incisor. J. Cell Biol. **56:** 838–845.
7. Li, S. T. & E. Katz. 1977. On the state of demineralized matrices of bone and dentin. Calcif. Tissue Res. **22:** 275–284.
8. McKee, M. D. & A. Nanci. 1994. Osteopontin at mineralized tissue interfaces in

bone, teeth and osteointegrated implants: ultrastructural distribution and implications for mineralized tissue formation, turnover, and repair. Microsc. Res. Tech. In press.

9. Endo, A. 1987. Potential role of phosphoprotein in collagen mineralization—an experimental study *in vitro*. J. Orthop. Assoc. **61:** 563–569.

10. Glimcher, M. J. 1989. Mechanisms of calcification in bone: role of collagen fibrils and collagen-phosphoprotein complexes *in vitro* and *in vivo*. Anat. Rec. **224:** 139–153.

11. Boskey, A. L., S. B. Doty, R. Mendelsohn & I. Binderman. 1994. ATP promotes mineralization in differentiating chick limb-bud mesenchymal cell cultures. Microsc. Res. Technique **28:** 492–504.

12. Boskey, A. L., I. Binderman, D. Stiner, R. Mendelsohn & S. B. Doty. 1994. Viable chondrocytes are essential for normal mineral deposition in culture. Trans. ORS **19:** 122.

13. Boskey, A. L., M. Maresca, W. Ullrich, S. B. Doty, W. T. Butler & C. W. Prince. 1993. Osteopontin-hydroxyapatite interactions *in vitro*: inhibition of hydroxyapatite formation and growth in a gelatin-gel. Bone Miner. **22:** 147–159.

14. Boskey, A. L. 1989. Hydroxyapatite formation in a dynamic gelatin gel system: effects of type I collagen, lipids, and proteoglycans. J. Phys. Chem. **93:** 1628–1633.

15. Hunter, G. K. & H. A. Goldberg. 1993. Nucleation of hydroxyapatite by bone sialoprotein. Proc. Natl. Acad. Sci. USA **90:** 8562–8565.

16. Boskey, A. L., M. Maresca, S. B. Doty, B. Sabsay & A. Veis. 1990. Concentration-dependent effect of dentin-phosphophoryn in the regulation of *in vitro* hydroxyapatite formation and growth. Bone Miner. **11:** 55–65.

17. Hunter, G. K., C. L. Kyle & H. A. Goldberg. 1994. Modulation of crystal formation by bone phosphoproteins: structural specificity of the osteopontin-mediated inhibition of hydroxyapatite formation. Biochem. J. **300:** 723–728.

18. Ashkar, S., D. B. Teplow, M. J. Glimcher & R. A. Saavedra. 1993. *In vitro* phosphorylation of mouse osteopontin expressed in *E. coli*. Biochem. Biophys. Res. Commun. **191:** 126–133.

19. Mann, S., J. Webb & R. J. P. Williams, Eds. 1988. Biomineralization: Chemical and Biochemical Perspectives. 35–60. VCH Publishers. New York.

20. Gorski, J. P. 1992. Acidic phosphoproteins from bone matrix: a structural rationalization of their role in biomineralization. Calcif. Tissue Int. **50:** 391–396.

21. Evans, J. A. & S. I. Chan. 1994. Phosphophoryn, a biomineralization template protein: pH-dependent protein folding experiments. Biopolymers **34:** 507–527.

22. Moreno, E. C., M. Kresak & D. I. Hay. 1991. Adsorption of molecules of biologic interest onto hydroxyapatite. Calcif. Tissue Int. **49:** 193–196.

23. Peterson, E., B. Luning, H. Mickos & D. Heinegård. Synthesis, NMR and function of an O-phosphorylated peptide comprising the RGD-adhesion sequence of osteopontin. Acta Chem. Scand. **45:** 604–608.

24. Baccaranti, C. M., F. Taparelli, F. Baraldi & R. I. Pasquali. 1994. Mineralization in PXE is induced by proteins with high affinity for calcium. Abstracts XIV Meeting of the Federation of European Connective Tissue Societies, N3.

25. Giachelli, C. M., N. Bue, M. Almeida, D. T. Denhardt, C. E. Alpeprs & S. M. Schwarts. 1993. Osteopontin is elevated during neontima formation in rat arteries and is a novel component of human atherosclerotic paques. J. Clin. Invest. **92:** 1686–1696.

26. Kohr, K., Y. Suzuki, K. Yoshida & K. Yamamoto. 1992. Molecular cloning and sequencing of cDNA encoding urinary stone protein which is identical to osteopontin. Biochem. Biophys. Res. Commun. **184:** 859–864.

27. Shiraga, H., W. Min, W. J. van Dusen, M. D. Clayman, D. Miner, C. L. T. Terrell, J. R. Sherbotie, J. W. Foreman, C. Pryzysiecki, E. G. Nelson & J. R. Hoyer. 1992. Inhibition of calcium oxalate crystal growth *in vitro* by uropontin: another member of the aspartic acid-rich protein superfamily. Proc. Natl. Acad. Sci. USA **89:** 426–430.

28. Worcester, E. M., S. S. Blumenthal, A. M. Beshensky & D. L. Lewand. 1992. The calcium oxalate crystal growth inhibitor protein produced by mouse kidney cortical cells in culture is osteopontin. J. Bone Miner. Res. **7:** 1029–1036.

29. GERSTENFELD, L. C., Y. GOTOH, M. D. MCKEE, A. NANCI, W. J. LANDIS & M. J. GLIMCHER. 1990. Expressions and ultrastructural immunolocalization of a major 66-kDa phosphoprotein synthesized by chicken osteoblasts *in vitro*. Anat. Res. **228:** 93–103.
30. IBARAKI, K., J. D. TERMINE, S. W. WHITSON & M. F. YOUNG. 1992. Bone matrix mRNA expression in differentiating fetal bovine osteoblasts. J. Bone Miner. Res. **7:** 743–754.
31. REINHOLT, F. P., K. HULTENBY, A. OLDBERG & D. HEINEGÅRD. 1990. Osteopontin—a possible anchor of osteoclasts to bone surface. Proc. Natl. Acad. Sci. USA **87:** 4473–4475.
32. ARAI, N., K. OHYAK & H. OGURA. 1993. Osteopontin mRNA expression during bone resorption: an *in situ* hybridization study of induced ectopic bone in the rat. Bone Miner. **22:** 129–145.
33. HAY, D. I., E. C. MORENO & D. H. SCHLESINGER. 1979. Phosphoprotein inhibitors of calcium phosphate precipitation from salivary secretions. Inorg. Perspect. Biol. Med. **2:** 271–285.

Osteopontin in Urinary Stone Formation[a]

JOHN R. HOYER,[b,d] LASZLO OTVOS, JR.,[c] AND LASZLO URGE[c]

[b]*Departments of Pediatrics and Medicine*
University of Pennsylvania
and
The Children's Hospital of Philadelphia
[c]*The Wistar Institute*
Philadelphia, Pennsylvania 19104

Calcium oxalate (CaOx) is the primary constituent of the majority of stones formed within the urinary space.[1,2] Since normal urine is frequently supersaturated with respect to calcium oxalate, urinary inhibitors of crystallization appear to have important roles in preventing urinary stone formation. The majority of the inhibition of CaOx crystallization in normal urine is provided by protein macromolecules. These proteins may be protective by slowing the growth and/or aggregation of crystals and/or raising the concentration of CaOx required for the spontaneous formation of new crystals. *In vitro* studies have shown that the urinary form of osteopontin from human urine (uropontin) is a potent inhibitor of CaOx crystallization.[3,4] Although the precise role of uropontin has not been defined, information concerning its urinary excretion and molecular features provides the basis for considering its potential role in stone formation.

Isolation of Uropontin by Immunoaffinity Purification

Uropontin was initially isolated by immunoaffinity purification from human urine on the basis of its capacity to inhibit calcium oxalate monohydrate crystal growth.[3,5] In these studies, our approach to purification and characterization of crystal inhibitor proteins used monoclonal antibodies produced by hybridomas that we obtained after immunization with an inhibitory fraction of urinary proteins. These antibodies were coupled to Sepharose beads and used to purify a protein that migrated in SDS-polyacrylamide gels at a position of ~50 Kd and inhibited calcium oxalate crystal growth. The N-terminal amino acid sequence of this urinary protein[3] is identical to that of human osteopontin from bone.[6] In addition, compositional analysis of uropontin revealed a very high percentage of aspartic acid residues and an overall distribution of amino acids comparable to that determined or predicted for human osteopontin from bone.[6,7] Furthermore, the nucleotide sequences of cDNAs from human kidney[3] and bone[7,8] that encode osteopontin

[a] This work was supported by Grant DK33501 from the National Institutes of Health.
[d] Address correspondence to: Dr. John R. Hoyer, Children's Hospital of Philadelphia, 34th St. & Civic Center Blvd., Philadelphia, PA 19104.

are identical. However, substantial variation in the migration of osteopontin from different tissue sources in polyacrylamide gels has been demonstrated. In addition, as discussed in detail elsewhere in this volume, the extent of posttranslational modifications may vary within the same tissue. Additional evidence for heterogeneity has been provided by our initial studies using proteins derived from different human sources. Osteopontin derived from human milk migrated in 10% PAGE gels at a position of ~75 kD, while osteopontin from urine (uropontin) migrated at a position of ~50 kD in adjacent lanes. In addition, when evaluated by ELISA using uropontin standards and our ZH monoclonal antibody to uropontin, the concentration of osteopontin from milk detected was less than 2% of the value determined by chemical protein determination. Comparable studies of rOP27, a recombinant human osteopontin using the ZH monoclonal antibody gave ELISA values that were only 20% of that predicted by chemical protein determination. In addition, the recombinant protein was not recovered by immunoaffinity purification using ZH beads. Further studies of the detection of MP using a panel of polyclonal antibodies to UP or bone osteopontin (OP) are shown in FIGURE 1. It is noteworthy that the polyclonal antibody prepared by immunization with osteopontin from bone (LF7) detected uropontin much better than it detected osteopontin from milk. These immunologic studies provide additional evidence for differences in the molecular characteristics of osteopontins derived from different

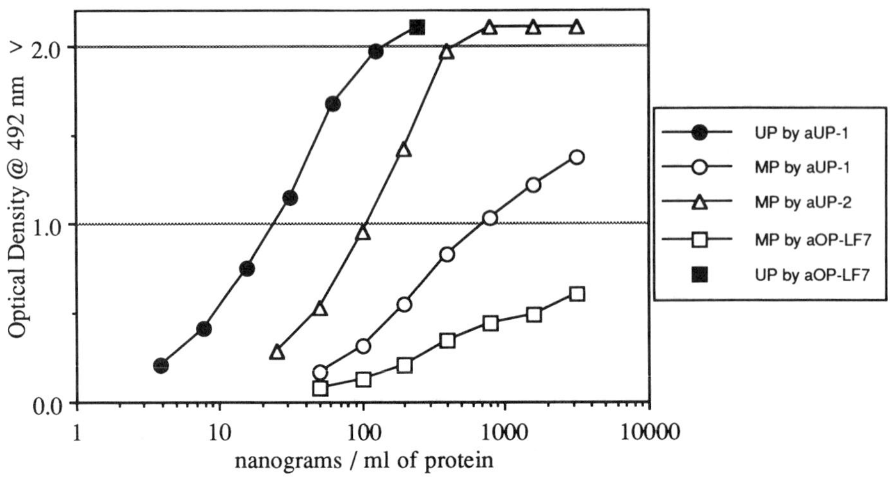

FIGURE 1. The horizontal axis is expressed as the concentrations based on chemical determination that were present in the dilutions of uropontin (UP) and osteopontin from milk (MP) used to coat a microtiter plate. Each polyclonal rabbit antisera was used on the same plate at an optimal concentration for detecting UP. The results showed that these polyclonal antibodies did not detect MP as well as UP. When the slope of the linear portion of the curve for the detection of UP by anti-UP-1 (*solid circles*) is used to calculate the concentration of MP (480 μg/ml) that is detected, only 27 μg/ml was detected by anti-UP-1 (*open circles*), while 102 μg/ml was detected by the other polyclonal antisera to uropontin (anti-UP-2) (*open triangles*), and <9 μg/ml was detected by LF7 (*open squares*), a polyclonal antibody to human osteopontin from bone prepared by Dr. Larry Fisher. The low level of detection of MP by LF7 was not related to the concentration of antibody used; the same dilution of this antisera readily detected 200 ng/ml of UP on the same plate (*closed square*).

sources. Thus, the term *uropontin* identifying the urinary origin of this form of osteopontin appears to be a useful designation.

Biomineralization

The roles of osteopontin in mineralization mediated through interactions of its cell-binding Arg-Gly-Asp-Ser (RGDS) sequence with bone cells are considered in detail elsewhere in this volume. Therefore, the following discusison will emphasize the potential role of another highly conserved domain, the aspartic acid-rich sequences. We previously concluded that the biologic distribution of aspartic acid-rich proteins in close association with mineralization events in a broad range of organisms and tissues, reflected the presence of a superfamily of aspartic acid-rich proteins (AARP).[3] A potential function of osteopontin through interaction with crystals was suggested by a series of elegant *in vitro* studies demonstrating tissue-specific patterns of modulation of crystal growth by aspartic acid-rich proteins from other mineralizing tissues. Specific interactions of an AARP from mollusk shells with selected faces on calcium dicarboxylic acid crystals promotes growth of these faces, while slowing simultaneously the overall rate of crystal growth.[9] While inhibitory in the fluid phase, prior adsorption of this mollusk AARP onto a solid substrate induced calcite crystal formation in an orientation characteristic of biological mineralization.[9] Substituting a less acidic AARP from the sea urchin demonstrated the specificity of this interaction. This AARP template produced biologically relevant differences in the internal texture and fracture properties of the calcite crystals formed.[10]

Although the abundance of carboxylic acid residues in AARPs supports the potential for interaction with crystal surfaces, the conformations adopted by these proteins and the resulting pattern of charge density also appear to be crucial for the specific interactions that lead to modulation of crystal growth. For example, the aspartic acid-rich proteins extracted with EDTA from mollusk shells have adopted a β-sheet conformation upon exposure to calcium.[11] By contrast, circular dichroism studies of the secondary structure of osteopontin extracted from rat bone showed no conformational changes from a random coil after exposure to 8 to 800-fold molar excesses of calcium, findings attributed to protein denaturation during extraction from bone.[12] More recently, we obtained evidence that the method of purification profoundly influences the results of studies of secondary structure. To allow direct comparison with the earlier study, rat osteopontin isolated from the conditioned media of rat ROS 17/2.8 cells by immunoaffinity purification was used in our circular dichroism studies. A column of immobilized MPIIIB10 monoclonal antibodies was used for immunopurification. As shown in FIGURE 2, exposure of rat osteopontin to a 100-fold molar excess of calcium ions shifted the CD spectra. Computer analysis by the spectra by the algorithm of Yang[13] indicated a significant shift from ~46% β-sheet to ~58% β-sheet structure after addition of calcium.

Uropontin is a very potent inhibitor of calcium oxalate monohydrate (COM) crystal growth *in vitro* with a Kd for inhibition of 2.5×10^{-8} M estimated from a Langmuir type isotherm plot.[4] Osteopontin secreted by mouse kidney cells in culture appears to be comparable to uropontin in potency as an inhibitor of COM crystal growth.[14] It should be noted that uropontin inhibited crystal growth and produced intracellular signal transduction after integrin-binding at concentrations within the physiologic range,[4,15] while relatively larger amounts (>10-fold higher) of osteopontin isolated from bone were required to achieve comparable effects.[16,17]

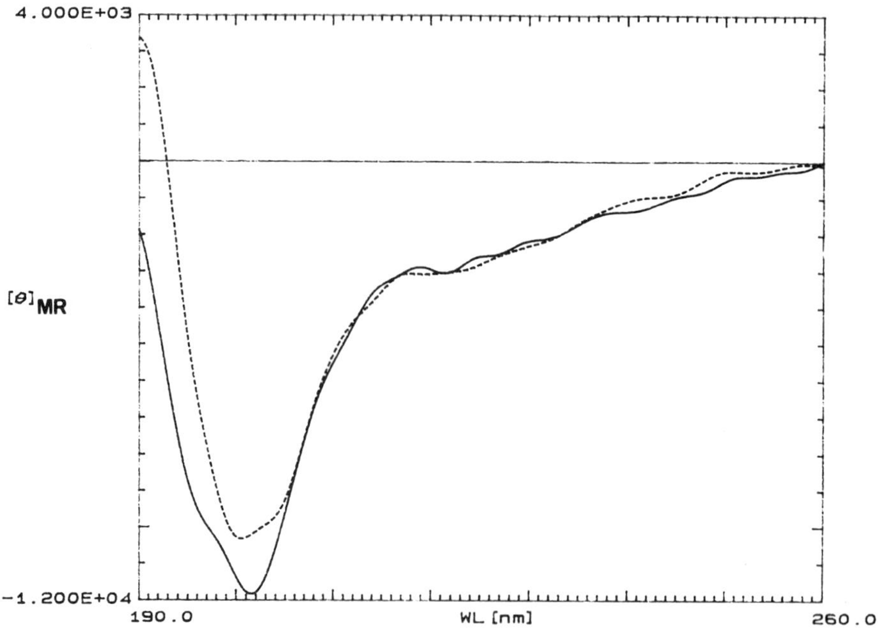

FIGURE 2. Circular dichroism spectra of rat osteopontin (80 μg/ml) isolated by monoclonal antibody immunoaffinity purification. Spectra obtained on a Jasco model J-720 CD spectrometer before (*solid line*) and after exposure to a 100-fold excess of calcium ions (*broken line*) are shown. The conformation calculated by the algorithm of Yang[13] shifted from ~46% β-sheet to ~58% β-sheet structure after the addition of calcium.

The relative contributions of differences in species, sensitivity of crystal growth system to inhibition or to denaturation of osteopontin during isolation to differences in potency are not known. More recent studies using osteopontin obtained by EDTA extraction[18] in an autotitration system showed inhibition of hydroxyapatite crystal growth at concentrations comparable to those at which uropontin is active. Both studies of the influence of osteopontin on hydroxyapatite crystal growth showed markedly decreased inhibition after the protein was dephosphorylated.[17,18]

Taken together, the above functional studies add support to our suggestion that the osteopontin may contribute to the regulation of mineralization within the fluid spaces of multicellular organisms that are supersaturated with respect to their mineral constituents.[3] AARPs, acting in concert with other proteins, may exert either restrictive or permissive effects on the formation or remodeling of crystal surfaces in a variety of body fluids. Regulatory roles and tissue-specific functional expression are suggested by evidence that the quantities and molecular forms of osteopontin are responsive to hormonal influences, including those of vitamin D,[19,20] parathyroid hormone,[21] estrogen, and progesterone.[22–24] In addition to the posttranslational modifications that are discussed in detail elsewhere in this volume, the functional effects of osteopontin may be influenced by alternative RNA splicing. A highly conserved sequence of 14 amino acids is encoded by an exon of the mouse and human osteopontin genes;[25,26] this sequence is deleted in one of the isoforms encoded by cDNAs from human bone, decidua and kidney.[3,7,8]

Uropontin in Calcium Oxalate Stones

Inhibition of CaOx crystal growth *in vitro* by uropontin suggested that it would also interact with calcium crystals *in vivo*. This hypothesis was tested by using a sensitive ELISA to evaluate EDTA extracts of a series of individual human urinary stones weighing approximately 100 mg and classified on the basis of the predominant mineral component.[27,28] In these initial studies, the uropontin content of CaOx monohydrate stones (mean of ~800 μg/100 mg of stone) was much greater than in hydroxyapatite of CaOx dihydrate stones (mean of <10 μg and <20 μg/100 mg of stone, respectively), and greater than in brushite stones (mean of ~90 μg/100 mg). By contrast, the content of Tamm-Horsfall protein, albumin and the urinary inter-α-trypsin inhibitor[29] detected in the same extracts were very low (all <10 μg/100 mg of CaOx monohydrate stone). This abundance of uropontin in CaOx monohydrate stones is consistent with the preponderance of acidic amino acids shown in previous studies of the overall amino acid composition of CaOx stones.[30–32] Cloning studies using a polyclonal antiserum to proteins extracted from CaOx stones have confirmed the presence of osteopontin in CaOx stones.[33]

Uropontin Excretion

Protection from stone formation provided by an inhibitor depends on the presence of sufficient concentrations within urine as well as its intrinsic inhibitory capacity. Our studies of inhibitory capacity showed that the relationship between uropontin concentration and inhibition of crystal growth follows a Langmuir-type isotherm with a dissociation constant of 2.5×10^{-8} M.[4] Thus, our analysis of the patterns of uropontin excretion of normal adult human volunteers using an affinity-purified polyclonal antiserum showed that the mean concentration of uropontin present in normal adult human urine ($\sim 6 \times 10^{-8}$ M)[34] is sufficient to significantly inhibit CaOx crystal growth. Uropontin concentration varied inversely with urine volume, a characteristic that favors the inhibition of CaOx crystallization. While other tissue sources for uropontin have not been excluded, renal synthesis of osteopontin has been well established. High levels of mRNA have been consistently observed in kidneys, and recent studies employing both *in situ* hybridization and immunohistochemical analysis demonstrated osteopontin expression in a subset of distal nephrons in normal adult mouse kidneys[35] and human kidneys.[36] Recent ultrastructural analysis provided evidence that osteopontin is also synthesized by thin limbs of Henle's loop and by the papillary surface epithelium of rat kidneys.[37] Thus, osteopontin is delivered to sites where urine is highly concentrated and the presence of inhibitors is most crucial.

Factors that alter uropontin excretion rates could decrease the capacity for inhibiting crystallization and thereby influence urinary stone formation. Although effects have not yet been directly evaluated, studies of osteopontin expression within tissues and cell culture provide clues regarding such effects. For example, a low incidence of stone formation prevails during pregnancy, despite the physiologically increased calcium excretion of pregnancy. In this setting, the increased OPN expression during pregnancy[24] may counterbalance hypercalciuria by increasing inhibitory capacity. Osteopontin expression has been shown to be increased by exposure to phorbol esters,[38] bacterial infection,[39] angiotensin II,[40] oxalate,[41] urinary obstruction,[42] and renal ischemia.[37,43] On the other hand, decreased quantities of uropontin secreted into urine may result from decreased responsiveness to hormones or other factors. Quantitative deficiencies could also result from intrinsic genetic defects such as those linked by mapping studies to resistance to Rickettsial infection.[39] However, detection of decreased quantities of uropontin in urine should not be taken as definitive evidence for a synthetic defect, since reduced quantities could also result from increased adsorption of uropontin onto increased quantities of crystals in the urine of some stone formers with marked CaOx supersaturation. Furthermore, functional deficiencies of uropontin could result from inactivation by urine conditions such as pH, ionic strength, and calcium ion concentration in a manner comparable to the increased self-aggregation of Tamm-Horsfall protein from some stone formers.[44] Interactions of osteopontin with other proteins[45,46] also have the potential for lowering the effective concentrations of uropontin in urine.

Uropontin Structure

A broad range of potential structural defects of peptide sequence and/or post-translational modifications of uropontin could also influence the inhibition of crystallization and the process of nephrolithiasis. Decreased phosphorylation, for example, may alter the patterns of charge density and/or conformation of uropontin and decrease its affinity for CaOx crystals. This would decrease its ability to inhibit crystallization. Alternatively, molecular abnormalities could lead to a net promotion, rather than inhibition of crystallization. In this regard, sulfation of osteopontin has been shown to be associated with mineralization in bone.[47] Thus, a net inhibition of CaOx nucleation by normal uropontin[4] might be reversed by a structural modification of this protein that led to a net promotion of crystallization.

Proteins may have dual roles in crystallization, exerting a net inhibitory effect in the fluid phase, but serving as nucleation sites after immobilization.[48] The binding of uropontin to the $\alpha_v\beta_3$ integrin that is present on the surfaces of some cells would provide a basis for immobilization of this protein. This integrin is not normally accessible to uropontin within the urinary space, since it is present on basolateral surfaces of tubular cells in the normal mouse kidney.[35] However, epithelial cell injury,[49] including that induced by crystalluria,[50] may disrupt cell polarity and redirect this integrin to an accessible site on luminal surfaces. Attachment of CaOx crystals to uropontin on the injuried epithelium could support stone formation by promoting the retention of crystals within the kidney.

The high content of uropontin in CaOx monohydrate stones strongly suggests a specific interaction of this aspartic acid-rich protein with CaOx crystals and an important role in calcium oxalate stone formation. However, its pathogenetic role is uncertain. Presence of uropontin in these stones could primarily reflect a protective fluid phase interaction that partially inhibited crystallization, and was overwhelmed by extremes of CaOx supersaturation. Alternatively, promotion, rather than inhibition of crystallization might be induced by a structurally abnormal uropontin. Uropontin is also a potential nucleation site if it is bound to the cell surface of diseased urinary epithelium. The elucidation of the primary structure of osteopontin provides the basis for investigative approaches at the molecular level to the structural basis for the modulation of biomineralization previously suggested by studies of other aspartic acid-rich proteins whose primary sequences are not known. Further investigations will seek to better understand the role of osteopontin in normal and pathologic mineralization.

ACKNOWLEDGMENTS

The murine monoclonal antibody to rat osteopontin, MPIIIB10 was obtained from the Developmental Studies Hybridoma Bank maintained by the Department of Pharmacology and Molecular Sciences, Johns Hopkins University School of Medicine, Baltimore, Maryland and the Department of Biological Sciences, University of Iowa, Iowa City, Iowa under contract NO1-HD-6-2915 from the NICHD. The polyclonal antiserum to osteopontin from human bone (LF7) was provided by Dr. Larry Fisher. Osteopontin from human milk was provided by Dr. Ronald Ingram and recombinant human osteopontin (rOP27) was produced by Drs. Dana Hu and Jeffrey Smith. Conditioned media from ROS 17/2.8 rat osteosarcoma cells was provided by Dr. Gideon Rodan.

REFERENCES

1. Coe, F. L., J. H. Parks & J. R. Asplin. 1992. The pathogenesis and treatment of kidney stones. N. Engl. J. Med. **327:** 1141–1152.
2. Worcester, E. M. & J. Lemann. 1994. Nephrolithiasis. Massry and Glassock's Textbook of Nephology. 3rd edit. S. G. Massry & R. J. Glassock, Eds. 1054–1078. Williams & Wilkins. Baltimore.
3. Shiraga, H., W. Min, W. J. VanDusen, M. D. Clayman, D. Miner, C. H. Terrell, J. R. Sherbotie, J. W. Foreman, C. Przysiecki, E. G. Neilson & J. R. Hoyer. 1992. Inhibition of calcium oxalate crystal growth *in vitro* by uropontin: another member of the aspartic acid-rich protein superfamily. Proc. Natl. Acad. Sci. USA **89:** 426–430.
4. Asplin, J., J. R. Hoyer, Y. Nakagawa & F. L. Coe. 1992. Uropontin (UP) inhibits calcium oxalate monohydrate (COM) crystal growth and nucleation. J. Am. Soc. Nephrol. **3:** 679.
5. Shiraga, H., M. D. Clayman, E. G. Neilson & J. R. Hoyer. 1989. Affinity purification of urinary crystal growth inhibitor (CGI). Kidney Int. **35:** 363.
6. Fisher, L. W., G. R. Hawkins, N. Tuross & J. D. Termine. 1987. Purification and partial characterization of small proteoglycans I and II, bone sialoproteins I and II, and osteonectin from the mineral compartment of developing human bone. J. Biol. Chem. **262:** 9702–9708.
7. Kiefer, M. C., D. M. Bauer & P. J. Barr. 1989. The cDNA and derived amino acid sequence for human osteopontin. Nucleic Acids Res. **17:** 3306.
8. Young, M. F., J. M. Kerr, J. D. Termine, U. M. Wewer, M. G. Wang, O. W. McBride & L. W. Fisher. 1990. cDNA cloning, mRNA distribution and heterogeneity, chromosomal location, and RFLP analysis of human osteopontin (OPN). Genomics **7:** 491–502.
9. Addadi, L. & S. Weiner. 1985. Interactions between acidic proteins and crystals; stereochemical requirements in biomineralizaiton. Proc. Natl. Acad. Sci. USA **82:** 4110–4114.
10. Berman, A., L. Addadi & S. Weiner. 1988. Interactions of sea-urchin skeleton macromolecules with growing calcite crystals; a study of intracrystalline proteins. Nature **331:** 546–548.
11. Worms, D. & S. Weiner. 1986. Mollusk shell organic matrix: Fourier transform infrared study of the acidic macromolecules. J. Exp. Zool. **237:** 11–20.
12. Prince, C. W. 1989. Secondary structure predicitons for rat osteopontin. Connect. Tissue Res. **21:** 15–20.

13. YANG, J. T., G.-S. WU & H. MARTINEZ. 1986. Calculation of protein conformation from circular dichroism. Methods Enzymol. **130:** 208–269.
14. WORCESTER, E. M., S. S. BLUMENTHAL & A. BESHENSKY. 1992. The calcium oxalate crystal growth inhibitor protein produced by mouse kidney cortical cells in culture is osteopontin. J. Bone Miner. Res. **7:** 1029–1036.
15. ZIMOLO, Z., G. WESOLOWSKI, H. TANAKA, J. HYMAN, J. R. HOYER & G. A. RODAN. 1994. Soluble $\alpha_v\beta_3$ integrin ligands increase intracellular [Ca^{++}] in rat osteoclasts and mouse osteoclast-like cells. Am. J. Physiol. **266:** C376–C381.
16. MIYAUCHI, A., J. ALVAREZ, E. M. GREENFIELD, A. TETI, M. GRANO, S. COLUCCI, A. ZAMBONIN-ZALLONE, F. P. ROSS, S. L. TEITELBAUM, D. CHERESH & K. A. HRUSKA. 1991. Recognition of osteopontin and related peptides by an $\alpha_v\beta_3$ integrin stimulates immediate cell signals in osteoclasts. J. Biol. Chem. **266:** 20369–20374.
17. BOSKEY, A. L., M. MARESCA, W. ULLRICH, S. B. DOTY, W. T. BUTLER & C. W. PRINCE. 1993. Osteopontin-hydroxyapatite interactions *in vitro;* inhibition of hydroxyapatite formation and growth in a gelatin-gel. Bone Miner. **22:** 147–159.
18. HUNTER, G. K., C. L. KYLE & H. A. GOLDBERG. 1994. Modulation of crystal formation by bone phosphoproteins: structural specificity of the osteopontin-mediated inhibition of hydroxyapatite formation. Biochem. J. **300:** 723–728.
19. YOON, K., R. BUENAGA & G. A. RODAN. 1987. Tissue specificity and developmental expression of rat osteopontin. Biochem. Biophys. Res. Commun. **148:** 1129–1136.
20. CHANG, P.-L. & C. W. PRINCE. 1991. 1α,25-dihydroxyvitamin D_3 stimulates synthesis and secretion of nonphosphorylated osteopontin (secreted phosphoprotein 1) in mouse JB6 epidermal cells. Cancer Res. **51:** 2144–2150.
21. NODA, M. & G. A. RODAN. 1989. Transcriptional regulation of osteopontin production in rat osteoblast-like cells by parathyroid hormone. J. Cell Biol. **108:** 713–718.
22. CRAIG, A. M. & D. T. DENHARDT. 1991. The murine gene encoding secreted phosphoprotein 1 (osteopontin): promoter structure, activity, and induction *in vivo* by estrogen and progesterone. Gene **100:** 163–171.
23. TURNER, R. T., D. S. COLVARD & T. C. SPELSBERG. 1990. Estrogen inhibition of periostial bone formation in rat long bones; down regulation of gene expression for bone matrix proteins. Endocrinology **127:** 1346–1351.
24. WATERHOUSE, P., R. S. PARHAR, X. GUO, P. D. LALA & D. T. DENHARDT. 1992. Regulated temporal and spatial expression of the calcium-binding proteins calycin and OPN (osteopontin) in mouse tissues during pregnancy. Mol. Reprod. Dev. **32:** 315–323.
25. MIYAZAKI, Y., M. SETOGUCHI, S. YOSHIDA, Y. HIGUCHI, S. AKIZUKI & S. YAMAMOTO. 1990. The mouse osteopontin gene; expression in monocytic lineages and complete nucleotide sequence. J. Biol. Chem. **265:** 14432–14438.
26. HIJIYA, N., M. SETOGUCHI, K. MATSUURA, Y. HIGUCHI, S. AKIZUKI & S. YAMAMOTO. 1994. Cloning and characterization of the human osteopontin gene and its promoter. Biochem. J. **303:** 255–262.
27. HOYER, J. R. & E. DAIKHIN. 1992. Uropontin is a major matrix component of urinary calcium oxalate monohydrate stones. Connect. Tissue Res. **27:** 188.
28. HOYER, J. R., W. WIKOFF & S. C. BOCK. 1992. Selective incorporation of uropontin into urinary calcium oxalate monohydrate stones. J. Am. Soc. Nephrol. **3:** 684.
29. HOCHSTRASSER, K., P. REISINGER, G. J. ALBRECHT, E. WACHTER & O. L. SCHONBERGER. 1984. Isolation of acid-resistant urinary trypsin inhibitors by high performance liquid chromatography and their characterization by N-terminal amino-acid sequence determination. Hoppe-Seyler's Z. Physiol. Chem. **365:** 1123–1130.
30. LIAN, J. B., E. L. PRIEN, M. J. GLIMCHER & P. M. GALLOP. 1977. The presence of protein-bound γ-carboxy-glutamic acid in calcium-containing renal calculi. J. Clin. Invest. **59:** 1151–1157.
31. SPECTOR, A. R., A. GRAY & E. L. PRIEN. 1976. Kidney stone matrix, differences in acid protein composition. Invest. Urol. **13:** 387–389.
32. WARPEHOSKI, M. A., P. J. BUSCEMI, D. C. OSBORN, B. FINLAYSON & E. P. GOLDBERG. 1981. Distribution of organic matrix in calcium oxalate renal calculi. Calcif. Tissue Int. **33:** 211–222.

33. KOHRI, K., Y. SUZUKI, K. YOSHIDA, K. YAMAMOTO, N. AMASAKI, T. YAMATE, T. UMEKAWA, M. IGUCHI, H. SINOHARA & T. KURITA. 1992. Molecular cloning and sequencing of cDNA encoding urinary stone protein, which is identical to osteopontin. Biochem. Biophys. Res. Commun. **184:** 859–864.
34. CHALKO, C., G. KRISHNA, J. R. HOYER & S. GOLDFARB. 1992. Characterization of urinary uropontin excretion in humans. J. Am. Soc. Nephrol. **3:** 681.
35. LOPEZ, C. A., J. R. HOYER, P. D. WILSON, P. WATERHOUSE & D. T. DENHARDT. 1993. Heterogeneity of osteopontin expression among nephrons in mouse kidneys and enhanced expression in sclerotic glomeruli. Lab. Invest. **69:** 355–363.
36. BROWN, L. F., B. BERSE, L. VAN DE WATER, A. PAPADOPOULOS-SERGIOU, C. A. PERRUZZI, E. J. MANSEAU, H. F. DVORAK & D. R. SENGER. 1992. Expression and distribution of osteopontin in human tissues; widespread association with luminal epithelial surfaces. Mol. Biol. Cell **3:** 1169–1180.
37. KLEINMAN, J. G., D. BROWN, J. V. BONVENTRE, A. M. BESHENSKY & E. M. WORCESTER. 1993. Distribution of osteopontin in rat kidney; upregulation in ischemia. J. Am. Soc. Nephrol. **4:** 611.
38. CRAIG, A. M., J. H. SMITH & D. T. DENHARDT. 1989. Osteopontin, a transformation-associated cell adhesion phosphoprotein, is induced by 12-O-tetradecanoylphorbol 13-acetate in mouse epidermis. J. Biol. Chem. **264:** 9682–9689.
39. PATARCA, R., G. J. FREEMAN, R. P. SINGH, F.-Y. WEI, T. DURFEE, F. BLATTNER, D. C. REGNIER, C. A. KOZAK, B. A. MOCK, H. C. MORSE, T. R. JERRELLS & H. CANTOR. 1989. Structural and functional studies of the early T lymphocyte activation 1 (*Eta-1*) gene; definition of a novel T cell-dependent response associated with genetic resistance to bacterial infection. J. Exp. Med. **170:** 145–161.
40. GIACHELLI, C., N. BAE, M. ALMEIDA, D. T. DENHARDT, C. E. ALPERS & S. SCHWARTZ. 1993. Osteopontin is elevated during neointima formation in rat arteries and is a novel component of human atherosclerotic plaques. J. Clin. Invest. **92:** 1686–1696.
41. KOHRI, K., S. NOMURA, Y. KITAMURA, T. NAGATA, K. YOSHIOKA, M. IGUCHI, T. YAMATE, T. UMEKAWA, Y. SUZUKI & H. SINOHARA. 1993. Structure and expression of the mRNA encoding urinary stone protein (osteopontin). J. Biol. Chem. **268:** 15180–15184.
42. KANETO, H., J. MORRISSEY, R. MCCRACKEN, A. A. REYES & S. KLAHR. 1994. Osteopontin expression in the kidney during ureteral obstruction. J. Am. Soc. Nephrol. **5:** 784.
43. DENHARDT, D. T. & X. GUO. 1993. Osteopontin: a protein with diverse functions. FASEB J. **7:** 1475–1482.
44. HESS, B., Y. NAKAGAWA, J. H. PARKS & F. L. COE. 1991. Molecular abnormality of Tamm-Horsfall glycoprotein in calcium oxalate nephrolithiasis. Am. J. Physiol. **260:** F569–F578.
45. NEMIR, M., M. W. DEVOUGE & B. B. MUKHERJEE. 1989. Normal rat kidney cells secrete both phosphorylated and non-phosphorylated forms of osteopontin showing different physiologic properties. J. Biol. Chem. **264:** 18202–18208.
46. RITTER, N. M., M. C. FARACH-CARSON & W. T. BUTLER. 1992. Evidence for the formation of a complex between osteopontin and osteocalcin. J. Bone Miner. Res. **7:** 877–885.
47. KASUGAI, S., R. TODESCAN, T. NAGATA, K. L. YAO, W. T. BUTLER & J. SODEK. 1991. Expression of bone matrix proteins associated with mineralized tissue formation by adult rat bone marrow cells *in vitro:* inductive effects of dexamethasone on the osteoblastic phenotype. J. Cell. Physiol. **147:** 111–120.
48. CAMPBELL, A. A., A. EBRAHIMPOUR, L. PEREZ, S. A. SMESKO & G. H. NANCOLLAS. 1989. The dual role of polyelectrolytes and proteins as mineralization promoters and inhibitors of calcium oxalate monohydrate. Calcif. Tissue Int. **45:** 122–128.
49. RIESE, R. J., N. S. MANDEL, J. H. WIESSNER, G. S. MANDEL, C. G. BECKER & J. G. KLEINMAN. 1992. Cell polarity and calcium oxalate crystal adherence to cultured collecting duct cells. Am. J. Physiol. **262:** F177–F184.
50. KHAN, S. R., B. FINLAYSON & R. HACKETT. 1982. Experimental calcium oxalate nephrolithiasis in the rat. Role of the renal papillae. Am. J. Pathol. **107:** 59–69.

Osteopontin Production by Cultured Kidney Cells[a]

ELAINE M. WORCESTER,[b] JACK G. KLEINMAN, AND ANN M. BESHENSKY

Nephrology Division
Department of Medicine
Department of Veterans Affairs Medical Center
and
The Medical College of Wisconsin
Milwaukee, Wisconsin 53295

INTRODUCTION

One of the roles that osteopontin is felt to play is as a regulator of crystallization. This protein has been localized to sites of both physiologic and pathologic calcifications, such as renal stones and atherosclerotic plaque,[1,2] and its structure bears similarities to proteins found in other mineralized tissue, such as mollusk shells.[3,4] Its role in modulating calcium crystal formation and growth is only partly understood. One site where this activity is of particular importance is in the kidney. Osteopontin purified both from urine[5] and from primary cultures of mouse kidney cells[6] has been found to inhibit the growth of calcium oxalate crystals in metastably supersaturated solutions. Studies of the nature of this inhibitory activity may shed light on the ability of osteopontin to modulate calcium crystallization in general.

Renal stones are a common occurrence—the lifetime prevalence of stone disease is approximately 10% in men by the seventh decade[7]—and 70–80% of renal stones are composed primarily of calcium salts.[8] Stones form because urine, and renal tubular fluid in the distal portions of the nephron,[8,9] are often supersaturated with respect to calcium oxalate (CaOx) and calcium phosphate. This results from the necessity of excreting these rather insoluble substances in small volumes of water. Supersaturation can lead to formation of crystals of these salts, which may provide the nucleus of a renal stone. However, it has long been recognized that urine also contains substances that inhibit calcification,[10,11] which may protect against the formation of crystals and stones. The tendency to form stones may be due to abnormalities that augment the level of supersaturation (such as an increased level of calcium or oxalate excretion into the urine) or ones that diminish the amount or activity of urinary inhibitors. While the fomer situation is well documented, the latter is still primarily hypothetical, largely because the urinary inhibitors of crystallization have only begun to be identified and studied in detail in the past decade.

[a] This work was supported by the Department of Veterans Affairs Medical Research Service and by Research Grants DK41725 and R55 DK48504 from the National Institutes of Health.

[b] Correspondence to: Elaine M. Worcester, M.D., Nephrology Section/111K, VA Medical Center, 5000 West National Avenue, Milwaukee, WI 53295.

Isolation of Osteopontin from Kidney Cell Cultures

We noted that supernatants of primary cultures of mouse kidney cortical cells contained an inhibitor of CaOx crystal growth, and set out to purify and study this inhibitor.[6] These cells, which are grown in serum-free medium, as a whole retain differentiated features of proximal tubule cells, such as sodium-dependent glucose and phosphate cotransport, and 25-hydroxy-vitamin D_3 1,α-hydroxylase activity, but they may contain some cells originating from other tubule segments as well.

Aliquots of media incubated with confluent cells were tested for the ability to inhibit CaOx crystal growth in a seeded crystal assay. Inhibitory activity increased rapidly over the first 12–24 hours, and more slowly thereafter (FIG. 1). When conditioned media were subjected to anion exchange chromatography, the inhibitory activity eluted from the column in two major peaks (FIG. 2). The first peak was further purified using gel filtration, and a single peak of inhibitory activity eluted at a molecular weight of approximately 78 kDa. The protein from this peak was sequenced, and the 21 N-terminal amino acids were identical to the sequence of mouse osteopontin/2ar.[12] The protein appeared as a single band at a molecular weight of 75 kDa on 10% SDS-PAGE (FIG. 3A), consistent with prior reports that osteopontin migrates at anomalously high molecular weights in such gels.[13,14] Antisera were raised to this protein in rabbits, and these antisera were compared in a Western blot with antisera raised to fusion proteins containing either the N-terminal (2arN) or C-terminal (2arC) portion of osteopontin/2ar (a kind gift of Dr. David Denhardt) (FIG. 3B). Both sets of antisera recognized the same bands, including higher molecular weight bands that may represent aggregated forms of osteopontin, which have also been noted by other investigators.[13,14]

The purified protein inhibits the growth of CaOx seed crystals in a metastably supersaturated solution of CaOx. In this assay, the calcium activity of the solution

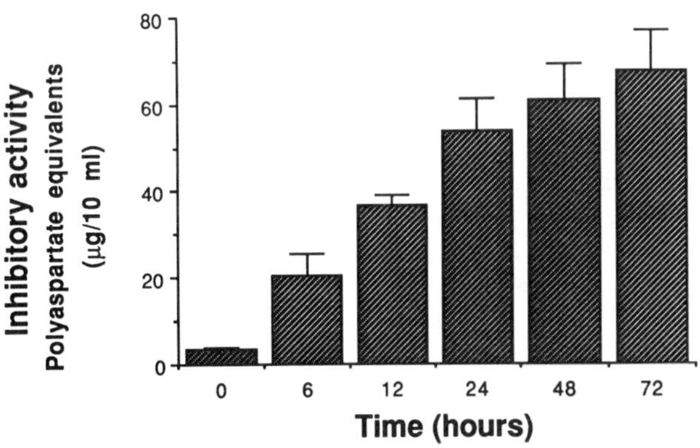

FIGURE 1. Appearance of CaOx crystal growth inhibitory activity in media incubated with cultured mouse kidney cortical cells. Fresh medium was placed on confluent cells, and aliquots removed from three to four plates at the times shown. Media were dialyzed, lyophilized, and assayed in a seeded growth system, where polyaspartic acid is used as a standard inhibitor. Values are mean ± SEM. (From Worcester *et al.*[6] Reprinted by permission from the *Journal of Bone and Mineral Research.*)

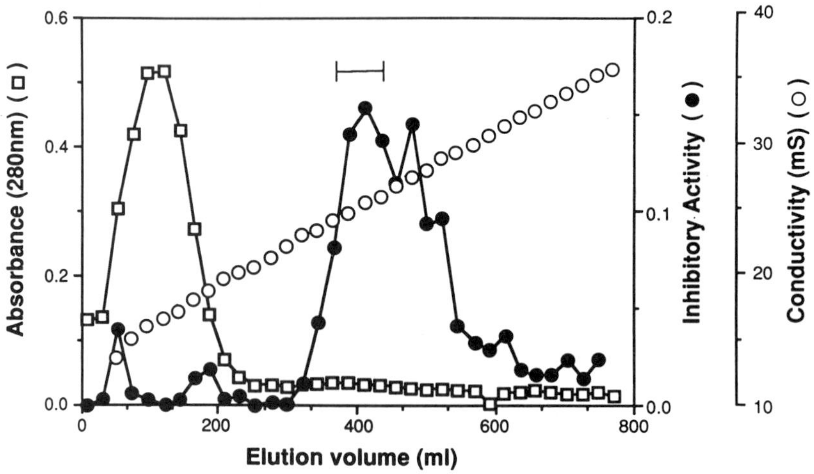

FIGURE 2. Isolation of the crystal growth inhibitor protein from conditioned media using DEAE-cellulose column chromatography (1.5 × 30 cm). Protein was eluted using a linear NaCl gradient from 0.1 to 0.4 M in 0.05 M Tris-HCl, pH 7.3. Protein elution was monitored at 280 nm, and CaOx crystal growth inhibition was assayed in a seeded crystal growth system. The main peaks of inhibitory activity eluted between 23 and 27 mS. The *bracket* indicates the fractions pooled for further purification. (From Worcester *et al.*[6] Reprinted by permission from the *Journal of Bone and Mineral Research.*)

is continuously monitored with a calcium selective electrode, and the concentrations of calcium and oxalate in the solution maintained constant by the potentiostatic addition of equimolar amounts of each ligand by automatic burettes. The rate of calcium addition is taken as the rate of crystal growth. The growth of the seed crystals is linear with time, and is slowed in the presence of osteopontin (FIG. 4). If the interaction between crystals and inhibitor protein follows a simple Langmuir adsorption isotherm, in which crystal growth rate is proportional to the number of free or unbound growth sites on the crystal surface, then the crystal growth rate at different inhibitor concentrations should reflect the affinity of the inhibitor protein for crystal surface, according to the equation

$$k_w/k_w - k_e = K/[I]$$

in which k_e is the crystal growth rate at a given osteopontin concentration, k_w represents the rate in the absence of inhibitor, K represents the dissociation constant for the crystal-inhibitor complex, and [I] is the inhibitor concentration. Under conditions in which the assumptions of the Langmuir adsorption isotherm are valid, a plot of 1/[I] versus $k_w/k_w - k_e$ yields a straight line with slope K. The data for crystal growth in the presence of osteopontin are linear when plotted in this form (FIG. 5), with a dissociation constant of 3.7×10^{-8} M (when the molecular weight of osteopontin is assumed to be 44,000 daltons, which is close to the experimentally determined values for osteopontin in the literature). This suggests that the protein binds reversibly to growth sites on the surface of the crystal, and prevents further growth at these sites. This general model seems to hold true for the known urinary crystal growth inhibitors.

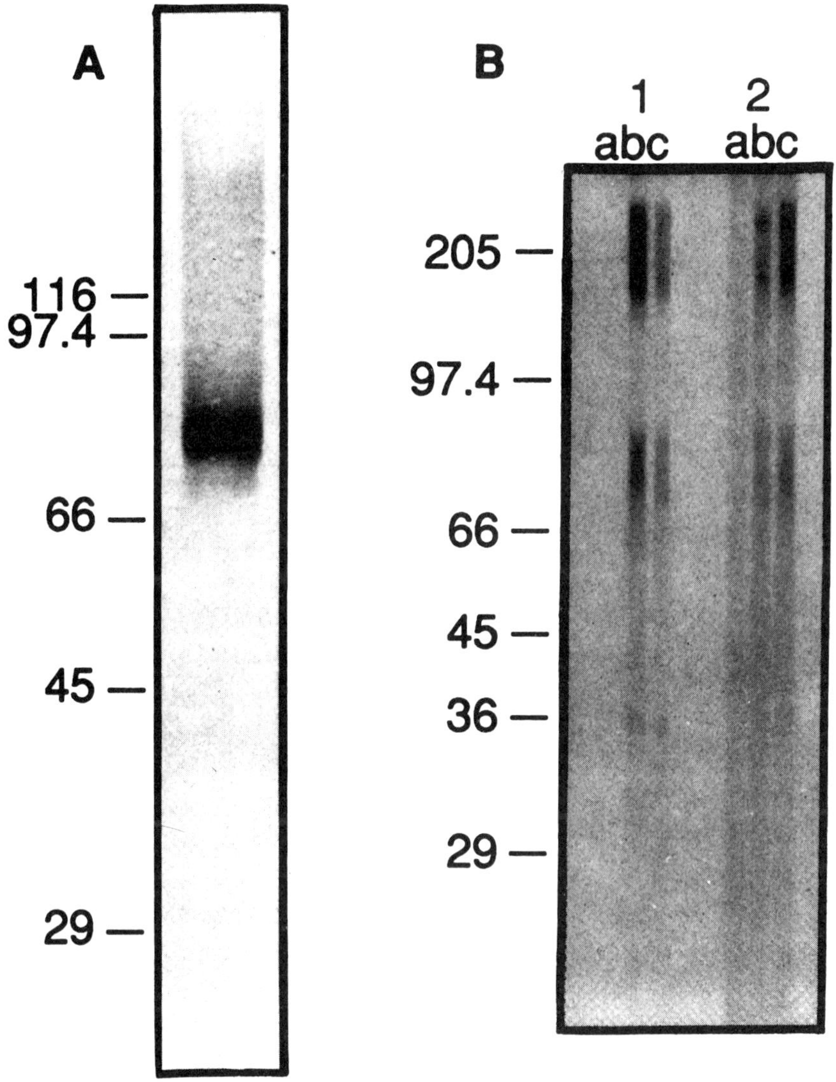

FIGURE 3. Purified osteopontin from culture media separated by 10% SDS-PAGE and blotted to nitrocellulose. **(A)** Colloidal gold stain of 1 μg protein; major band is seen at 80 kDa. **(B)** Western blot using 3 μg protein per lane, showing staining with antisera raised to the purified protein, and antisera raised to fusion proteins produced by plasmids containing the N-terminal (2arN) or C-terminal (2arC) coding sequences of 2ar cDNA. *Lane 1,* antisera to purified inhibitor: *a*, nonimmune rabbit serum 1:1000; *b*, serum H7 1:1000; *c*, serum H6 1:1000. *Lane 2,* antisera to 2ar fusion proteins: *a*, nonimmune serum 1:40; *b*, 2arN antiserum 1:40; *c*, 2arC antisera 1:40. (From Worcester *et al.*[6] Reprinted by permission from the *Journal of Bone and Mineral Research.*)

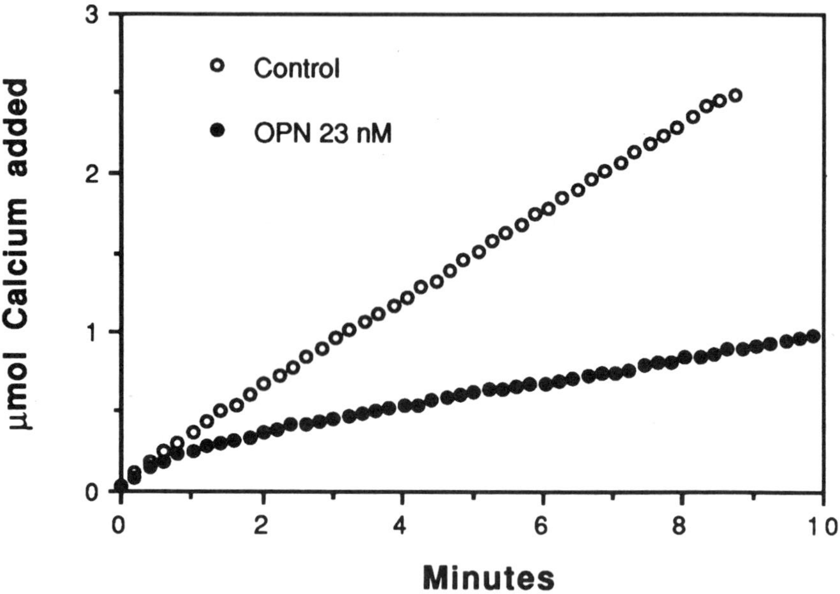

FIGURE 4. The effect of osteopontin on CaOx crystal growth in the constant composition assay. Crystal growth is decreased as shown by a slower uptake of calcium from solution in the presence of 1 μg/ml osteopontin (●) compared to control (○).

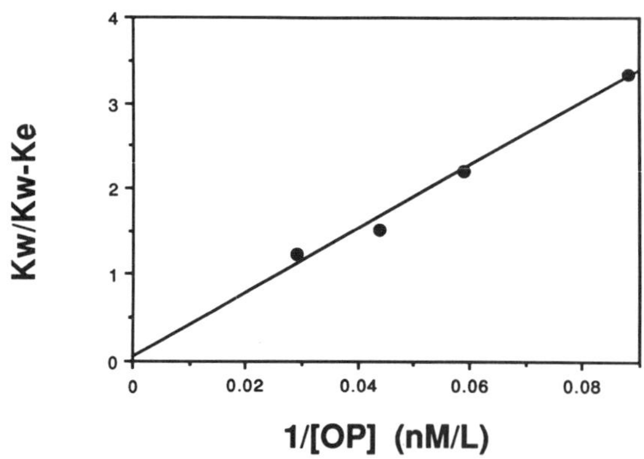

FIGURE 5. The effect of osteopontin from mouse kidney cortical cell culture media on growth rate of CaOx crystals, plotted in the form of a Langmuir isotherm. k_e and k_w refer to crystal growth in the presence and absence of purified inhibitor, respectively. The slope of the regression line, 3.7×10^{-8} M, is the dissociation constant for the protein with respect to CaOx crystal. (From Worcester *et al.*[6] Reprinted by permission from the *Journal of Bone and Mineral Research.*)

Osteopontin has been isolated from human urine by immunoaffinity chromatography, where it has been called uropontin;[5] we have identified it in the urine of mice and rats as well, at concentrations of 6–8 μg/ml.

Molecular Features Involved in Crystal Growth Inhibition

The amino acid composition of osteopontin isolated from culture media is similar to that predicted for 2ar (TABLE 1). Some similarities to the composition of nephrocalcin, the currently best studied urinary CaOx crystal growth inhibitor, are also seen. Both proteins contain a high percentage of acidic amino acids and relatively few basic amino acid residues; both nephrocalcin and osteopontin are also reported to be phosphorylated and glycosylated. It is not yet clear which features are critical for the inhibitory activity of these molecules. The presence of a large number of acidic residues is a common finding in macromolecular crystallization inhibitors.[3] In osteopontin, the highly conserved series of 9 to 10 consecutive Asp residues found in the N-terminal portion of the molecule has been thought to play an important role in mineral binding. Recent data from our lab suggest that this portion of the molecule may not be required for crystal growth inhibition, however.

TABLE 1. Amino Acid Composition of Inhibitor Isolated from Mouse Kidney Cortical Cell Tissue Culture Media (Weight %): Comparison with Composition of Previously Isolated Crystal Growth-Inhibitor Proteins

Amino Acid	Mouse Culture Medium	2ar (Predicted)	Nephrocalcin	
			Human Urine[a]	Rat Kidney[b]
Lysine	3.8	6.1	3.6	2.8
Histidine	4.3	5.4	2.0	1.8
Arginine	1.5	2.2	4.1	2.8
Asp/Asn	22.0	17.6	10.6	11.0
Threonine	5.5	4.3	8.8	8.3
Serine	12.4	17.6	9.7	11.0
Glu/Gln	14.6	14.4	12.2	14.7
Proline	5.3	5.4	5.9	6.4
Glycine	3.8	3.2	10.8	8.3
Alanine	6.5	4.3	7.6	5.5
Valine	6.2	5.4	6.4	6.4
Methionine	0.9	1.1	0.6	0.9
Isoleucine	1.3	1.8	2.4	3.7
Leucine	5.0	7.6	6.3	7.3
Tyrosine	1.0	1.4	1.0	1.8
Phenylalanine	2.4	1.8	3.1	3.7
Tryptophan	nm	0.4	0.9	nr
Cysteine	nm	0	1.7	1.8
Gla	0	0	2.0	1.8

[a] From Nakagawa *et al.*[31]
[b] From Nakagawa *et al.*[32]
nm, not measured; nr, not reported.
(From Worcester *et al.*[6] Reprinted by permission from the *Journal of Bone and Mineral Research.*)

During purification of osteopontin from culture media, a smaller protein, with an apparent molecular weight of approximately 29,000 daltons on 10% SDS-PAGE, was also seen in some preparations. This peptide was further isolated, and the major band sequenced after blotting to Immobilon. The N-terminal sequence of this band was Ser-Phe-Gln-Val-Ser-Asp-Glu-Gln-Tyr-Pro-Asp-Ala-Thr-Asp-Glu. It is derived by cleavage of intact osteopontin between Arg_{141}–Ser_{142} by a serine-type protease, and the sequence is identical to that predicted for 2ar at this site. The peptide is recognized by the antisera which was produced to the intact molecule, as well as by 2arC, the antisera made against a fusion protein containing the C-terminal portion of the molecule, but not by 2arN, the antisera directed against the N-terminal portion of osteopontin (FIG. 6). The peptide preparation does not contain any appreciable amount of intact osteopontin, though some reactivity to 2arN is seen at a molecular weight slightly above that of the peptide, which may indicate a small amount of an N-terminal peptide is present. The peptide preparation is able to inhibit CaOx crystal growth, in a seeded growth assay using [^{14}C]oxalate depletion from solution as a marker of crystal growth. The dissociation constant of the protein with respect to crystal surface was 4×10^{-8} M, if the molecular weight of the peptide is taken as 16 kDa, the approximate weight of the amino acid backbone. The dissociation constant for the intact protein, using the predicted polypeptide molecular weight of 32,350, is 5.1×10^{-8} M. Thus, the crystal growth inhibiting ability of this peptide preparation is not different from that for the intact protein. As this peptide preparation is able to inhibit crystal growth effectively, despite the fact that the major peptide present lacks the Asp repeat sequence found in the N-terminal portion of the molecule, the Asp repeat motif may not be required for crystal binding and growth inhibition. The intact molecule seems not to be required for growth inhibition, in any case. Although the Asp repeat sequence is missing from this portion of the molecule, it does possess a putative calcium-binding site, and potential phosphorylation sites.[14,15]

It is not known whether the peptide is produced by proteolytic cleavage during purification, or is a product of intracellular processing. The cultures are grown in the presence of 200 μg/ml leupeptin, a serine protease inhibitor, and additional leupeptin, as well as 200 μM PMSF, 1 μM pepstatin, and 100 μM EDTA were added during purification to inhibit proteases. Similar peptides have been noted in other cells, such as MDCK cells,[16,17] where they are thought to represent tissue-specific processing of the intact molecule.

As noted above osteopontin shares certain features commonly found in calcium crystallization inhibitors, both those in urine, and those found in other mineralized tissues. Most are anionic macromolecules, and are frequently phosphorylated and glycosylated. Phosphorylation is felt to be important to the inhibitory activity of nephrocalcin.[18] In osteopontin as well, phosphorylation has been shown to be important in mediating its effects on formation of hydroxyapatite *in vitro*. Osteopontin is able to inhibit the spontaneous formation of hydroxyapatite both in supersaturated solutions[19] and in steady-state gel systems,[19,20] and dephosphorylation by treatment of the protein with alkaline phosphatase significantly attenuates this activity.[19,20] However, phosvitin, a highly phosphorylated protein from egg yolk, was a less effective inhibitor than osteopontin,[19] suggesting that other factors such as protein sequence and conformation must be of importance as well.

Abnormalities in the structure and function of some crystallization inhibitors have been found in calcium stone-formers. Both nephrocalcin[21] and Tamm-Horsfall protein[22] have been found to have structural abnormalities, and to exhibit defects in their abilities to inhibit crystal growth, nucleation, or aggregation. For nephrocal-

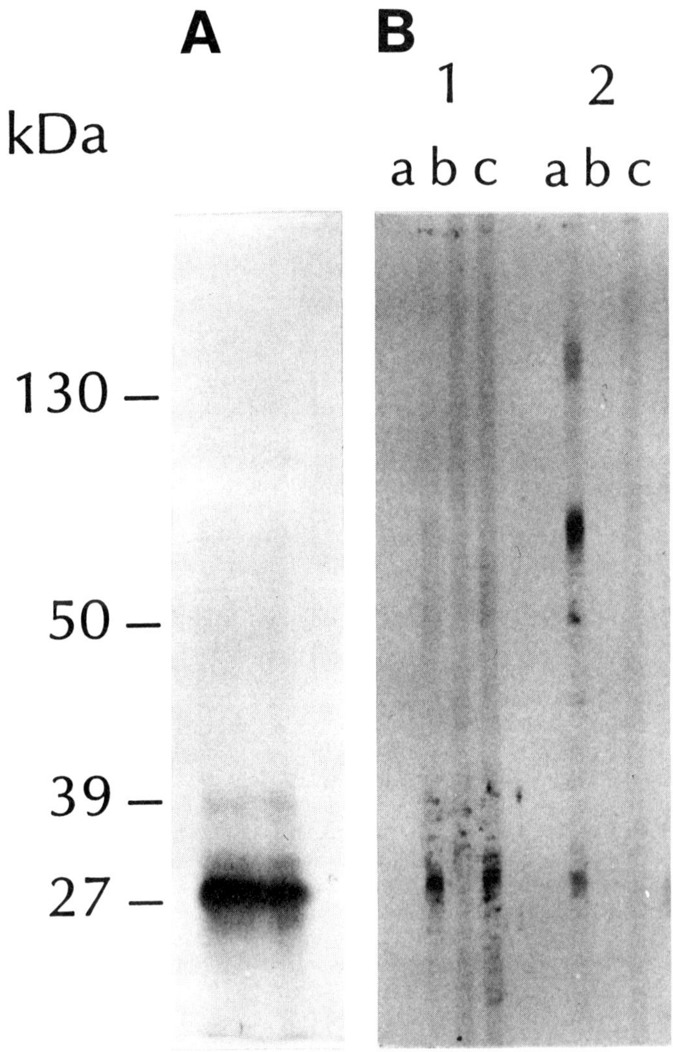

FIGURE 6. Peptide isolated from mouse kidney cortical cell culture media, electrophoresed on a 10% SDS-polyacrylamide gel and blotted to nitrocellulose. **(A)** Gold stain of 1 μg of peptide. The major band is at 29,000 daltons. **(B)** *Lane 1*: Western blot using 1 μg peptide. *a*, antiserum to intact osteopontin (H7) 1 : 1000; *b*, 2arN antiserum 1 : 50; *c*, 2arC antiserum 1 : 50. Staining of the peptide is seen with the antisera to the intact protein and to the C-terminal portion of osteopontin, but not with antiserum against the N-terminal portion of the molecule. *Lane 2:* 1 μg peptide and 0.5 μg intact osteopontin. *a*, H7 1 : 1000; *b*, nonimmune serum 1 : 1000; *c*, nonimmune serum 1 : 50.

cin, this includes absence of gamma-carboxylation in the protein isolated from stone-formers, and formation of less stable films at an air-water interface, with lower collapse pressures. Defects in phosphorylation have also been reported in some individuals.[23] Thus far, no abnormalities have been identified in osteopontin isolated from the urine of calcium stone-formers.

Effects of Osteopontin on Crystal Nucleation

It has been suggested that osteopontin may function as a nucleator of calcium salts in some situations. Anionic proteins found in mollusk shells are able to promote nucleation in a preferred orientation of crystals from solutions, and a similar role has been proposed for osteopontin in bone matrix. The ability to nucleate crystals may depend on whether the protein is free in solution or immobilized on a surface, however. Recent studies performed in steady-state gel systems suggest that osteopontin is not a nucleator of hydroxyapatite under these conditions, and in fact may inhibit nucleation of crystals.[19,20,24] At no concentration did osteopontin promote formation of hydroxyapatite, and at high concentrations inhibition of mineral deposition was seen.

Osteopontin has been identified in the matrix of calcium stones,[1] and this has led to speculation that it may be involved in stone formation by promoting nucleation of minerals from urine or renal tubular fluid. Using a metastably supersaturated solution of calcium and oxalate in which the calcium activity was continuously monitored by a calcium selective electrode, we have investigated the effect of osteopontin on spontaneous calcium oxalate nucleation. The calcium and oxalate concentrations were each 0.6 mM, and the calcium-oxalate ion activity product (calculated with the EQUIL program[25]) was 8.2 times the solubility product for the salt, a level of supersaturation that may be found in urine. Crystal nucleation was indicated by a fall in the calcium electrode reading, which occurred in control experiments after a lag phase of 18.2 ± 6.9 minutes. Addition of 0.24 μg/ml osteopontin to the solution prolonged the lag phase to 38.3 ± 11.9 minutes, and addition of osteopontin in concentrations ranging from 0.96–3.9 μg/ml prevented any fall in the calcium concentration for at least 60 minutes, at which point the experiment was stopped. These are concentrations of osteopontin that are lower than those found in rodent urine, and which might be present in the renal tubules. These experiments suggest that osteopontin in solution does not promote nucleation of calcium oxalate, but does not rule out the possibility that immobilized osteopontin might do so.

In addition to crystal nucleation and growth, the process of stone formation requires the immobilization of crystals at a site in the kidney where the crystals can grow into a stone. Aggregation of small crystals into larger masses might lead to entrapment of the crystals if the size of the aggregate exceeds that of the renal tubule.[26] Alternatively, crystals may attach to sites on renal epithelial cells.[27] Urinary macromolecules may be able to affect both these processes.[10,28,29] No data exist about the effect of osteopontin on calcium oxalate crystal aggregation, but recent studies suggest that the protein may be able to inhibit binding of preformed calcium oxalate crystals to the apical surface of renal epithelial cells *in vitro*.[28]

Localization of Osteopontin in the Kidney

The site of osteopontin production and secretion in the nephron is of interest in understanding its role as a potential crystallization inhibitor. The antibodies

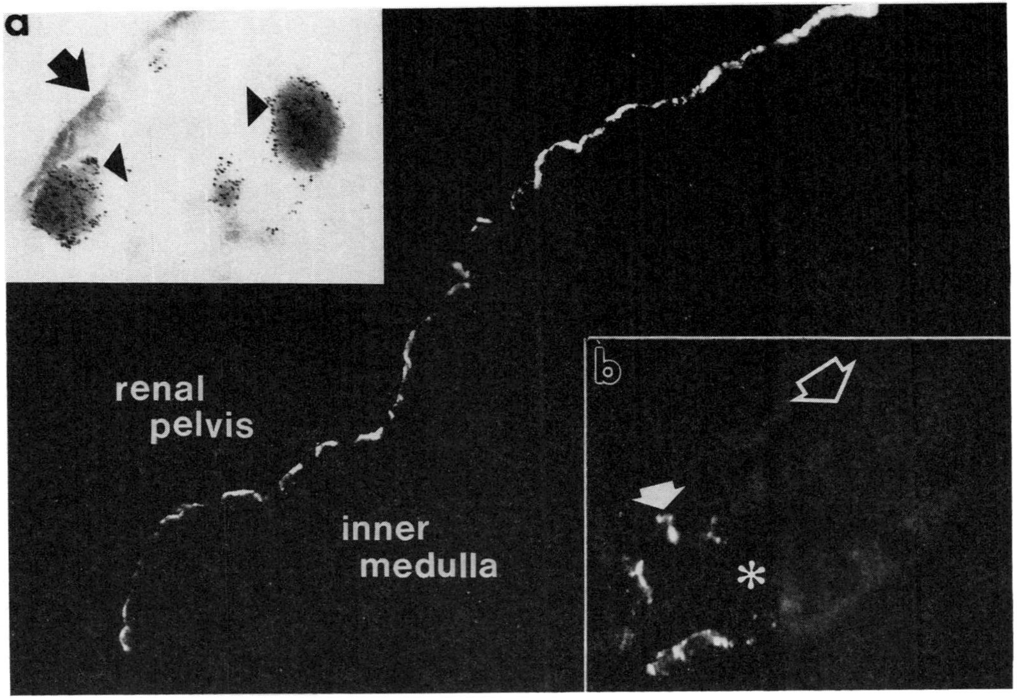

FIGURE 7. Osteopontin in the papillary surface epithelium. (Original magnification ×320.) Indirect immunofluorescence staining for osteopontin in a 5-μm cryosection showing staining in the subapical region of cells lining the renal pelvis. *Inset a,* indirect immunogold staining of an ultrathin frozen section of the papillary surface epithelium using affinity purified antiserum. Gold particles are associated with large, electron dense bodies (*arrowheads*) lying just below the plasma membrane (*arrow*). (Original magnification ×21,000.) *Inset b,* osteopontin in descending thin limbs. A confocal image of indirect immunofluorescence on a 5-μm cryosection showing the transition (*asterisk*) between the S_3 segments of the proximal tubule (*open arrow*) and the thin limbs (*filled arrow*). Staining is seen in the apical region of the thin limb epithelial cells. (Original magnification ×320.)

raised against osteopontin purified from the culture media were used to localize the protein in rat kidneys that were fixed by perfusion *in situ* with paraformaldehyde-periodate-lysine fixative.[30] Using fluorescence, confocal and electron microscopy, we localized osteopontin primarily to the descending thin limb of Henle and the papillary surface epithelium in the calyceal fornix (FIG. 7). In both sites, staining was most intense along the luminal border of the cell. Immunogold label localized primarily to a population of intracellular dense vesicles, which were seen mainly in the apical region of the cells. The nature of these vesicles was investigated to determine if the staining observed could be explained by uptake of osteopontin from tubule fluid or urine. The vesicles were shown to be distinct from endocytotic vesicles by studies using kidneys isolated from animals that were perfused with fluorescein-labeled dextran. The kidneys were then stained for osteopontin using rhodamine-labeled second antibody. There was little if any uptake of FITC-Dextran in the thin descending limbs, and thus no colocalization with osteopontin. In the papillary surface epithelium, FITC-Dextran uptake into endosomic vesicles was observed in some cells, but did not colocalize with osteopontin staining. Osteopontin-containing vesicles were distinguished from lysosomes by incubating tissue sections with an antiserum against the lysosomal protein LGP 120 as well as the osteopontin antiserum. Again, the patterns of staining with the two antibodies were sufficiently different as to make it unlikely that osteopontin was present in lysosomes. Osteopontin mRNA was localized by *in situ* hybridization using probes made from 2ar (a kind gift of Dr. David Denhardt). Osteopontin message was found in the same tubule segments at which protein was identified by immunocytochemistry, while no specific message was detected in other tubule segments.

Osteopontin turnover was examined in the kidneys of rats that were injected with the protein synthesis inhibitor cycloheximide 6 hours prior to fixation. These kidneys no longer demonstrated osteopontin in the descending thin limbs, and the immunofluorescence in the papillary surface was attenuated. Thus, the protein appears to have a relatively rapid turnover.

CONCLUSIONS

Osteopontin is produced by renal tubular epithelial cells and secreted into the tubular lumen. *In vitro* studies demonstrate that it has the ability to inhibit the formation of new calcium oxalate crystals, and prevent the growth of existing crystals. It is synthesized by nephron segments proximal to sites where supersaturation with respect to stone mineral constituents would be expected. In addition, it is also present along the renal papilla, the site of attachment of many stones. The molecular features needed for inhibition of crystal growth are not yet well understood, although phosphorylation appears to be important. What role osteopontin plays in the pathogenesis of renal stones is not clear, but its ability to inhibit crystal nucleation and growth suggest that it could protect against crystallization along the nephron, where supersaturation with respect to calcium oxalate and calcium phosphate is common.

REFERENCES

1. KOHRI, K., Y. SUZUKI, K. YOSHIDA, K. YAMAMOTO, N. AMASAKI, T. YAMATE, T. UMEKAWA, M. IGUCHI, H. SINOHARA & T. KURITA. 1992. Molecular cloning

and sequencing of cDNA encoding urinary stone protein, which is identical to osteopontin. Biochem. Biophys. Res. Commun. **184:** 859–864.

2. GIACHELLI, C. M., N. BAE, M. ALMEIDA, D. T. DENHARDT, C. E. ALPERS & S. M. SCHWARTZ. 1993. Osteopontin is elevated during neointima formation in rat arteries and is a novel component of human atherosclerotic plaques. J. Clin. Invest. **92:** 1686–1696.
3. ADDADI, L. & S. WEINER. 1985. Interactions between acidic proteins and crystals: sterochemical requirements in biomineralization. Proc. Natl. Acad. Sci. USA **82:** 4110–4114.
4. WHEELER, A. P., J. W. GEORGE & C. A. EVANS. 1981. Control of calcium carbonate nucleation and crystal growth by soluble matrix of oyster shell. Science **212:** 1397.
5. SHIRAGA, H., W. MIN, W. J. VANDUSEN, M. D. CLAYMAN, D. MINER, C. H. TERRELL, J. R. SHERBOTIE, J. W. FOREMAN, C. PRZYSIECKI, E. G. NEILSON & J. R. HOYER. 1992. Inhibition of calcium oxalate crystal growth *in vitro* by uropontin: another member of the aspartic acid-rich protein superfamily. Proc. Natl. Acad. Sci. USA **89:** 426–430.
6. WORCESTER, E. M., S. S. BLUMENTHAL, A. M. BESHENSKY & D. L. LEWAND. 1992. The calcium oxalate crystal growth inhibitor protein produced by mouse kidney cortical cells in culture is osteopontin. J. Bone Miner. Res. **7:** 1029–1036.
7. SOUCIE, J. M., M. J. THUN, R. J. COATES, W. MCCLELLAN & H. AUSTIN. 1994. Demographic and geographic variability of kidney stones in the United States. Kidney Int. **46:** 893–899.
8. COE, F. L., J. H. PARKS, & J. R. ASPLIN. 1992. The pathogenesis and treatment of kidney stones. N. Engl. J. Med. **327:** 1141–1152.
9. DEGANELLO, S., J. ASPLIN & F. L. COE. 1990. Evidence that tubule fluid in the thin segment of the loop of Henle normally is supersaturated and forms a poorly crystallized hydroxyapatite that can initiate renal stones. Kidney Int. **37:** 472 (abstract).
10. RYALL, R. L., R. M. HARNETT & V. R. MARSHALL. 1981. The effect of urine, pyrophosphate, citrate, magnesium and glycosaminoglycans on the growth and aggregation of calcium oxalate crystals *in vitro*. Clin. Chim. Acta **112:** 349–356.
11. ROBERTSON, W. G., M. PEACOCK, R. W. MARSHALL, D. H. MARSHALL & B. E. C. NORDIN. 1976. Saturation-inhibition index as a measure of the risk of calcium oxalate stone formation in the urinary tract. N. Engl. J. Med. **294:** 249–252.
12. CRAIG, A. M., J. H. SMITH & D. T. DENHARDT. 1989. Osteopontin, a transformation-associated cell adhesion phosphoprotein, is induced by 12-*O*-tetradecanoylphorbol 13-acetate in mouse epidermis. J. Biol. Chem. **264:** 9682–9689.
13. BUTLER, W. T. 1989. The nature and significance of osteopontin. Connect. Tissue Res. **23:** 123–136.
14. DENHARDT, D. T. & X. GUO. 1993. Osteopontin: a protein with diverse functions. FASEB J. **7:** 1475–1482.
15. PRINCE, C. W. 1989. Secondary structure predictions for rat osteopontin. Connect. Tissue Res. **21:** 15.
16. ULLRICH, O., K. MANN, W. MAASE & C. KOCH-BRANDT. 1991. Biosynthesis and secretion of an osteopontin-related 20-kDa polypeptide in the Madin-Darby Canine Kidney cell line. J. Biol. Chem. **266:** 3518–3525.
17. ZHANG, Q., C. DOMENICUCCI, H. A. GOLDGERG, J. L. WRANA & J. SODEK. 1990. Characterization of fetal porcine bone sialoproteins, secreted phosphoprotein I (SPPI, osteopontin), bone sialoprotein, and a 23-kDa glycoprotein: demonstration that the 23-kDa glycoprotein is derived from the carboxyl terminus of SPPI. J. Biol. Chem. **265:** 7583–7589.
18. NAKAGAWA, Y., T. OTSUKI & F. L. COE. 1989. Elucidation of multiple forms of nephrocalcin by ^{31}P-NMR spectrometer. FEBS Lett. **259**(2): 187–190.
19. HUNTER, G. K., C. L. KYLE & H. A. GOLDBERG. 1994. Modulation of crystal formation by bone phosphoproteins: structural specificity of the osteopontin-mediated inhibition of hydroxyapatite formation. Biochem. J. **300:** 723–728.
20. BOSKEY, A. L., M. MARESCA, W. ULLRICH, S. B. DOTY, W. T. BUTLER & C. W. PRINCE. 1993. Osteopontin-hydroxyapatite interactions *in vitro:* inhibition of hydroxyapatite formation and growth in a gelatin-gel. Bone Miner. **22:** 147–159.

21. NAKAGAWA, Y., V. ABRAM, J. H. PARKS, H. S. H. LAU, J. K. KAWOOYA & F. L. COE. 1985. Urine glycoprotein crystal growth inhibitors: evidence for a molecular abnormality in calcium oxalate nephrolithiasis. J. Clin. Invest. **76:** 1455–1462.
22. HESS, B., Y. NAKAGAWA, J. H. PARKS & F. L. COE. 1991. Molecular abnormality of Tamm-Horsfall glycoprotein in calcium oxalate nephrolithiasis. Am. J. Physiol. **260:** F569–F578.
23. NAKAGAWA, Y., S. C. REINHART, J. H. ASPLIN, F. L. COE & S. J. SCHEINMAN. 1993. Characterization of abnormal nephrocalcin in X-linked recessive nephrolithiasis. J. Am. Soc. Nephrol. **4:** 712.
24. HUNTER, G. K. & H. A. GOLDBERG. 1993. Nucleation of hydroxyapatite by bone sialoprotein. Proc. Natl. Acad. Sci. USA **90:** 8562–8565.
25. WERNESS, P. G., C. M. BROWN, L. H. SMITH & B. FINLAYSON. 1986. Equil2: a basic computer program for the calculation of urinary saturation. J. Urol. **134:** 1242–1244.
26. KOK, D. J. & S. R. KHAN. 1994. Calcium oxalate nephrolithiasis, a free or fixed particle disease. Kidney Int. **46:** 847–854.
27. RIESE, R. J., J. G. KLEINMAN, J. H. WIESSNER, G. S. MANDEL & N. S. MANDEL. 1988. Specificity in calcium oxalate adherence to papillary epithelial cells in culture. Am. J. Physiol. **255:** F1025–F1032.
28. LIESKE, J. C. Personal communication.
29. HESS, B., Y. NAKAGAWA & F. L. COE. 1989. Inhibition of calcium oxalate monohydrate crystal aggregation by urine proteins. Am. J. Physiol. **257:** F99–F106.
30. KLEINMAN, J. G., A. BESHENSKY, E. M. WORCESTER & D. BROWN. 1994. Expression of osteopontin, a urinary inhibitor of stone mineral crystal growth, in rat kidney. Kidney Int. In press.
31. NAKAGAWA, Y., V. ABRAM, F. J. KEZDY, E. T. KAISER & F. L. COE. 1983. Purification and characterization of the principal inhibitor of calcium oxalate monohydrate crystal growth in human urine. J. Biol. Chem. **258:** 12594–12600.
32. NAKAGAWA, Y., V. ABRAM & F. L. COE. 1984. Isolation of calcium oxalate crystal growth inhibitor from rat kidney and urine. Am. J. Physiol. **247:** F765–772.

Expression of Osteopontin in the Inner Ear[a]

ROBIN L. DAVIS,[b] CECILIA A. LOPEZ, AND KEWA MOU

Department of Biological Sciences
Rutgers University
Piscataway, NJ 08855

INTRODUCTION

The inner ear is a functionally diverse structure in which osteopontin (OPN) is localized to discrete regions. Within this single sensory endorgan, both auditory and vestibular transduction takes place, neural coding is initiated, and a complex fluid homeostasis is maintained. The contribution of OPN in these processes is as yet unknown; however, its localization within specialized inner ear cell types with known functions may give insights into its role.

Considering that OPN was first purified from mineralized bone matrix,[1] and its expression was induced in cultured JB6 cells by the tumor-promoting phorbol ester TPA,[2] *opn* expression within the membranous labyrinth of the inner ear was unexpected. Since OPN was initially detected early in otocyst morphogenesis, it is possible that the protein has a role in inner ear ontogeny.[3,4] This hypothesis is consistent with studies that have shown a relationship between OPN gene expression or protein production with exposure to growth factors[5,6] and upregulation of the *ras* protooncogene.[7,8] However, the recent localization of OPN to differentiated cells within adult inner ear tissues[9–11] indicates that OPN is more than a developmentally-regulated protein, but is also a constituent protein in the fully-functioning auditory and vestibular periphery.

OPN has not been observed in other sensory endorgans, such as the retina or olfactory mucosa,[12] so its presence is most likely associated with processes that are unique to the inner ear. Beyond the signal transduction mechanisms that differ for each sensory modality, the inner ear is further specialized in that it consists of three separate fluid compartments that contain two different fluid types. The ionic composition of one of these fluids, the endolymph, is regulated by the marginal cells of the stria vascularis and the dark cells of the vestibular labyrinth.[13,14] These specialized epithelia are structurally and functionally analogous to cells within the kidney and choroid plexus. The endolymph, however, has a distinctive ionic composition when compared to urine or cerebrospinal fluid. Like the intracellular milieu, endolymph contains high levels of potassium (150–165 mM) and low levels of sodium (0.5–2.0 mM); the concentration of calcium (25 μM), however, is between intracellular and extracellular levels.[13,14] The regulation of calcium ion concentration is particularly critical in the inner ear since it is involved in sensory transduction and is required for the process of adaptation.[15] Another

[a] This work was supported by NIH Grant DC01295 and a Charles and Joanna Busch Memorial Fund award to R. L. Davis.

[b] Please address all correspondence to: Dr. Robin L. Davis, Nelson Biological Laboratories, Rutgers University, Piscataway, N.J. 08855-1059.

specialization that is associated with the inner ear fluid compartments is the presence of the otolith, a membrane that contains calcium carbonate crystals. This structure is confined to the region above the sensory hair cells within the saccule and utricle but is not found in the ampullæ of the semicircular canals or the cochlea, despite the interconnection of these fluid spaces.

In order to understand the function of OPN and how it relates to the specific workings of the inner ear, it is critical to determine its expression patterns precisely. In this report we show the distribution of OPN in the inner ear of the adult mouse, with particular attention to the vestibular endorgans. A previous study focusing on the auditory periphery was reported elsewhere.[9] Comparisons with embryonic *opn* expression will be made and functional implications will be discussed. Furthermore, a model culture system of the inner ear tissues is presented in which manipulations may be made *in vitro* to test the role of OPN *in vivo*.

METHODS

Immunohistochemistry

Temporal bones from adult (CBA/J) mice and neonatal Mongolian gerbils and Sprague-Dawley rats were immunostained for OPN using either monoclonal or polyclonal anti-OPN (described below). The animals were deeply anesthetized with urethane (1.5 gm/kg, i.p.) and perfused intracardially with phosphate buffered saline (PBS) containing 0.1% sodium nitrite as a vasodilator. This was followed by approximately 100 ml of fixative (usually 10% formalin in PBS, but in some cases also containing zinc chloride or zinc dichromate salts). Following this, temporal bones from adult animals were removed from the skull and placed in 0.12 M EDTA or 4 N formic acid and gently stirred at room temperature until they were decalcified. Tissues were dehydrated in ethanol, cleared in xylene, embedded in paraffin, cut at 6 μm, and mounted on slides coated with poly-L-lysine. For immunostaining, selected sections were dewaxed in xylene, hydrated, and incubated for 1 hour in a 5% solution of serum of the species in which the secondary antibody was raised. The primary antibody was then applied and left overnight at room temperature. The following day a biotinylated secondary antiserum was applied (Vector) for one hour, followed by an avidin-biotin-HRP complex (ABC or VIP kit, Vector) for one hour, and then diaminobenzidine/H_2O_2 for 2–10 minutes. Between each step the preceding solution was removed by washing three times in PBS. In some cases the immunoreaction was amplified using the biotinylated tyramine (BT) reaction.[16] Control sections included tissue treated identically except that either nonimmune or preimmune sera or affinity-purified anti-β-galactosidase were substituted for anti-OPN.

Antibodies

Two OPN antibodies were used: polyclonal anti-mouse-OPN and monoclonal anti-human-uropontin. The polyclonal antiserum (2arC) was raised against a β-galactosidase fusion protein synthesized in *E. coli,* which contained the carboxyl proximal sequence composed of 80 amino acids.[7] Control antiserum was raised against β-galactosidase. Both the control and polyclonal antisera were affinity purified. The monoclonal antibody to human uropontin (ZH)[33] was provided by Dr. John R. Hoyer of Children's Hospital, Philadelphia, PA.

In situ *Hybridization*

Inner ear tissues were dehydrated in a graded series of ethanol solutions, cleared with two changes of xylene for 20 min, and embedded in paraffin. Sections were cut at 5 μm, mounted on subbed microscope slides,[17] and dried thoroughly at 42°C for two hours. The slides were deparaffinized and rehydrated in solutions of decreasing ethanol concentrations and air-dried. The tissue sections were incubated with self-digested pronase[18] (1 mg/ml) at room temperature, washed with PBS containing glycine (2 mg/ml), and fixed in 4% formaldehyde in PBS for 20 min. The slides were washed twice in glycine/PBS for 5 min, immersed in 0.1 M triethanolamine buffer (pH 8.0) for 5 min, and incubated in a freshly prepared mixture of 0.25% acetic anhydride in triethanolamine buffer for 10 min. They were washed twice in 2 × SSC (3 M sodium chloride, 0.3 M sodium citrate) for 5 min, dehydrated in increasing concentrations of ethanol, and air-dried.[19]

Prehybridization buffer containing 50% formamide, 20 mM piperazine N,N′-bis[2-ethane]-sulfonic acid (PIPES), 0.1% ficoll, 0.1% polyvinylpyrrolidone, 0.1% bovine serum albumin, 0.2% SDS, 40 mM DTT, 250 mg/ml sheared salmon sperm DNA, and 250 mg/ml yeast tRNA, pH 6.8, was applied to each slide (1 ml/slide). The hybridization buffer was identical to the prehybridization buffer except for the addition of 10% dextran sulfate. For *in situ* hybridization, a 990-bp *Hind*III-fragment[7] (nt -89 to -1079) of murine osteopontin, subcloned into pGEM3, was linearized by *Eco*RI. [^{35}S]UTP-labeled single-stranded sense and antisense RNA probes were prepared,[20] reduced to an average size of 150 bases by limited alkaline hydrolysis,[21] and used in the hybridization at a final concentration of 10^6 cpm/100 μl/slide. Prehybridization and hybridization steps were carried out at 50°C for 1 and 12 h, respectively. The sections were covered with a baked coverslip and sealed with 1 : 1 ratio of Carter's rubber cement and petroleum ether. After hybridization, the slides were immersed in 4 × SSC, 10 mM 2-mercaptoethanol, and the sections were soaked for 5 min. They were then washed twice in 4 × SSC for 10 min each and incubated in RNase A (20 mg/ml) and RNase T1 (2 mg/ml) in 0.5 M NaCl, 10 mM Tris-HCl, 5 mM EDTA (pH 8.0) for 30 min. The slides were rinsed in the same buffer without RNase A, in 2 × SSC for 30 min at 25°C, in 0.1 × SSC for 30 min at 45°C, and air-dried. They were then dipped in diluted Kodak NTB-2 emulsion (1 : 1 dilution with water) and air-dried for 2 h at room temperature in the dark. Exposure was carried out in a light-tight box with a dessicant for 4 days at 4°C. The exposed slides were developed in Kodak D-19 developer and fixed in Kodafix solution (Eastman Kodak Company, Rochester, NY). They were counterstained with hematoxylin-eosin and mounted.

Western Blot Analysis

Conditioned medium collected from tissue cultures serum-starved for 16 h was resolved by 12% SDS-PAGE and transferred to nitrocellulose membranes. Purified human recombinant OPN (produced by transfecting a human OPN cDNA into a human embryonic kidney cell line (NO1L)[22] was used as a positive control. Prior to SDS-PAGE analysis, conditioned medium was dialyzed against water and lyophilized. All of the samples were diluted with 2 × Laemmli's buffer and boiled for 5 min before electrophoresis.

The blotted membrane was blocked for 1 h in 5% nonfat milk in TTBS (0.1 M Tris-HCl, pH 7.5, 0.9% NaCl, 0.1% Tween 20) and immunoblotted with affinity-purified 2arC (1 : 200 dilution) for 1 h at 25°C. The blots were washed with three changes of TTBS for 10 min. This was followed by incubation with a secondary antibody conjugated to horseradish peroxidase (goat anti-rabbit IgG at 1 : 3000 dilution, BioRad). The blots were then washed with three changes of TTBS and developed using enhanced chemiluminescence (ECL, Amersham) according to the manufacturer's protocol.

Tissue Culture Preparation

Primary cultures of inner ear tissue were grown from explants of selected regions of the neonatal inner ear. The cochleæ were removed from the temporal bone and placed into L-15 Leibovitz medium (Sigma L-4386) for microdissection. The spiral ligament with its attached stria vascularis was removed from the outer regions of the cochlea. In some cultures the stria vascularis was peeled away from the spiral ligament and placed into a separate culture dish. Often the stria vascularis was removed as an intact thin strip of tissue with the vascularization still evident under the dissecting microscope. Explants of the stria vascularis with the attached spiral ligament were also placed in culture. The tissue was then divided into eight to twelve small pieces which were plated on polyornithine-coated culture dishes. Cultures were maintained at 37°C, 5% CO_2, humidified atmosphere, in DMEM culture medium with 10% fetal calf serum.

The procedures involving animals in this study were approved by the Rutgers University Institutional Review Board for the Use and Care of Animals (IRB-UCA), protocol #90-073 and #92-062.

RESULTS

OPN Distribution in the Inner Ear

OPN was segregated into different functional regions within the adult inner ear. FIGURE 1 shows the distribution of OPN protein revealed by immunohistochemistry. Protein epitopes were localized to each sensory endorgan: the cochlea, saccule, utricle and the ampullæ of the semicircular canals. Within the vestibular endorgans, OPN was distributed within the subepithelial stroma (FIG. 1B,C) and in patches along the epithelial lining (FIG. 1C) that correspond to the distribution patterns of vestibular dark cells.[23] The dark cells are analogous in function to the marginal cells of the stria vascularis—both are thought to produce the unique ionic composition of the endolymph. There were scattered OPN epitopes noted within the vestibular sensory epithelium, the intensity being significantly reduced from the stroma. Similar staining patterns were also detected with ZH, a monoclonal antibody made to uropontin. Control sections that substituted nonimmune or preimmune sera or affinity-purified anti-β-galactosidase for the primary antibody at similar protein levels showed no specific staining above background levels.

In situ hybridization confirmed these observations: the antisense probe hybridized to patches along the vestibular epithelial lining, the sensory epithelium, and the subepithelial stroma of adult tissues. Control sections exposed to sense probe showed no enhanced hybridization to any region within the inner ear above the background levels.

Immunostaining and *opn* expression have been reported within the auditory region of the inner ear.[9] Some of these areas can be seen in panel A of FIGURE 1.

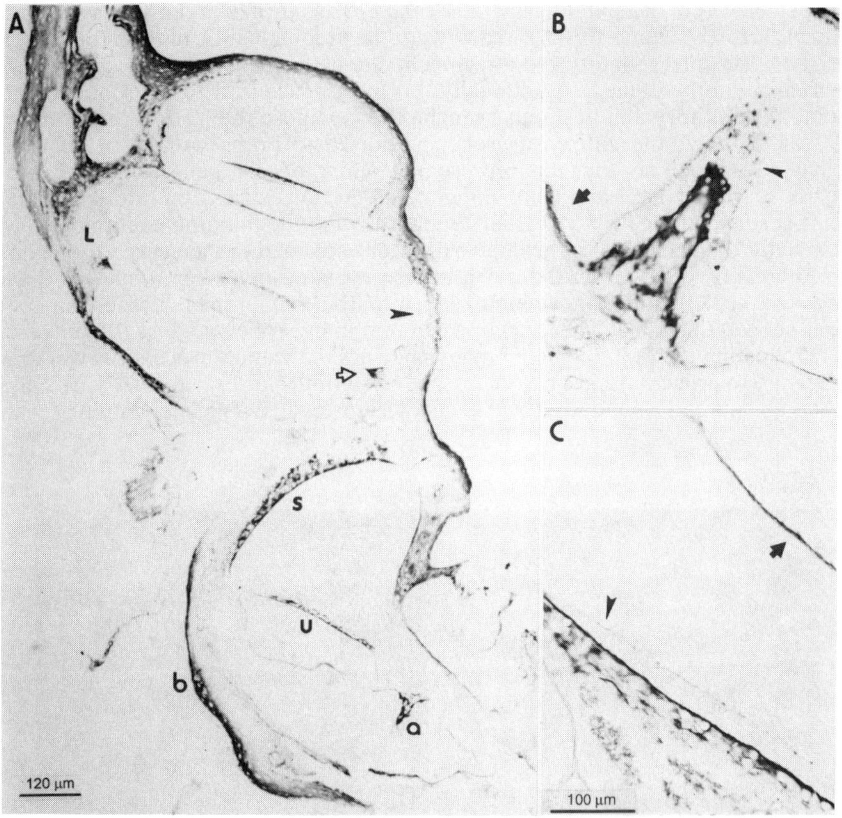

FIGURE 1. OPN was localized to a number of functionally disparate regions within the inner ear. **(A)** Low magnification of the auditory and vestibular endorgans. The *upper half* of the photograph shows OPN immunostaining in the cochlea. The outermost regions consist of the bony labyrinth (b) and as expected show a high degree of staining. Other areas in which OPN antibody staining can be detected are the limbus (L), the luminal border of the stria vascularis (*arrowhead*) and the attachment site of the basilar membrane to the spiral ligament (*open arrow*). The *lower portion* of the figure shows the saccular macula (s), utricular macula (u) and the crista ampullaris (a). The region below the sensory epithelium in each vestibular endorgan shows dark reaction product. **(B)** High magnification of the crista ampullaris. The *arrowhead* indicates the region of the sensory hair cells where light, punctate staining is observed. The subepithelial stroma below this region is darkly stained. Stained patches along the epithelial lining are also shown (*filled arrow*) which corresponds to regions that contain the vestibular dark cells. **(C)** High magnification of the utricle shows the same features demonstrated in the crista ampullaris. Patches of membrane (*arrow*), punctate staining over the vestibular hair cells (*arrowhead*), and the dark reaction product in the underlying stroma are characterizing features. The *calibration bar* in the lower left hand corner also applies to panel (B). Anti-2arC was used with the BT technique at a concentration of 1 : 10,000.

Fibrocytes within the spiral ligament (L) and epithelial cells of the stria vascularis (FIG. 1, arrowhead and FIG. 4, arrow) were the predominant cell types observed. Regions that correspond to the attachment site of the basilar membrane (arrow) were also darkly stained. Additionally, OPN was localized to neuronal somata within the auditory and vestibular ganglia that compose the VIIIth cranial nerve. FIGURE 2 shows the antisense probe hybridized to primary-auditory neuronal somata. It is unlikely that this represents staining of glial cells, since it can be clearly seen that Schwann cells along axons do not express the message. This point is substantiated by an earlier report showing the immunoreaction product from anti-OPN localized to the large ovoid cells confined to the neuronal ganglia.[9]

Although we have focused our attention on adult inner ear tissue, we have also observed OPN immunostaining of the otolithic membrane in tissues removed from neonatal animals. These sections were optimal for evaluating the presence of OPN within the otolithic membrane, since decalcification procedures were not necessary to process tissues before the bony labyrinth was fully formed. FIGURE 3 shows an anti-2arC immunostained neonatal gerbil cochlea. Accumulation of reac-

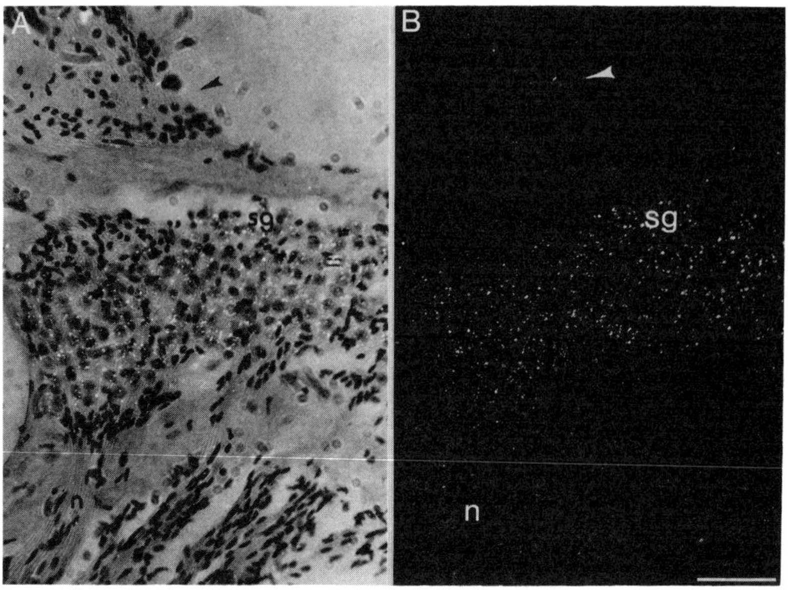

FIGURE 2. The OPN probe hybridized to neuronal somata of the auditory neurons within the spiral ganglion. **(A)** Brightfield view of hematoxylin and eosin-stained tissue showing the overlying silver grains which demarcate the 2ar probe. One can detect flattened Schwann cell nuclei along the myelinated nerve fibers which project into the CNS (n). The spiral limbus located above the spiral ganglion (*filled arrowhead*) often shows a slight amount of hybridized probe above background levels, although this is not evident in this tissue section. **(B)** Darkfield view of the same section shown in panel (A). The silver grains are confined to the regions which contained the neuronal cell bodies and do not extend to the nerve fibers (n). Sense *2ar* probe showed very low background levels that corresponded to the regions outside of the spiral ganglion. The *calibration bar,* which represents 125 μm, applies to both panels. (Adapted from Lopez *et al.*[9])

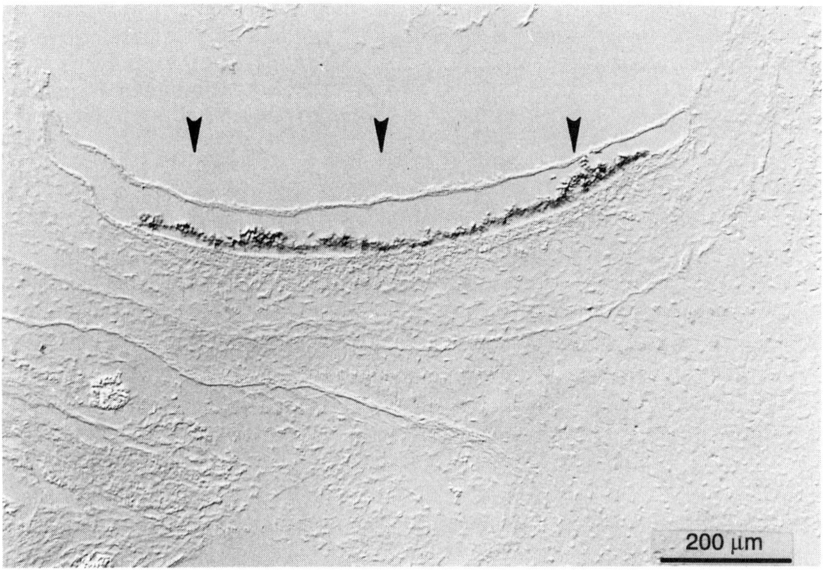

FIGURE 3. Stained otolithic membrane in tissue processed from a 7-day-old gerbil. Anti-2arC was used at a concentration of 1:200 with the ABC (Vector) technique, without applying the BT method. Since decalcification procedures were not used, the otolithic membrane was retained (*arrowheads*), whereas it was not detected in the decalcified adult tissues that were used for immunostaining (compare FIG. 1C). The lack of sensory epithelium staining was likely due to the low amplification immunostaining technique used.

tion product was observed within the otolithic membrane, yet the remaining tissues were devoid of the chromagen. We have noted, however, that *in situ* hybridization of neonatal tissue (rat, 4 days postnatal) showed *opn* expression localized to the sensory epithelium of the utricle, saccule and ampullæ of the semicircular canals. This is in agreement with the findings of Mark *et al.*[12] on rat neonatal inner ear tissues. Despite the lack of immunostaining in the gerbil tissue, which could be due to the relatively low amplification detection method that was used for this section, we conclude that *opn* was indeed present in the sensory epithelia of neonatal vestibular endorgans based upon *in situ* hybridization results from rat tissue.

Cell Culture Model of Stria Vascularis

In order to examine the expression of OPN dynamically in inner ear tissues, we developed a tissue culture system of the stria vascularis. This tissue was chosen because both immunocytochemistry and *in situ* hybridization showed homogeneous localization of OPN to the apical regions of the marginal cells. In addition, we hypothesized that the marginal cells were the source of OPN that was detected within the cochlear fluids withdrawn from the scala media.[9] When examined *in vivo,* the apical edge of the marginal cells was stained uniformly (FIG. 4), although this belies the complex shape of these luminal epithelia. The highly convoluted basolateral membrane of these cells invaginates into two other cell layers that consist of intermediate and basal cells.[24] Therefore, of the trilaminate stria vascularis, only the luminal regions of the stria vascularis contain a preponderance of

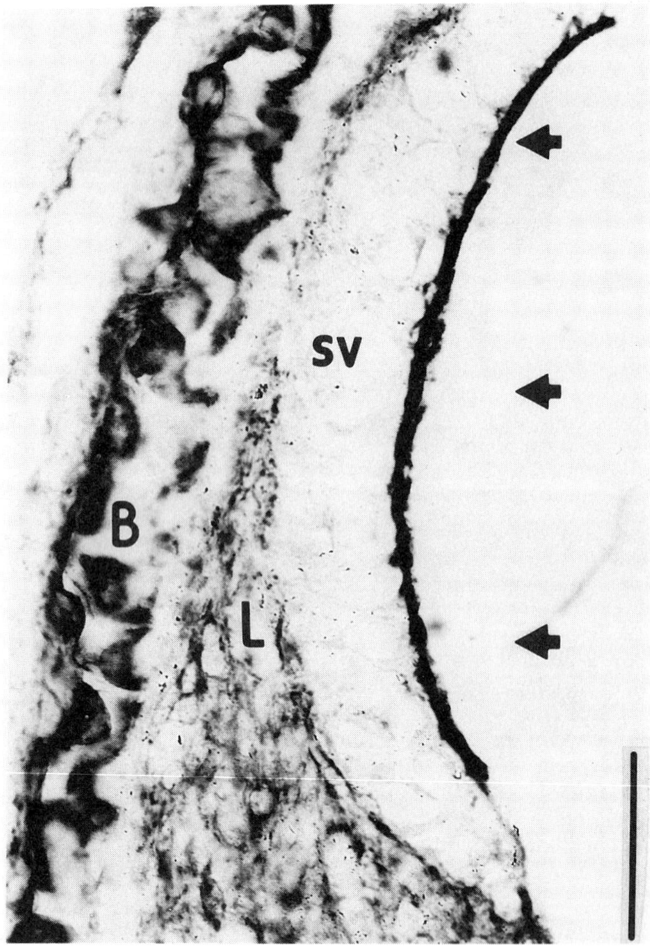

FIGURE 4. Immunoreaction product was confined to the apical region of stria vascularis marginal cells. The basolateral membrane of these same epithelia is extremely convoluted, extending well below the lumen of the stria vascularis and interdigitating with the underlying cell layers. Only the apical membrane is evident with the anti-OPN antibody staining, indicating regional localization of the protein within the marginal cells. The stria vascularis (sv) is a three-layered tissue that extends from the luminal surface (*arrows*) to the spiral ligament (L); the bony labyrinth is also stained (B). Anti-uropontin (ZH) was used with the BT technique at a concentration of 1 : 10,000. The *calibration bar* (*lower right*) represents 40 μm.

OPN epitopes. The unstained intermediate and basal cells are located between the marginal epithelia (FIG. 4, arrows) and stained spiral ligament (L).

Explants removed from mouse neonatal inner ear and grown for several days formed complex structures along the surface of the tissue culture dish. The explants contained a number of cell types which were identified by intermediate filament content, gross morphology, and ultrastructure.[25] We detected features associated with the three cell layers of the stria vascularis (marginal, intermediate and basal cells) and concluded that each of the three cell types was present *in vitro*.

As determined by *in situ* hybridization (FIG. 5) and immunohistochemistry (unpublished data), OPN was not expressed in a cell-specific manner, but rather was localized to cells along tissue borders and cells that had not yet been incorporated into the multicellular layer elaborated from the explant. We have not yet determined whether OPN is produced exclusively by epithelial cells, yet it is clear that it is not produced by all the epithelia that are present *in vitro*. For example, the epithelial cells were the top layer of the multicellular structures that were formed in culture (as determined by cytokeratin distribution, polygonal shape, and distinctive ultrastructure features, such as microvilli, membrane interdigitation, tight junctions, and electron-dense cytoplasm), yet only the epithelia along the cell borders expressed *opn*.

To determine whether the cells that express *opn* also secrete the protein, we collected conditioned media from cultures grown from stria vascularis and spiral ligament explants. In all cases examined to date (n = 4), Western blot analysis has revealed the presence of OPN in the conditioned media from both tissue culture types. Two OPN isoforms with an apparent molecular weight of 57 and 66 kDa were revealed with anti-2arC; these isoforms were identical in molecular weight to the purified human recombinant OPN derived from NO1L cells (FIG. 6). Moreover, these isoforms match those observed in cochlear fluids that were removed from the scalæ tympani and media of adult gerbil cochleæ.[9]

DISCUSSION

The Role of OPN in Inner Ear Ontogeny

OPN has been widely implicated as a participant in growth, development, and metastasis. For example, mouse *opn* was induced by the tumor-promoting phorbol ester TPA in JB6 cells,[2] was enhanced in metastatic disease[26] and can be regulated by *ras*.[7,8] Growth factors are also effective stimulators of OPN; enhanced OPN production was observed in arterial smooth muscle cells exposed to bFGF, TGF-β and angiotensin II.[5,6] Since OPN is a constitutive protein in different cell types of the fully formed inner ear, it is clear that this protein performs more than a developmental function. This does not, however, rule out an additional function during ontogeny. Swanson *et al.*[4] showed that OPN is one of thc earliest identified genes observed in the otocyst, appearing at embryonic day (E) 16.5 in the mouse inner ear. Since hearing is not fully developed until one to two weeks postnatally,[27] it is clear that OPN is present before the inner ear can transduce sound.

OPN is expressed during embryogenesis in the primary-afferent neurons, the stria vascularis, the utricular and saccular maculæ, and patches of epithelia lining the vestibular endorgans and endolymphatic sac. OPN expression was initially detected within the vestibular membranous labyrinth and the endolymphatic sac at E16.5. At this time, the utricular and saccular maculæ have differentiated into multilayered structures consisting of two to three layers of supporting cell nuclei

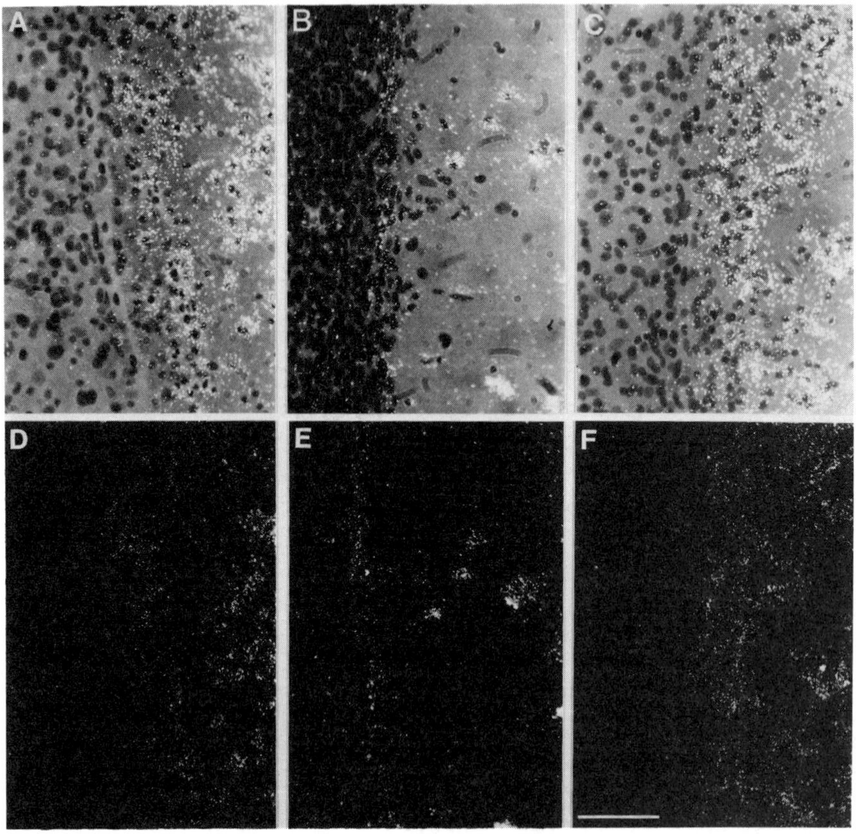

FIGURE 5. *Opn* expression was restricted to cells within and surrounding tissue borders formed *in vitro*. Three different stria vascularis/spiral ligament cultures stained with hematoxylin and eosin are shown in brightfield **(A–C)**, and in darkfield **(D–F)**. Comparison of the two views shows that OPN probe hybridized to cells that formed the edge of the proliferating tissues and to single cells that had migrated beyond the tissue border. Hybridization close to background levels was observed in regions that comprised the center of these tissue structures. The sense probe showed low background levels, without any specific localization to tissue borders or single cells. The explants were removed from neonatal mice (4 days postpartum) and maintained in culture for 16 days. The *calibration bar* in the *lower left* of panel (F), which represents 125 μm, applies to all panels.

beneath tall columnar cells.[28] *Opn* expression is not coincident with the formation of the maculæ, since differentiation of these structures begins at E13–14. However, *opn* expression does appear at a critical time for otoconia development. Sher[28] showed that calcium carbonate crystals were initially detected at E17 and their numbers increased substantially in the next two days (E18–19). Anniko[29] showed that the otolithic matrix was formed as early as E14, yet found that at times corresponding to *opn* expression (E16), high concentrations of Ca^{++} were localized to the newly-formed hexagonal matrix.

Other evidence indicating that OPN may contribute to otoconia formation is its presence within the otolithic membrane and the differential distribution of its expression in the vestibular endorgans. During embryogenesis, the appearance of OPN epitopes was confined to the utricle and saccule.[3,4] However, *opn* expression was not observed during the same period of time in the ampullæ of the semicircular canals: vestibular endorgans without an otolithic membrane. There was an altered distribution in the fully developed structure (TABLE 1). In adult tissues OPN was no longer preferentially localized to the utricle and saccule; it was also detected within the ampullæ of the semicircular canals. Furthermore, the subepithelial stroma, as well as the sensory epithelium, of each vestibular endorgan contained both the OPN immunoreaction product and hybridized probe.

At E17.5 *opn* was initially detected within the statoacoustic ganglion and the stria vascularis; interestingly, this corresponds to different stages of development in these two regions. Initial *opn* expression in the statoacoustic ganglion occurred after neuronal differentiation and outgrowth, yet prior to synaptogenesis. Neural ganglia are present as early as E12[28] and their terminal mitosis occurred at E14[30] in CBA/J mice. The central processes were present at E12, while peripheral projections reached the periphery at E18.[28] Therefore, *opn* expression, which is retained in adult tissues (TABLE 1), occurred after initial neural outgrowth, but before the final innervation patterns and synaptogenesis were complete. In con-

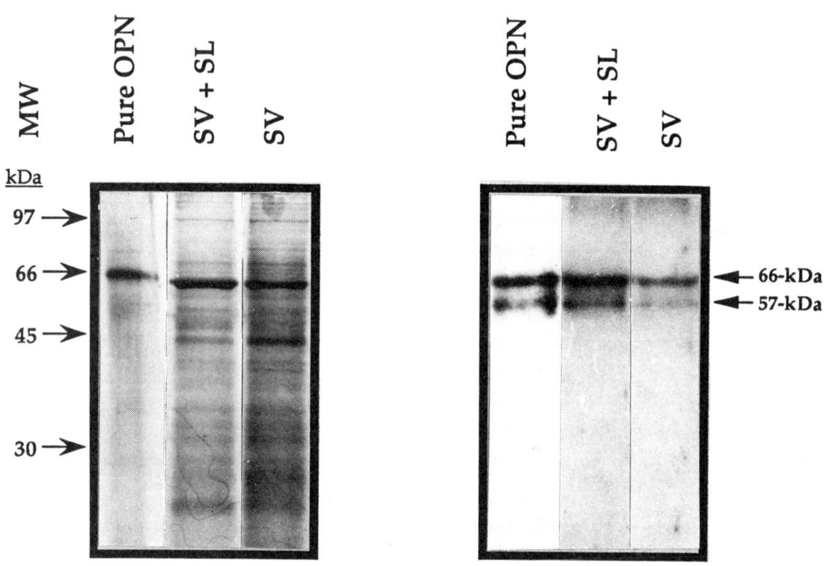

FIGURE 6. OPN species present in conditioned media of inner ear cells were identical to purified human recombinant OPN. *Left panel.* Silver-staining of total proteins in conditioned medium from stria vascularis (SV) cultures, stria vascularis and spiral ligament (SV & SL) cultures, and purified human recombinant OPN. Molecular weight markers are indicated on the *left side* of the panel. *Right panel.* Western blot of conditioned medium from SV, and SV & SL cultures, and purified recombinant human OPN. The apparent molecular weights of the OPN species migrating as a doublet at 57 and 66 kDa, are indicated with *arrows*. Samples were immunoblotted with affinity-purified 2arC antiserum and detected by enhanced chemiluminescence.

TABLE 1. Comparison of OPN Distribution in Embryonic, Neonatal, and Adult Animals[a]

		Embryonic	Neonatal	Adult
Age		E16.5, E17.5[1,2]	1–4,[3,6] 7 day old[6]	adult[4,5,6,7]
Animal		mouse[1,2]	rat,[3,6] gerbil[6]	mouse,[5,6] guinea pig,[4,5] rat,[5,7] gerbil[5,6]
Location				
Cochlea:	Organ of Corti	0[1,2]	0[3]	
	Stria vascularis	developing stria vascularis[1]	apical membrane of marginal cells[3]	apical membrane of marginal cells[4–7]
	Spiral ligament			type I fibrocytes,[4,5] suprastriatal fibrocytes,[4] area of type IV fibrocytes[5]
	Spiral limbus			connective tissue cells,[4] fibrocytes[5]
	Auditory nerve	ganglion and nerve[1,2]	neuronal somata[3]	neuronal somata,[5,7] 0[4]
Vestibular:	Vestibular nerve	ganglion and nerve[1]	neuronal somata[3]	neuronal somata,[5] 0[4]
	Utricular macula	sensory epithelium[1,2]	sensory epithelium[3,6]	subepithelial stroma,[4,6], sensory epithelium[4,6,7]
	Saccular macula	sensory epithelium[1,2]	sensory epithelium[3,6]	subepithelial stroma,[4,6], sensory epithelium[4,6,7]
	Crista ampullaris	0[1]	sensory epithelium[3,6]	subepithelial stroma,[4,6], sensory epithelium[4,6,7]
	Epithelial lining	epithelial patches (dark cells)[1]	epithelial patches (dark cells)[3,6]	epithelial patches (dark cells),[6,7] 0[4]
Related areas:	Endolymphatic sac	epithelial lining[1]		distal epithelia, subepithelial tissues[4]
	Endolymphatic duct	0[1]		equivocal[4]
	Otolithic membrane		otolithic membrane[6]	otolithic membrane[7]
	Cochlear fluids			two OPN isoforms at 57 and 66 kDa[5]
	Choroid plexus		core of villi[3]	apical region of the cuboidal epithelium[5]
	Cerebrospinal fluid			three to four OPN isoforms at 57, 66, 76 and possibly 45 kDa[5]

[a] 0 denotes that OPN was not detected in a particular area.

References:

1. Swanson *et al.*, 1989[4]
2. Nomura *et al.*, 1988[3]
3. Mark *et al.*, 1988[12]
4. Ichimiya *et al.*, 1994[10]
5. Lopez *et al.*, 1995[41]
6. Present study
7. Sakagami *et al.*, 1994[11]

trast, the initiation of *opn* expression in the stria vascularis at E17.5 was coincident with the differentiation of this tissue. During this time in otocyst ontogeny, *opn* was expressed within the stria vascularis at the early stages of morphogenesis, when only the rudimentary mesenchymal structure was present.[28]

OPN appeared to be differentially distributed during ontogeny within the choroid plexus, a tissue type which produces the cerebrospinal fluid that gives rise to the perilymph (fluid contained within the scalæ tympani and vestibuli). Expression was not detected in embryonic tissue,[4] but the cores of the villi were strongly immunostained in neonatal rat tissue.[12] In adult mouse tissue, however, the protein distribution was localized instead to the luminal epithelia of the choroid plexus.[9] Whether the difference in OPN epitope localization was due to species differences or a true developmental sequence of events remains to be determined. By comparison, OPN distribution remained constant throughout ontogeny within the endolymphatic sac and duct, areas of the inner ear that are involved in fluid regulation.

In summary, *opn* expression was detected early in embryogenesis and was coincident with different developmental processes. Initial *opn* expression occurred within the utricular and saccular maculæ during the formation of the otolithic membrane, within the stria vascularis during differentiation, and within the primary-afferent neurons preceding synaptogenesis. This pattern of expression indicates that OPN may contribute to very different developmental processes; however, since OPN's contribution to the underlying mechanisms has not yet been elucidated, it is not clear whether OPN contributes to a single function or to multiple functions.

The Role of OPN in Inner Ear Homeostatic Mechanisms

This and previous reports[9–11] have shown that OPN is present in the adult inner ear and therefore most likely participates in homeostatic auditory and vestibular functions. The protein has a widespread distribution in functionally unrelated areas, and therefore may either play a role that is common to these disparate regions, or may serve in a multifunctional capacity. For example, ion transport-mediating enzymes are frequently found in widely dissimilar regions within the cochlea (such as the spiral ligament, stria vascularis and primary-sensory neurons), yet the basic function of the enzyme remains constant. There is, however, extensive evidence that OPN may indeed subserve different functions.[31] Strong indications that support this hypothesis reside within the protein structure since the molecule contains distinctive domains that may serve different purposes. Classic examples are the presence of an aspartic acid-rich area and GRGDS motif. The polyaspartic acid stretch has been correlated with enhanced Ca^{++} binding potential[32] (~20 Ca^{++} ions with a relatively low affinity; $K_d \sim 1$ mM in 150 mM NaCl) and inhibition of calcium oxalate crystal growth,[33,34] whereas the GRGDS motif classically mediates cell attachment, through interactions with $\alpha_v\beta_3$, $\alpha_v\beta_1$ and $\alpha_v\beta_5$ integrin receptors.[35,36] Therefore, from its structure and ability to bind to multiple receptors, one might speculate that OPN has the capacity to mediate a variety of events. Furthermore, this speculation has been supported by experiments that have provided convincing evidence for OPN's role in many different functions such as ion regulation/crystal formation,[32–34] cell adhesion,[35,37–39] mediation of the defense mechanism in response to infection[40] as well as observations that OPN-receptor binding mediates multiple cell-specific signalling cascades.[41]

In our studies, *opn* expression and protein distribution were localized to three broad classes of cells: neurons, fibrocytes, and epithelia. The patchy distribution of the protein along the epithelial lining of the semicircular canals and the utricle was consistent with the distribution of dark cell epithelia that perform a function similar to the marginal cells of the stria vascularis.[23] Both the marginal and the dark cells most likely are responsible for secreting the two isoforms of OPN which have been detected in fluid withdrawn from the scala media.[9] It is likely, therefore, that OPN produced within this region of the inner ear may be critical for some aspect of ion regulation. The fibrocytes of the limbus, subepithelial stroma and spiral ligament are rich in ion-transport proteins and related enzymes. Thus, it may not be surprising that a protein with the ability to bind multiple calcium ions is also produced by these cells.

The auditory and vestibular neurons showed the highest expression of *opn* and the protein appeared to be confined to the neuronal somata. This is a more difficult finding to explain since OPN is a secreted protein, yet the vast majority of these neuronal somata are myelinated and unable to secrete proteins directly into the extracellular environment. It is possible that OPN is transported into the neurites to be released at terminals in such a way that it is essentially undetectable outside of the cell soma. Moreover, OPN is not localized to all neurons, only to a subset of neuronal ganglia,[42] suggesting a specialized function of the protein. Recent evidence has shown that OPN binding can inhibit the induction of nitric oxide synthase (NOS) mRNA by γ-interferon and lipopolysaccharide in primary cultures of proximal tubular epithelial cells.[39] Since NOS is present within the spiral ganglion neurons[43] it is possible that one function of OPN in neurons is to regulate nitric oxide production.

OPN Expression in Stria Vascularis Explants In Vitro

The potential multifunctionality of OPN, its widespread distribution within the inner ear, and its possible contribution to both ontogeny and homeostatic mechanisms render it difficult to establish a clear role for this protein in the auditory and vestibular periphery. For this reason we have developed a culture system that consists solely of cell types isolated from the stria vascularis or from the stria vascularis in combination with the spiral ligament. We focused our efforts on the stria vascularis because OPN is homogeneously expressed consistently within a specific cell type from the onset of morphogenesis to adult tissue. Furthermore, the timing of initial *opn* expression correlates with stria vascularis tissue differentiation and therefore may contribute to the morphogenesis that occurs *in vitro*.

Opn expression was retained in these tissue cultures, and surprisingly, appeared to occur exclusively within cells that are located at or close to the borders of structures that were formed. This finding was intriguing since the pattern of OPN expression *in situ* was replicated *in vitro*. Furthermore, this feature of OPN distribution along cell borders was not unique. For example, OPN was recently localized to transitional regions of growing tissues and tumors.[44,45] From these observations we have speculated that OPN may be involved in the formation or retention of cell borders that were formed *in vitro*. In this regard, secreted OPN may not only play a role within fluids regulated by the marginal cells, but may also participate in stria vascularis morphogenesis. We hope to exploit this tissue culture system further in order to manipulate levels of constitutively-produced OPN to determine the effects on growth, differentiation and adhesion.

ACKNOWLEDGMENT

We wish to extend our thanks to Dr. John R. Hoyer of Children's Hospital, Philadelphia, PA for his contribution of monoclonal anti-uropontin (ZH).

REFERENCES

1. PRINCE, C. W., T. OOSAWA, W. T. BUTLER, M. TOMANA, A. S. BHOWN, M. BHOWN & R. E. SCHROHENLOHER. 1987. Isolation, characterization and biosynthesis of a phosphorylated glycoprotein from rat bone. J. Biol Chem. **262:** 2900–2907.
2. SMITH, J. H. & D. T. DENHARDT. 1987. Molecular cloning of a tumor promoter-inducible mRNA found in JB6 mouse epidermal cells: induction is stable at high, but not at low densities. J. Cell. Biochem. **34:** 13–22.
3. NOMURA, S., A. J. WILLS, D. R. EDWARDS, J. K. HEATH & B. L. M. HOGAN. 1988. Developmental expression of 2ar (osteopontin) and SPARC (osteonectin) RNA as revealed by *in situ* hybridization. J. Cell Biol. **106:** 441–450.
4. SWANSON, G. J., S. NOMURA & B. L. M. HOGAN. 1989. Distribution of expression of *2AR* (osteopontin) in the embryonic mouse inner ear revealed by *in situ* hybridization. Hearing Res. **41:** 169–178.
5. GIACHELLI, C., N. BAE, D. LOMBARDI, M. MAJESKY & S. SCHWARTZ. 1991. Molecular cloning and characterization of 2B7, a rat mRNA which distinguishes smooth muscle cell phenotypes *in vitro* and is identical to osteopontin (secreted phosphoprotein I, 2ar). Biochem. Biophys. Res. Commun. **177:** 867–873.
6. GIACHELLI, C., R. PICHLER, D. LOMBARDI, D. T. DENHARDT, C. ALPERS, S. SCHWARTZ & R. J. JOHNSON. 1994. Osteopontin expression in angiotensin II-induced tubulointerstitial nephritis. Kidney Int. **45:** 515–524.
7. CRAIG, A. M., J. H. SMITH & D. T. DENHARDT. 1989. Osteopontin, a transformation-associated cell adhesion phosphoprotein, is induced by 12-o-tetradecanoylphorbol-13-acetate in mouse epidermis. J. Biol. Chem. **264:** 9682–9689.
8. CHAMBERS, A. F., E. I. BEHREND, S. M. WILSON & D. T. DENHARDT. 1992. Induction of expression of osteopontin (OPN; secreted phosphoprotein) in metastatic, *ras*-transformed NIH3T3 cells. Anticancer Res. **12:** 43–48.
9. LOPEZ, C. A., E. S. OLSON, J. C. ADAMS, K. MOU, D. T. DENHARDT & R. L. DAVIS. 1995. Osteopontin expression detected in adult cochleæ and inner ear fluids. Hearing Res. In press.
10. ICHIMIYA, I., J. C. ADAMS & R. S. KIMURA. 1994. Immunolocalization of Na^+, K^+-ATPase, Ca^{++}-ATPase, calcium-binding proteins, and carbonic anhydrase in the guinea pig inner ear. Acta Otolaryngol. (Suppl.) **114:** 167–176.
11. SAKAGAMI, M., T. TAKEMURA, M. UMEMOTO, K. FUKAZAWA & S. NOMURA. 1994. Distribution of osteopontin in the adult rat inner ear by non-radioisotopic *in situ* hybridization and immunohistochemistry. Assoc. Res. Otolaryngol. **17:** 557.
12. MARK, M. P., C. W. PRINCE, S. GAY, R. L. AUSTIN & W. T. BUTLER. 1988. 44-kDa bone phosphoprotein (osteopontin) antigenicity at ectopic sites in newborn rats: kidney and nervous tissues. Cell Tissue Res. **251:** 23–30.
13. MARCUS, D. C. 1986. Nonsensory electrophysiology of the cochlea: stria vascularis. *In* Neurobiology of Hearing: the Cochlea. R. A. Altschuler, D. W. Hoffman & R. P. Bobbin, Eds. 123–137. Raven Press. New York.
14. SALT, A. N. & T. KONISHI. 1986. The Cochlear Fluids: Perilymph and Endolymph. *In* Neurobiology of Hearing: the Cochlea. R. A. Altschuler, D. W. Hoffman & R. P. Bobbin, Eds. 109–122. Raven Press. New York.
15. HUDSPETH, A. J. 1989. How the ear's works work. Nature **341:** 397–404.
16. ADAMS, J. C. 1992. Biotin amplification of biotin and horseradish peroxidase signals in histochemical stains. J. Histochem. Cytochem. **40:** 1457–1463.
17. BRAHIC, M. & A. T. HAASE. 1978. Detection of viral sequences of low reiteration frequency by *in situ* hybridization. Proc. Natl. Acad. Sci. USA **75:** 6125–6129.
18. HAFEN, E., M. LEVINE, R. L. GARBER & W. J. GEHRIG. 1983. An improved *in situ*

hybridization method for the detection of cellular RNAs in drosophila tissue sections and its application for localizing transcripts of the homeotic antennapedia gene complex. EMBO J. **2:** 617–623.

19. Hayashi, S., I. C. Gillam, A. D. Delaney & G. M. Tener. 1978. Acetylation of chromosome squashes of *Drosophila melanogaster* decreases the background in autoradiographs from hybridization with [^{125}I]-labeled RNA. J. Histochem. Cytochem. **26:** 677–679.
20. Melton, D., P. Krieg, M. Rebagliati, T. Maniatis, K. Zinn & M. R. Green. 1984. Efficient *in vitro* synthesis of biologically active RNA and RNA hybridization probes from plasmids containing a bacteriophage SP6 promoter. Nucleic Acids Res. **12:** 7035–7056.
21. Cox, K. H., D. V. Deleon, L. M. Angerer & R. C. Angerer. 1984. Detection of mRNAs in sea urchin embryos by *in situ* hybridization using asymmetric RNA probes. Dev. Biol. **101:** 485–502.
22. Lopez, C. A. & D. T. Denhardt. Activation of a signal transduction pathway by osteopontin. In preparation.
23. Kimura, R. S. 1969. Distribution, structure, and function of dark cells in the vestibular labyrinth. Ann. Otol. Rhinol. Laryngol. **78**(3): 542–561.
24. Kimura, R. S. & H. F. Schuknecht. 1970. The ultrastructure of the human stria vascularis. Part I. Acta Otolaryngol. **69:** 415–427.
25. Davis, R. L. & K. Mou. Growth patterns of stria vascularis cells *in vitro*. In preparation.
26. Senger, D. R., C. A. Perruzzi & A. Papadopoulos. 1989. Elevated expression of secreted phosphoprotein (osteopontin, 2ar) as a consequence of neoplastic transformation. Anticancer Res. **9:** 1291–1300.
27. Shnerson, A. & R. Pujol. 1983. Development: Anatomy, electrophysiology and behavior. *In* The Auditory Psychobiology of the Mouse. J. F. Willot, Ed. 395–425. C.C. Thomas, Springfield, IL.
28. Sher, A. E. 1971. The embryonic and postnatal development of the inner ear of the mouse. Acta Otolaryngol. (Suppl.) **285:** 3–29.
29. Anniko, M. 1980. Development of otoconia. Am. J. Otolaryngol. **1**(5): 400–410.
30. Ruben, R. J. 1967. Development of the inner ear of the mouse: a radioautographic study of terminal mitoses. Acta Otolaryngol. (Suppl.) **220:** 4–43.
31. Denhardt, D. T. & X. Guo. 1993. Osteopontin: a protein with diverse functions. FASEB J. **7:** 1475–1482.
32. Chen, Y., B. S. Bal & J. P. Gorski. 1992. Calcium and collagen binding properties of osteopontin, bone sialoprotein, and bone acidic glycoprotein-75 from bone. J. Biol. Chem. **267:** 24871–24878.
33. Shiraga, H., W. Min, W. J. VanDusen, M. D. Clayman, D. Miner, C. H. Terrell, J. R. Sherbotie, J. W. Foreman, C. Przysiecki, E. G. Neilson & J. R. Hoyer. 1992. Inhibition of calcium oxalate crystal growth *in vitro* by uropontin, a new member of the aspartic acid-rich protein superfamily. Proc. Natl. Acad. Sci. USA **89:** 426–430.
34. Worchester, E. M., S. S. Blumenthal, A. M. Beshensky & D. L. Leward. 1992. The calcium oxalate crystal growth inhibitor protein produced by mouse kidney cortical cells in culture is osteopontin. J. Bone Miner. Res. **7:** 1029–1036.
35. Senger, D. R., C. A. Perruzzi, A. Papadopoulos-Sergiou & L. Van De Water. 1994. Adhesive properties of osteopontin: regulation by a naturally occurring thrombin-cleavage in close proximity to the GRGDS cell-binding domain. Mol. Biol. Cell **5:** 565–574.
36. Giachelli, C. M., L. Liaw, C. E. Murry, S. M. Schwartz & M. Almeida. 1995. Osteopontin expression in cardiovascular diseases. This volume.
37. Flores, M. E., M. Norgard, D. Heinegård, F. P. Reinholt & G. Andersson. 1992. RGD-directed attachment of isolated rat osteoclasts to osteopontin, bone sialoprotein, and fibronectin. Exp. Cell Res. **201:** 526–530.
38. Liaw, L., M. Almeida, C. E. Hart, S. M. Schwartz & C. M. Giachelli. 1994. Evidence that a non-RGD domain in rat osteopontin is involved in cell attachment. Circ. Res. **74**(2): 214–224.

39. HWANG, S.-M., C. A. LOPEZ, D. E. HECK, G. R. GARDNER, D. L. LASKIN, J. D. LASKIN & D. T. DENHARDT. 1994. Osteopontin inhibits nitric oxide synthase gene expression by inflammatory mediators in mouse kidney epithelial cells. J. Biol. Chem. **269:** 711–715.
40. SINGH, R. P., R. PATARCA, J. SCHWARTZ, P. SINGH & H. CANTOR. 1990. Definition of a specific interaction between the early T lymphocyte activation 1 (Eta-1) protein and murine macrophages *in vitro* and its effect upon macrophages *in vivo*. J. Exp. Med. **171:** 1931–1942.
41. LOPEZ, C. A., R. L. DAVIS, K. MOU & D. T. DENHARDT. 1995. Activation of a signal transduction pathway by osteopontin. This volume.
42. BROWN, L. F., B. BERSE, L. VAN DE WATER, A. PAPADOPOULOS-SERGIOU, C. A. PERRUZZI, E. J. MANSEAU, H. F. DVORAK & D. R. SENGER. 1992. Expression and distribution of osteopontin in human tissues, widespread association with luminal epithelial surfaces. Mol. Biol. Cell **3:** 1169–1180.
43. ZDANSKI, C. J., J. PRAZMA, P. PETRUSZ, G. GROSSMAN, E. RAYNOR & H. C. PILLSBURY. 1994. Nitric oxide synthase is an active enzyme in the spiral ganglion cells of the rat cochlea. Assoc. Res. Otolaryngol. **17:** 548.
44. MCKEE, M. D. & A. NANCI. 1995. Osteopontin and the bone remodeling sequence: colloidal-gold immunocytochemistry of an interfacial extracellular matrix protein. This volume.
45. SENGER, D. R., L. F. BROWN, C. A. PERRUZZI, A. PAPADOPOULOS-SERGIOU & L. VAN DE WATER. 1995. Osteopontin at the tumor/host interface: functional regulation of thrombin-cleavage and consequences for cell adhesion. This volume.

Phosphorylation of Osteopontin by Golgi Kinases[a]

S. ASHKAR, J. L. SCHAFFER, E. SALIH,
L. C. GERSTENFELD, AND M. J. GLIMCHER

*Laboratory for the Study of Skeletal Disorders and Rehabilitation
Department of Orthopaedic Research
Children's Hospital and Harvard Medical School
Boston, Massachusetts 02115*

INTRODUCTION

Considerable evidence has accumulated to suggest that the multiple isoforms of osteopontin, secreted by several cells types, are the product of differential posttranscriptional modifications (for review see Refs. 1, 2). These isoforms differ in their extent of phosphorylation, glycosylation, or sulfation[1,2] and appear to be specifically induced by several factors including cytokines and tumor promoters.[1,2] To determine the molecular mechanisms controlling the differential posttranslational modification of osteopontin, we localized the cellular compartment where osteopontin is phosphorylated and have begun to characterize the protein kinases in it.

MATERIALS AND METHODS

Chicken calvarial osteoblasts were isolated and cultured as previously described.[3]

RESULTS AND DISCUSSION

This study examined the effects of pharmacological agents, known to affect the posttranslational modifications of proteins, on the phosphorylation of osteopontin synthesized by mineralizing cultures of chicken osteoblasts. Colchicine, which inhibits the assembly of microtubules and causes the dissociation of the Golgi, inhibits the secretion and the intracellular phosphorylation of osteopontin, suggesting that the phosphorylation of osteopontin occurs after its exit from the endoplas-

[a] This study was supported by NIH Grant PO1 AR34078 and by a grant from the Orthopaedic Research and Education Foundation.

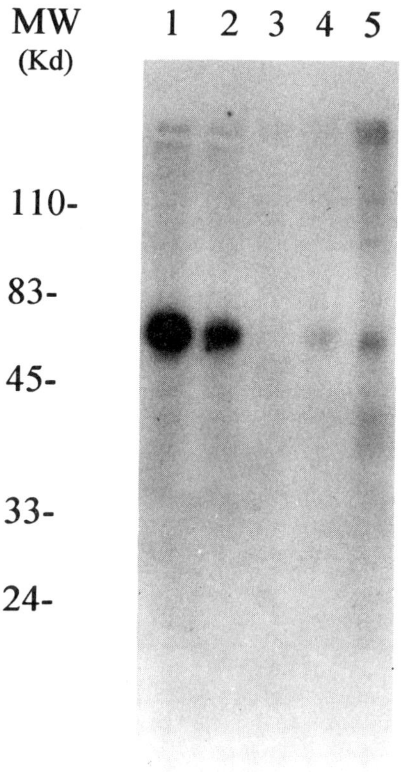

FIGURE 1. Immunoprecipitation of ^{32}P-labeled osteopontin from osteoblasts treated with Brefeldin A, colchicine, monensin and tunicamycin. Mineralized cultures were incubated in BGJ_b media containing either 1 μg/ml of Brefeldin A or 1 μM monensin. After 4 h, the media were exchanged for phosphate-free media containing either 1 μg/ml of Brefeldin A or 1 μM monensin. The cells were pulsed for 30 min with 200 μCi/ml ^{32}Pi and then chased with 20 mM phosphate. After 2 h, the osteoblasts were lysed, for 5 min, at 4°C in 1 ml of a 10 mM HEPES buffer, pH 7.2, containing 0.5% triton X-100, 300 mM sucrose, 100 mM KCl, 5 mM $MgCl_2$, 10 mM EGTA, 1 mM NaF, 0.1 mM PPi, 0.1 mM sodium orthovanadate, 1 mM levamisole, 0.1 M ε-amino-n-caproic acid, 5 mM benzamidine, 1 mM p-hydroxymercuribenzoate, 5 mg/l pepstatin, 1 mg/l leupeptin. Osteopontin was immunoprecipitated with rabbit anti-osteopontin IgG from 100 μl of clarified lysate after pretreating the lysate with protein A-sepharose. The osteopontin-antibody complex was precipitated with protein A-sepharose beads and osteopontin was eluted from the washed beads by boiling in laemmli buffer. ^{32}P-labeled osteopontin was immunoprecipitated from untreated osteoblasts (1) or osteoblasts treated for 4 h with either 1 μg/ml Brefeldin A (2), 10 μM colchicine (3), 1 μM monensin (4), or 8 μg/ml tunicamycin (5).

mic reticulum (FIG. 1). Similarly tunicamycin, which inhibits N-glycosylation, also blocks the intracellular phosphorylation of osteopontin (FIG. 1). Monensin, a sodium ionophore known to block the transport of proteins from medial to the trans cisternae of the Golgi stack, inhibited both the sulfation and phosphorylation of osteopontin, indicating that these modifications occur in the trans Golgi (FIG. 1). To differentiate between the trans Golgi cisternae and the trans Golgi network, osteoblasts were treated with Brefeldin A, an antibiotic which causes the redistribution to the endoplasmic reticulum of membrane-bound enzymes of the trans cisternae of the Golgi stack, but not of the trans Golgi network. Treatment of osteoblasts with Brefeldin A resulted in the intracellular accumulation of phosphorylated but not sulfated osteopontin, indicating that the phosphorylation of osteopontin occurs in the trans cisternae of the Golgi stack (FIG. 1).

To identify the protein kinases responsible for the phosphorylation of osteopontin, Golgi membranes were isolated from mineralizing osteoblasts by differential centrifugation. These membranes, which were devoid of any alkaline phosphatase activity, phosphorylate recombinant mouse osteopontin *in vitro* (FIG. 2). The activity of Golgi protein kinases was partially inhibited by either 50 μg/ml heparin (an inhibitor of independent kinase I and II), or by 10 μg of H7 (an inhibitor of protein kinase C and protein kinase A). Furthermore, the phosphorylation of osteopontin by Golgi membranes was completely inhibited by a combination of

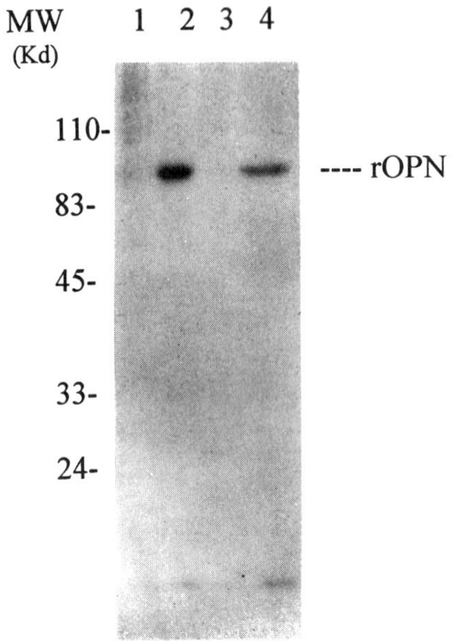

FIGURE 2. *In vitro* phosphorylation of recombinant osteopontin by Golgi kinases. 10 μg of recombinant osteopontin were incubated with 0.1 μg of isolated Golgi membranes in 50 μl of tris-HCl buffer, pH 7.4, containing 150 mM NaCl, 10 mM $MgCl_2$ and 10 μM ^{32}P-ATP (10 mCi/mmol) at 25°C for 30 min. The reactions were stopped by boiling in laemmli buffer and processed through SDS-PAGE and fluorography. (1) Control (no rOPN added), (2) phosphorylation of rOPN by Golgi kinases, (3) phosphorylation of rOPN in the presence of 50 μg/ml heparin and 10 μg H7, (4) phosphorylation of rOPN in the presence of 50 μg/ml heparin.

both compounds, indicating that there are at least two different types of protein kinase which are responsible for the phosphorylation of the recombinant osteopontin by Golgi membranes. These results are consistent with the sequence of the phosphorylated sites on osteopontin (see Salih *et al.* and Gerstenfeld *et al.* in this volume) which show that at least two different classes of protein kinases are responsible for the phosphorylation of osteopontin *in vivo*.

The results on the localization and identity of the protein kinases responsible for the phosphorylation of osteopontin suggest that the protein is probably phosphorylated in a specific intracellular compartment along its secretory path and that this phosphorylation can be modulated by either regulating the activity of the protein kinases within the Golgi or their translocation into the Golgi compartment.

REFERENCES

1. Denhardt, D. T. & X. Guo. 1993. FASEB J. **7:** 1475–1482.
2. Patarca, R., R. Saavedra & H. Cantor. 1993. Crit. Rev. Immunol. **13:** 225–246.
3. Gerstenfeld, L. C., S. D. Chipman, J. Glowacki & J. Lian. 1987. Dev. Biol. **122:** 49–60.

Expression of Antisense Osteopontin RNA in Metastatic Mouse Fibroblasts Is Associated with Reduced Malignancy

E. I. BEHREND,[a,b,e] A. M. CRAIG,[c] S. M. WILSON,[a]
D. T. DENHARDT,[d] AND A. F. CHAMBERS[a,b]

[a]*London Regional Cancer Centre*
[b]*Department of Oncology and Microbiology & Immunology*
University of Western Ontario
London, Ontario N6A 4L6, Canada

[c]*Department of Neuroscience*
University of Virginia School of Medicine
Charlottesville, Virginia, 22908

[d]*Department of Biological Sciences*
Rutgers University
Piscataway, New Jersey 08855-1059

Osteopontin (OPN) is a secreted, calcium-binding phosphoprotein that binds to cells via interaction with the $\alpha_v\beta_3$ integrin.[1] OPN has been associated with the transformed phenotype on a number of occasions.[1] To investigate whether elevated secretion of OPN by *ras*-transformed NIH 3T3 cells functionally contributes to the malignant properties of these cells, we expressed antisense *opn* RNA in metastatic, *ras*-transformed NIH 3T3 cells (PAP2) and assayed the effects of antisense *opn* RNA expression on the tumorigenic and metastatic properties of the cells.[2]

Cells were transfected, using the polycation polybrene, with the mammalian expression vector pNMH-asOPN, which contains the *opn* cDNA in an inverted orientation and under control of the heavy metal-inducible metallothionein (Mt)-I promoter. This expression vector can generate antisense *opn* RNA complementary to the *opn* mRNA. Four Geneticin-resistant transfectants were selected and screened by Southern blot, Northern blot, and sodium dodecyl sulfate polyacrylamide gel electrophoresis (SDS-PAGE) analysis. Experimental metastatic ability was assayed in chick embryos and nude mice, and tumorigenicity was assayed in nude mice.

Southern blot analysis showed that the transfectants asOPN1, asOPN2, asOPN3, and asOPN4 had incorporated 1, >10, <1, and 1 copy/cell, respectively, of the transfected antisense *opn* DNA.[2] Analysis of the transfectants by Northern blotting demonstrated that asOPN1 and asOPN2 express antisense *opn* RNA when grown *in vitro* and induced with cadmium chloride ($CdCl_2$) (FIG. 1). Clones asOPN3 and asOPN4 did not express antisense *opn* RNA (FIG. 1). The results in FIGURE 1 also show that the levels of *opn* mRNA expression are still high in all 4 clones. Expression levels of *ras* were analyzed, because loss of *ras* expression can result in loss of *opn* expression and loss of tumorigenic and metastatic ability. Analysis

[e] Address for correspondence: Ms. Elke Behrend, London Regional Cancer Centre, 790 Commissioners Road East, London, Ontario N6A 4L6, Canada.

of the levels of metabolically labeled OPN protein secreted by the transfectants grown *in vitro* and induced with $CdCl_2$ showed that OPN protein secretion was not reduced in the clones expressing antisense *opn* RNA (asOPN1 and asOPN2) relative to controls.[2]

When cells were tested for their tumorigenic and metastatic properties *in vivo,* asOPN1 and asOPN2 had reduced primary tumor growth rates in nude mice (FIG. 2) and formed fewer experimental metastases in both a chick embryo metastasis assay and in nude mice.[2] Analysis of the tumors that did eventually arise from asOPN2 by Southern and Northern blotting showed that in all 5 tumors from mice injected with asOPN2 expression of antisense *opn* RNA was lost, and that this was due in part to a loss in copy number of the transfected antisense *opn* DNA.[2]

We speculate that the tumors from asOPN2, which formed only after an extended lag period relative to controls (PAP2 and vector) and had significantly reduced growth rates, could only have grown after asOPN2 cells had eliminated expression of the antisense *opn* RNA, in part by reducing the copy number of the transfected *opn* DNA. It is possible that expression of antisense *opn* RNA by these cells in the nude mice are quite high, since factors known to drive expression of the Mt-I promoter used here are present in the mice. We have excluded the possibility that the observed reduction in malignancy of these cells is due to the action of a protein encoded in the antisense *opn* DNA. Furthermore, it is unlikely

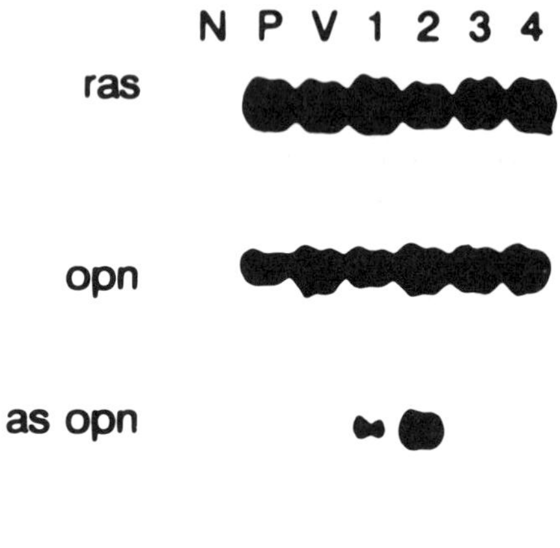

FIGURE 1. Northern blot analysis of RNA from cells induced with 20 μM cadmium chloride for 4 hours. Total cytoplasmic RNA (10 μg/lane) was separated in a 1.8% agarose gel containing 6.8% formaldehyde, blotted onto a nylon membrane, and hybridized with 5′ end-labeled oligonucleotide probes at 1 M NaCl, 50% formamide, and 42°C. *Lane N,* NIH 3T3; *lane P,* PAP2; *lane V,* Vector; *lane 1,* asOPN1; *lane 2,* asOPN2; *lane 3,* asOPN3; *lane 4,* asOPN4. The same blot was sequentially stripped and reprobed with the following oligonucleotide probes: ras, opn, as opn, and 18S. The oligonucleotide probe opn specifically recognizes and hybridizes to OPN mRNA, whereas as opn specifically recognizes and binds to OPN antisense RNA. Transcript sizes are; ras, 1.3 kb; opn, 1.6 kb; as opn, 1.6 kb; and 18 S, 2.1 kb. (From Behrend *et al.*[2] Reprinted by permission from *Cancer Research.*)

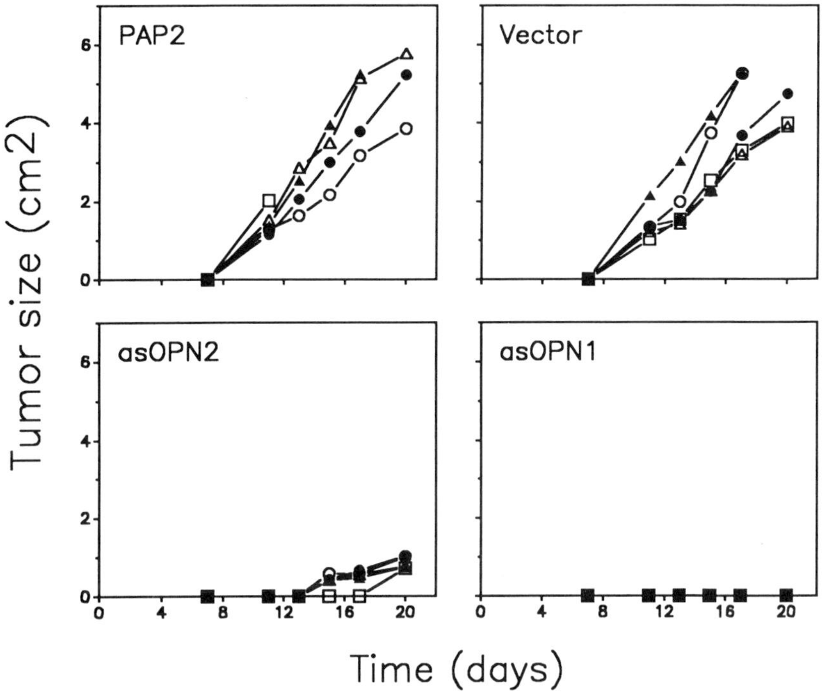

FIGURE 2. Tumorigenicity in nude mice. Tumor growth rates were assayed by subcutaneous injection of 1×10^5 cells/mouse in 0.2 ml Hanks' balanced salt solution. Tumors were measured every second day (length and width) starting on day 7 postinjection. Mice were killed after the tumor reached a cross-sectional area of ~5 cm^2. The primary tumors were dissected under sterile conditions and used for preparation of RNA and DNA. Five mice were injected per cell line. Each symbol represents the growth of a tumor in a single mouse. For asOPN1 the 5 symbols overlap since no tumor grew in any of the 5 mice. (From Behrend *et al.*[2] Reprinted by permission from *Cancer Research.*)

that the reduced malignancy is due to clonal variation based on our experience with clones of PAP2 cells.

REFERENCES

1. DENHARDT, D. T. & X. GUO. 1993. Osteopontin: a protein with diverse functions. FASEB J. **7:** 1475–1482.
2. BEHREND, E. I., A. M. CRAIG, S. M., WILSON, D. T. DENHARDT & A. F. CHAMBERS. 1994. Reduced malignancy of *ras*-transformed NIH 3T3 cells expressing antisense osteopontin RNA. Cancer Res. **54:** 832–837.

HL-60 Cell Differentiation and Osteopontin Expression

JANICE E. BERRY,[a,d] MARTHA J. SOMERMAN,[a,b]
ZHILA KHALKHALI-ELLIS,[c] PHIL OSDOBY,[c] AND
ROBERT U. SIMPSON[b]

[a]*Department of Periodontics/Prevention/Geriatrics*
[b]*Department of Pharmacology*
University of Michigan
Ann Arbor, Michigan 48109

[c]*Department of Biology*
and
Division of Bone and Mineral Disease
Washington University
St. Louis, Missouri 63130

INTRODUCTION

HL-60 cells, a human promyelocytic leukemic cell line, grow in suspension, but when induced to differentiate along the monocyte pathway they adhere to tissue culture plates and undergo morphological changes.[1,2] In contrast, when HL-60 cells are induced to differentiate along the granulocyte pathway cells remain in suspension. There is increasing evidence that adhesion molecules and their associated receptors play a key role in regulating numerous biological and pathological events, including cell differentiation.[3] Adhesion molecules either increase or decrease depending on cell type during, or as a result of, differentiation.[4–7] Also, they and their associated cell surface receptors are expressed in a unique fashion during development, both temporally and spatially, suggesting a role for such molecules in cell differentiation,[8–10] although the precise roles for adhesion molecules in these events remain to be determined. We hypothesized that differentiation of HL-60 cells along the monocyte/macrophage pathway is related to increased expression of specific adhesion molecules and integrin receptors. Furthermore, such adhesion molecules would not be regulated in the same fashion during differentiation along the granulocyte pathway.

METHODS

HL-60 cells obtained from ATCC were maintained in RPMI, and media of test cultures were treated with phorbol 12-myristate 13-acetate (PMA), a monocyte-inducing agent, at a concentration of 1.5 nM in ethanol, retinoic acid (RA), a granulocyte-inducing agent, at a concentration of 10^{-5} M in ethanol, or an equiva-

[d] Correspondence: Ms. Janice E. Berry, Department of Periodontics/Prevention/Geriatrics, University of Michigan, 1011 N. University Ave., Ann Arbor, MI 48109-1078.

lent amount of ethanol alone. Flasks were harvested at 6, 24, 48, and 72 hours after treatment, and RNA was isolated.

For Northern analysis, total RNA (30 μg/lane) was denatured and electrophoresed on a 1.2% agarose gel containing 18% of a 37% formaldehyde solution. RNA was transferred to a nylon membrane and UV cross-linked. The membrane was probed sequentially with α^{32}-P dCTP-labeled cDNAs specific for α_v integrin, osteopontin (OPN) and 18S/28S RNA.

ELISA analysis was performed using partially purified monoclonal antibodies to osteoclast-specific cell surface markers 121F and 75B. ELISA values were normalized for total sample protein.

RESULTS

Cells exposed to PMA showed a dramatic change in morphology, associated with adhesion, when compared to cells treated with RA or vehicle. As seen in TABLE 1, nearly 100% of PMA-treated cells were attached to flasks by 24 hr of treatment, while cells treated with vehicle or RA continued to grow in suspension over time.

Northern analysis showed an induction of OPN mRNA with as little as 6 hr of PMA treatment. OPN levels peaked at 24 hr of treatment, and levels remained high for further time points. Levels of α_v mRNA also increased with PMA treatment, but this increase occurred more slowly with continued increase in expression up to 72 hr. This result contrasted with Northern analysis of control and RA-treated cells (FIG. 1), where no OPN or α_v mRNA could be detected after 48 hr of treatment.

Induction of 121F and 75B, osteoclast-specific cell-surface antigens, with PMA treatment paralleled increases in mRNA levels of OPN and α_v.

CONCLUSIONS

OPN levels peak at approximately 24 hr following induction of differentiation along the monocyte/macrophage pathway. In contrast, levels of α_v integrin increase more gradually. Differentiation along the granulocyte pathway resulted in no apparent induction of OPN or α_v, and cells did not adhere to flasks. An increase in osteoclast-specific cell-surface antigens accompanied adhesion molecule induction.

TABLE 1. Effects of Agents on Cell Attachment *In Vitro*[a]

	% Adherent Cells			
Treatment	6 Hr	24 Hr	48 Hr	72 Hr
Control	0	0	0	0
RA	0	0	0	0
PMA	0	75	90	93

[a] Cells were plated at a concentration of 2×10^5 cells/ml, and treated with retinoic acid (RA), 10^{-5} M, 12-myristate 13-acetate (PMA), 1.5 nM or ethanol (control) and percent of cells attached versus suspended determined at designated times by Coulter Counter.

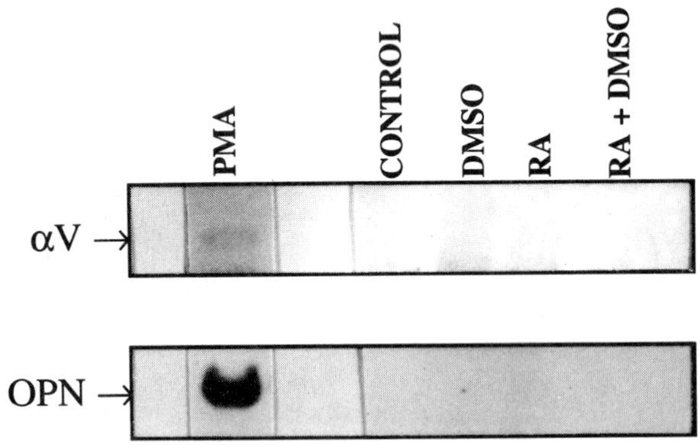

FIGURE 1. Effect of experimental treatments on mRNA levels of α_v and OPN in HL-60 cells. Northern hybridization of RNA from cells treated with retinoic acid or PMA. The *first lane* shows mRNA levels with PMA treatment for 48 hr, which is compared with control and 48-hr treatment with 1% DMSO, 10^{-5} M retinoic acid, or 10^{-5} M retinoic acid + 1% DMSO.

These findings indicate an association between the adhesion of differentiated HL-60 cells to tissue culture dishes and expression of integrins and OPN, an adhesion molecule considered to be instrumental in regulation of differentiation and function of many cell types, including osteoclasts.[11]

REFERENCES

1. TANAKA, H., E. ABE, C. MIYAURA, T. KURIBAYASHI, K. KONNO, T. NISHII & T. SUDA. 1982. Biochem. J. **204:** 713–719.
2. MARTELL, R. E., R. U. SIMPSON & J. M. TAYLOR. 1987. J. Biol. Chem. **262:** 5570–5575.
3. HYNES, R. O. 1992. Cell **69:** 11–25.
4. PATEL, V. P. & H. T. LODISH. 1984. Science **224:** 996–998.
5. PLANTIFABER, L. C. & R. U. HYNES. 1989. Cell **56:** 281–290.
6. IGNOTZ, R., J. HEINO & J. MASSAGUE. 1989. J. Biol. Chem. **264:** 389–392.
7. FERREIRA, O. C., J. E. VALINSKY, D. SHERIDAN, E. A. WAYNER, C. BIANCO & A. GARCIA-PARDO. 1991. Exp. Cell Res. **193:** 20–26.
8. VAINIO, S., M. JALKANEN & I. THESLEFF. 1989. J. Cell Biol. **108:** 1945–1954.
9. SLAVKIN, H. C. 1990. Crit. Rev. Oral Biol. Med. **1:** 1–16.
10. MACNEIL, R. L. & M. J. SOMERMAN. 1993. J. Periodont. Res. **28:** 550–559.
11. DENHARDT, D. T. & X. GUO. 1993. FASEB J. **7:** 1475–1482.

The Inhibitory Activity of Osteopontin on Hydroxyapatite Formation *In Vitro*[a]

HARVEY A. GOLDBERG[b] AND GRAEME K. HUNTER

Division of Oral Biology and Department of Biochemistry
Faculty of Dentistry
The University of Western Ontario
London, Ontario, Canada N6A 5C1

INTRODUCTION

The major sialoproteins of bone, osteopontin (OPN) and bone sialoprotein (BSP), are synthesized by osteoblasts and deposited in bone matrix.[1] While they are similar in many respects (size and chemical modifications), BSP and OPN differ in a number of important aspects. In bone the temporal expression of these proteins differs, consistent with distinct roles in the mineralization process (see Ref. 1). *In vitro* studies also suggest major differences in the functional properties of the sialoproteins. We have demonstrated that BSP is a potent nucleator of hydroxyapatite (HA) formation in steady-state agarose gels,[2] while OPN is a potent inhibitor of HA formation in both steady-state agarose gels and calcium phosphate solutions.[3] These latter data confirm the findings of others that OPN is a potent inhibitor of HA formation[4] and calcium oxalate formation.[5] To determine which features on OPN contribute to this inhibitory activity, enzymatically and chemically modified OPN and model synthetic polypeptides were studied for their effects on HA formation.

METHODS

Osteopontin Purification and Modification

OPN was extracted from adult porcine calvaria bone by sequential extraction using guanidine hydrochloride followed by demineralization and extraction of the sialoproteins in buffer containing EDTA.[6] OPN was purified on a fast protein liquid chromatography (FPLC) system using Q sepharose fast flow, hydroxyapatite, gel filtration and mono Q columns.[6] The gel filtration step was important for separating native OPN from OPN-derived peptides. Alkaline phosphatase was used to dephosphorylate OPN (85% effective). Glycine ethyl ester in the presence of carbodiimide was used to block COOH groups in OPN (86% effective).

Functional Studies

Studies on the *de novo* formation of HA were performed by autotitration,[3] using 2.5 mM $CaCl_2$, 1.5 mM sodium phosphate, 150 mM NaCl, pH 7.4. Since

[a] This study was supported by the J. P. Bickell Foundation and the Medical Research Council of Canada.

[b] To whom correspondence should be addressed.

the formation of HA results in the loss of OH ions from the solution, the amount of NaOH added to maintain the pH at 7.4 is a measure of the amount of HA formed. The effects of OPN and the synthetic polypeptides polyaspartic acid (poly-Asp) and polyglutamic acid (poly-Glu) on HA formation were also studied in steady-state agarose gels as previously described.[2] The calcium phosphate concentrations used were above the threshold for spontaneous precipitation of HA (see FIG. 1).

Conformational Studies

The secondary structures of the synthetic polypeptides were determined by circular dichroism (CD) spectroscopy using a JASCO J500C spectropolarimeter. The polypeptides were analyzed in 10 mM Tris/HCl buffer containing 150 mM NaCl, pH 7.4.

RESULTS AND DISCUSSION

The inhibitory effects of native OPN and various compounds on HA formation in metastable calcium phosphate solution are shown in TABLE 1. Native OPN is the most potent of the agents tested, causing half-maximal inhibition (IC_{50}) at a concentration of 1.9 nM. Dephosphorylation of OPN caused a greater than 40-fold decrease in potency. Similarly, a recombinant GST-OPN fusion protein, which is nonphosphorylated, exhibited reduced inhibitory potency (data not shown).

Studies on the relative importance of the carboxylate groups of OPN were performed using modification with glycine ethyl ester in the presence of carbodiimide. This resulted in a 6-fold decrease in potency. Since OPN contains a conserved region of contiguous aspartic acids, which is believed to be important for HA binding, the effects of synthetic polypeptides on mineral formation were studied. Poly-Asp (29 kDa; IC_{50} of 3.8 nM) was almost as potent an inhibitor as native OPN, while poly-Glu (34 kDa) was 1000-fold less potent than poly-Asp.

TABLE 1. Inhibition Constants for Osteopontin and Model Compounds[a]

Test Substances	IC_{50} (μg/ml)	IC_{50} (μM)
Osteopontin	0.06	0.0019
Poly(Asp) (28.8 kDa)	0.11	0.0038
Poly(Glu) (34 kDa)	155	4.56
AP-OPN	>4	>0.125
AP-OPN (control)	0.15	0.0047
GEE-OPN	1.2	0.031
GEE-OPN (control)	0.19	0.006
Phosvitin	1.9	0.054
Dextran sulphate	>250	>5.21
Bovine serum albumin	>250	>3.73

[a] Effector concentrations giving half-maximal inhibition of HA formation (IC_{50}) using the autotitrator system. The data were derived by plotting the relative inhibition of HA formation over a range of concentrations for each test substance. Each concentration used is the mean of three determinations, with a coefficient of variation of approximately 10%. AP, alkaline phosphatase; GEE, glycine ethyl ester.

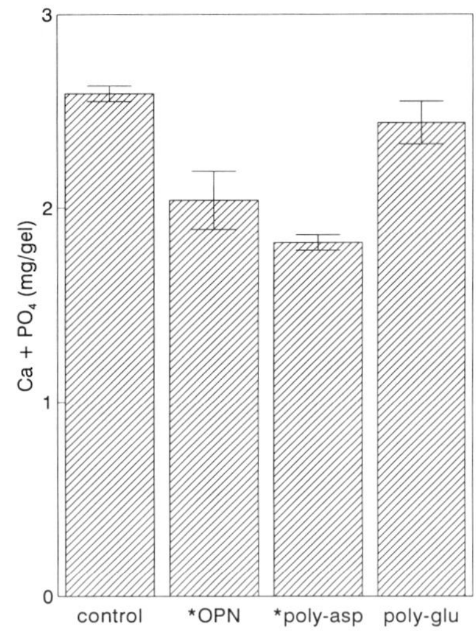

FIGURE 1. Effects of osteopontin and synthetic polypeptides on hydroxyapatite formation in steady state-agarose gels. Solutions containing 1% agarose in buffer (10 mM Tris/HCl, 150 mM NaCl, 0.01% azide, pH 7.4) and the indicated test substances at a concentration of 100 μg/ml were aliquoted (1 ml) into specially designed cells, with dialysis membrane on each side. After the agarose had set, the cells were linked in series and buffer solutions containing 7.5 mM $CaCl_2$ or 7.5 mM sodium phosphate were pumped along opposite ends of cells at a flow rate of 1 ml/hr per cell without recirculation. After five days at 37°C the amount of HA formed was measured by determination of Ca (atomic absorption spectrometry) and phosphate (colorimetric assay). Results shown are mean ± standard deviation of three determinations. Statistical significance was determined using analysis of variance and the Tukey multiple correlations test. Control, no addition to gel; OPN, osteopontin; poly-asp, the synthetic homopolymer poly-aspartic acid (28.8 kDa); poly-glu, the synthetic homopolymer poly-glutamic acid (34 kDa). * Significantly different from control ($p < 0.05$).

The effects of these model compounds in a steady-state agarose gel system also showed significant differences in their ability to inhibit HA formation (FIG. 1). OPN and poly-Asp but not poly-Glu caused significant inhibition. In fact, poly-Glu but not OPN or poly-Asp was shown to be effective as a nucleator of HA formation in this system run under conditions of subthreshold calcium and phosphate concentration.[7]

As aspartic acid and glutamic acid are chemically very similar, the functional differences between poly-Asp and poly-Glu are difficult to explain unless these polypeptides assume different conformations in solution. Using CD spectroscopy, such differences became apparent. Poly-Asp appears to lack ordered secondary structure while poly-Glu, with a positive peak maximum at 216 nm and negative peak maximum at 206 nm, appears to exist as a left handed 'extended helix.' Differences in the CD spectra in these model compounds have been extensively documented.[8] Our data are consistent with these previous observations.

OPN-derived peptides from porcine calvaria have been isolated and characterized. These peptides, which are not thrombin-derived, include a 45-kDa peptide (containing the N-terminus), a 30-kDa peptide (containing the C-terminus) and

20- and 23-kDa peptides (encompassing the contiguous poly-Asp region). In preliminary experiments, these peptides were less potent inhibitors of HA formation than intact OPN (data not shown). Further studies on these peptides and peptides generated from recombinant OPN may give insight into the structure of the region of OPN involved in mineral modulation.

The data derived from these studies indicate that maximal inhibitory activity of OPN requires an intact, phosphorylated molecule. The endogenous degradation of OPN in bone and the presence of different phosphorylated forms in various tissues suggest a complex role for this protein in physiological and pathological calcification processes.

ACKNOWLEDGMENTS

We thank Dr. Martin J. Stillman, Department of Chemistry, University of Western Ontario, for assistance in circular dichroism spectroscopy.

REFERENCES

1. Sodek, J., J. Chen, S. Kasugai, T. Nagata, Q. Zhang, H. S. Shapiro, J. L. Wrana & H. A. Goldberg. 1992. *In* Biological Mechanisms of Tooth Movement and Craniofacial Adaptation. Z. Davidovitch, Ed. 127–136. Ohio State Univ. Press. Columbus.
2. Hunter, G. K. & H. A. Goldberg. 1993. Proc. Natl. Acad. Sci. USA **90:** 8562–8565.
3. Hunter, G. K., C. L. Kyle & H. A. Goldberg. 1994. Biochem. J. **300:** 723–728.
4. Boskey, A. L., M. Maresca, W. Ullrich, S. B. Doty, W. T. Butler & C. W. Prince. 1993. Bone Miner. **22:** 147–159.
5. Shiraga, H., W. Min, W. J. Van Dusen, M. D. Clayman, D. Miner, C. H. Terrell, J. R. Sherbolie, J. W. Foreman, C. Przysiecki, E. G. Nielson & J. R. Hoyer. 1992. Proc. Natl. Acad. Sci. USA **89:** 426–430.
6. Goldberg, H. A. & J. Sodek. 1994. J. Tissue Culture Meth. In press.
7. Hunter, G. K. & H. A. Goldberg. 1994. Biochem. J. **302:** 175–179.
8. McDiarmid, R. & P. Doty. 1966. J. Phys. Chem. **70:** 2620–2627.

A Monoclonal Antibody against Osteopontin Inhibits RGD-Mediated Cell Adhesion to Osteopontin

D. S. BAUTISTA,[a] J-W. XUAN,[a] C. HOTA,[a]
A. F. CHAMBERS,[a,b] AND J. F. HARRIS[a,b,c]

[a]*London Regional Cancer Center*
and
[b]*Departments of Oncology and Microbiology and Immunology*
University of Western Ontario
London, Ontario, Canada N6A 4L6

Osteopontin (OPN) contains a conserved Gly-Arg-Gly-Asp-Ser (GRGDS) sequence that facilitates cell binding via an integrin-mediated mechanism, and a conserved thrombin-cleavage site in close proximity. In order to study the structure-function relationships of OPN, we developed five murine monoclonal antibodies (mAb) that recognize both recombinant and native human OPN. We used as immunogen recombinant human OPN-glutathione-S-transferase fusion protein[1] (GST-hOPN) and our improved hybridoma fusion partner, Sp2/mIL6,[2] to raise these mAb. TABLE 1 summarizes the properties of these mAb. Using GST-hOPN inhibition immunoassays, we found that these mAb are highly specific for human OPN since they did not cross-react with GST-mouse OPN and bovine bone OPN, and that all of the mAb have relatively high affinities. Cross-competition assays using biotinylated mAb indicated that all mAb recognize nonoverlapping epitopes except for mAb87-B and mAb87-C.

To further characterize these mAb, we used Western blot analyses.[3] We found that mAb53 and mAb87-B were efficient in detecting recombinant human OPN, native OPN secreted by human MDA-MB-435 breast cancer cells and that expressed by a transfected human OPN cDNA in kidney cells. When recombinant and native human OPN were cleaved by thrombin, the protein fragments were detected efficiently by mAb87-B but not by mAb53.[3] This suggested that the epitope of mAb53 was localized to the RGD-thrombin cleavage region of human OPN.

We previously showed that GST-hOPN can mediate cell adhesion in an *in vitro* assay by a RGD-dependent mechanism.[1] We examined the ability of mAb53 and mAb87-B to inhibit the cell adhesion function of OPN (FIG. 1). Since the thrombin site is only six residues away from the RGD site, we predicted that mAb53, but not mAb87-B, would inhibit cell adhesion. Adhesion inhibition assays (FIG. 1) indicated that mAb53 (10 μg/ml) strongly inhibited GST-hOPN-mediated cell adhesion of both PAP2 (*ras*-transformed NIH 3T3 cells, inhibition >90%) and MDA-MB-435 (>90%) cells. In addition, the inhibition by mAb53 of PAP2 cell attachment was dose dependent with 50% inhibition of approximately 0.2 μg/ml. The inhibition was specific for GST-hOPN, since cell adhesion to GST-mouse

[c] Address correspondence to: Dr. John F. Harris, London Regional Cancer Centre, 790 Commissioners Road East, London, Ontario, Canada N6A 4L6.

TABLE 1. Properties of Monoclonal Antibodies against Osteopontin

	GST-hOPN[a]	GST-hOPN[b] + Thrombin	Activity when[c] Biotinylated	Affinity (M) (50% Inhibition)[d]	Cell Adhesion Inhibition[e]
MAb53	+	−	+	3×10^{-10}	+
MAb87-A	+	+	−	10^{-8}	−
MAb87-B	+	+	+/−	2×10^{-9}	−
MAb87-C	+	+	+/−	2×10^{-9}	−
MAb87-D	+	+	−	10^{-8}	−

[a,b] Activity determined by western blotting;[3] (+) positive, (−) negative; GST-hOPN, glutathione-S-transferase-human OPN fusion protein.

[c] Binding activity of antibody against GST-hOPN in solid phase ELISA: (+) full activity, (+/−) intermediate activity, and (−) no activity.

[d] Affinities were determined by solution phase inhibition ELISA using GST-hOPN on solid phase and as competitor.[3]

[e] Inhibition of cell adhesion was determined by cell attachment to GST-hOPN using 10 μg/ml of mAb. Cell lines: PAP2, *ras*-transformed NIH 3T3 cells; MDA-MB-435, human mammary carcinoma.

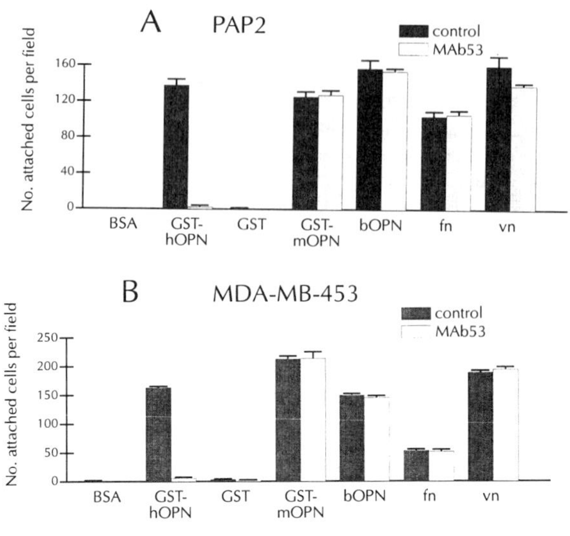

FIGURE 1. Antibody mAb53 inhibits cell adhesion to GST-hOPN. Adhesion of PAP2 **(A)**, an H-*ras*-transformed derivative of NIH 3T3, or MDA-MB-435 **(B)**, a human mammary carcinoma, in an *in vitro* cell attachment assay. Cell lines used were grown as monolayers to about 75% confluence before harvesting. Microtitre plates for the cell attachment assay were coated with one of the following: BSA, bovine serum albumin; GST-hOPN, GST-human OPN fusion protein; GST, GST control protein; GST-mOPN, GST-mouse OPN fusion protein; bOPN, bovine bone OPN; fn, fibronectin; vn, vitronectin. After coating with the substrate, mAb53 (*light hatch*) or a control antibody (*dark hatch*) of the same isotype as mAb53 was added to the wells prior to addition of cells. Unattached cells were washed off, and the remaining adherent cells were fixed and stained. The bar graphs represent the mean and the error bars represent the standard error of the mean of four or five counts of separate wells.

OPN, bovine OPN, fibronectin, and vitronectin were not significantly affected by mAb53. Also, mAb87-B and a control IgG_{2a}/κ antibody failed to inhibit cell adhesion.[3]

Primary sequence analysis of OPN from seven species reveals that the RGD cell binding sequence and the thrombin cleavage site are highly conserved structural domains, consistent with this region being functionally important.[3] The conserved GRGDS domain is identical in all species, suggesting a common receptor for OPN. Since mAb53 interacts with the conserved RGD/thrombin cleavage region, it must recognize an epitope that is unique to human OPN. Such an epitope may depend on molecular conformation but may also arise from the variable species-specific differences in the amino acid sequence surrounding this domain. Our results strengthen the idea that the conserved RGD cell binding domain and the adjacent thrombin-cleavage site of OPN are important functionally. Since the epitope of mAb53 maps in this domain, both by structural and functional criteria, it will be a useful monoclonal antibody in further clarifying the role of this domain of OPN in various biological processes.

REFERENCES

1. Xuan, J. W., C. Hota & A. F. Chambers. 1994. J. Cell. Biochem. **54:** 247–255.
2. Harris, J. F., R. G. Hawley & G. C. Crawford-Sharpe. 1992. J. Immunol. Methods **148:** 199–207.
3. Bautista, D. S., J.-W. Xuan, C. Hota, A. F. Chambers & J. F. Harris. 1994. J. Biol. Chem. **269:** 23280–23285.

Characterization of the Interaction between Integrins and Recombinant Human Osteopontin[a]

DANA D. HU,[b] JOHN R. HOYER,[c] AND JEFFREY W. SMITH[b]

[b]*Department of Vascular Biology (VB-1)*
The Scripps Research Institute
10666 North Torrey Pines Road
La Jolla, California 92037
and
[c]*Children's Hospital of Philadelphia*
34th and Civic Center Boulevard
Philadelphia, Pennsylvania 19104

Osteopontin (OPN) is a multifunctional extracellular matrix protein that supports osteoclast adhesion to the bone by binding to integrin $\alpha_v\beta_3$.[1] We characterized the interaction between OPN and integrin $\alpha_v\beta_3$ with recombinant human osteopontin (rOPN) and the urinary form of human OPN, uropontin (UP). The cDNA of human OPN was modified and engineered into an expression vector pGEX-5x-1 (Invitrogen). The recombinant osteopontin was expressed in *E. coli* (BL26, Novagen) as a fusion protein with glutathione S-transferase (GST) and cleaved from GST with factor Xa. The protein was purified by DEAE chromatography and HPLC. rOPN migrated at 60 kD on a 10% SDS-PAGE, but mass spectroscopic measurement showed its mass to be 27,046 daltons. The amino acid sequencing analysis reported this form of OPN has an intact amino terminus. Therefore rOPN is truncated, presumably by bacterial proteases, at arginine residue 228, 70 amino acids short of the native C-terminus. Thus it is termed rOP27. Human UP was purified under nondenaturing conditions by monoclonal antibody affinity chromatography.[2]

UP and rOP27 are functionally identical in their ability to support cell adhesion and to bind purified integrin $\alpha_v\beta_3$. Both events are RGD dependent. In addition, an $\alpha_v\beta_3$-specific antibody, Fab9,[3] abolished cell adhesion to rOP27, proving that the adhesion is mediated by $\alpha_v\beta_3$. Further study showed that neither rOP27 nor UP binds platelet integrin $\alpha_{IIb}\beta_3$ with detectable affinity. Thus, osteopontin is the first naturally occurring RGD-containing protein found to have a substantially greater affinity for the α_v integrin than for the α_{IIb} integrin.

The major finding from this study is that divalent cations regulate the interaction between integrins and rOP27. Cell adhesion was most efficient in Mn^{2+} and Mg^{2+}. Moreover, cell adhesion to rOPN supported by either Mn^{2+} or Mg^{2+} is blocked

[a] This study was supported by National Institutes of Health Grants CA 56483 and AR 42750 to JWS and DK33501 to JRH. D. Hu was supported initially by a postdoctoral grant from Monsanto/Searle and subsequently by a postdoctoral grant from the California Affiliate of the American Heart Association.

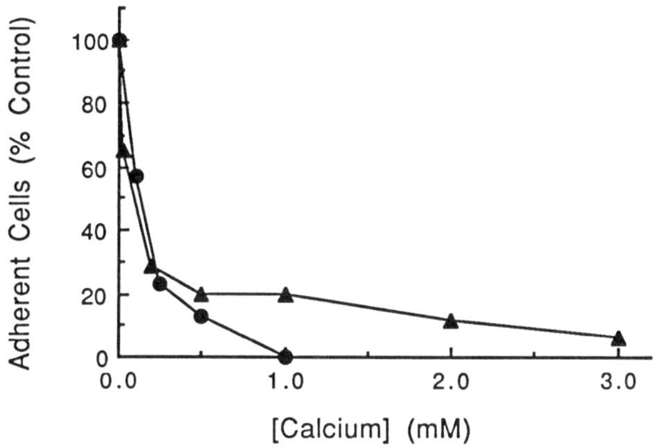

FIGURE 1. Ca^{2+} blocks cell adhesion to OPN. To ascertain whether Ca^{2+} can inhibit cell adhesion to osteopontin, an experiment was performed in which M21 cells (1.5×10^6 cells/ml) were allowed to adhere to GST-OPN in adhesion buffer containing a range of Ca^{2+} in addition to either Mg^{2+} (1.0 mM, ▲) or Mn^{2+} (0.2 mM, ●). The data are expressed as a percentage of control adhesion in the absence of Ca^{2+}. Nearly identical results were obtained with MG63 osteosarcoma cells and when UP was used as immobilized ligand. (From Hu *et al.*[5] Reprinted by permission from the *Journal of Biological Chemistry*.)

by physiologic Ca^{2+} (FIG. 1). The inhibition of adhesion to OPN by Ca^{2+} is due to the ability of this cation to lower binding affinity for integrin $\alpha_v\beta_3$. The effects of Ca^{2+} on the binding affinity were measured by surface plasmon resonance (SPR) using BIAcore (pharmacia).[4] We found that calcium downregulates the affinity of OPN for $\alpha_v\beta_3$ by reducing the association rate and increasing the dissociation rate of the interaction. The affinity between rOPN and $\alpha_v\beta_3$ is 26-fold lower in Ca^{2+} than in Mn^{2+} and 9-fold lower than in Mg^{2+} (TABLE 1). This observation could be particularly pertinent to bone where the resorbing osteoclast generates elevated levels of extracellular Ca^{2+}. Consequently, the findings presented here suggest a previously unappreciated mechanism for the modulation of bone resorption by extracellular Ca^{2+}.

TABLE 1. Affinity Constants between GST-OPN and Integrin $\alpha_v\beta_3$ in Different Divalent Cations[a]

Cation	k_1 ($M^{-1}s^{-1}$)	k_{-1} (s^{-1})	K_d (M)
Ca^{2+}	3.5×10^4	3.8×10^{-4}	1.1×10^{-8}
Mg^{2+}	1.1×10^5	1.4×10^{-4}	1.3×10^{-9}
Mn^{2+}	1.8×10^5	7.8×10^{-5}	4.3×10^{-10}

[a] The association (k_1) and dissociation (k_{-1}) rate constants between GST-OPN and $\alpha_v\beta_3$ were determined with surface plasmon resonance. All SPR measurements were obtained at 25°C. The levels of each divalent ion used for these measurements are Mn^{2+}, 0.2 mM; Mg^{2+}, 2 mM; Ca^{2+}, 2 mM. These divalent cation concentrations have been shown to be optimal for ligand binding to integrin $\alpha_v\beta_3$. (From Hu *et al.*[5] Reprinted by permission from the *Journal of Biological Chemistry*.)

We also report the identification of two new OPN receptors: $\alpha_v\beta_1$ and $\alpha_v\beta_5$. Solid phase binding assays showed that UP can bind to purified $\alpha_v\beta_5$ with nanomolar affinity. Cells lacking $\alpha_v\beta_3$ but expressing mainly either β_1 or β_5 integrins are highly adhesive to immobilized rOPN, and the adhesion is RGD and divalent cation dependent. The adhesion of β_1-expressing human kidney carcinoma 293 cells was blocked by a monoclonal antibody directed to the β_1 subunit. In addition, the adhesion was boosted tenfold by a monoclonal antibody known to "activate" β_1-integrins. These are the first data indicating that integrins β_1 and β_5 are also receptors for osteopontin and that cell adhesion to osteopontin may be activation dependent. Our results provide insight into understanding the complex mechanism of interaction between integrins and osteopontin.

REFERENCES

1. Flores, M. E., M. Norgard, D. Heinegård, F. P. Reinholt & G. Andersson. 1992. Exp. Cell Res. **201:** 526–530.
2. Shiraga, H., W. Min, W. J. VanDusen, M. D. Claymen, D. Miner, C. H. Terrell, J. R. Sherbotie, J. W. Foreman, C. Przysiecki, E. G. Neilson & J. R. Hoyer. 1992. Proc. Nat. Acad. Sci. USA **89:** 426–430.
3. Barbas, C. F. III, L. R. Languino & J. W. Smith. 1993. Proc. Natl. Acad. Sci. USA **90:** 10003–10007.
4. Altschuh, D., M-C. Dubs, E. Weiss, G. Zeder-Lutz & M. H. V. Van Regenmortel. 1992. Biochemistry **31:** 6298–6304.
5. Hu, D., J. R. Hoyer & J. W. Smith. 1995. J. Biol. Chem. In press.

Osteopontin: a Ligand for the $\alpha_v\beta_3$ Integrin of the Osteoclast Clear Zone in Osteopetrotic (ia/ia) Rats[a]

K. HULTENBY,[b,c] F. P. REINHOLT,[c,f] D. HEINEGÅRD,[d]
G. ANDERSSON,[c] AND S. C. MARKS, JR.[e]

[b]Clinical Research Center
[c]Division of Pathology
Karolinska Institutet
Huddinge Hospital
S-141 86 Huddinge, Sweden
[d]Department of Medical and Physiological Chemistry
University of Lund
S-221 00 Lund, Sweden
[e]Department of Anatomy
University of Massachusetts Medical School
Worcester, Massachusetts 01605

A key event in bone resorption is the attachment of the osteoclast to the bone surface. Upon binding, the cell develops two specialized areas facing the bone surface: a clear zone and a ruffled border. The active bone resorption is confined to the ruffled border which is sealed off by the firm interaction between cell and matrix at the clear zone. Data are accumulating supporting our notion that the attachment involves a receptor of the integrin family ($\alpha_v\beta_3$).[1,2] The integrins are transmembrane heterodimeric glycoproteins with noncovalently linked α- and β-subunits. The family consists of at least 15 α- and 8 β-subunits.[3] We previously demonstrated that osteopontin (OPN) is accumulated at the clear zone of osteoclasts, colocalizing with its receptor.[1,4] This suggests a role as a ligand in anchoring osteoclasts to the bone matrix. Osteopetrosis is a group of skeletal disorders characterized by excessive accumulation of bone due to impaired bone resorption. The inherited disease occurs in man, rat and mouse. Among rats the ia/ia is a well characterized mutant developing osteopetrosis. The impaired osteoclast function in these rats is due to the inability of the cells to develop ruffled borders and release lysosomal enzymes.[5]

In the present study ultrathin cryosections from metaphyseal bone of ia/ia rats and their normal littermates were incubated with polyclonal antibodies against integrin subunits α_v or β_3.[4] Sections from Lowicryl-embedded metaphyseal bone of the same groups of animals were incubated with polyclonal antibodies against OPN.[1] In both sets of experiments the bound antibodies were detected with 10-nm colloidal gold coated with protein A and quantitated. Tissue OPN mRNA levels

[a] This study was supported by the Swedish Medical Research Council, the Bank of Sweden Tercentenary Foundation, Karolinska Institutet, and the Axel and Margareta Ax:son Johnsons Foundation.

[f] Correspondence to: Dr. Finn P. Reinholt, Karolinska Institutet, Division of Pathology, Huddinge Hospital F42, S-141 86 Huddinge, Sweden.

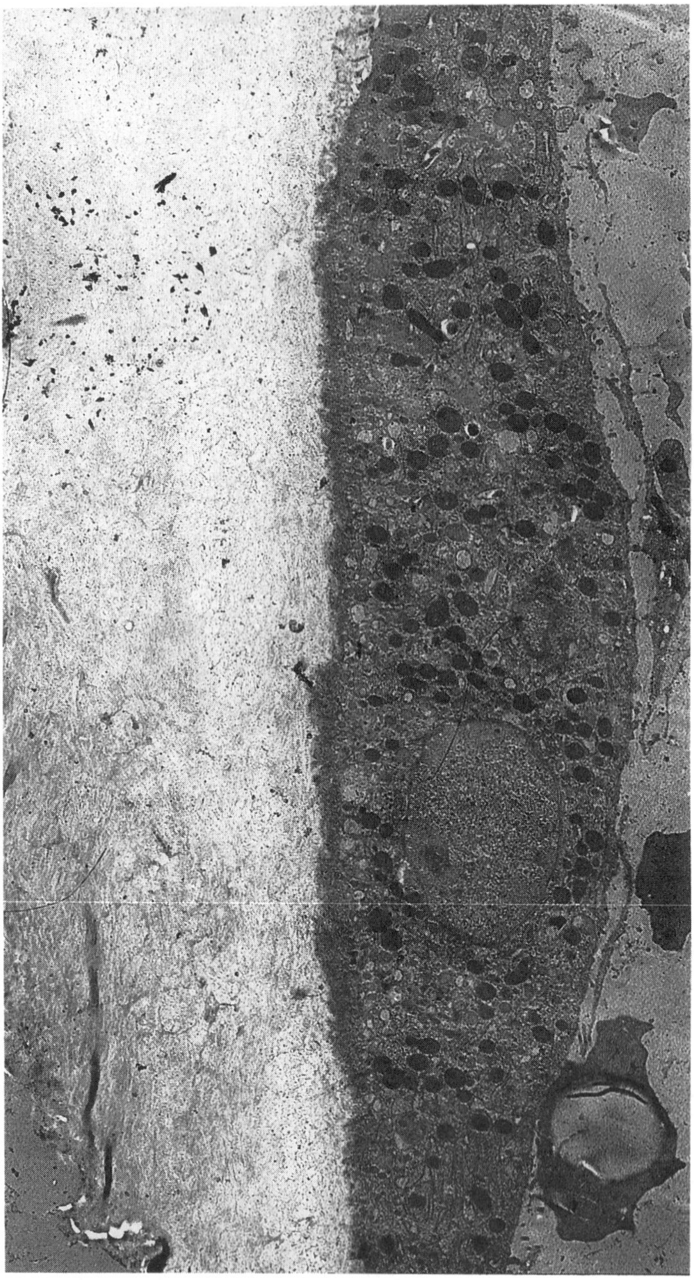

FIGURE 1. Osteoclast at the surface of a bone trabecula in an ia/ia rat. Note the extensive but thin clear zone facing the bone surface. Low temperature embedding, magnification 4,200×.

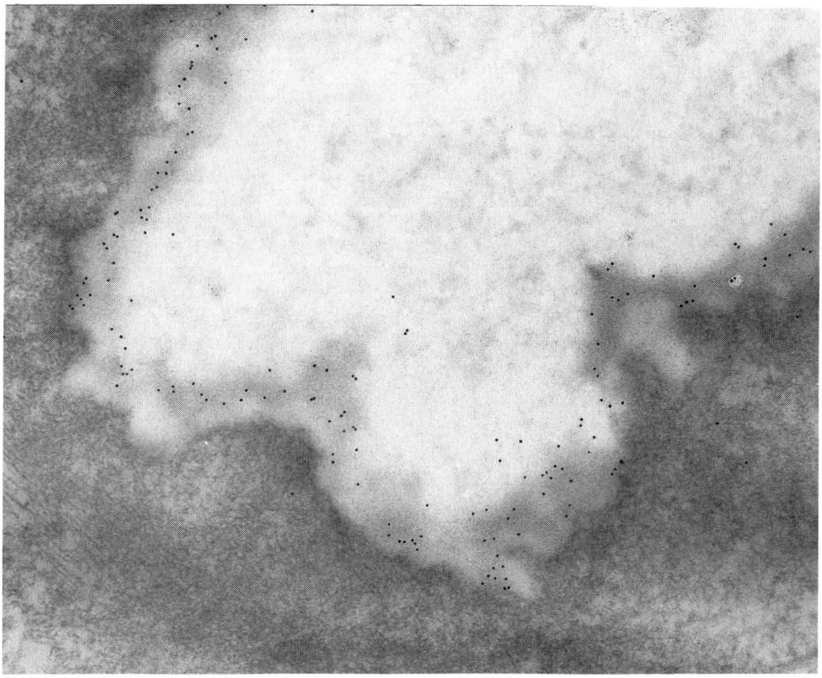

FIGURE 2. Immunolocalization of osteopontin in a bone surface facing the clear zone of an osteoclast in an ia/ia rat. The localization and amount of immunoreactivity is normal. Low temperature embedding, immunogold detection, magnification 52,000×.

were measured by solution hybridization. The results show that osteoclasts cover an increased part of the trabecular bone surface. The large osteoclasts are characterized by an extensive clear zone (FIG. 1). Immunohistochemistry of the ultracryosections demonstrates that the α_v- and β_3-subunits are enriched at the clear zone of osteoclasts in both groups of animals. Moreover, the total amount of receptor is slightly increased in clear zones of the ia/ia animals compared with their normal littermates. However, the ia/ia animals show the same distribution and level of OPN (FIG. 2) as their normal littermates. In the cytoplasm under the extended clear zones of the osteoclasts in ia/ia animals a large number of vesicles were observed. Such vesicles are likely transport vesicles. Notably, no immunoreactivity for OPN was observed in these vesicles. The present results suggest that ia/ia osteoclasts attach to the bone matrix through the same mechanisms as the osteoclasts of their normal littermates. Thus, the dysfunction of osteoclastic resorption in ia/ia animals is not due to a defective attachment. Interestingly, our results indicate that the low concentration of OPN observed at the ruffled border region of normal osteoclasts is a result of enzymatic activity.

REFERENCES

1. REINHOLT, F. P., K. HULTENBY, Å. OLDBERG, & D. HEINEGÅRD. 1990. Osteopontin—a possible anchor of osteoclast to bone. Proc. Natl. Acad. Sci. USA **87:** 4473–4475.

2. Flores, M., M. Norgård, D. Heinegård, F. P. Reinholt & G. Andersson. 1992. RGD-directed attachment of isolated rat osteoclasts to osteopontin, bone sialoprotein and fibronectin. Exp. Cell Res. **201:** 526–530.
3. Rouslahti, E., N. A. Noble, S. Kagami & W. A. Border. 1994. Integrins. Kidney Int. **45:** 17–22.
4. Hultenby, K., F. P. Reinholt & D. Heinegård. 1993. Distribution of integrin subunits on rat metaphyseal osteoclasts and osteoblasts. Eur. J. Cell Biol. **62:** 86–93.
5. Marks, S. C., Jr. 1973. Pathogenesis of osteopetrosis in the ia rat: reduced bone resorption due to reduced osteoclast function. Am. J. Anat. **138:** 165–190.

The Puerap Motif in the Promoter of the Mouse Osteopontin Gene[a]

KEVIN S. KIMBRO AND RAUL A. SAAVEDRA[b]

Harvard Medical School
and
The Children's Hospital
Enders Pediatric Research Center
300 Longwood Avenue
Boston, Massachusetts 02115

Osteopontin is a secreted glycosylated phosphoprotein expressed in bone, kidney, the immune system, carcinomas, and sarcomas.[1,2] Within the promoter of the mouse osteopontin gene, we have identified a DNA sequence that is located -686 base pairs from the initiation of transcription site.[1] This DNA sequence contains a Pu.1/Ets,[3] a steroid receptor element (SRE),[4] an acute phase responsive element (APRE.1),[5] and AP3-binding sites[6] (FIG. 1). In addition to the Pu.1/Ets site of this sequence, there are other potential Ets sites within the osteopontin promoter, which attest to the possible importance of these sites in the overall regulation of the osteopontin gene. The identification of a potential SRE site is consistent with the observation that the expression of osteopontin can be regulated by steroids.[7,8] The APRE.1 site of the above sequence was one of several potential APRE.1 sites found throughout the mouse osteopontin promoter. The presence of these sites within the osteopontin promoter correlates with the induction of this gene by tumor necrosis factor-α, interleukin-1, lymphocyte inhibitory factor, interleukin-6 and lipopolysaccharides.[2,9] The AP3 site could be a potential binding site for a transcription factor that is involved in the signal transduction pathway elicited by treatment with phorbol-esters,[6] and which may be involved in the regulation of osteopontin expression.

We synthesized a double-stranded DNA fragment that contained these motifs of the mouse osteopontin promoter, and we called it Puerap (FIG. 1). We found proteins in nuclear extracts from mouse kidney, osteoblast-like (MC3T3-E1),[10] and osteosarcoma (MGH-OGS)[11] cell lines that bound specifically to the Puerap DNA fragment and caused a retardation of its electrophoretic mobility in gel-shift assays. The Puerap DNA fragment after being incubated with kidney nuclear extracts generated two DNA-protein complexes (X and Y). The Puerap fragment after incubation with extracts from osteoblast-like or osteosarcoma cells, however, generated complex X and two other complexes different from complex Y. The expression of the proteins responsible for the formation of complexes X and Y appeared to be developmentally regulated in mouse kidney. We also synthesized shorter double-stranded DNA fragments which contained specific motifs of the

[a] This work was supported by Grant No. IR29AR41046 from the National Institutes of Health (NIH), No. RG24458A2 from the National Multiple Sclerosis Society, and by a grant from the Milton Fund (Harvard Medical School) to RAS, and Fellowship No. F32AR08302 from NIH to KSK.

[b] Corresponding author.

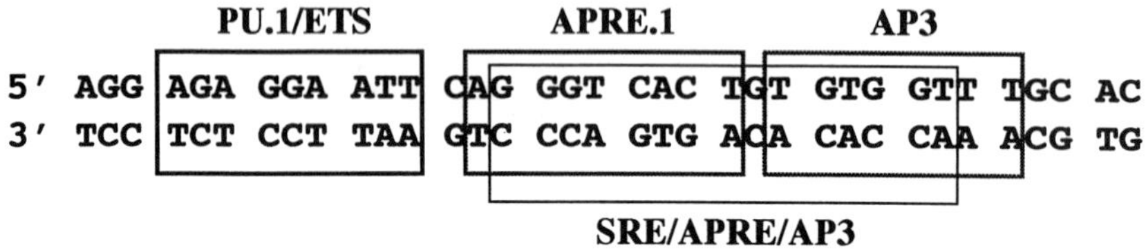

FIGURE 1. The sequence of the Puerap DNA fragment from the mouse osteopontin promoter. The sequences of the Pu.1/Ets, APRE.1, AP3, and SRE/APRE/AP3 motifs are indicated by *boxes*.

Puerap DNA. A fragment that contained the Pu.1/Ets motif incubated with MC3T3-E1 cell extracts generated only one complex. Another DNA fragment which contained the SRE/APRE/AP3 motifs incubated with kidney or MC3T3-E1 cell extracts generated two different patterns with three complexes each. DNA fragments containing either the Pu.1/Ets, APRE.1 or AP3 motifs incubated with kidney extracts generated no detectable complexes. These results indicate that the SRE and Pu.1/Ets motifs of the Puerap DNA fragments are targets for the specific binding of nuclear proteins from kidney tissue and/or cells of bone origin. Furthermore, these DNA-protein interactions could play an important regulatory role in the expression of the osteopontin gene *in vivo*.

REFERENCES

1. Denhardt, D. T. &. X. Guo. 1993. FASEB J. **7:** 1475–1482.
2. Patarca, R., R. A. Saavedra & H. Cantor. 1993. Crit. Rev. Immunol. **13:** 225–246.
3. Klemsz, M. J., S. R. McKercher, A. Celada, C. VanBeverer & R. A. Maki. 1990. Cell **61:** 113–124.
4. Martinez, E., F. Givel & W. Wahli. 1987. EMBO J. **6:** 3719–3727.
5. Wegenka, U. M., J. Bushmann, C. Lutticken, P. C. Heinrich & F. Horn. 1993. Mol. Cell. Biol. **13:** 276–288.
6. Mercurio, F. & M. Karin. 1989. EMBO J. **8:** 1455–1460.
7. Yoon, K., R. Buenaga & G. A. Rodan. 1987. Biochem. Biophys. Res. Commun. **148:** 1129–1136.
8. Craig, A. M. & D. T. Denhardt. 1991. Gene **100:** 163–171.
9. Jin, C. H., C. Miyaura, I. Yoshiko, M. H. Hong, S. Toshiyuki, E. Abe & T. Suda. 1990. Mol. Cell. Endocrinol. **74:** 221–228.
10. Kodama, H., Y. Amagai, H. Sudo, S. Kasai & S. Yamanto. 1981. Jpn. J. Oral Biol. **23:** 899–901.
11. Choi, C. H., R. S. Sedlacek & H. D. Suit. 1979. Eur. J. Cancer **15:** 433–442.

Upregulation of Osteopontin Expression by Ischemia in Rat Kidney

J. G. KLEINMAN,[a] E. M. WORCESTER,[a]
A. M. BESHENSKY,[a] A. M. SHERIDAN,[b]
J. V. BONVENTRE,[b] AND D. BROWN[b]

[a]*Nephrology Division*
VA Medical Center
Medical College of Wisconsin
Milwaukee, Wisconsin 53295
and
[b]*Renal Unit*
Massachusetts General Hospital
Harvard Medical School
Boston, Massachusetts 02129

In normal rat kidney, osteopontin protein and mRNA has been observed in thin descending limbs of the loop of Henle and in papillary surface epithelium away from the papillary tip.[1] Enhanced expression of the protein has also been described in models of renal injury.[2,3] The current study was done to investigate the expression of osteopontin during recovery from ischemic injury and to determine whether the ATP depletion that characterizes ischemia can stimulate expression of this protein.

Sprague-Dawley rats (225–235 g) were anesthetized and the left renal pedicle clamped. After 40 minutes, the clamp was removed and the rats allowed to recover. After various intervals of reperfusion, rats were again anesthetized and the kidneys fixed *in situ* as previously described.[4] Immunocytochemistry was performed on 5 μm frozen sections using a polyclonal rabbit antiserum.[1]

To examine the effect of ATP depletion on osteopontin expression, mouse cortical tubule cells in culture were exposed to 5 mM cyanide in the presence of 5 mM 2-deoxyglucose, a maneuver shown to decrease cellular ATP levels by approximately 95%.[5] After 2 hours, cyanide was removed and D-glucose restored. After an additional 2 hours, cells were immersion fixed and permeabilized, and immunocytochemistry was performed as on kidney sections.

No differences in expression between the kidneys subjected to ischemia and the contralateral nonischemic kidneys were observed after 1 and 3 hours of reflow. At 24 hours, however, osteopontin was markedly induced in thick ascending limbs of the loop of Henle (FIG. 1, *left*), and there was also a moderate amount of osteopontin observed in proximal convoluted tubules. However, no staining was detected in proximal straight tubules, the segment with the most severe histological damage. By 48 hours of reperfusion some immunofluorescence was observed in distal tubules and in proximal tubule S_3 segments, but the proximal convoluted tubule osteopontin expression had wanted. After 5 days of reperfusion, the S_3 segment osteopontin was increased even more (FIG. 1, *right*). At this time point, osteopontin was still observed in the distal tubule, but had faded almost completely from the proximal convoluted tubule.

Two hours after the cyanide was removed from the medium of cultured mouse proximal tubules and D-glucose ws restored, the fluorescent signal for osteopontin was markedly increased compared to controls (FIG. 2).

The ischemia-reperfusion injury produced by clamping and releasing the rat renal artery is particularly severe in the S_3 region of the proximal tubule.[4] However, expression of osteopontin is delayed in this segment. On the other hand, in the proximal convoluted tubule and the thick ascending limb of Henle, where the damage appears less severe, osteopontin is more briskly upregulated and then quickly wanes. Although osteopontin expression may be a feature of normal cellular proliferation, it is unlikely that the widespread expression observed here is a manifestation of the proliferation induced by ischemia-reperfusion injury. Such proliferation is observed in the S_3 segment at an earlier time point and is largely completed by 5 days.[4] In this segment, osteopontin does appear to follow the pattern of expression of vimentin, a marker of dedifferentiation in proximal tubules.[4] However, in contrast to osteopontin, at no time after reperfusion is vimentin detected in S_1 or S_2 segments.[4]

Thus, the temporal segment-specific expression of osteopontin in this model appears to be different from that observed with other protein markers of injury or repair. The signal for osteopontin upregulation in this and other types of renal

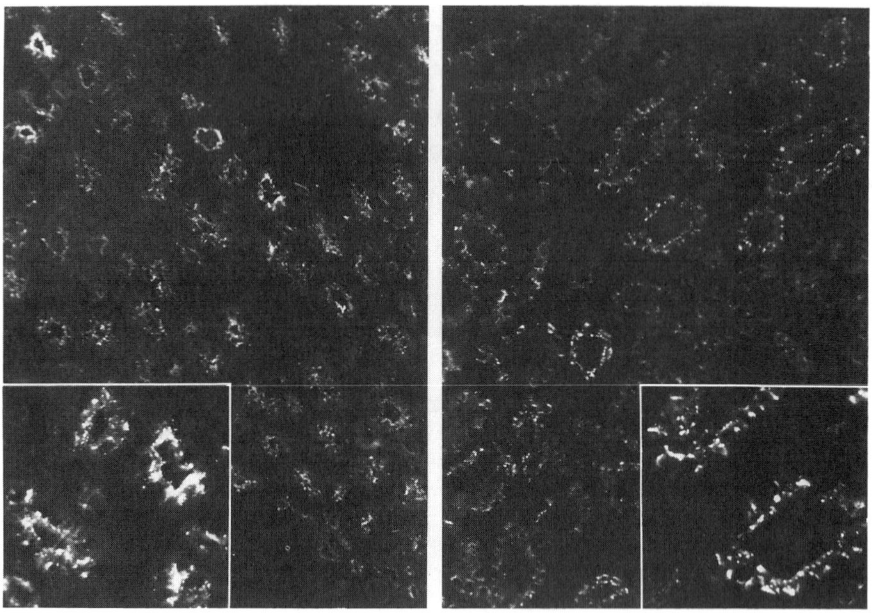

FIGURE 1. Osteopontin in rat kidney after ischemia and reperfusion. Immunocytochemistry was performed on 5-μm cryosections of rat kidney perfused *in situ* with paraformaldehyde-periodate-lysine fixative. *Left.* Inner stripe of outer medulla shows staining of the thick ascending limbs after 24 hours of reperfusion. *Right.* Outer stripe of outer medulla shows staining of the S_3 segments of proximal tubules after 5 days of reperfusion. *Insets.* Higher magnifications showing largely subapical and perinuclear distribution of staining. Original magnifications ×320 and ×640, respectively.

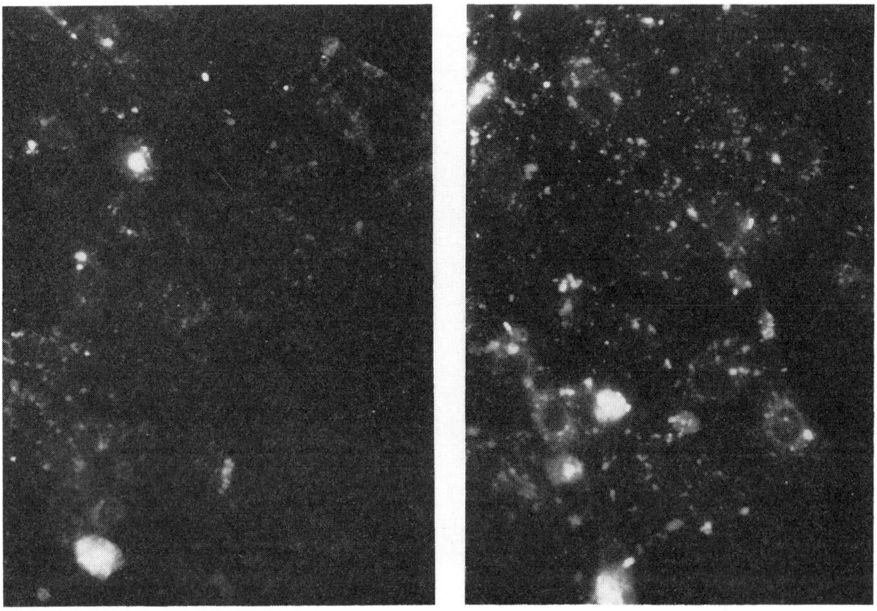

FIGURE 2. Osteopontin in energy-depleted, cultured mouse cortical tubule cells. Cells were immersed in paraformaldehyde-periodate-lysine fixative, permeabilized with Triton X-100 and immunocytochemistry performed as described in text and FIGURE 1. *Left.* Control cultures show staining in cytoplasmic vesicles of some cells. *Right.* Cultures 2 hours after restoring glucose and removal of cyanide show a much larger amount of staining in most of the cells. Magnification ×770.

injury is not known. However, the results in the cell culture model suggest that ATP depletion alone can stimulate osteopontin expression. These findings suggest that osteopontin subserves a protective or reparative function in renal tissue.

REFERENCES

1. KLEINMAN, J. G., A. M. BESHENSKY, E. M. WORCESTER & D. BROWN. Kidney Int. In press.
2. KOHRI, K., S. NOMURA, Y. KITAMURA, T. NAGATA, K. YOSHIOKA, M. IGUCHI, T. YAMATE, T. UMEKAWA, Y. SUZUKI, H. SINOHARA & T. KURITA. 1993. J. Biol. Chem. **268:** 15180–15184.
3. GIACHELLI, C. M., R. PICHLER, D. LOMBARDI, D. T. DENHARDT, C. E. ALPERS, S. M. SCHWARTZ & R. J. JOHNSON. 1994. Kidney Int. **45:** 515–524.
4. WITZGALL, R., D. BROWN, C. SCHWARZ & J. V. BONVENTRE. 1994. J. Clin. Invest. **93:** 2175–2188.
5. SHERIDAN, A. M., J. H. SCHWARTZ, V. M. KROSHIAN, A. M. TERCYAK, J. LARAIA, S. MASINO & W. LIEBERTHAL. 1993. Am. J. Physiol. **265:** F342–F350.

Activation of a Signal Transduction Pathway by Osteopontin[a]

CECILIA A. LOPEZ, ROBIN L. DAVIS, KEWA MOU, AND DAVID T. DENHARDT

Department of Biological Sciences
P.O. Box 1059
Rutgers University
Piscataway, NJ 08855-1059

Numerous groups have shown that upon integrin binding extracellular matrix components such as fibronectin and vitronectin can stimulate tyrosine phosphorylation of focal adhesion proteins such as paxillin and p125 focal adhesion kinase ($pp125^{FAK}$).[1-4] To determine the functional implications of osteopontin (OPN)-receptor binding, we have characterized alterations in tyrosine phosphorylation of specific intracellular proteins using purified recombinant OPN expressed in an immortalized human embryonic kidney cell line.

OPN is a sialic acid-rich secreted phosphoprotein containing a GRGDS sequence capable of binding to the vitronectin receptor ($\alpha_v\beta_3$ integrin). It functions in cell adhesion, migration, and chemoattraction.[5,6] Binding of OPN to osteoclasts via the $\alpha_v\beta_3$ integrin receptor can effect changes in intracellular Ca^{++} levels.[7,8] It can activate phospholipase C[9] and can inhibit the induction of iNOS mRNA in gamma-interferon- and lipopolysaccharide-stimulated primary proximal tubule epithelial cells[10] and in primary cultures of inner ear explants (unpublished observations).

As illustrated in FIGURE 1, addition of OPN to serum-starved subconfluent cells resulted in an increase in tyrosine phosphorylation of several proteins in T24 Ha- *ras*-transformed NIH3T3 cells (PAP2). In epithelial cells subcultured from a stria vascularis explant (inner ear tissue that has an ion regulatory role in the cochlea, SVE) a sustained decrease of $pp125^{FAK}$ was observed (Denhardt *et al.*, this volume). A transient increase of tyrosine phosphorylation of $pp60^{\underline{src}}$ was seen in all the cell lines tested. Characterization of the nature of tyrosine phosphorylation of $pp60^{\underline{src}}$ by OPN (1 nM) was performed in PAP2 cells. It was time and dose dependent. Phosphorylation of $pp60^{\underline{src}}$ peaked between 3 and 6 min after OPN addition, and disappeared after 9 min. Maximal stimulation was observed at 0.1–1.0 nM OPN; at 4 nM OPN, phosphorylation was decreased. Control experiments were done to determine the specificity of our findings. Phosphorylation was inhibitable by anti-mouse OPN antibody. A GST-OPN fusion protein (1 nM, a gift from A. Chambers, London Regional Cancer Centre, Canada) made in *E. coli* was also able to stimulate tyrosine phosphorylation of $pp60^{\underline{src}}$ while GST (glutathione *S*-transferase) alone did not. OPN also enhanced for at least 16 min the tyrosine phosphorylation of paxillin and tensin.

Our results suggest that OPN behaves as a cell signalling molecule and that the pathway involves transient and sustained changes in tyrosine phosphorylation

[a] This work was supported by NIH Grant DC01295.

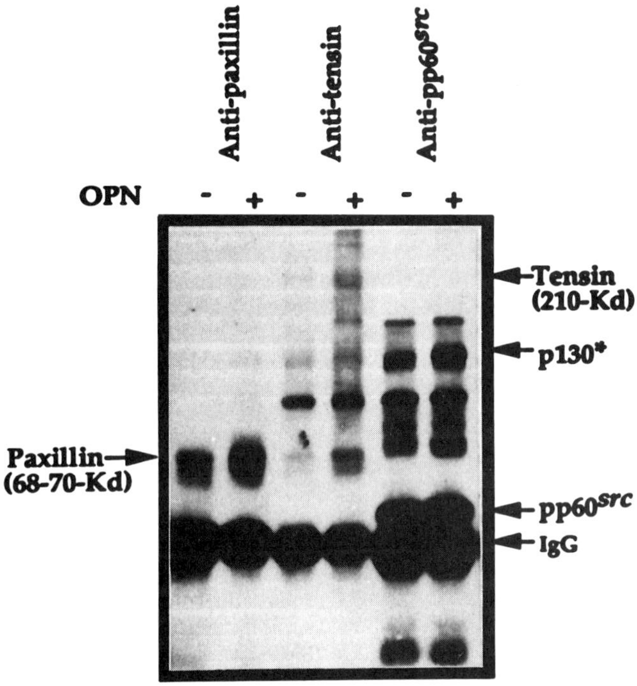

FIGURE 1. Detection of tyrosine-phosphorylated proteins with anti-phosphotyrosine antibody (PY-54) in PAP2 cells. An increase in tyrosine phosphorylation of paxillin, tensin, pp60^{src}, and possibly p130, was observed at 4 min. Serum-starved, semiconfluent PAP2 cells were incubated with 100 μM sodium orthovanadate for 1 h, exposed to OPN (50 ng/ml; 1 nM) for 4 min, quickly washed with ice-cold phosphate-buffered saline, and lysed with RIPA buffer. Sedimented cell lysates were first precleared with protein A-Sepharose beads and immunoprecipitated with anti-paxillin, anti-tensin, and anti-pp60^{src} (mAb327) overnight at 4°C. Goat anti-mouse IgG bound to protein A Sepharose beads was then added to these samples. The beads were washed with lysis buffer three times, electrophoresed on an SDS-7.5% polyacrylamide gel, and immunoblotted with a monoclonal antibody to phosphotyrosine (PY-54, Oncogene Science). Detection was performed by enhanced chemiluminescence (ECL; Amersham). *p130 is a *src* transformation-associated protein of unknown identity.[11]

of focal adhesion proteins. This regulation is cell-type specific and possibly mediated by the $\alpha_v\beta_3$ integrin.

ACKNOWLEDGMENTS

We would like to thank Drs. T. Maciag, X. Zhan, E. Yurkow, and X. Hu for their valuable time and useful suggestions regarding immunoprecipitation and immunoblotting. We are thankful to Dr. D. J. Fujita for a generous supply of mAb327 (anti-pp60^{src}); Dr. T. J. Parsons for mAb2A7 (anti-p125^{FAK}); Dr. K.

Burridge for anti-tensin and anti-paxillin; and Dr. L. W. Fisher for LF-7 (anti-human OPN antibody) and human bone OPN cDNA.

REFERENCES

1. GUAN, J.-L., J. E. TREVITHICK & R. O. HYNES. 1991. Cell Regulation **2:** 951–964.
2. GUAN, J.-L. & D. SHALLOWAY. 1992. Nature **358:** 690–692.
3. SCHALLER, M. D., C. A. BORGMAN, B. S. COBB, R. R. VINES, A. B. REYNOLDS & J. T. PARSONS. 1992. Proc. Natl. Acad. Sci. USA **89:** 5192–5196.
4. BURRIDGE, K., C. E. TURNER & L. H. ROMER. 1992. J. Cell Biol. **119:** 893–903.
5. DENHARDT, D. T. & X. GUO. 1993. FASEB J. **7:** 1475–1482.
6. PATARCA, R., R. A. SAAVEDRA & H. CANTOR. 1993. Crit. Rev. Immunol. **13:** 225–246.
7. ZIMOLO, G., G. WESOLOWSKI, H. TANAKA, J. L. HYMAN, J. R. HOYER & G. A. RODAN. 1994. Am. J. Physiol. **266:** C376–C381.
8. MIYAUCHI, A., J. ALVAREZ, E. M. GREENFIELD, A. TETI, M. GRANO, S. COLUCCI, A. ZAMBONIN-ZALLONE, F. P. ROSS, S. M. TEITELBAUM, D. CHERESH & K. A. HRUSKA. 1991. J. Biol. Chem. **266:** 20369–20374.
9. ROLNICK, F., A. GUPTA, M. HUSKEY, U. ALVAREZ & K. HRUSKA. 1993. J. Bone Miner. Res. **8:** S240 (abstract 495).
10. HWANG, S.-M., C. A. LOPEZ, D. E. HECK, C. R. GARDNER, D. L. LASKIN, J. D. LASKIN & D. T. DENHARDT. 1994. J. Biol. Chem. **269:** 711–715.
11. KANNER, S. B., A. B. REYNOLDS, R. R. VINES & T. J. PARSONS. 1990. Proc. Natl. Acad. Sci. USA **87:** 3328–3332.

The Mutagenesis of the RGD Sequence of Recombinant Osteopontin Causes It to Lose Its Cell Adhesion Ability

ROSSOLYN J. McFARLAND, SONYA GARZA,
WILLIAM T. BUTLER,[a] AND MAGNUS HÖÖK

*Albert B. Alkek Institute of Biosciences and Technology
Center for Extracellular Matrix Biology
Texas A&M University
2121 West Holcombe Boulevard
Houston, Texas 77030*

[a]*Texas Medical Center Dental Branch
Department of Biological Chemistry
Houston, Texas 77025*

INTRODUCTION

The adhesion of cells to the extracellular matrix (ECM) plays a crucial role in embryonic development and normal physiology of a multicellular organism. Adherent cells use specific cell surface receptors, integrins, to bind to adhesive extracellular macromolecules.[1] Several bone ECM proteins have been shown to support the adhesion of osteoclasts, osteoblasts, and fibroblasts.[2] These adhesive proteins often contain the amino acid sequence RGD which is recognized by the $\alpha_v\beta_3$ integrin.[3] Osteopontin (OPN), which contains an RGD sequence, has been proposed to contain additional cell adhesion signals.[4]

We expressed wild type as well as mutated OPN in a prokaryotic system and evaluated the resulting recombinant proteins for their ability to support cell adhesion in an *in vitro* adhesion assay. Specifically, the importance of the RGD sequence was studied by comparing OPN mutants where this sequence had been changed to RAD and RGE, respectively. Our data suggest that the recombinant form of OPN contains one single cell attachment site comprised of the RGD sequence.

MATERIALS AND METHODS

Protein Expression

Rat or human osteopontin cDNA was PCR amplified and cloned into the pQE vector. Recombinant fusion protein was produced with 6 histidines at the amino terminal end. The histidine tail allows for a one-step purification using metal chelating chromatography.

Cell Culture

Human skin fibroblast (ATCC) and osteoblast cells (a gift from Dr. Tony Mikos, Rice University, Houston, TX) were maintained in Dulbecco's modified Eagle's medium (DMEM) (GIBCO, Grand Island, NY) supplemented with 10% fetal bovine serum (Intergen, Purchase, NY). Single cell suspensions for experiments were obtained by trypsinization with 0.05% trypsin and 0.5 mM EDTA (GIBCO), followed by 0.5% soybean trypsin inhibitor (Sigma, St. Louis, MO).

Adhesion Assay

Falcon (#3912) 96-well microtiter plates were coated with 5 μg/ml of OPN, OPN mutants, or recombinant fragment dissolved in phosphate buffered saline for one hour at room temperature. The wells were incubated with 100 μl of 1% heat denatured BSA in PBS to saturate all unoccupied protein binding sites for one hour at room temperature. Wells were washed three times with PBS. Cells were trypsinized as previously stated and the suspension centrifuged at 500 g for five minutes. Cell pellets were washed with DMEM and resuspended in DMEM to yield 2×10^4 cells/50 μl, which were transferred to each well and incubated for one hour at 37°C, 5% CO_2. The wells were washed three times with PBS, fixed with 50% glutaraldehyde, and washed three times with dH_2O. The attached cells were then stained for twenty minutes with 0.1% crystal violet in 200 mM MES, pH 6.0. The wells were washed three times with dH_2O to remove unbound dye. The bound samples were solubilized with 10% acetic acid and the absorbance read at 570 nm (Molecular Devices, Menlo Park, CA). All experiments were performed in triplicate.

RESULTS

FIGURE 1A shows the various constructs produced with the pQE-30 expression system. The full length rat OPN (rOPN) with amino acids 1–301, the truncated form (rOPN-1) representing amino acids 42–301, and human OPN (hOPN) all have the RGD motif. A recombinant fragment (rOPN-2) representing the C-terminal domain (residue 142–301) lacks the RGD attachment sequence. OPN mutants where the RGD sequence had been replaced with RGE and RAD, respectively, is shown in FIGURE 1B. As shown in FIGURES 2A and 2B, hOPN, rOPN and rOPN-1 promoted attachment of human skin fibroblasts and osteoblasts. However, cells plated on rOPN-2 and mutants that underwent site-directed mutagenesis were unable to support cell attachment in this assay.

DISCUSSION

The data presented in this study indicate that there is a single cell attachment site, the RGD site, present in recombinant OPN. A truncated recombinant fragment (OPN-2) lacking this site and mutated recombinant OPN where RGD has been changed to RGE or RAD did not support cell attachment. A previous study demonstrated that a 28-kD peptide (comprising the C-terminal domain of OPN and corresponding to OPN-2) obtained after digestion of OPN with endoproteinase Arg-C supported cell attachment.[4] This fragment does not contain the RGD sequence and it was therefore suggested that another site in this native fragment may support OPN-mediated cell attachment. It was also suggested that the cell adhesion properties of OPN are dependent on posttranslational modifications.[5]

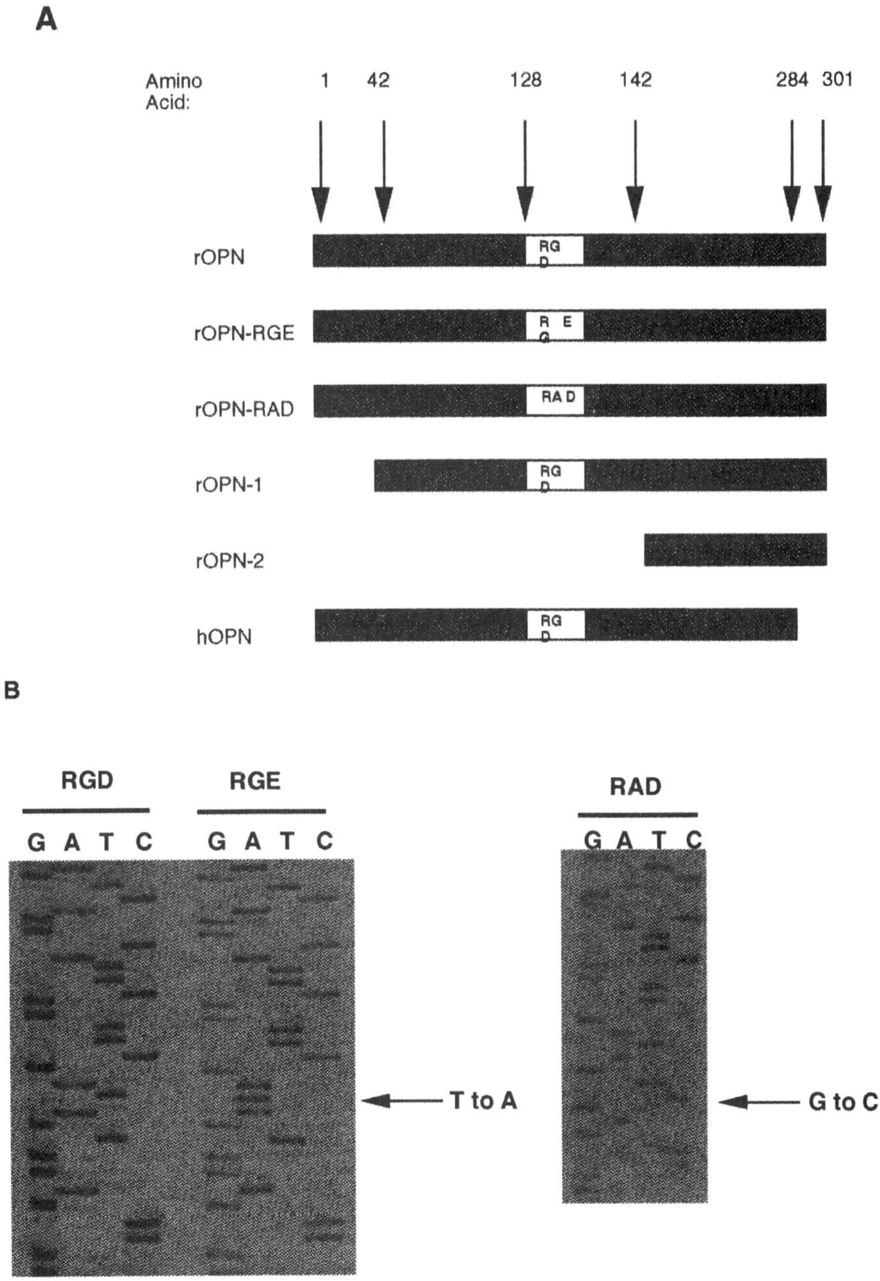

FIGURE 1. (A) Various constructs produced with the pQE-30 expression system. The full-length rat OPN (rOPN) with amino acids 1–301, the truncated form (rOPN-1) representing amino acids 42–301, and human OPN (hOPN) all have the RGD motif. A recombinant fragment (rOPN-2) representing the C-terminal domain (residue 142–301) lacks the RGD attachment sequence. **(B)** DNA sequence showing OPN mutants where the RGD sequence had been replaced with RGE and RAD, respectively.

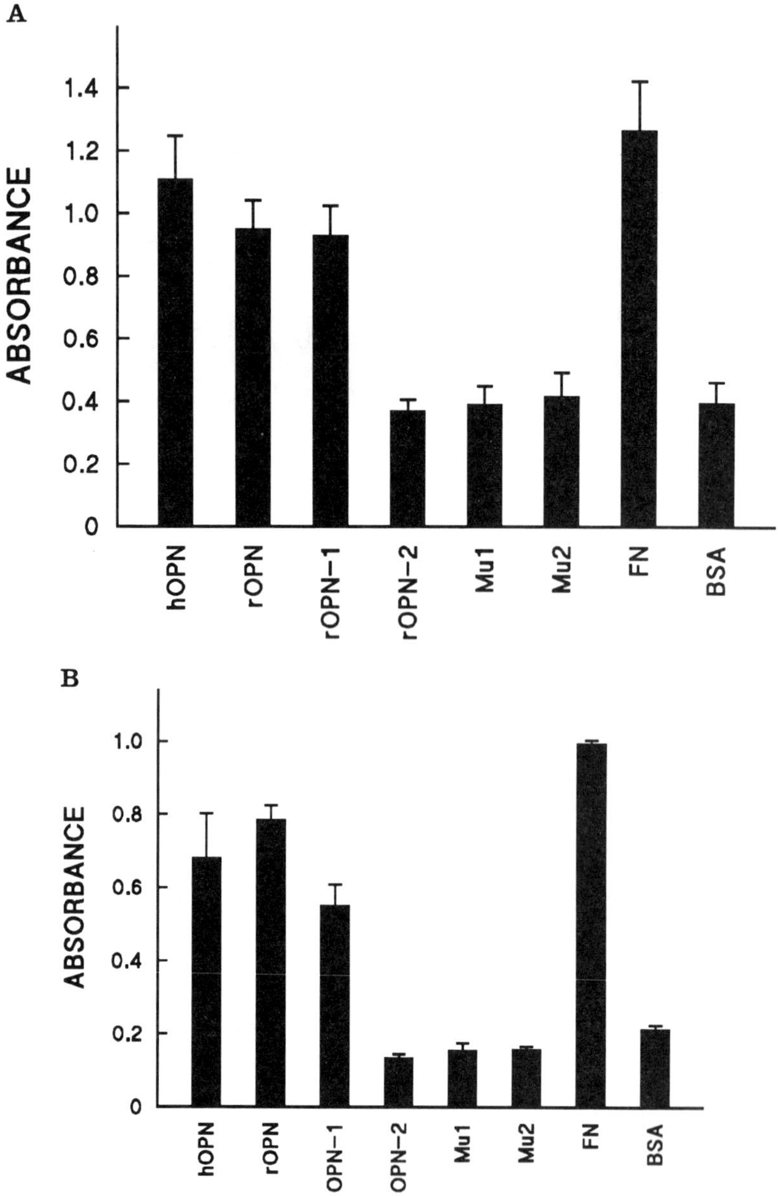

FIGURE 2. Attachment of human skin fibroblasts **(A)** and osteoblasts **(B)** to coated surfaces. Cells were added to 96-well plates that had been coated with 5 μg/ml of OPN, OPN mutants (Mu1: RGD > RGE, Mu2: RGD > RAD), or truncated fragments. Fibronectin (FN) and BSA controls at 10 μg/ml. Cells adhered for 1 h, were fixed, and were stained. The attached cells were solubilized and the absorbance read at 570 nm. All experiments were performed in triplicate.

For example, osteoclasts were found to adhere to phosphorylated OPN.[5] If in fact the posttranslational modifications of OPN contribute to the cell adhesion inducing activity we would not detect this activity using our prokaryotic recombinant proteins. The protein core does only seem to contain one cell adhesion-inducing site composed of the RGD sequence.

REFERENCES

1. YAMADA, K. M. 1991. J. Biol. Chem. **266:** 12809–12812.
2. SOMERMAN, M. J., C. W. PRINCE, W. T. BUTLER, R. A. FOSTER, M. MOEHRING & J. J. SAUK. 1989. Matrix **9:** 49–54.
3. DENHARDT, D. T. & X. GUO. 1993. FASEB J. **7:** 1475–1482.
4. VAN DIJK, S., J. A. D'ERRICO, M. J. SOMERMAN, M. C. FARACH-CARSON & W. T. BUTLER. 1993. J. Bone Miner. Res. **8:** 1499–1505.
5. EK-RYLANDER, B., M. FLORES, WENDEL, MIKAEL, D. HEINEGÅRD & G. ANDERSSON. 1994. J. Biol. Chem. **269:** 14853–14856.

Divergent Expressions of Two Cell Adhesion Molecules at the Leading Edge of a Migrating Cell Sheet

R. N. MASCARDO,[a] J. KSIAZEK,[b] AND R. W. BARTON[b]

[a]*Division of Endocrinology*
University of Connecticut School of Medicine
Farmington, Connecticut 06032
and
[b]*Department of Inflammatory Disease*
Boehringer Ingelheim Pharmaceuticals
Ridgefield, Connecticut 06877

The intercellular adhesion molecule 1 (ICAM-1)[1] and endothelial leukocyte adhesion molecule 1 (ELAM-1)[2] are two distinct cell surface glycoproteins that mediate leukocyte adhesion to endothelial cells and cytokine-induced leukocyte locomotion. Utilizing monoclonal antibodies against ICAM-1 and ELAM-1, and immunoperoxidase microscopy, we localized these linker glycoproteins in several human fibroblastic, epithelial, and endothelial cell monolayers, including a human umbilical vein endothelial cell (HUVEC) culture.

We previously observed that the expression and cellular localization of ICAM-1 changed in actively migrating cells at the edge of an artificial wound path of a confluent human fibroblast (CCD76SKP5) monolayer exposed to the cytokines interleukin-1, tumor necrosis factor-alpha (TNF), and interferon-gamma.[3] To determine if the cellular expression and localization of ICAM-1 and ELAM-1 differed in actively moving cell sheets, we induced an artificial wound stimulus[4] in a confluent HUVEC culture that was exposed to TNF (500 U/ml × 24 hr), and observed the changes in the cell expression of the two adhesion molecules at the wound edge by immunoperoxidase microscopy.

Wounding induced cell polarization and migration at the wound edge within 3 hr after disrupting the continuity of the monolayer. Cells at the wound edge exhibited a progressive decline in ICAM-1 expression (from 55% of cells at 0 hr to 25% at 3 hr), whereas confluent cells distal to the wound edge showed no change (55%). In contrast, the expression of ELAM-1 at the wound border increased significantly (from 25% of cells at 0 min to 65% at 30 min), then progressively decreased to its basal level (25%) at 3 hr after wounding. The expression of ELAM-1 in cells distal to the wound remained unchanged (25%).

Thus, the initial cellular events observed in directed cell migration of wound closure, including centrosomal orientation and cellular locomotion at the wound edge, were associated with an increase in the expression of ELAM-1 and a decline in ICAM-1 expression. These observations suggest that these two pathfinding adhesion glycoproteins play opposite functional roles in cell locomotion induced by disrupting the continuity of the cell monolayer.

REFERENCES

1. DUSTIN, M. L., R. ROTHLEIN, A. K. BHAN, C. A. DINARELLO & T. A. SPRINGER. 1986. Induction by IL 1 and interferon-gamma: tissue distribution, biochemistry, and function of a natural adherence molecule (ICAM-1). J. Immunol. **137:** 245–254.
2. ROSEN, S. D. 1990. The LEC-CAMs: an emerging family of cell-cell adhesion receptors based upon carbohydrate recognition. Am. J. Respir. Cell Mol. Biol. **3:** 397–402.
3. MASCARDO, R. N., R. ROTHLEIN & R. W. BARTON. 1990. Immunofluorescence localization of the intercellular adhesion molecule-1 in confluent cells subjected to an artificial wound stimulus. Ann. N.Y. Acad. Sci. **580:** 518–520.
4. MASCARDO, R. N., W. THOMPSON, M. A. GALLO, T. C. CHAMBERS & G. EILON. 1990. A novel dipeptide stimulates directed cell migration by modulating protein kinase. C. J. Cell. Physiol. **142:** 41–49.

Expression of Osteopontin in a Macrophage Cell Line and in Transgenic Mice with Pulmonary Fibrosis Resulting from the Lung Expression of a Tumor Necrosis Factor-α Transgene

YOSHITAKA MIYAZAKI,[a] TAKAYOSHI TASHIRO,[a] YASUNORI HIGUCHI,[b] MIHOKO SETOGUCHI,[b] SHUNSUKE YAMAMOTO,[b] HIROYUKI NAGAI,[a] MASARU NASU,[a,d] AND PIERRE VASSALLI[c]

[a]*Second Department of Internal Medicine*
[b]*Department of Pathology*
Oita Medical University
Hasama-machi
Oita 879-55, Japan

[c]*Department of Pathology*
University of Geneva
1, rue Michel-Servet
CH1211 Geneva 4, Switzerland

INTRODUCTION

OPN is a phosphoprotein whose cDNA was originally cloned from a rat osteoblast cell line.[1] Among other effects OPN is thought to be involved in host defense mechanisms, in particular against *Rickettsia tsutsugamushi, Listeria monocytogenes,* and *Flavivirus*.[2] OPN mRNA has been detected in a macrophage cell line,[3] in which its level of expression was significantly increased by stimulation with crude lymphokines.[4] Recently, Denhardt *et al.* reported that OPN can inhibit nitric oxide synthase (NOS) production by activated macrophages, as a result of its calcium-binding property. NOS activity can be induced in macrophages by exposure to tumor necrosis factor (TNF)-α.[5] In order to understand the relationship between TNF-α and OPN in macrophages, we explored the expression level of OPN mRNA in macrophages stimulated *in vitro* with TNF-α and in the pulmonary lesions of transgenic mice overexpressing TNF-α in the lung.

The transgenic model has been created to explore the possible pathogenesis of idiopathic pulmonary fibrosis, usually a fatal disorder of the lung.[6] It is generally assumed that the infiltration of inflammatory cells, degeneration of alveolar epithelial cells and proliferation of interstitial cells with the development of fibrosis are

[d] Corresponding author.

related to an overproduction of certain cytokines.[7,8] Presence of TNF-α or its mRNA has been detected in cases of idiopathic pulmonary fibrosis in man[9] and in the lung of mice with pulmonary fibrosis elicited by exposure to bleomycin[10] or silica.[11] That the pulmonary release of TNF-α can induce the whole spectrum of lesions associated with pulmonary fibrosis has been shown by generating transgenic mice in which the TNF-α was placed under the control of the lung specific surfactant protein C (SP-C) promoter (Y. Miyazaki *et al.*, in preparation). Since pulmonary macrophages are associated with the fibrotic lesions created by the permanent release of TNF in the lung of these transgenic mice, it was of interest to study osteopontin expression in these lesions.

MATERIALS AND METHODS

Transgenic Mice

An animal model for pulmonary fibrosis was generated by creating transgenic methods (Y. Miyazaki *et al.*, in preparation). Briefly, a chimeric DNA was constructed by introducing the entire region of the murine TNF-α gene[12] into a pBluescript II SK- plasmid (STRATAGENE, La Jolla, CA) containing the human SP-C promoter.[13] The transgene was microinjected into the male pronucleus of C57BL/6xDBA2 F1 fertilized eggs. Transgenic offsprings were identified by polymerase chain reaction (PCR) analysis of genomic DNA isolated from tail biopsies, using a 5′ primer specific to the SP-C promoter and a 3′ primer specific to the TNF-α gene. Several transgenic found mice were obtained, and those with the highest level of TNF mRNA and the most severe progressive pulmonary disease selected for back crossing with C57BL/6.

Cell Culture

The aHINS-B3, a murine macrophage cell line, has been described elsewhere.[14] aHINS-B3 cells were cultured in Dulbecco's modified Eagle medium (DMEM) supplemented with 10% endotoxin-free fetus calf serum (FCS) (HyClone, Logan, Utah). After trypsinisation, the cells were harvested and plated on plastic plates (1.4×10^5 cells/ϕ10-cm plate) in the culture medium to which recombinant murine TNF-α (Genzyme, Kent, UK) was added (100 U/ml).

Northern Blot Analysis

RNAs were prepared from lungs of control or transgenic mice by the acid-guanidine-phenol-chloroform method,[15] and from aHINS-B3 cells using ISOGEN (Nippongene, Toyama, Japan). Northern blot hybridizations were performed as previously described.[16] Equal RNA loading was controlled by staining the filters with methylene blue in 0.5 M soldium acetate PH4.8. Filters were hybridized with 32P-labeled OPN cRNA probe or with an OPN probe labeled by the use of a Megaprime DNA labeling system (Amersham International, Burlinghamshire, UK). The filters hybridized with the OPN cDNA probe were exposed to Imaging

Plate BAS1000 (Fuji Photo Film, Tokyo, Japan), and data analysis was performed using the software for the BAS1000 supplied by Fuji Photo Film.

In Situ *Hybridization*

33P-labeled mouse TNF-α and OPN cRNA probes and control sense probes were used for *in situ* hybridization, which was described elsewhere.[17]

RESULTS

Induction of Enhanced Expression of OPN mRNA by TNF-α in a Macrophage Cell Line

OPN mRNA was detected in the aHINS-B3 macrophage cell line before stimulation with TNF-α (FIG. 1), and its amount showed an average 3-fold increase when the cells were exposed for 2 hours to TNF, remaining in a plateau (2.3-fold) after 5 and 9 hr, and being still slightly elevated (1.6-fold) after 24 hr.

Expression of OPN mRNA in Interstitial Alveolitis and Pulmonary Fibrosis of SP-C/TNF Transgenic Mice

OPN mRNA was detected on Northern blots of lung RNAs of normal and transgenic mice. The expression level of OPN in transgenic mice, observed between 2 and 6 months of age, was consistently higher with lesions and culminating in severe fibrosis (FIG. 2) than that found in lung of normal mice (FIG. 3). *In situ* hybridization revealed that TNF-α mRNA was expressed in epithelial alveolar cells as a result of the cell specificity of the SP-C promoter (FIG. 4a),[18] with a distribution contrasting to that observed for OPN cRNA, which is found in alveolar macrophages (FIG. 4b).

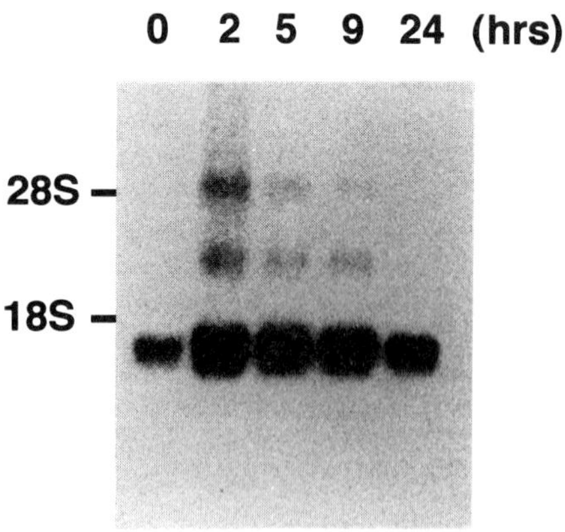

FIGURE 1. Northern blot of aHINS-B3 mRNAs. RNAs were prepared by using ISOGEN from confluent plates (10 cmϕ) of aHINS-B3 stimulated with TNF-α for various durations. Equal loading of RNA was verified by methylene blue staining of the filters, which was hybridized with a 32P-labeled DNA probe in rapid hybridization buffer (Amersham, UK). 0, unstimulated aHINS-B3; 2, 2 hr after TNF-α addition; 5, 5 hr; 9, 9 hr; 24, 24 hr.

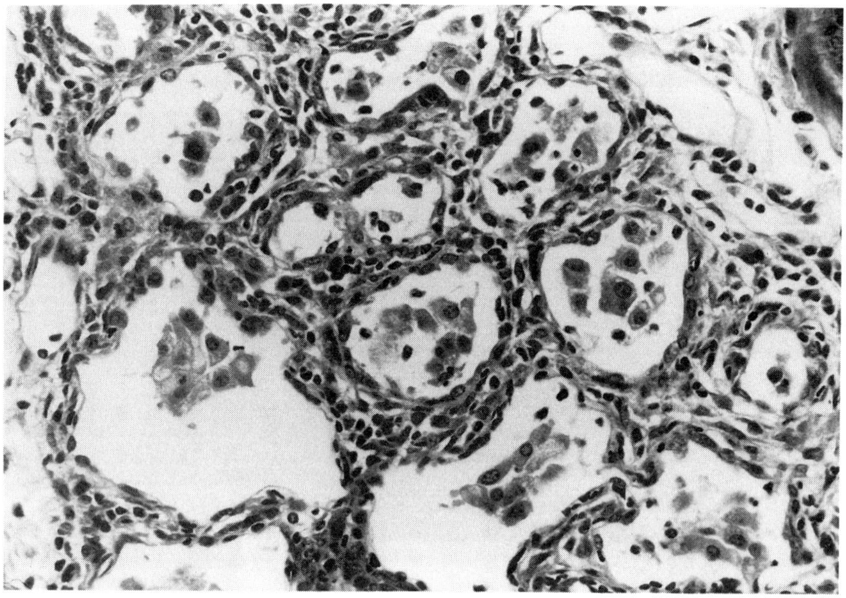

FIGURE 2. Histologic section of the lung of a SP-C/TNF transgenic mouse. Hematoxylin-eosin stain. These lesions were observed in a two-month-old mouse and represent a relatively early lesion of alveolitis progressing towards fibrosis. The thickened alveolar septa are hypercellular, and infiltrated mostly by lymphocytes. This experimental pulmonary disease appears indistinguishable from idiopathic pulmonary fibrosis in man.

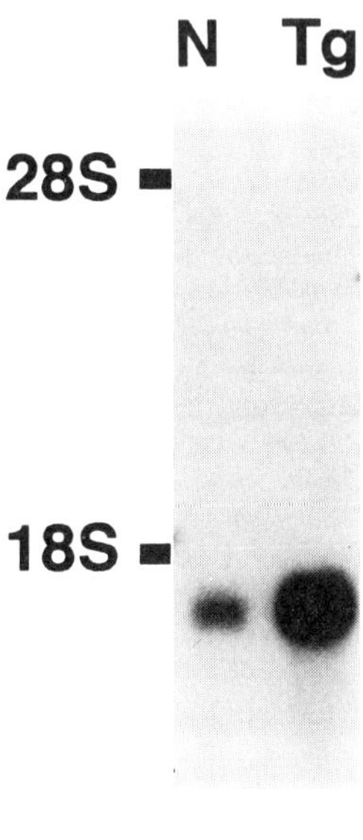

FIGURE 3. Northern blot of lung mRNAs. RNAs were prepared from lungs and hybridized with a 32P-labeled OPN cRNA probe. Equal amounts of lung RNA were placed in each track, and the evenness of loading was judged by methylene blue staining of the filters showing the ribosomal RNA bands. N, normal mouse; Tg, transgenic mouse, comparable to that shown in FIGURE 2.

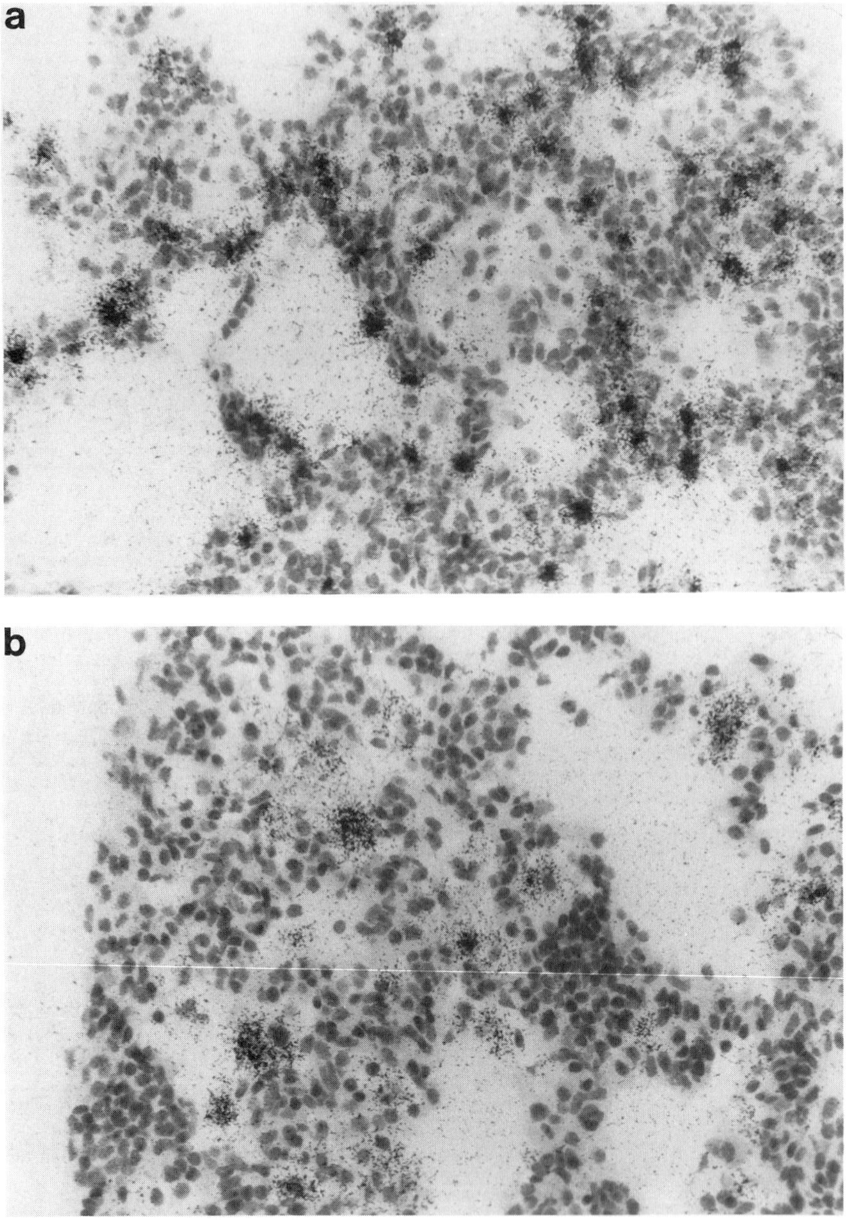

FIGURE 4. *In situ* hybridization with 33P-labeled TNF-α **(a)** and OPN **(b)** cRNA probes. **(a)** Silver grains corresponding to TNF mRNA are seen in cells located at the air-alveolar border. Lamellar bodies were identified in these cells by electron microscopy ascertaining their type II pneumocyte identity. **(b)** Foamy cells in the alveolar space bear grains corresponding to OPN mRNA. These cells were stained by the F4/80[20] monoclonal antibody, indicating their macrophage nature.

DISCUSSION

OPN, explored in the present report, and found to be induced in macrophages by TNF stimulation, is not only an adhesion molecule characterized by an RGD motif, but also a cytokine motif released by activated T cells.[19] In trying to understand if a change in OPN level may be involved in some types of pathology, we focused our attention on pulmonary lesions which are elicited by lung TNF release in transgenic mice.

In man, TNF-α is mainly synthesized in alveolar macrophages in normal lungs, although type II epithelium becomes a predominant source of TNF-α production in interstitial pneumonia leading to pulmonary fibrosis.[9] In our transgenic mouse model of pulmonary fibrosis, there was no significant increase in the lungs of mRNA for various cytokines (*e.g.*, GM-CSF, TGF-β) or adhesion molecules (*e.g.*, ICAM-1), except for a moderate increase in VCAM-1 mRNA. We show here that OPN mRNA is expressed in alveolar macrophages, in which the expression of TNF-α cannot be detected (or at least is rather insignificant, in comparison to the amount of TNF mRNA expressed in type II epithelial pneumocytes in which the SP-C promoter leads to the transgene expression).

The role of enhanced expression of OPN in the macrophages of associated interstitial pneumonitis and progressive pulmonary fibrosis is not clear. It is not known whether TNF-α leads to fibrosis in a direct or indirect way, *i.e.*, through the production of other cytokines. TNF-α stimulation of macrophages induces both OPN and nitric oxide synthase (NOS), but the relationship of these two events is unknown. The contrast production of NOS requires the accumulation of the intracellular calcium ion; on the other hand, the OPN reduces this concentration in cells, and thus may eventually lead to inhibition of NOS production. We have not yet tested NOS expression in these fibrotic lungs; it must be pointed out that NOS, leading to the production of NO, is a potent damaging agent for cells surrounding the macrophages, and is able to elicit the proliferation of interstitial cells. Thus, if increased OPN decreases NOS synthesis, this may represent a protective mechanism against fibrosis: on the other hand, OPN enhances type IV collagenase expression, which might play a role in fibrosis.

The combined observation that TNF increases OPN expression in macrophages, and that increased expression of OPN is found in macrophages associated with severe pulmonary fibrosis, indicates that the role of OPN and possibly its interaction with NOS in developing pulmonary fibrosis should be further explored. This is the aim of our current work.

SUMMARY

The expression level of osteopontin (OPN) mRNA was found to be increased in a macrophage cell line in the presence of recombinant tumor necrosis factor-α (TNF-α). OPN mRNA level was also explored in the lungs of transgenic mice which were expressing TNF-α in type II pneumocytes, a condition leading to pulmonary alveolitis and progressive fibrosis. OPN mRNA was significantly increased in the lungs of these transgenic mice. *In situ* hybridization showed that it was localized mostly in alveolar macrophages. Since OPN can be induced in macrophages by TNF-α stimulation and since on the other hand osteopontin appears to decrease the level of nitric oxide synthase, and thus the production of nitric oxide, osteopontin might also indirectly have an antifibrotic effect. The role played by osteopontin in fibrotic lesions resulting from the release of TNF-α deserves further study, since it may be involved in the balance of opposite effects eventually leading to local tissue damage ending in fibrosis.

REFERENCES

1. OLDBERG, A., A. FRANZEN & D. HEINEGÅRD. 1986. Cloning and sequence analysis of rat bone sialoprotein (osteopontin) cDNA reveals an Arg-Gly-Asp cell-binding sequence. Proc. Natl. Acad. Sci. USA **83:** 8819–8823.
2. PATARCA, R., R. A. SAAVEDRA & H. CANTOR. 1993. Molecular and cellular basis of genetic resistance to bacterial infection: the role of the early T-lymphocyte activation-1/osteopontin gene. Crit. Rev. Immunol. **13:** 225–246.
3. MIYAZAKI, Y., M. SETOGUCHI, S. YOSHIDA, Y. HIGUCHI, S. AKIZUKI & S. YAMAMOTO. 1989. Nucleotide sequence of cDNA for mouse osteopontin-like protein. Nucleic Acids Res. **17:** 3298.
4. MIYAZAKI, Y., M. SETOGUCHI, S. YOSHIDA, Y. HIGUCHI, S. AKIZUKI & S. YAMAMOTO. 1990. The mouse osteopontin gene expression in the monocytic lineages and complete nucleotide sequence. J. Biol. Chem. **265:** 14432–14438.
5. BUSSE, R. & A. MULSCH. 1990. Induction of nitric oxide synthase by cytokines in vascular smooth muscle cells. FEBS Lett. **275:** 87–90.
6. CRYSTAL, R. G., J. D. FULMER, W. C. ROBERTS, M. L. MOSS, B. R. LINE & H. Y. REYNOLDS. 1976. Idiopathic pulmonary fibrosis: clinical, histological, radiographic, physiologic, scintigraphic, cytologic, and biochemical aspects. Ann. Intern. Med. **85:** 769–788.
7. WESTERGREN-THORSSON, G., J. HERNNAS, B. SARNSTRAND, A. OLDBERG, D. HEINEGÅRD & A. MALMSTROM. 1993. Altered expression of small proteoglycans, collagen, and transforming growth factor-β1 in developing bleomycin-induced pulmonary fibrosis. J. Clin. Invest. **92:** 632–637.
8. ANTONIADES, H. N., M. A. BRAVO, R. E. AVILA, T. GALANOPOLOUS, J. NEVILLE-GOLDEN, M. MAXELL & M. SELMAN. 1990. Platelet-derived growth factor in idiopathic pulmonary fibrosis. J. Clin. Invest. **86:** 1055–1064.
9. PIGUET, P. F., C. RIBAUX, V. KARPUZ, G. E. GRAU & Y. KAPANCI. 1993. Expression and localization of tumor necrosis factor α and its mRNA in idiopathic pulmonary fibrosis. Am. J. Pathol. **143:** 651–655.
10. PIGUET, P. F., M. A. COLLART, G. E. GRAU, Y. KAPANCI & P. VASSALLI. 1989. Tumor necrosis factor/cachectin plays a key role in bleomycin-induced pneumopathy and fibrosis. J. Exp. Med. **170:** 655–663.
11. PIGUET, P. F., M. A. COLLART, G. E. GRAU, A. P. SAPPINO & P. VASSALLI. 1990. Requirement of tumor necrosis factor for development of silica-induced pulmonary fibrosis. Nature **344:** 245–247.
12. SEMON, D., E. KAWASHIMA, C. V. JONGENEEL, A. N. SHAKHOV & S. A. NEDOSPASOV. 1987. Nucleotide sequence of the murine TNF locus, including the TNF-α (tumor necrosis factor) and TNF-β (lymphotoxin) genes. Nucleic Acids Res. **15:** 9083–9084.
13. GLASSER, S. W., T. R. KORFHAGEN, C. M. PERME, T. J. PILOT-MATIAS, S. E. KISTER & J. A. WHITSETT. 1988. Two SP-C genes encoding human pulmonary surfactant proteolipid. J. Biol. Chem. **263:** 10326–10331.
14. SETOGUCHI, M., S. YOSHIDA, Y. HIGUCHI, S. AKIZUKI & S. YAMAMOTO. 1988. Molecular analysis of expression of parental cell properties in hybrids between monocytes and a myeloma cell line. Somat. Cell Mol. Genet. **14:** 427–438.
15. CHOMCZYNSKI, P. & N. SACCHI. 1987. Single-step method of RNA isolation by acid guanidinium thiocyanate-phenol-chloroform extraction. Anal. Biochem. **162:** 156–159.
16. COLLART, M. A., D. BELIN, J. D. VASSALLI, S. DE KOSSODO & P. VASSALLI. 1986. Gamma interferon enhances macrophage transcription of the tumor necrosis factor/cachectin, interleukin-1, and urokinase genes which are controlled by short lived repressor. J. Exp. Med. **164:** 2113–2118.

17. SAPPINO, A. P., J. HUARTE, D. BELIN & J. D. VASSALLI. 1989. Plasminogen activators in tissue remodeling and invasion: messenger RNA localization in mouse ovaries and implanting embryos. J. Cell Biol. **109:** 2471–2479.
18. GLASSER, S. W., T. R. KORFHAGEN, S. E. WERT, M. D. BRUNO, K. M. MCWILLIAMS, D. K. VORBROKER & J. A. WHITSETT. 1991. Genetic element from human surfactant protein SP-C gene confers bronchiolar-alveolar cell specificity in transgenic mice. Am. J. Physiol. **261:** L349–L356.
19. PATARCA, R., G. J. FREEMAN, R. P. SINGH, F.-Y. WEI, T. DURFEE, F. BLATTNER, D. C. REGNIER, C. A. KOZAK, B. A. MOCK, H. C. MORSE III, T. R. JERRELLS & H. CANTOR. 1989. Structural and functional studies of the early T lymphocyte activation 1 (Eta-1) gene definition of a novel T cell-dependent response associated with genetic resistance to bacterial infection. J. Exp. Med. **170:** 145–161.
20. AUSTYN, J. M. & S. GORDON. 1981. F4/80, a monoclonal antibody directed specifically against the mouse macrophage. Eur. J. Immunol. **11:** 805.

Osteopontin Expression in Clonal Dental Pulp Cells

TOSHIHIKO NAGATA, MIKA YOKOTA, SEIJI NISHIKAWA, HIROSHI ISHIDA, AND YOICHI WAKANO

Department of Periodontology and Endodontology
Tokushima University School of Dentistry
3-18-15 Kuramoto
Tokushima 770, Japan

Dental pulp, a primitive loose connective tissue surrounded by dentin, has an important lifelong role in the maintenance of dentin. Although primary dentin is formed by odontoblasts, products of supplementary mineralization such as reparative dentin and pulp stone are thought to be associated not with odontoblasts but with other dental pulp cells such as pulp fibroblasts and mesenchymal cells. Phosphoprotein is generally required during the formation of calcified tissue. Odontoblasts are known to synthesize phosphophoryn, a major noncollagenous protein in dentin, although it is unclear whether or not dental pulp cells produce phosphoproteins. The clonal cell lines, RPC-C2A and RDP4-1, derived from rat dental pulp from which odontoblasts have been removed, both exhibit high alkaline phosphatase activity.[1,2] The present study revealed that osteopontin (OPN) is a major phosphoprotein synthesized by these dental pulp cells, and that OPN expression is regulated by 1α,25-dihydroxyvitamin D_3 (Vit-D_3).

RPC-C2A and RDP4-1 cells were maintained in Eagle's MEM and α-MEM as reported previously.[1,2] The newly synthesized proteins were studied by radiolabeling with [$^{32}PO_4$] or [^{35}S]methionine. After labeling, the media were collected, desalted, freeze-dried, and then analyzed by sodium dodecyl sulfate-polyacrylamide gel electrophoresis (SDS-PAGE). The proteins were then detected by autoradiography. Some samples were followed by thrombin digestion or immunoprecipitation, as described previously,[3] before SDS-PAGE. For Northern hybridization analysis, total RNA was obtained by the guanidinium thiocyanate-CsCl method, separated by 1% agarose gel electrophoresis, transferred to a nylon membrane, and hybridized with a [^{32}P]-labeled OPN cDNA probe.

Phosphorylated proteins from the media of RPC-C2A and RDP4-1 cells are shown in FIGURE 1. Major bands were observed between 50 and 60 kDa on 10% gels (lanes 1 and 4). These bands appeared at 56 and 60 kDa on a 10% gel (lane 1) or at 44 and 55 kDa on a 15% gel (lane 2) in RPC-C2A cells, and at 56 kDa on a 10% gel (lane 4) or 44 kDa on a 15% gel (lane 5) in RDP4-1 cells. It is known that thrombin susceptibility is a characteristic of OPN. The digestion resulted in the appearance of 26- and 28-kDa bands on 15% gels

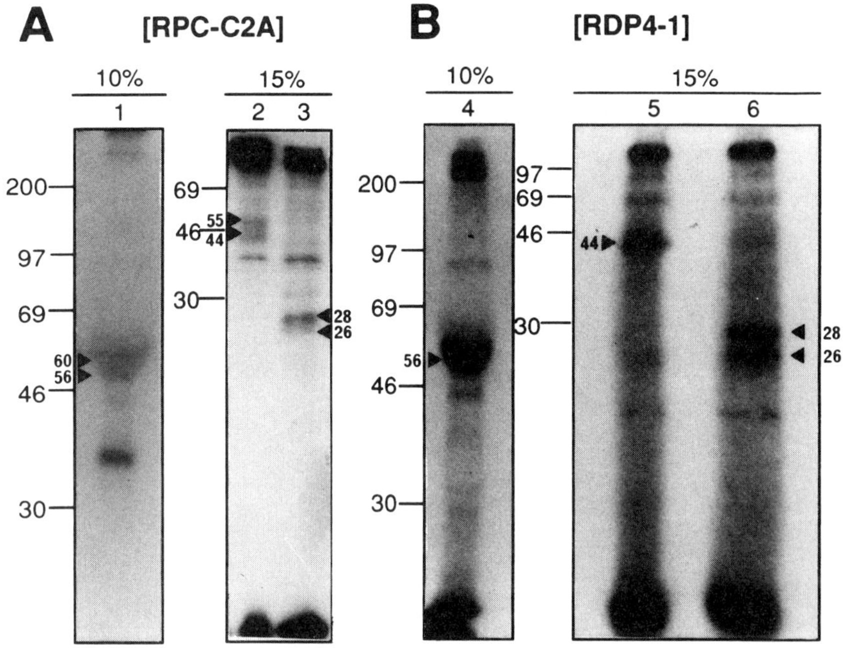

FIGURE 1. Phosphorylated proteins synthesized by the clonal dental pulp cell lines RPC-C2A (**A**) and RDP4-1 (**B**). Confluent cells were labeled with 3.7 MBq/ml [$^{32}PO_4$] for 24 h. Aliquots of culture media were analyzed by SDS-PAGE on 10% (*lanes 1, 4*) and 15% gels (*lanes 2, 3, 5, 6*); then the gel was processed for autoradiography. Thrombin-digested samples are shown in *lanes 3 and 6*. The *arrowheads* indicate the positions of the major phosphorylated proteins. The molecular weight marker sizes are shown in kDa on the *left*.

(lanes 3 and 6), demonstrating 56- and 60-kDa OPN proteins. This anomalous behavior and this thrombin digestion pattern seen on the gels were consistent with those of OPN synthesized by bone cells.[3] Thus, OPN was a major phosphoprotein synthesized by the clonal dental pulp cells, but no phosphophoryn-like protein was observed. Immunoprecipitation analysis also confirmed that the 56- and 60-kDa proteins were OPN when radiolabeled by [$^{32}PO_4$] and [^{35}S]methionine (FIG. 2A,B). Since Vit-D_3 is a potent regulatory factor of OPN production, its effect on biosynthesis and mRNA expression was examined, as shown in FIGURE 2. Dose-dependent stimulation by Vit-D_3 was apparent in both phosphorylated-OPN synthesis and the expression of its mRNA in RPC-C2A cells. When 10^{-7} M Vit-D_3 was added, densitometric analysis revealed a 4.4-fold increase in [$^{32}PO_4$]-labeled OPN and a 1.3-fold increase in [^{35}S]methionine-labeled OPN, suggesting that Vit-D_3 stimulated both phosphorylation and protein synthesis. The expression of OPN mRNA was increased 2.2-fold by 10^{-7} M Vit-D_3 (FIG. 2C). These findings indicate that Vit-D_3 may play a key role in determining the action of osteopontin in dental pulp.

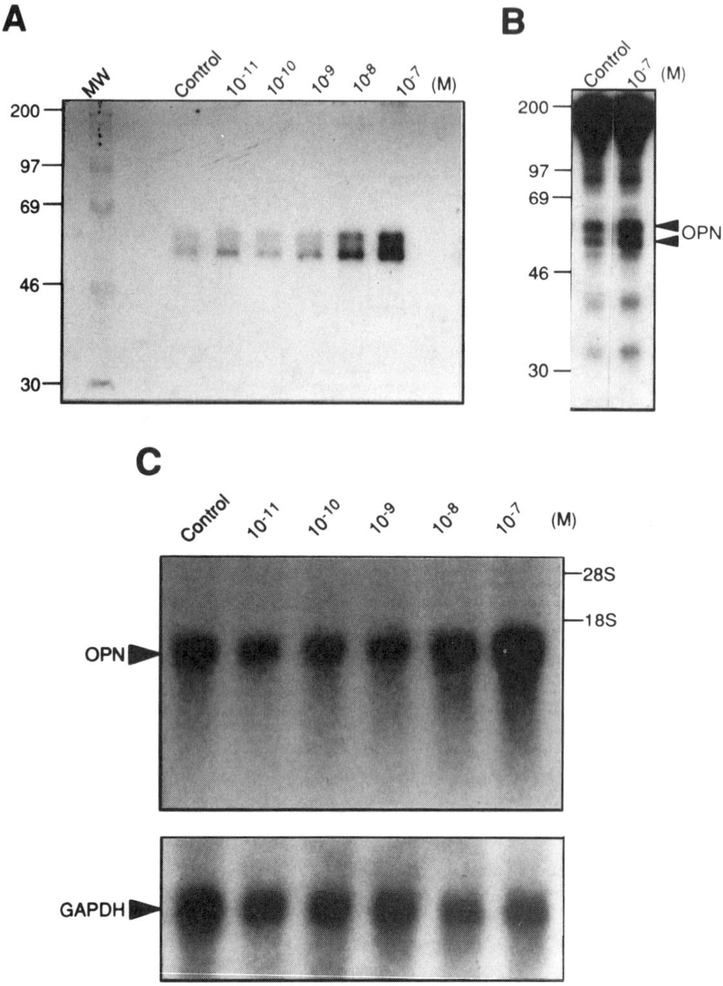

FIGURE 2. Immunoprecipitation and Northern hybridization analysis of OPN and the effect of Vit-D_3 on OPN synthesis and the expression of its mRNA in RPC-C2A cells. The cells were radiolabeled for 6 h with 3.7 MBq/ml [$^{32}PO_4$] **(A)** or [^{35}S]methionine **(B)**, and immunoprecipitated samples were anlayzed by SDS-PAGE. Total RNA extracted from the cell layers was separated, transferred, and hybridized with a [^{32}P]-labeled OPN or GAPDH cDNA probe **(C)**. Note the dose-dependent stimulatory effects of Vit-D_3 both on protein and mRNA levels.

REFERENCES

1. Kasugai, S., M. Adachi & H. Ogura. 1988. Establishment and characterization of a clonal cell line (RPC-C2A) from dental pulp of the rat incisor. Arch. Oral Biol. **33:** 887–891.

2. KAWASE, T., M. ORISAKA & A. SUZUKI. 1988. A clonal prostaglandin-responsive cell line (RDP4-1) derived from rat dental pulp. Bone Miner. **11:** 163–175.
3. NAGATA, T., C. G. BELLOWS, S. KASUGAI, W. T. BUTLER & J. SODEK. 1991. Biosynthesis of bone proteins [SPP-1 (secreted phosphoprotein-1, osteopontin, BSP (bone sialoprotein) and SPARC (osteonectin)] in association with mineralized-tissue formation by fetal-rat calvarial cells in culture. Biochem. J. **274:** 513–520.

Regulated Expression of Osteopontin in Human Trophoblasts

AKINYINKA OMIGBODUN,[a] ERIC DAITER,[a]
DENISE WALINSKY,[a] LARRY FISHER,[b]
MARIAN YOUNG,[b] JOHN HOYER,[a]
AND CHRISTOS COUTIFARIS[a,c]

[a]*University of Pennsylvania Medical Center*
Philadelphia, Pennsylvania 19104
and
[b]*Bone Research Branch*
National Institute of Dental Research
Bethesda, Maryland

INTRODUCTION

Osteopontin (OPN), a secreted phosphoprotein which binds cells via the $\alpha_V\beta_3$ integrin, mediates cell adhesion and signaling in many mammalian cells.[1,2] Although the expression of OPN in various human tissues has been demonstrated,[3,4] its regulation has not been elucidated. In addition a model for the study of OPN expression in the human has not been fully established. We recently showed that human trophoblasts express OPN and that its expression is under regulatory control by cyclic-AMP.[5] In the present short communication, we report the regulation of OPN expression in human trophoblasts by phorbol esters.

RESULTS AND DISCUSSION

Human chorionic villi obtained from the first trimester and third trimester of pregnancy were prepared for immunohistochemistry. OPN was localized to the cytotrophoblast layer (FIG. 1). At no time of gestation was OPN protein immunolocalized within the syncytiotrophoblasts.

In order to study the regulation of OPN expression, mononuclear cytotrophoblasts were isolated from chorionic villi as previously described.[6] Under monolayer culture conditions these cells aggregate and subsequently fuse to form multinucleated syncytia.[7] This *in vitro* morphogenetic process recapitulates the *in vivo* differentiation of these cells. Cells were harvested at timed intervals and total RNA was isolated using standard techniques. Some cells were exposed to 100 ng/ml of the phorbol ester 12-*O*-tetradecanoyl-phorbol-13-acetate (TPA).

[c] Address all correspondence to: Christos Coutifaris, M.D., Ph.D., Divisions of Human Reproduction and Reproductive Biology, Department of Obstetrics and Gynecology, 106 Dulles Pavilion, Hospital of the University of Pennsylvania, 3400 Spruce Street, Philadelphia, PA 19104.

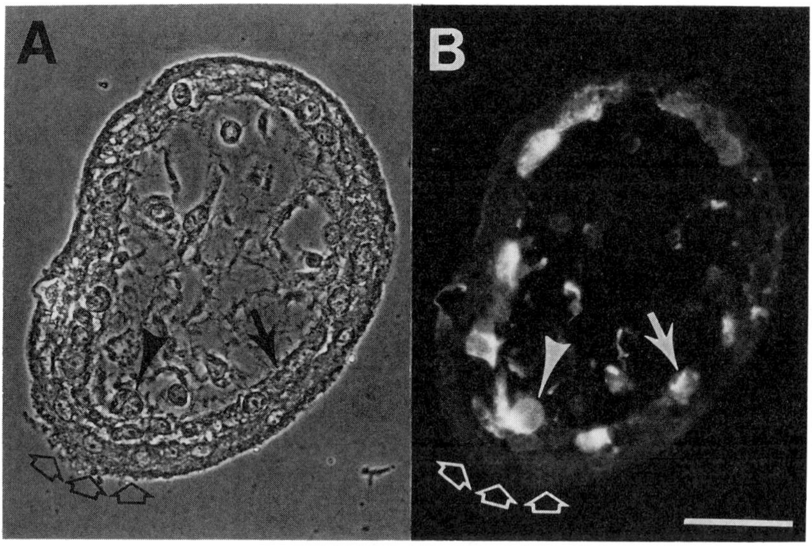

FIGURE 1. Phase contrast and OPN indirect immunofluorescence of a first trimester human chorionic villus. Note the positive staining of mononuclear cytotrophoblasts (*arrow*) and the negative staining of the overlying syncytial trophoblast layer (*open arrows*). Some cells in the core of the villus, presumably macrophages (*arrowhead*), also stain positively for OPN. *Bar* = 50 μm.

Results were confirmed with the use of 10 nM 4β-phorbol 12,13-didecanoate and using the inactive α-analog as control. The expression of OPN at six hours was significantly enhanced by TPA (FIG. 2). In contrast, at later times in culture when the predominant cellular forms were multinucleated syncytia, there was no upregulation of OPN expression after exposure to TPA. Thus, the observed effect appears to be cell differentiation dependent: mononuclear cytotrophoblasts respond to phorbol ester treatment while syncytiotrophoblasts do not.

The expression of OPN in various nonhuman cells has been shown to be regulated by hormones, growth factors and tumor promoters.[2] In rodent cell lines, increases in intracellular cAMP or activation of protein kinase C (PKC) by the phorbol ester TPA inhibit or enhance OPN expression, respectively.[8,9] It is well recognized that cAMP is the second messenger in the protein kinase A (PKA) pathway and its activity in downregulating OPN expression in trophoblasts[5] suggests that this pathway is involved in the control of OPN expression in these cells. On the other hand, the phorbol ester TPA is known to activate the protein kinase C (PKC) pathway, and this may be the mechanism by which it upregulates OPN expression in trophoblasts. The data to date suggest that osteopontin expression in human trophoblasts is under dual regulatory control through both PKA-mediated inhibitory and PKC-mediated stimulatory pathways. In addition, we propose that primary cultures of human trophoblasts provide a good experimental model for the study of the regulation of OPN expression.

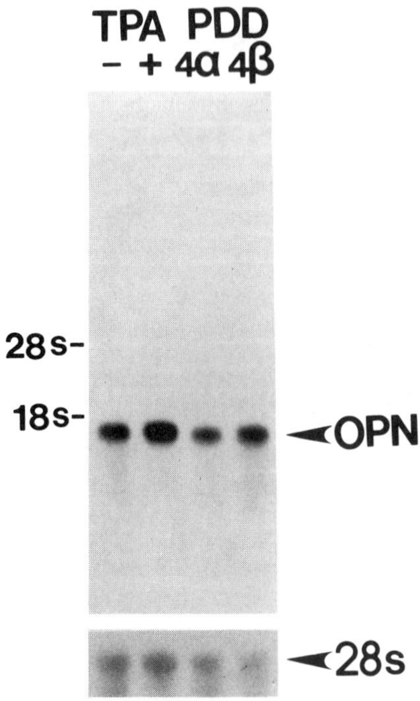

FIGURE 2. Effect of phorbol esters on OPN mRNA expression in human trophoblasts. Note that both TPA and the 4β active PDD isomer upregulated OPN mRNA. Loading is normalized by probing the same membrane with a 28s ribosomal RNA probe.

REFERENCES

1. MIYAUCHI, A., J. ALVAREZ, E. M. GREENFIELD, A. TETI, M. GRANO, S. COLUCCI, A. ZAMBONIN-ZALLONE, F. P. ROSS, S. L. TEITELBAUM, D. CHERESH & K. A. HRUSKA. 1991. Recognition of osteopontin and related peptides by an $\alpha_v\beta_3$ integrin stimulates immediate cell signals in osteoclasts. J. Biol. Chem. **266:** 20369–20374.
2. DENHARDT, D. T. & X. GUO. 1993. Osteopontin: a protein with diverse functions. FASEB J. **7:** 1475–1482.
3. BROWN, L. F., B. BERSE, L. VAN DE WATER, A. PAPADOPOULOS-SERGIOU, C. A. PERRUZZI, E. J. MANSEAU, H. F. DVORAK & D. R. SENGER. 1992. Expression and distribution of osteopontin in human tissues: widespread association with luminal epithelial surfaces. Mol. Biol. Cell **3:** 1169–1180.
4. YOUNG, M. F., J. M. KERR, J. D. TERMINE, U. M. WEWER, M. G. WANG, O. W. MCBRIDE & L. W. FISHER. 1990. cDNA cloning, mRNA distribution and heterogeneity, chromosomal location and RFLP analysis of human osteopontin (OPN). Genomics **7:** 491–502.
5. DAITER, E., A. OMIGBODUN, S. WANG, D. HAWKINS, J. F. STRAUSS III, L. FISHER, M. YOUNG, J. HOYER & C. COUTIFARIS. Expression of osteopontin in human trophoblast cells: regulation by endogenous cyclic AMP and cell differentiation. Submitted.
6. KLIMAN, H. J., J. E. NESTLER, E. SERMASI, J. M. SANGER & J. F. STRAUSS III. 1986. Purification, characterization, and *in vitro* differentiation of cytotrophoblasts from human term placentae. Endocrinology **118:** 1567–1582.
7. COUTIFARIS, C., L.-C. KAO, H. SEDHEV, U. CHIN, G. BABALOLA, O. BLASCHUK & J. F. STRAUSS III. 1991. E-cadherin expression during the differentiation of human trophoblasts. Development **113:** 767–777.

8. NODA, M. & G. A. RODAN. 1989. Transcriptional regulation of osteopontin production in rat osteoblast-like cells by parathyroid hormone. J. Cell Biol. **108:** 713–718.
9. CRAIG, A. M., J. H. SMITH & D. T. DENHARDT. 1989. Osteopontin, a transformation-associated cell adhesion phosphoprotein, is induced by 12-*O*-tetradecanoylphorbol 13-acetate in mouse epidermis. J. Biol. Chem. **264:** 9682–9689.

Osteopontin Gene Expression in Avian Tibial Dyschondroplasia[a]

V. KNOPOV,[b] R. M. LEACH,[c] T. BARAK-SHALOM,[b]
S. HURWITZ,[b] AND M. PINES[b,d]

[b]*Institute of Animal Science*
The Volcani Center
Bet Dagan, Israel
and
[c]*Department of Poultry Science*
Pennsylvania State University
Pennsylvania

During the process of bone growth, proliferating chondrocytes within the growth plate differentiate into hypertrophic chondrocytes which, in turn, initiate matrix mineralization. Chondrocytes and the cartilaginous matrix are consequently replaced by osteoblasts and bone matrix, respectively. The rate of division of the stem cells, proliferation rate of the chondrocytes, the size of the proliferation zone and the degree of cellular hypertrophy are controlled precisely, and any deviation from that will cause abnormal bone growth. Tibial dyschondroplasia (TD) is one of the most prevalent skeletal abnormalities in birds, which results in deformed bones and lameness. TD is a disease of rapid growth and occurs especially in broilers[1] and turkeys[2] at their maximal genetic potential. TD is characterized by the appearance of a plug of nonvascularized, unmineralized, white opaque cartilage dominating the proximal metaphysis of the tibiatarsus and occasionally the tarsometatarsus[3] (FIG. 1). The various morphological and biochemical manifestations of the TD lesion, such as changes in carbonic anhydrase activity[4] and collagen type II and X production[5,6] suggested that TD-chondrocytes fail to undergo the complete differentiation which normally leads to cartilage vascularization and mineralization.

To investigate the relationship between osteopontin (OPN) synthesis and chondrocyte differentiation, we compared OPN gene expression in the growth plate of normal chickens with that of chickens afflicted with TD. TD was induced in broiler chickens (Peterson X Arbor Acre) by the following diets: (a) a copper-deficient diet;[7] (b) a basal diet supplemented with 16.7 mg/kg $CuSO_4$; and (c) a basal diet supplemented with the fungicide tetramethylthiuram disulfide.[8] At four weeks of age, the broilers were sacrificed, and the epiphyseal growth plates were taken for evaluation of OPN gene expression by *in situ* hybridization using avian OPN specific probe. Alkaline phosphatase (AP) activity, the appearance of which marks the onset of hypertrophy and calcification, was determined as a marker for chondrocyte differentiation.

[a] This study was supported by a grant from the United States-Israel Binational Agricultural Research and Development Fund (US-2117-92).

[d] Corresponding author and address for correspondence: Mark Pines, Ph.D., Beth Israel Hospital, Department of Medicine, Division of Bone and Mineral Metabolism, 330 Brookline Ave., Boston, MA 02115.

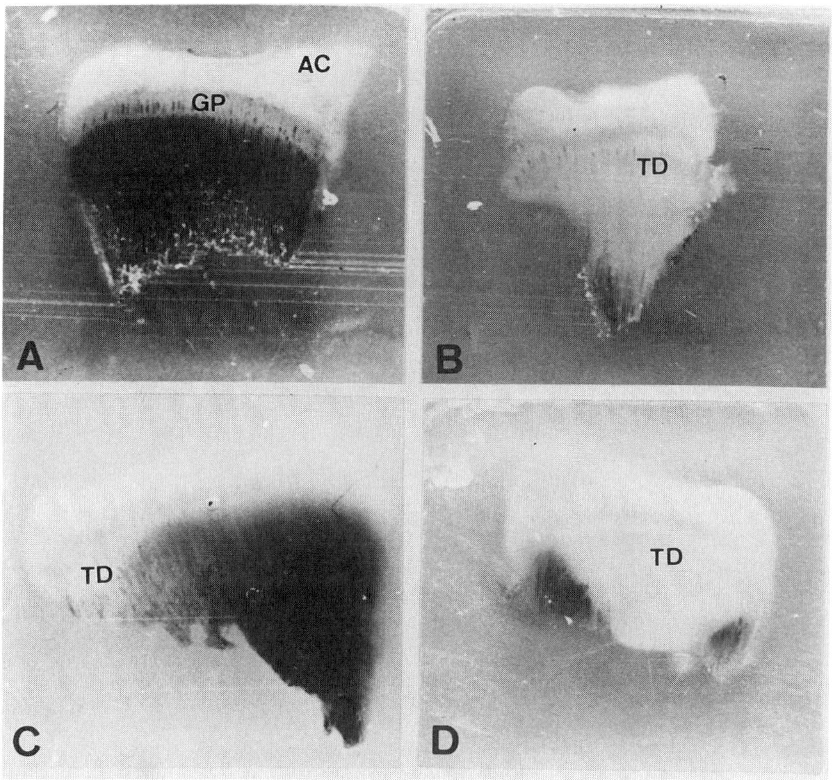

FIGURE 1. Low-magnification morphology of normal and TD tibiae. **(A)** Normal chicken; **(B)** chicken fed a copper-deficient diet; **(C)** chicken fed the basal diet supplemented with 16.7 mg/kg $CuSO_4$ $5H_2O$; and **(D)** chicken fed basal diet supplemented with 16.7 mg/kg $CuSO_4$ $5H_2O$ and 25 mg/kg of the fungicide tetramethylthiuram disulfide. AC, articular cartilage; GP, growth plate; TD, tibial dyschondroplasia lesion. Note the appearance of a plug of nonvascularized, unmineralized, white opaque cartilage (B–D) replacing the normal GP and the zone of mineralization (A).

In the normal growth plate, OPN gene was expressed by (a) cells of the subperichondrial zone surrounding the articular cartilage; (b) a narrow layer of hypertrophic chondrocytes at the hypertrophic zone, and (c) lower hypertrophic chondrocytes at the zone of matrix calcification and endochondral bone formation (FIG. 2B). The latter two layers were separated by OPN-negative chondrocytes (FIG. 2D). Osteopontin gene was not expressed throughout the zone of articular cartilage, in the nonhypertrophic or upper hypertrophic portions of the growth-plate cartilage. Only at sites of calcification of the lower hypertrophic zone was the expression of the OPN gene associated with AP activity. In all TD lesions, regardless of the induction procedure, the layer of chondrocytes of the lower hypertrophic zone expressing the OPN gene, and the layer of OPN-negative cells

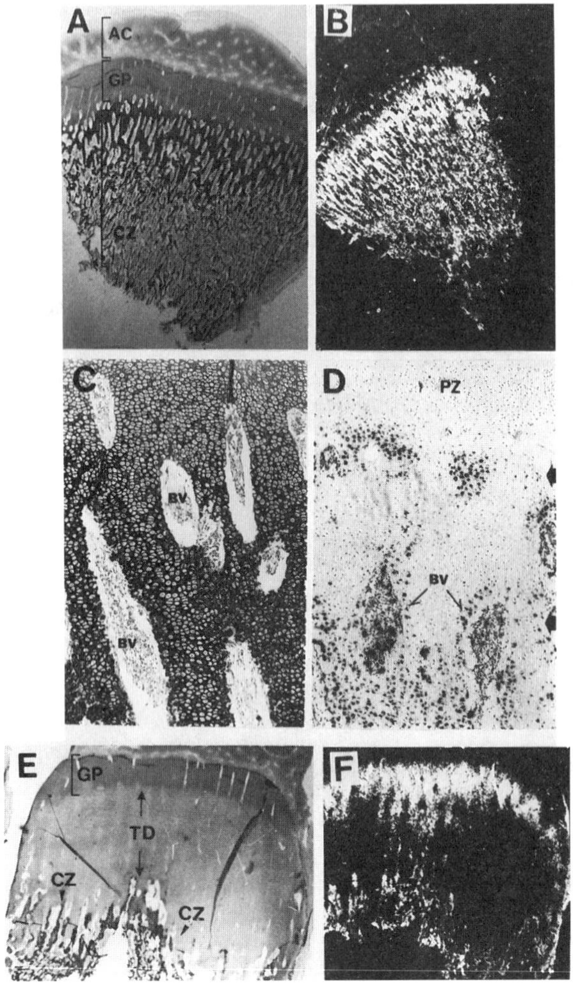

FIGURE 2. Osteopontin gene expression in normal and TD tibia. **(A,C,E)** Hematoxylin and eoxin staining; **(B,D,F)** dark field *in situ* hybridization with avian OPN probe. In the normal tibia (magnification × 2; A,B) OPN gene expression was mainly observed at the lower hypertrophic chondrocytes in the zone of matrix calcification and at the zone of endochondral bone formation. In addition, OPN gene was expressed also at the subperichondrial zone surrounding the articular cartilage. At higher magnification (×25; C,D), OPN gene was expressed by two layers of hypertrophic chondrocytes (indicated by the *arrows*) separated by nonlabeled chondrocytes. Within the upper layer, the chondrocytes expressing the OPN gene were arranged in clusters. In the tibia of chicken afflicted with TD after having been fed the fungicide tetramethylthiuram disulfide (E,F) OPN gene was expressed in two layers: one by chondrocytes at the lower hypertrophic zone and the second by chondrocytes, which were associated with mineralization. These two layers were widely separated. AC, articular cartilage; GP, growth plate; CZ, calcification zone; PZ, proliferative zone; BV, blood vessels.

separating the two areas of OPN-expressing cells, were grossly enlarged (FIG. 2F). This resulted in a wide discontinuity between the chondrocytes of the lower hypertrophic zone expressing the OPN gene and the cells expressing the OPN gene which are associated with mineralization. In TD, no AP activity was detected within the growth plate cartilage, but normal OPN gene expression was observed at the subperichondrium zone and at the zone of endochondral bone formation.

The results of this study suggest that in the epiphyseal growth plate, OPN gene expression is not restricted to sites of bone calcification. In addition, chickens afflicted with tibial dyschondroplasia may serve as a useful model for understanding the involvement of various genes in growth-plate chondrocyte differentiation, which is prerequisite for calcification, endochondral bone formation, and longitudinal bone growth.

REFERENCES

1. LEACH, M. R. & M. S. LILBURN. 1992. Poult. Sci. Rev. **4:** 57–65.
2. WYERS, M., Y. CHEREL & G. PLASSIART. 1991. Avian Dis. **35:** 408–414.
3. HARGEST, T. E., R. M. LEACH & C. V. GAY. 1984. Am. J. Pathol. **119:** 175–190.
4. GAY, C. V., R. E. ANDERSON & R. M. LEACH. 1985. Avian Dis. **29:** 812–821.
5. BASHEY, R., R. M. LEACH, C. V. GAY & S. A. JIMENEZ. 1989. Lab. Invest. **60:** 106–112.
6. CHEN, Q., E. GIBNEY, R. M. LEACH & T. F. LINSENMAYER. 1993. Dev. Dyn. **196:** 54–61.
7. ROSSELOT, G., C. SOKOL & R. M. LEACH. 1994. Poult. Sci. **73:** 452–456.
8. VARGAS, M., M. J. LAMAS & V. ALVARENGA. 1983. Poult. Sci. **62:** 1195–1200.

Osteopontin mRNA Expression in a Chemically-Induced Model of Atherogenesis

ALAN R. PARRISH AND KENNETH S. RAMOS[a]

Faculty of Toxicology
and
Department of Physiology and Pharmacology
College of Veterinary Medicine
Texas A&M University
College Station, Texas 77843

Repeated cycles of vascular injury by allylamine (AAM), an atherogenic amine, modulate rat aortic smooth muscle cells (SMCs) from a quiescent to a proliferative/synthetic phenotype.[1] This switch in phenotypic modulation is maintained during serial passage and involves specific changes in the secretion and deposition of extracellular matrix proteins.[2] In view of previous observations suggesting that osteopontin (OPN) might be overexpressed in allylamine cells (TABLE 1), the present studies were conducted to examine gene expression of osteopontin, a thrombin-sensitive 46-kDa acidic phosphoprotein which regulates cell-matrix interactions through the $\alpha_v\beta_3$ integrin.[5] Giachelli *et al.* have demonstrated that OPN levels are elevated in SMCs following injury.[6] Furthermore, OPN accumulates in atherosclerotic plaques in humans,[7] and is thought to play an important role in the migration of smooth muscle cells to the intima during the early stages of atherosclerosis.[8] To date, little attention has been given to the role of OPN in the regulation of smooth muscle cell proliferation.

Expression of OPN was examined at the protein and mRNA levels. 2-D gel electrophoresis of proteins secreted by AAM and control cells confirmed the presence of a small acidic protein (26 kDa, pI 5.2) exclusively in AAM cells, as well as overexpression of a 46-kDa protein.[9] Northern analysis of the gene expression of OPN demonstrated that serum-inducible mRNA levels were elevated in AAM cells as compared to controls (FIG. 1). To test the hypothesis that the small protein secreted by AAM cells was a cleavage fragment of OPN, exogenous thrombin, a protease inhibitor known to cleave OPN, was added to the conditioned media of [^{35}S]methionine-labeled AAM cells. This manipulation was associated with a decrease in the intensity of a 45-kDa protein and enhancement in the intensity of a 30-kDa band.[9]

In a series of experiments designed to further characterize the biochemical properties of proteins secreted into the media and sequestered in the extracellular matrix by AAM cells, both cell types were plated on several different matrices. Plating of AAM and control cells on culture dishes coated with type I collagen nullified the proliferative advantage of AAM cells by decreasing [^{3}H]thymidine incorporation in AAM cells, without affecting control cultures.[9] Plating of cultures on dishes coated with Pronectin, a synthetic fibronectin substrate, reduced DNA

[a] Corresponding author.

TABLE 1. Characteristics of Aortic Smooth Muscle Cells Harvested from Rats following Repeated Cycles of Chemical Injury by Allylamine[a]

Observation	Reference
Enhanced secretion of a 46-kDa protein, novel secretion of a 30-kDa protein.	Ramos *et al.*[2]
Enhanced PKC-mediated phosphorylation of a 45-kDa and a 30-kDa protein.	Ramos and Ou[3]
Enhanced levels of c-Ha-ras mRNA in proliferating cells.	Bowes and Ramos[4]
Regulation of the proliferative capacity of AAM/control cells by factors secreted by AAM cells and sequestered in the extracellular matrix.	Ramos *et al.*[2]

[a] These observations led us to investigate the expression of osteopontin at the protein and mRNA levels in allylamine cells. A phosphoprotein corresponding to the reported size of osteopontin is secreted by allylamine cells, as well as a fragment corresponding to a thrombin-produced cleavage product. Of particular significance is the observation that a protein secreted and sequestered in the extracellular matrix by allylamine cells affords control cells increased proliferative capacity, which suggests cell-matrix interactions are central to allylamine-induced smooth muscle cell proliferation.

synthesis in both cell types. These data suggest that the enhanced proliferative capacity afforded to control cells when plated on a matrix synthesized by AAM cells is associated with a novel protein produced by AAM cultures.

These findings implicate OPN overexpression in the phenotypic modulation of SMCs by AAM in a chemically induced model of atherogenesis. Of particular interest is the fact that enhanced levels of OPN are sustained after removal of the

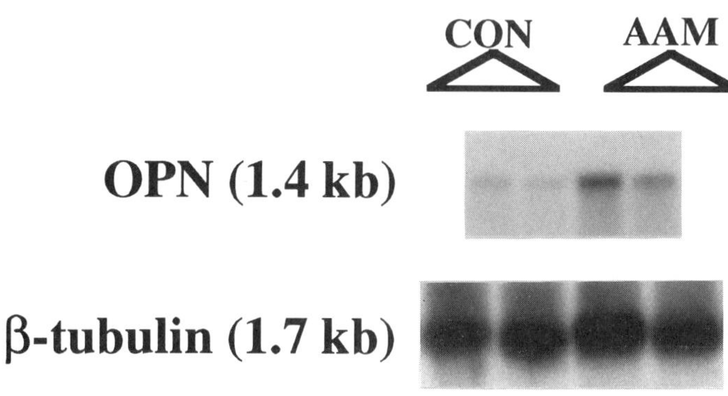

FIGURE 1. Serum-inducible mRNA levels of osteopontin in synchronized cultures of allylamine and control cells. Cells were plated at 75 cells/mm^2 on 100-mm plastic culture dishes and growth-arrested in M199 (0.1% FBS) for 72 hr following a 24-hr attachment period. Total RNA was extracted following mitogenic (10% FBS) stimulation, and 7.5 μg of RNA was separated by gel electrophoresis and transferred to a nylon membrane. [^{32}P]dCTP-labeled osteopontin (1.4 kB) or β-tubulin (1.7 kB) cDNA probes were hybridized to the membrane and visualized by autoradiography.

injurious agent and during serial passage. These data emphasize the fidelity of this *in vivo/in vitro* model and suggest that allylamine cells may be useful in defining the role of OPN in SMC proliferation.

REFERENCES

1. Cox, L. R. & K. Ramos. 1990. J. Exp. Pathol. **71:** 11–18.
2. Ramos, K. S., T. J. Weber & G. Liau. 1993. Biochem. J. **289:** 57–63.
3. Ramos, K. S. & X. Ou. 1994. Toxicol. Lett. **73:** 123–133.
4. Bowes, R. C. III & K. S. Ramos. 1993. Toxicol. Lett. **66:** 263–272.
5. Denhardt, D. T. & X. Guo. 1993. FASEB J. **7:** 1475–1482.
6. Giachelli, C., N. Bae, D. Lombardi, M. Majesky & S. Schwartz. 1991. Biochem. Biophys. Res. Commun. **177:** 867–873.
7. Giachelli, C. M., N. Bae, M. Almeida, D. T. Denhardt, C. E. Alpers & S. M. Schwartz. 1993. J. Clin. Invest. **92:** 1686–1696.
8. Yue, T-L., P. J. McKenna, E. H. Ohlstein, M. C. Farach-Carson, W. T. Butler, K. Johanson, P. McDevitt, G. Z. Feuerstein & J. M. Stadel. 1994. Exp. Cell Res. **214:** 459–464.
9. Parrish, A. R., T. J. Weber, D. M. Mitchell & K. S. Ramos. 1994. Submitted.

Identification of the *In Vivo* Phosphorylated Sites of Secreted Osteopontin from Cultured Chicken Osteoblasts

ERDJAN SALIH, SAMY ASHKAR,
LOUIS C. GERSTENFELD, AND MELVIN J. GLIMCHER

Laboratory for the Study of Skeletal Disorders and Rehabilitation
Department of Orthopaedic Surgery
Harvard Medical School and Children's Hospital
Boston, Massachusetts 02115

INTRODUCTION

It has been well established that the processes of phosphorylation and dephosphorylation of proteins, catalyzed by protein kinases and phosphatases, respectively, are important in the initiation, regulation and termination of a wide range of intracellular biochemical processes with significant functional consequences. The secretions of phosphoproteins into the extracellular matrix of which they become a part[1,2] have been postulated to play a major role in hydroxyapatite crystal formation.[3,4] One of the most studied and characterized of these secreted phosphoproteins is osteopontin, which is found to be secreted in a heavily phosphorylated form from cultured chicken osteoblasts.[5,6]

In order to understand the structure-function of the extracellular phosphoproteins, we have isolated and purified *in vivo* ^{32}P-labeled secreted osteopontin from cultured chicken osteoblasts and identified the phosphorylated peptide regions as well as specific residues on which phosphate resides.

RESULTS AND DISCUSSION

Quantitative secreted ^{32}P-labeled osteopontin (OPN) from cultured chicken osteoblasts was isolated and purified by reverse-phase HPLC on C-4 column (FIG. 1). The N-terminal sequencing showed a single sequence of WPVSKRQHAISA, completely consistent with the deduced cDNA sequence of chicken OPN.[7] Thrombin digest of the purified OPN followed by reverse-phase HPLC led to isolation of three ^{32}P-labeled fragments and on N-terminal sequencing, each thrombin fragment gave a single sequence with sequences: fragment 1: WPVSKRQHAIS; fragment 2: SHHTHRYHGDHVD; and fragment 3: ASKLRKAARKL, with apparent M_rs ~5 kDa, 30 kDa, and 35 kDa, respectively, on SDS-PAGE. Thus, thrombin specifically cleaved chicken OPN at two sites, one between Arg(22)-Ser(23) and the other between Lys(138)-Ala(139), leading to generation of a small N-terminal end fragment starting with Trp-1 and extending to Arg-22; a middle fragment starting with Ser-23 and extending to Lys-138; and a C-terminal end fragment starting with

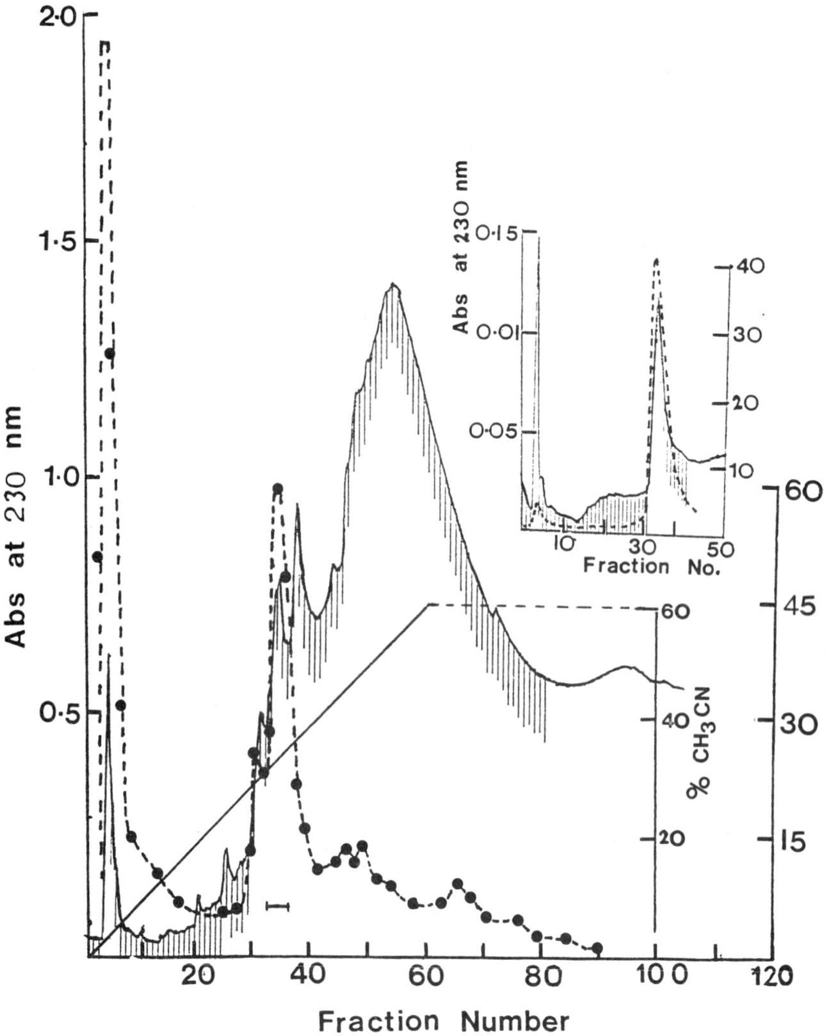

FIGURE 1. Reverse-phase HPLC on a C-4 column of secreted ^{32}P-labeled osteopontin from cultured chicken osteoblasts. Chicken calvarial osteoblasts from 17-day chicken embryos were isolated and cultured as described previously.[8] ^{32}P-labeling was carried out by incubation (18 hr) of the mineralized cultures with DME phosphate-free media containing 5 mg/ml ascorbic acid + mCi $^{32}P_i$. The cells were lysed and washed and the mineral layer was extracted with 0.1 M HCl (pH ~2.0) for 3 hr. The extract was lyophilized, resuspended in 100% H_2O + 0.1% TFA and separated by HPLC on a Vydac C-4 column (15 × 0.38 cm) by a linear gradient from 100% H_2O + 0.1% TFA to 60% CH_3CN + 0.55% TFA in 60 min. The ^{32}P-radioactive peak (|—|) was rechromatographed (*inset*). *Solid line* Abs. = 230 nm, and *dashed line* ^{32}P-count.

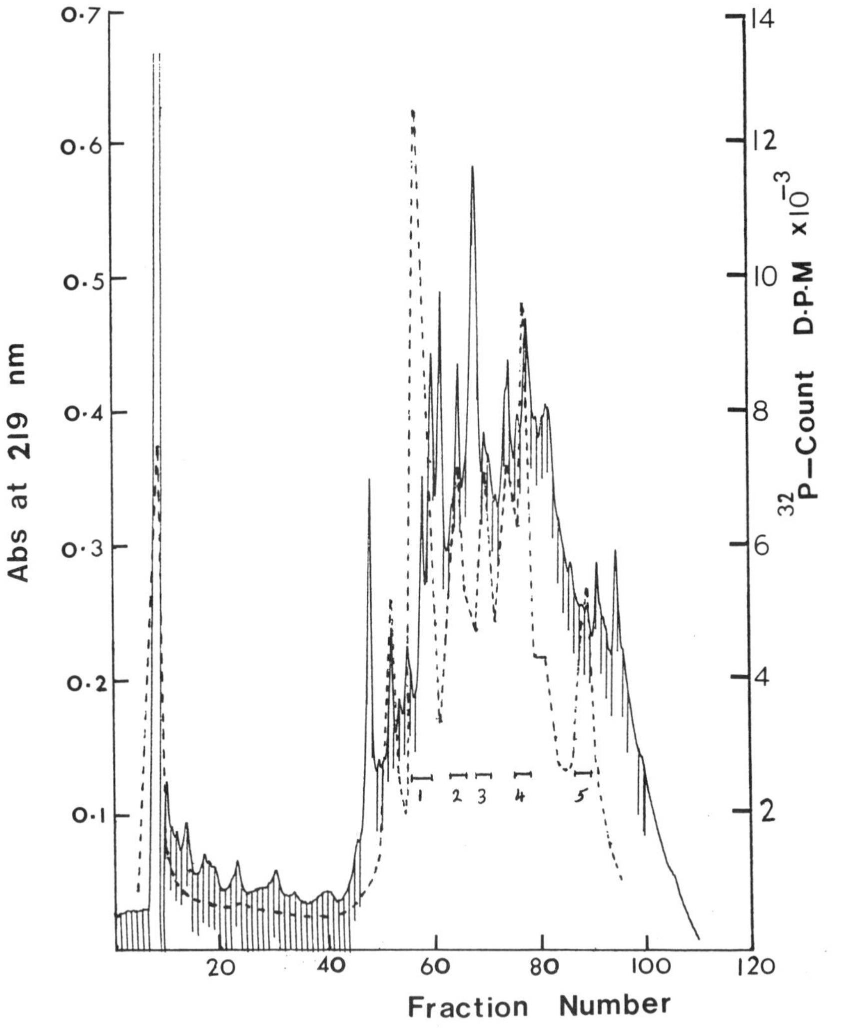

FIGURE 2. Isolation and purification of tryptic peptides of ^{32}P-labeled osteopontin by reverse-phase HPLC on a C-18 column. The purified ^{32}P-labeled osteopontin (0.5 mg) was digested with trypsin (2% w/w) in 0.2 mls of 50 mM NH_4HCO_3, pH 8.0, for 20 hr at 37°C. The peptides were separated by HPLC using a linear gradient from 100% H_2O + 0.1% TFA to 60% CH_3CN + 0.055% TFA in 120 min. Fractions of 0.5 ml were collected and aliquots of 10 μl counted for ^{32}P-. *Solid line* Abs. = 219 nm, and *dashed line* ^{32}P-count. Each ^{32}P-peak was rechromatographed and sequenced.

Ala-139 and extending to Arg-248. The tryptic digest followed by reverse-phase HPLC on a C-18 column (FIG. 2) led to isolation of five ^{32}P-labeled peptides. N-terminal sequencing using the solid-phase a.a. sequencing technique[8] identified clearly the phosphorylated peptide regions of osteopontin to be: peptides with a.a.s: 8–18: QHAISASSEEK; peptide 39–54; LASQQTHYSSEENAD; peptide 150–171; LIEDDATAEVGDSQLAGLWLPK; peptide 194–205; FDSPEVGGG-DSK; and peptide 214–219: ESLASR. The specific a.a. on which the phosphate was residing (*asterisk,* *) was determined from the release of ^{32}P-radioactivity by counting 2/3 of the Edman degradation products of each cycle during sequencing. Of the five identified phosphorylated peptide regions, two reside close to the N-terminal end of the OPN molecule, whereas the remaining three peptides are concentrated towards the C-terminal half of the molecule. The first two phosphorylated peptides contain the recognition amino-acid sequences for message-independent protein kinase (-SSEE-), while the third and fourth peptides contain sequences around the phosphorylated residues (-EDDATAE) and (-PRFDSPEV) with possible recognition sequence for message-independent protein kinase, and/or possible likelihood of another kinase. The fifth phosphorylated peptide (SRESLASR) has no clear classical recognition site for message-independent kinase. The presence of an Arg(R) residue at the third position N-terminal to the phosphorylated residue in the fourth peptide suggests possible phosphorylation by calmodulin-dependent protein kinase II.

REFERENCES

1. UCHIYAMA, A., M. SUZUKI, B. LEFTERIOU & M. J. GLIMCHER. 1988. Biochemistry **24:** 6363–6369.
2. PRINCE, C. W., T. OOSAWA, W. T. BUTLER, M. TOMAMA, A. S. BROWN & R. E. SCHROHENLOHER. 1989. J. Biol. Chem. **262:** 2900–2907.
3. GLIMCHER, M. J. 1984. Phys. Trans. R. Soc. London B **304:** 479–508.
4. VEIS, A. & B. SABSAY. 1983. *In* Biomineralization and Biological Metal Accumulation. P. Westhroeke & E. S. DeJohn, Eds. 273–284. Reidal Publishing Co.
5. GERSTENFELD, L. C, J. B. LIAN, Y. GOTOH, D. D. LEE, W. J. LANDIS, M. D. MCKEE, A. NANCI & M. J. GLIMCHER. 1989. Connect. Tissue Res. **21:** 215–225.
6. GOTOH, Y., L. C. GERSTENFELD & M. J. GLIMCHER. 1990. Eur. J. Biochem. **187:** 49–58.
7. MOORE, M. A., Y. GOTOH, K. RAFIDI & L. C. GERSTENFELD. 1991. Biochemistry **30:** 2501–2508.
8. GERSTENFELD, L. C., S. D. CHIPMAN, J. GLOWACKI & J. B. LIAN. 1987. Dev. Biol. **122:** 49–60.
9. WETTENHALL, R. E. H., R. H. AEBERSOLD & L. E. HOOD. 1991. Methods Enzymol. **201:** 186–199.

Identification of Osteopontin mRNA and Protein in Rat Epididymis and of Protein on Epididymal Sperm[a]

JON E. SIITERI AND DAVID W. HAMILTON

Department of Cell Biology and Neuroanatomy
University of Minnesota
Minneapolis, Minnesota 55455

We previously reported the presence of osteopontin (OPN) mRNA and protein in the male reproductive tract.[2] Our data indicated that adult rat Sertoli cell extracts, but not testicular interstitial cell or germ cell preparations, contained OPN mRNA and protein. In this report, we describe the presence of OPN mRNA and protein in the rat epididymis and on the surface of rat epididymal sperm.

Polyadenylated RNAs isolated from rat liver, testis, or whole rat epididimydes were amplified by reverse transcription-polymerase chain reaction (RT-PCR) using primers specific for OPN (nucleotides 515–532 and 1198–1215).[3] The products of RT-PCR were separated by agarose electrophoresis and visualized under ultraviolet light. A PCR product of the appropriate size (700 base pairs) was amplified from OPN mRNA present faintly in rat liver, more so in testis, and markedly in epididymal tissue.

By using monoclonal antibody MPIII B_{10}, indirect immunofluorescence confirmed the presence of OPN in all regions of the rat epididymis with the exception of the distal cauda epididymidis. Immunofluorescence was apparent throughout the epithelial cell cytoplasm, and was occasionally seen in interstitial connective tissue. Tissue sections treated with control antibody were negative.

The presence of OPN in epididymal epithelial tissue suggested the possibility that this organ could secrete OPN into the lumen where it could interact with spermatozoa. The presence of OPN on epididymal sperm, previously detected by immunoblots, was confirmed by indirect immunofluorescence. On both nonfixed and fixed caput epididymal sperm, fluorescence was restricted principally to the dorsal surface of the head region in the majority of the cells examined. Staining of the midpiece was detected in some of the cells observed, and a small percentage of cells had no staining at all. Nonfixed cauda epididymal sperm stained faintly on the head, but more strongly on the midpiece and principal piece of the sperm tail. In contrast, staining of fixed cauda epididymal sperm was intense along the midpiece and principal piece of the sperm tail. Fluorescence staining of cells after incubation with control ascites fluid was negative.

We have extended our initial observations[2] that the OPN mRNA and protein are present in adult rat testis and epididymis tissues. We are the first to report the presence of OPN in the epididymal epithelium and on the surface of rat epididymal sperm.

The physiological function of OPN in the epididymis and on epididymal sperm is unclear at this time. The presence of an -RGDS- cell attachment sequence within

[a] A complete description of the work presented here was subsequently published in Ref. 1.

OPN suggests that cell adhesion is one important function of OPN. Hence, OPN could play a role in potential epithelial-germ cell interactions in the epididymis.

An alternative function could involve the presumed calcium binding properties of OPN. The presence of a calcium-binding protein in epididymal fluid may serve to act as a "decapacitation" factor to prevent premature activation of epididymal sperm motility and/or the acrosome reaction. Finally, OPN bound to the surface of epididymal sperm could play a role in fertilization, either during transit through the female reproductive tract or at the time of sperm-egg contact.

REFERENCES

1. Siiteri, J. E., K. M. Ensrud, A. Moore & D. W. Hamilton. 1995. Identification of osteopontin (OPN) mRNA and protein in the rat testis, and epididymis, and on sperm. Mol. Reprod. Dev. **40:** 16–28.
2. Siiteri, J. E. & D. W. Hamilton. 1993. Secreted phosphoprotein-1 (osteopontin) localization in the adult rat testis. Mol. Biol. Cell **4:** 137a.
3. Oldberg, Å., A. Franzén & D. Heinegård. 1986. Cloning and sequence analysis of rat bone sialoprotein (osteopontin) cDNA reveals an Arg-Gly-Asp cell-binding sequence. Proc. Natl. Acad. Sci. USA **83:** 8819–8823.
4. Brown, L. F., B. Berse, L. Van De Water, A. Papadopolous-Sergiou, C. A. Perruzzi, E. J. Manseau, H. F. Dvorak & D. R. Senger. 1992. Expression and distribution of osteopontin in human tissues: widespread association with luminal epithelial surfaces. Mol. Cell. Biol. **3:** 1169–1180.

Phosphorylation, Glycosylation, and Transglutaminase Sites in Bovine Osteopontin[a]

ESBEN S. SØRENSEN AND TORBEN E. PETERSEN

Protein Chemistry Laboratory
University of Aarhus
Science Park
Gustav Wieds Vej 10-C
DK-8000 Aarhus C, Denmark

The multiple functions of osteopontin (OPN) are all dependent on the unique structure of the protein, and especially the posttranslational modifications have been assigned significant importance. OPN isolated from rat bone has been shown to contain 12 phosphoserines, one phosphothreonine, and one N-linked and 5–6 O-linked glycosylations.[1] Furthermore, there are reports of sulphated tyrosine residues in OPN synthesized by rat bone marrow cells.[2]

To extend the studies of the posttranslational modifications of the protein, we investigated OPN isolated from bovine milk.[3] By a combination of mass spectrometric analysis and amino acid sequencing of S-ethylcysteine-derivatized phosphopeptides we have determined the complete *in vivo* phosphorylation pattern of bovine OPN, containing a total of 27 phosphoserines and one phosphothreonine.[4–6] Furthermore, three O-glycosylations were located in the sequence.[6] The sites of modification are indicated in the schematic model given in FIGURE 1. The phosphorylation and glycosylation pattern reported here is distinctly different from that described for rat bone OPN. This difference could be ascribed to species variation, or more likely the existence of tissue-specific isoforms that differ in their phosphorylation and glycosylation pattern. Examination of the phosphorylation pattern of bovine OPN reveals that the phosphorylations occur in clusters of approximately three phosphate groups, spanned by stretches of unmodified amino acids (FIG. 1). This phosphorylation pattern is probably important in the interaction with calcium ions and developing salt crystals. Interestingly, the regions containing the glycosylations, the RGD-cell adhesion sequence, the thrombin cleavage site, and the transglutaminase reactive glutamines are devoid of phosphorylations in the near proximity.

Investigations of bovine OPN as a substrate for transglutaminase (TG) revealed the presence of two reactive acceptor glutamines; Gln-34 and Gln-36.[7] Analysis of the rat OPN sequence places Gln-34 and Gln-36 in a region with no predicted structure linking an α-helix and a β-sheet in the N-terminal part of the protein.[8] Alignment of the TG-reactive region in bovine OPN with the corresponding region in other species shows that the glutamines corresponding to Gln-34 and Gln-36 in the bovine sequence are conserved in all known OPN sequences (FIG. 2). The conservation of these glutamines in OPN implies that this motif is critical to

[a] The work summarized in this paper is part of the FØTEK programme supported by the Danish Dairy Board and the Danish Government.

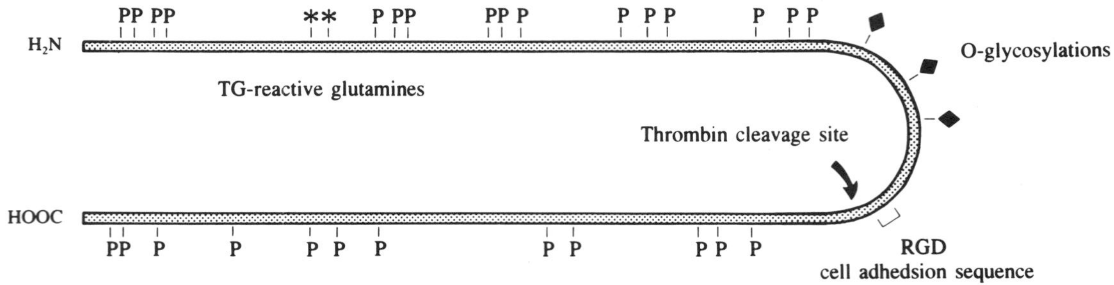

FIGURE 1. Schematic model of the posttranslational modifications of bovine osteopontin.[7] P's symbolize identified phosphorylations. The transglutaminase reactive glutamines are marked with *asterisks*. *Filled diamonds* symbolize O-linked glycosylations. The positions of the RGD-cell adhesion sequence and the thrombin cleavage site are as indicated.

```
           27                              43
BOVINE     W L K P D P S Q K Q T F L A P Q N
             : : : : : : : : : : : : : : : :
PIG        L L K P D P S Q K Q T F L A P Q N
           : :   : : : : : : :     : : : : :
HUMAN      W L N P D P S Q K Q N L L A P Q N
           : : : : : : : : : :     : : : : :
RAT        W L K P D P S Q K Q N L L A P Q N
           : :   : : : : : : :     : : : : :
MOUSE      W L V P D P S Q K Q N L L A P Q N
           : :   : : : : : : :     :   : : :
RABBIT     W L N P D P S Q K Q N L L T P Q N
                       : :   :     :
CHICKEN    H Q D H V D S Q S Q E H L Q Q T Q
```

FIGURE 2. Alignment of the transglutaminase reactive region of bovine OPN with the corresponding region in other species. All sequences are aligned to the bovine sequence. *Two dots* indicate a perfect match. The transglutaminase reactive glutamines in bovine OPN and the corresponding glutamines in the aligned sequences are shown in *bold*.

functional aspects of the protein. The proteins to which OPN cross-links *in vivo* have not been identified. In bone, a TG-catalyzed formation of complexes between OPN and other matrix proteins could be involved in the initiation and control of the mineralization processes.[9] Supporting this hypothesis, Western analysis of bovine bone reveals the presence of covalently linked OPN in high molecular mass complexes.[7]

ACKNOWLEDGMENTS

We wish to acknowledge Dr. Peter Højrup, University of Odense, Denmark, for performing the mass spectrometric analysis and Lise Møller for excellent technical assistance.

REFERENCES

1. Prince, C. W., T. Oosawa, W. T. Butler, M. Tomana, M. Bhown, A. S. Bhown & R. E. Schrohenloher. 1987. J. Biol. Chem. **262:** 2900–2907.
2. Nagata, T., R. Todescan, H. A. Goldberg, Q. Shang & J. Sodek. 1989. Biochem. Biophys. Res. Commun. **165:** 234–240.
3. Sørensen, E. S. & T. E. Petersen. 1993. J. Dairy Res. **60:** 189–197.
4. Sørensen, E. S., L. K. Rasmussen, P. Højrup & T. E. Petersen. *In* Methods in Protein Structure Analysis. M. Z. Atassi & E. Appella, Eds. Plenum Press. New York. In press.
5. Sørensen, E. S. & T. E. Petersen. 1994. Biochem. Biophys. Res. Commun. **198:** 200–205.
6. Sørensen, E. S., P. Højrup & T. E. Petersen. In preparation.
7. Sørensen, E. S., L. K. Rasmussen, L. Møller, P. H. Jensen, P. Højrup & T. E. Petersen. 1994. Biochem. J. **304:** 13–16.

8. Denhardt, D. T. & C. W. Prince. 1993. *In* Guidebook to the Extracellular Matrix and Adhesion Proteins. T. Kreis & R. Vale, Eds. 76–79. Oxford University Press. Oxford.
9. Aeschlimann, D., A. Wetterwald, H. Fleish & M. Paulsson. 1993. J. Cell Biol. **120:** 1461–1470.

Localization of Osteopontin during Mouse Development[a]

D. N. STERN,[b,c,d,e] M. J. GLIMCHER,[b,c]
AND R. A. SAAVEDRA[b,c]

[b]*Harvard Medical School*
[c]*Laboratory for the Study of Skeletal Disorders and Rehabilitation*
Children's Hospital
and
[d]*Forsyth Dental Center*
Boston, Massachusetts 02115

Osteopontin (OPN) is a secreted glycosylated phosphoprotein expressed by several cell types including osteoblasts and osteoclasts of bone,[1-3] hypertrophic chondrocytes of cartilage,[3,4] cells of the convoluted tubules of the kidney,[4,5] and cells of various epithelia.[6] The tissue distribution of OPN and its primary structure, including many acidic and phosphorylatable residues, have led to the proposal that OPN plays a major role in calcium and phosphate regulation.[7,8] Several other functions have been proposed for the protein,[8,9] but at present the functional roles of OPN are not clear.

Establishing the pattern of OPN expression during development is essential to understanding its function. We have initiated a project to study the localization of OPN during mouse development by using the Vectastain ABC immunocytochemical technique and a polyclonal antibody that was raised against mouse OPN expressed in *Escherichia coli*.[9,10] This antibody is expected to yield a minimal amount of nonspecific labeling, since it recognizes epitopes located on the backbone of OPN but not epitopes added during posttranslational modifications of the protein.

Pregnant mice were obtained from Taconic (Germantown, NY). Embryos were removed at 14, 15, 16, and 18 days of gestation and fixed in any of the following solutions: 0.1 M phosphate buffered saline (PBS), pH 7.4, containing 2% paraformaldehyde, or 4% paraformaldehyde, or 2% paraformaldehyde and 1% glutaraldehyde, for three hours. One neonate mouse was sacrificed at 1.5 days after birth, and various organs were dissected and fixed as indicated above. Tissues were rinsed in PBS containing increasing concentrations of sucrose (from 0 to 30%) and embedded in TISSUE TEK O.C.T. for freeze-sectioning.

In the 14-day-old embryo OPN was localized to chondrocytes of the cranium, and to chondrocytes of the vertebrae at the late stages of maturation prior to

[a] This work was supported by Grants IR29AR41046 and PO1AR34078 from the National Institutes of Health (NIH) to RAS and MJG, respectively, Grant RG24458A2 from the National Multiple Sclerosis Society, and a grant from the Milton Fund (Harvard Medical School) to RAS, and Training Grant T32ARO7112 (DNS).

[e] Correspondence to: Dr. Doris N. Stern, Children's Hospital, Laboratory for the Study of Skeletal Disorders and Rehabilitation, Enders Pediatric Research Center, 300 Longwood Avenue, Boston, MA 02115.

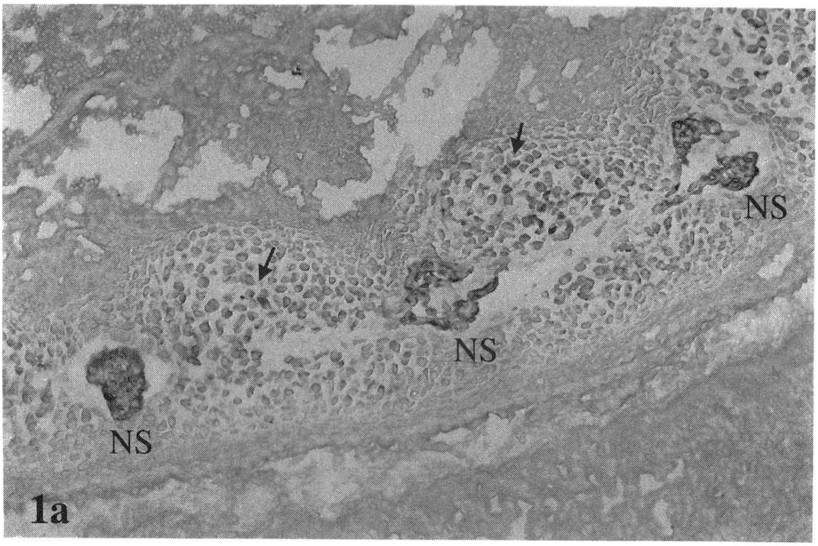

FIGURE 1a. Spine of a 14-day-old mouse embryo. The notochordal segments (NS) are intensely stained with the OPN antibody (diluted 1 : 150). Chondrocytes of the vertebrae (*arrows*) are also stained but cells surrounding the NS are not. NS is the future nucleus pulposus of the intervertebral disc, and the surrounding cells will secrete the fibrocartilage comprising the annulus fibrosus. Tissue was fixed in 4% paraformaldehyde. Original magnification = 200×.

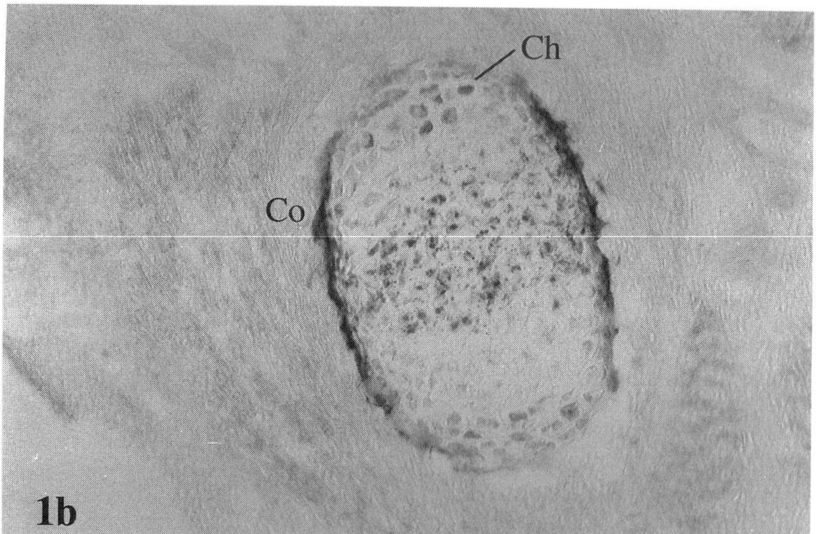

FIGURE 1b. Initial site of mineralization in the body of a 15-day-old mouse embryo. The bony collar (Co) of a developing rib stains intensely with the OPN antibody (diluted 1 : 150). Some chondrocytes (Ch) also appear stained, as do the cells and matrix in the center of the bone. Tissue was fixed in 4% paraformaldehyde. Original magnification = 200×.

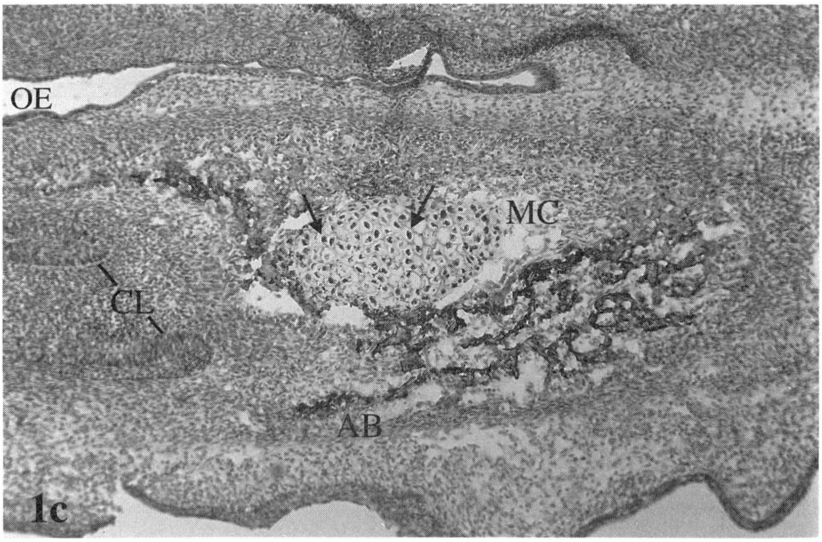

FIGURE 1c. Mineralization site in the mandible of a 16-day-old mouse embryo. The alveolar bone (AB) forms around Meckel's cartilage (MC) and stains with the antibody to OPN (diluted 1 : 100). Some of the chondrocytes in MC are stained (*arrows*). The cervical loops (CL) of the incisal tooth germ are seen at the *left* of the photograph. A portion of the oral epithelium (OE) in the same region is also stained. Tissue was fixed in 2% paraformaldehyde and the section was counterstained with hematoxylin. Original magnification = 100×.

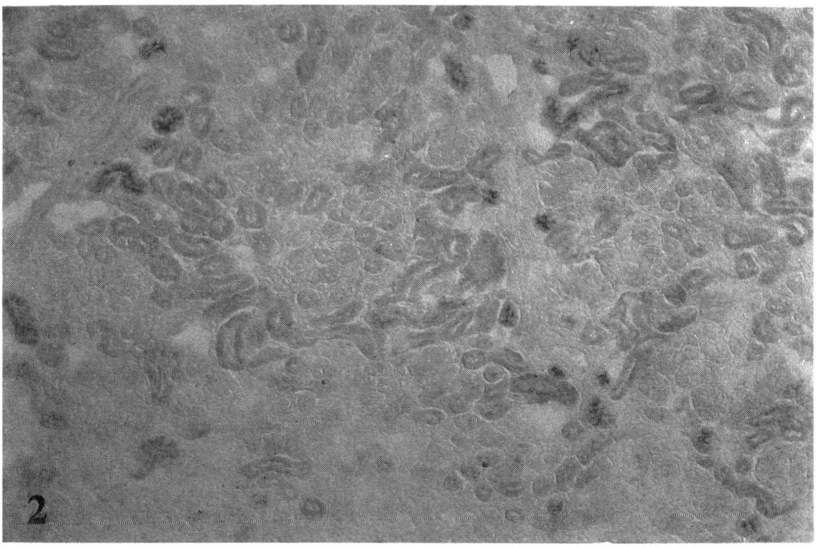

FIGURE 2. Kidney from an 18-day-old mouse embryo. OPN is localized at the luminal aspects of the convoluted tubules of some nephrons in the 18-day-old fetal kidney shown here, but it appears more homogeneously distributed in the convoluted tubules of the 1.5-day-old neonate (antibody diluted 1 : 100) (see text). Tissue was fixed in 4% paraformaldehyde. Original magnification = 200×.

evidence of calcification (FIG. 1a). Chondrocytes in the head appeared to stain more intensely (not shown) and in larger numbers than those in the vertebrae. In these embryos, the protein was also seen in the notochordal segments (FIG. 1a), the lip furrow and the maxillary oral epithelium (D. Stern, M. Glimcher, R. Saavedra, submitted). In the 15-day-old embryo, OPN was localized to primary centers of ossification (FIG. 1b), as previously reported by others.[3,4] In the 16-day-old embryo, OPN was also localized to the intramembranous alveolar bone of the mandible (FIG. 1c). In the 18-day-old embryo, OPN was detected in the kidney (FIG. 2) but the pattern of staining differed from that seen in the same organ of the 1.5-day-old neonate. The cells of the convoluted tubules in the embryonic kidney stained more heavily at their luminal aspects, whereas in the neonate the staining appeared more evenly distributed throughout the cells of the tubules (D. Stern, M. Glimcher, R. Saavedra, unpublished results). The results indicate that OPN is involved in mineralization but may also have important function(s) in the development of epithelia and other nonmineralizing tissues.

REFERENCES

1. OLDBERG, A., A. FRANZEN & D. HEINEGÅRD. 1986. Cloning and sequence analysis of rat bone sialoprotein (osteopontin) cDNA reveals an Arg-Gly-Asp cell-binding sequence. Proc. Natl. Acad. Sci. USA **83:** 8819–8823.
2. YOON, K., R. BUENAGA & G. A. RODAN. 1987. Tissue specificity and developmental expression of rat osteopontin. Biochem. Biophys. Res. Commun. **148:** 1129–1136.
3. MCKEE, M. D., M. J. GLIMCHER & A. NANCI. 1992. High-resolution immunolocalization of osteopontin and osteocalcin in bone and cartilage during endochondral ossification in the chicken tibiae. Anat. Rec. **234:** 479–492.
4. MARK, M. P., W. T. BUTLER, C. W. PRINCE, R. D. FINKELMAN & J.-V. RUCH. 1988. Developmental expression of 44-kDa bone phosphoprotein (osteopontin) and bone γ-carboxyglutamic acid (Gla)-containing protein (osteocalcin) in calcifying tissue of rat. Differentiation **37:** 123–136.
5. NOMURA, S., A. J. WILLS, D. R. EDWARDS, J. K. HEATH & B. L. M. HOGAN. 1988. Developmental expression of 2ar (osteopontin), and SPARC (osteonectin) RNA as revealed by *in situ* hybridization. J. Cell Biol. **106:** 441–450.
6. BROWN, L. F., B. BERSE, L. VAN DE WATER, A. PAPADOPOULUS-SERGIOU, C. A. PERRUZZI, E. J. MANSEAU, H. F. DVORAK & D. R. SENGER. 1992. Expression and distribution of osteopontin in human tissues: widespread association with luminal epithelial surfaces. Mol. Biol. Cell **3:** 1169–1180.
7. GLIMCHER, M. J. 1989. Mechanism of calcification: role of collagen fibrils and collagen-phosphoprotein complexes *in vitro* and *in vivo*. Anat. Rec. **224:** 139–153.
8. PATARCA, R., R. A. SAAVEDRA & H. CANTOR. 1993. Molecular and cellular basis of genetic resistance to bacterial infection: the role of the early T-lymphocyte activation-1/osteopontin gene. Crit. Rev. Immunol. **13:** 225–246.
9. DENHARDT, D. T. & X. GUO. 1993. Osteopontin: a protein with diverse functions. FASEB J. **7:** 1475–1482.
10. ASHKAR, S., D. TEPLOW, M. J. GLIMCHER & R. A. SAAVEDRA. 1993. *In vitro* phosphorylation of mouse osteopontin expressed in *E. coli*. Biochem. Biophys. Res. Commun. **191:** 126–133.

Functional Analysis of the Osteopontin Molecule

YOHKO KATAGIRI, KIYOSHI MORI, TOYOMICHI HARA, KUMIKO TANAKA, MASAAKI MURAKAMI, AND TOSHIMITSU UEDE[a]

Section of Immunopathogenesis
Institute of Immunological Science
Hokkaido University
Kita-15, Nishi-7, Kita-ku
Sapporo 060, Japan

Among various Arg-Gly-Asp (RGD) containing extracellular matrix and serum proteins such as fibronectin, vitronectin, collagen, thrombospondin, fibrinogen, osteopontin (OP), and von Willebrand factor, OP is very unique in the sense that it can be classified not only as extracellular matrix but also as a lymphokine, since it possesses signal sequence and is secreted by activated T cells.[1,2] We and others previously reported that OP can be secreted by abnormal double negative (DN) T cells and the induction of OP mRNA is detected in DN T cells of MRL-lpr/lpr and other lpr mice.[3,4] It was also shown that OP stimulated polyclonal activation of B cells and induced production of immunoglobulin when murine B cells and macrophages were cultured with OP.[3,5] These data collectively suggested that OP is involved in the pathogenesis of autoimmune disease. Therefore, we first tested whether the elevated level of serum OP can be detected not only in autoimmune-prone mice but also in patients of autoimmune disease. As shown in FIGURE 1, OP was detected in the serum of autoimmune-prone mice, but not of normal mice. The appearance of OP in lpr mice was age dependent and preceded the appearance of lymphoadenopathy as well as increase in the serum level of interleukin-6. OP was also detected in the serum of rheumatoid arthritis (RA) and systemic lupus erythematosus (SLE) patients, but not of healthy donors. It should be noted that two species of OP molecule are detected in autoimmune-prone mice and RA and SLE patients: the molecular weight of 54,000 (64,000 in human) as well as 28,000 (32,000 in human) daltons. The increase of serum OP correlated very well with the increase of the 28,000 (32,000)-dalton form of OP. The difference between the large and small form of OP is not known, but the small form of OP may be derived from the large form since it can be cleaved by thrombin. It is also possible that two species of OP may be due to alternative splicing of the OP gene.[6]

To better understand the role of OP in autoimmune disease, the characterization of cell binding domains of OP and their cellular receptors are essential, although it is known that RGD tripeptides within the OP molecule interact with the cell surface $\alpha_v\beta_3$ integrin receptor. Recently, additional cell binding domains were

[a] Corresponding author.

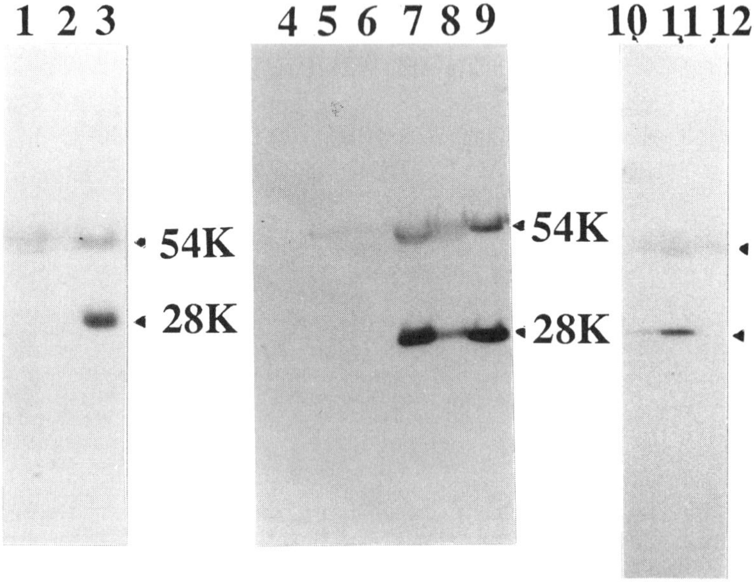

FIGURE 1. Detection of osteopontin in serum of autoimmune patients and autoimmune-prone mice. One hundred μl of serum obtained from mice and human were absorbed to a DE52 gel equilibrated with 50 mM Tris HCl buffer, pH 7.2, containing 6 M urea and 0.14 M NaCl. After extensive washing, the proteins were eluted with the same buffer containing 0.7 M NaCl and electrophoresed with Laemmli's system. Then, the proteins were blotted to PVDF membranes and immunostained with the monospecific antibody raised against synthetic peptide EQYPDATDEDLTSHM corresponding to internal sequence of mouse osteopontin. Serum were obtained from 7-week (w)-old MRL-+/+ (1), NZB/W F1 (2), and MRL-lpr/lpr (3) mice. MRL-lpr/lpr mouse serum was also obtained at the age of 11 (4) and 17 (5) days, 4w (6), 7w (7), and 5 (8) and 12 months (9). Human serum was obtained from an RA patient (10), an SLE patient (11), and a healthy donor (12).

identified in various RGD containing extracellular matrix such as fibronectin, fibrinogen, von Willebrand factor and thrombospondin.

We purified OP from human colostrum and examined the binding of murine melonoma cells (B16-BL6) bearing $\alpha_v\beta_3$ integrin to GRGDS peptides and OP. As shown in FIGURE 2A, the binding of B16-BL6 cells to GRGDS peptides coupled to BSA (bovine serum albumin) was almost completely abrogated by GRGDS peptides, while the binding was not inhibited by GRGES peptides. In addition, the binding of B16-BL6 to GRGDS peptides coupled to BSA was also strongly inhibited by monoclonal antibodies (mAb) against murine α_v integrin, but not by isotype-matched control mAb (FIG. 2B), indicating that $\alpha_v\beta_3$ integrin on B16-BL6 cells can interact with RGD peptides. However, the binding of B16-BL6 cells to purified OP was not inhibited by GRGDS or GRGES peptides (FIG. 2C). Furthermore, this binding was not inhibited by mAb reacting to murine α_v integrin (FIG. 2D). Taken together, our present data suggest that human OP possesses an additional cell binding site (different from RGD tripeptides) and its receptor may be distinct from $\alpha_v\beta_3$ integrin.

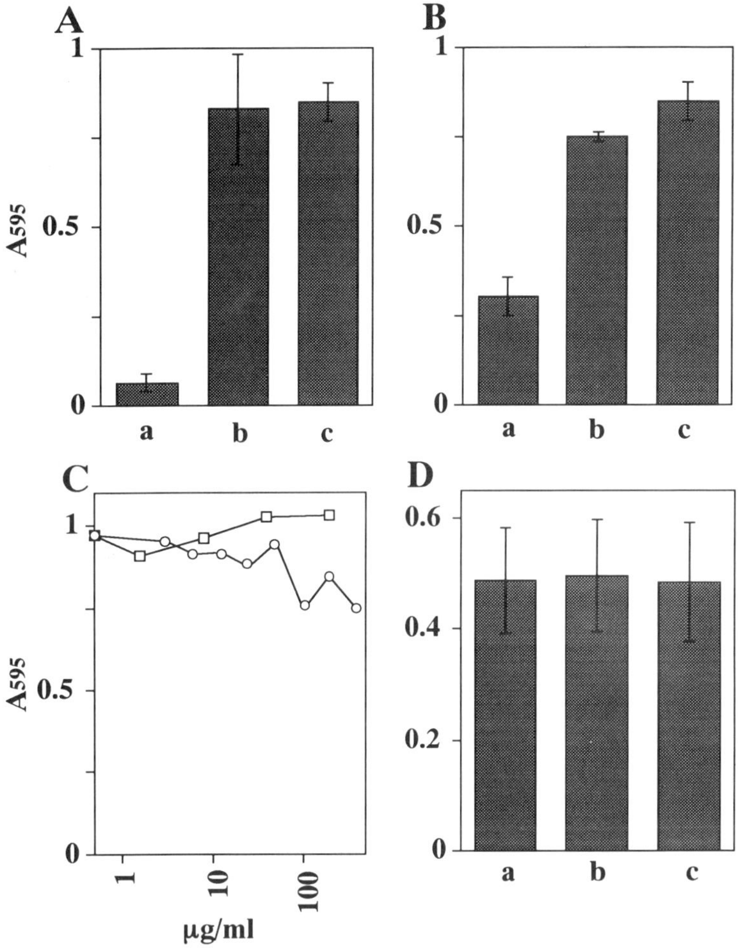

FIGURE 2. Involvement of an additional cell binding site within the osteopontin molecule to melanoma cells. The 5×10^4 B16-BL6 murine melanoma cells in 200 μl DMEM containing 0.25% BSA were plated to 96-well ELISA plates precoated with GRGDS peptides coupled to BSA (40 μg/ml) (A and B) or purified osteopontin (5 μg/ml) (C and D) in the presence or absence of inhibitory peptides (A and C) or mAb (B and D) and incubated for 1 hr at 37°C. The medium was ejected from the plates and all wells were washed twice with the same medium. The adherent cells were fixed and stained with 0.5% crystal violet in 20% methanol for 30 min at room temperature. All wells were rinsed with H_2O and adherent cells were lysed with 20% acetic acid. A_{595} of resulting supernatants from each well was read to determine the relative number of adherent cells. **(A)** The binding study was performed in the presence of 200 μg/ml of GRGDS (a), GRGES (b), or medium (c). (**B** and **D**) The binding study was performed in the presence of saturating amount of anti-α_v mAb (a), isotype-matched mAb (b), or medium (c). **(C)** GRGDS (○—○). GRGES (□—□).

REFERENCES

1. Oldberg, A., A. Franzen and D. Heinegård. 1986. Proc. Natl. Acad. Sci. USA **83:** 8819–8823.
2. Patarca, R., G. J. Freeman, R. P. Singh, F.-W. Wei, T. Durfee, F. Blattner, D. C. Regnier, C. A. Kozak, B. A. Mock, C. A. Morse III, T. R. Jerrells & H. Cantor. 1989. J. Exp. Med. **170:** 145–161.
3. Patarca, R., F.-W. Wei, P. Singh, M. I. Morasso & H. Cantor. 1990. J. Exp. Med. **172:** 1177–1183.
4. Mori, K., S. Kobayashi, M. Inobe, W.-Y. Jia, M. Tamakoshi, T. Miyazaka & T. Uede. 1994. Autoimmunity **17:** 49–57.
5. Lampe, M. A., R. Patarca, M. Iregui & H. Cantor. 1991. J. Immunol. **147:** 2902–2906.
6. Young, M. F., J. M. Kerr, J. D. Termine, U. M. Wewer, M. G. Wang, O. W. McBride & L. W. Fisher. 1990. Genomics **7:** 491–502.

Osteopontin Inhibits Nucleation of Calcium Oxalate Crystals[a]

ELAINE M. WORCESTER[b] AND ANN M. BESHENSKY

Nephrology Division
Department of Medicine
Department of Veterans Affairs Medical Center
and
Medical College of Wisconsin
Milwaukee, Wisconsin 53295

INTRODUCTION

Urine is supersaturated with respect to calcium oxalate (CaOx), which may lead to spontaneous nucleation of CaOx crystals. The CaOx concentration at which crystals begin to form is higher in urine than in simple salt solutions, implying the presence of nucleation inhibitors in urine. Osteopontin (OPN) is made by renal tubular cells, secreted into tubular fluid, and inhibits growth of CaOx crystals *in vitro*.[1,2] The concentration in rodent urine is 6–8 μg/ml; it has also been found in human urine.[3] OPN's effect on crystal nucleation is unknown; because it is found in the matrix of Ca stones, it has been suggested that OPN may promote nucleation of CaOx.

We developed an assay to measure the effect of OPN and other urinary proteins on spontaneous CaOx nucleation.

METHODS

The assay was performed in a 5-ml water-jacketed vessel maintained at 37°C, bubbled with nitrogen, and stirred continuously. All solutions were filtered through a 0.2-μm filter prior to use. The solution contained 0.01 M HEPES, pH 7.5, and .15 M NaCl. Ca and Ox concentrations were each 0.6 mM. The CaOx ion activity product in the solution was 1.84×10^{-8} M, which exceeded the solubility product by a factor of 8.21. This level of supersaturation may be seen in human urine. The Ca concentration of the solution was continuously monitored with a Ca-specific electrode for 60 min, and data were collected and analyzed using an A-to-D converter interfaced with a Macintosh SE30, using SuperScope II software. Proteins tested for inhibitory activity included OPN, purified from culture supernatants of primary cultures of mouse kidney cortical cells; Tamm-Horsfall protein (THP), a urinary glycoprotein that inhibits CaOx crystal aggregation but not growth; human serum albumin (HSA), an acidic protein found in urine that has

[a] This work was supported by the Department of Veterans Affairs Medical Research Service, and by Research Grants DK41725 and 1R55 DK48504 from the National Institutes of Health.

[b] Correspondence to: Elaine M. Worcester, M.D., Nephrology Section/111K, VA Medical Center, 5000 West National Avenue, Milwaukee, WI 53295.

weak CaOx growth inhibitory activity; and urine collected from a non-stone-forming subject, which was buffer-exchanged into 0.01 M HEPES, 0.15 M NaCl, pH 7.5, and tested immediately.

RESULTS

The Ca concentration of the solution containing 0.6 mM Ca and Ox fell after a lag phase, averaging 18.2 ± 6.9 min. This indicates spontaneous formation of crystals in the solution, which grow and deplete the solution of Ca. By contrast, when the Ca and Ox concentrations of the solution were each 0.5 mM, (relative supersaturation ratio 5.87) mV readings remained stable for as long as 6 hr, at which time the experiment was terminated.

OPN at concentrations of 5.5 and 11 nM (0.24 and 0.48 μg/ml) significantly prolonged the lag phase (TABLE 1), and at concentrations of 22–88 nM (0.96–3.9 μg/ml) there was no fall in Ca concentration. THP at concentrations ranging from 22 to 176 nM (1.8–14 μg/ml) had little effect on the lag phase, except for a modest prolongation induced by 44 nM THP. HSA at 88 and 176 nM (5.8 and 11.6 μg/ml) had no effect on the lag phase. However, 10% human urine from which many small MW substances had been removed prevented any fall in Ca for at least 60 min.

DISCUSSION

Purified mouse OPN slows the onset of evident spontaneous CaOx crystal nucleation from a supersaturated solution; inhibition occurs at OPN concentrations lower than those in urine. Soluble OPN does not promote crystal formation at any concentration tested. THP and albumin, two acidic proteins found in urine, do not slow the onset of evident crystal formation, while the macromolecular

TABLE 1. The Effect of Urine and Urinary Proteins on Spontaneous Nucleation of Calcium Oxalate[a]

Protein	n	Lag Phase (Min)	p vs Control
None (control)	22	18.2 ± 6.9	—
OPN— 5.5 nM	3	38.3 ± 11.9	<0.001
11 nM	3	45.7 ± 26.6	<0.0001
22 nM	2	>60	<0.0001
44 nM	10	>60	<0.0001
THP— 22 nM	3	14.3 + 2.1	0.349
44 nM	4	28.3 ± 9.7	0.019
88 nM	6	22.0 ± 13.5	0.348
176 nM	3	18.0 ± 3.0	0.956
HSA— 88 nM	4	19.3 + 7.2	0.789
176 nM	3	22.3 ± 6.4	0.340
10% human urine	3	>60	<0.0001

[a] Times are mean ± SD. OPN, osteopontin; THP, Tamm-Horsfall protein; HSA, human serum albumin.

fraction of normal urine does. We conclude that OPN is able to inhibit very early phases of CaOx crystallization at concentrations that may be present in the renal tubule, and at levels of supersaturation found in urine. This inhibitory activity is not a general property of anionic urinary proteins, as neither THP nor albumin slows the onset of evident nucleation. OPN secreted by renal tubular cells may help protect against pathological crystallization in the nephron.

REFERENCES

1. WORCESTER, E. M., S. S. BLUMENTHAL, A. M. BESHENSKY & D. L. LEWAND. 1992. J. Bone Miner. Res. **7**(9): 1029–1036.
2. KLEINMAN, J. G., D. BROWN, J. V. BONVENTRE, A. M. BESHENSKY & E. M. WORCESTER. 1993. J. Am. Soc. Nephrol. **4**(3): 611.
3. SHIRAGA, H., W. MIN, W. J. VANDUSEN *et al.* 1992. Proc. Natl. Acad. Sci. USA **89:** 426–430.

Effect of Recombinant Osteopontin on Adhesion and Migration of P388D1 Cells

SHUNSUKE YAMAMOTO, KAEI NASU,
TETSUYA ISHIDA, MIHOKO SETOGUCHI,
YASUNORI HIGUCHI, NAOKI HIJIYA,
AND SHIN'ICHIRO AKIZUKI

Department of Pathology
Oita Medical University
Hasama-machi
Oita 879-55, Japan

Osteopontin (OPN) is the smallest natural Gly-Arg-Asp (RGD) protein yet identified, and its matrix adhesiveness may be much more limited than that of fibronectin (FN) which is approximately six times larger. The OPN structure around RGD amino acids is closely related to that of FN, and OPN has been reported to bind to a receptor integrin, $\alpha_v\beta_3$, that is also bound by FN, although FN also binds to other integrins.[1,2] Therefore, it is possible that OPN modulates the function of FN. Here, we report the role of OPN in adhesion and haptotactic migration of a murine macrophage cell line, P388D1, using recombinant rabbit osteopontin (rerabOPN).

Recombinant wild type rabbit osteopontin and the protein with an Asp-to-Glu transposition induced by a point mutation in the rabbit osteopontin (rabOPN) cDNA within the RGD sequence were expressed in *E. coli* and purified to homogeneity. P388D1 cells bind rerabOPN in a saturable manner. rerabOPN induced adhesion and haptotaxis of P388D1 cells, whereas mutated rabOPN (rOPmut) did not. Anti-rerabOPN IgG F $(ab')_2$ and synthetic GRGDS peptide inhibited rerabOPN-mediated adhesion and haptotaxis of P388D1 cells (TABLES 1 and 2). FN-mediated adhesion of P388D1 cells was markedly inhibited in the presence of fluid-phase rerabOPN. Adhesion of P388D1 cells to rerabOPN was significantly inhibited by anti-VLA_4 α (α_4) and VLA_5 α (α_5) monoclonal antibodies (mAbs), but not anti-vitronectin (VN) receptor α (α_v) or Mac-1α (α_M) mAbs. Adhesion of P388D1 cells to FN and VN was significantly inhibited by anti-α_v mAb but not anti-α_4, -α_5 or -α_M mAbs. Haptotaxis of P388D1 cells to rerabOPN and VN was significantly inhibited by anti-α_v, but not by anti-α_4, -α_5 and -α_M mAbs, whereas that to FN showed no inhibition with all three mAbs. Haptotaxis of P388D1 cells to VN was significantly inhibited with anti-α_5 and anti-α_v but not by anti-α_4 and -α_M mAb. rerabOPN showed no chemotactic effect on P388D1 cells.

Receptors other than $\alpha_v\beta_3$ such as $\alpha_4\beta_1$ and $\alpha_5\beta_1$ are likely to mediate the binding of rerabOP to P388D1 cells, although others may be involved, since inhibition with anti-α_4 and α_5 mAbs was relatively weak. FN shows specific affinities for collagen, fibrin, heparin, DNA, glycosaminoglycans, and cell surfaces receptors. Although OPN has been suggested to bind positively charged matrix

TABLE 1. Inhibition of P388D1 Cell Adhesion to RGD Protein-Coated Wells by Various Proteins and Antibodies

	Percent Adhesion to Wells Coated with		
Samples	rerabOPN (5 μg/ml)	FN (5 μg/ml)	VN (5 μg/ml)
None	100	100	100
GRGDS (5 μg/ml)	6.1 ± 3.3	13.2 ± 1.2	nt
FN (5 μg/ml)	10.0 ± 1.2	nt	nt
VN (5 μg/ml)	100.0 ± 3.0	nt	nt
rerabOPN (5 μg/ml)	nt	6.4 ± 2.9	100.0 ± 3.0
rOPNmut (5 μg/ml)	93.1 ± 3.9	nt	nt
anti-rerabOPN IgG F(ab′)$_2$	10.6 ± 2.4	nt	nt
anti-α_4	59.8 ± 3.8	100.0 ± 5.6	100.0 ± 3.0
anti-α_5	83.3 ± 4.1	100.0 ± 4.7	100.0 ± 4.7
anti-α_v	100.0 ± 3.5	72.7 ± 3.6	22.7 ± 2.7
anti-α_M	100.0 ± 3.2	100.0 ± 2.7	100.0 ± 3.3

nt, not tested.

such as hydroxyapatite, its binding activity to other materials has not been investigated. OPN, therefore, could paradoxically play a role in inhibition of macrophage adhesion mediated by FN in tissue other than bone. Inhibition of FN-mediated cell adhesion by an RGD-containing extracellular matrix protein tenascin has been reported.[3]

Receptor use of P388D1 cells for rerabOPN was unexpected because haptotaxis by rerabOPN was not inhibited by anti-α_4 or -α_5 mAbs but was markedly inhibited by anti-α_v mAb, which had no effect on adhesion of P388D1 cells to rerabOPN. Thus haptotaxis of P388D1 induced by rerabOPN could be an α_v-mediated phenomenon.

TABLE 2. Inhibition of Haptotactic Migration of P388D1 Cells to RGD Protein-Coated Nuclepore Filters by Various Proteins and Antibodies

	Percent Haptotaxis to Filters Coated with		
Samples	rerabOPN (2 μg/filters)	FN (2 μg/filters)	VN (2 μg/filters)
None	100	100	100
GRGDS (2 μg/ml)	22.9 ± 3.5	nt	nt
FN (2 μg/ml)	96.4 ± 3.0	nt	nt
VN (2 μg/ml)	100.0 ± 4.1	nt	nt
rerabOPN (2 μg/ml)	nt	96.4 ± 3.3	98.4 ± 4.3
rOPNmut (2 μg/ml)	98.0 ± 1.6	nt	nt
anti-rerabOPN IgG F(ab′)$_2$	21.9 ± 1.7	nt	nt
anti-α_4	100.0 ± 4.7	100.0 ± 4.3	100.0 ± 6.3
anti-α_5	100.0 ± 2.8	100.0 ± 2.8	81.9 ± 4.7
anti-α_v	24.9 ± 2.2	100.0 ± 6.3	9.9 ± 3.8
anti-α_M	100.0 ± 3.3	100.0 ± 4.7	100.0 ± 3.5

nt, not tested.

REFERENCES

1. HYNES, R. O. 1992. Cell **69:** 11–25.
2. REINHOLT, F. P., K. HULTENBY, A. OLDBERG & D. HEINEGÅRD. 1990. Proc. Natl. Acad. Sci. USA **87:** 4473–4475.
3. CHIQUET-EHRISMANN, R., P. KALLA, C. A. PEARSON, K. BECK & M. CHIQUET. 1988. Cell **53:** 383–390.

High Expression of Osteopontin mRNA in Human Macrophages but Not Human Vascular Smooth Muscle Cells in Culture

C. M. NEWMAN, B. C. BRUUN, P. K. MISTRY,
P. L. WEISSBERG, AND C. M. SHANAHAN

Department of Medicine
Addenbrooke's Hospital
Cambridge CB2 2QQ, United Kingdom

Inappropriate proliferation of vascular smooth muscle cells (VSMCs) is involved in the pathogenesis of a diverse range of vascular pathologies, including atherosclerosis, restenosis following angioplasty, accelerated coronary atherosclerosis following cardiac transplantation, and late failure of saphenous vein coronary bypass grafts. Intimal VSMCs within vascular lesions exhibit a markedly different ("synthetic") phenotype compared to quiescent, contractile, VSMCs within the media of the blood vessel.[1] A transition from the contractile to the synthetic phenotype occurs when aortic smooth muscle cells are grown in culture.[2] Using differential screening of a cDNA library prepared from cultured rat aortic VSMCs, we have identified genes which are upregulated in cultured VSMCs compared to freshly dispersed cells.[3] One such gene is osteopontin. Osteopontin mRNA is also highly expressed in the neointima of the rat carotid artery following experimental balloon injury.[4,5] These data have led to the suggestion that osteopontin expression may be a marker for proliferating VSMCs, both *in vitro* and *in vivo*. More recently, we demonstrated by *in situ* hybridization that osteopontin mRNA is highly expressed in human atherosclerotic plaques.[6] In these lesions, however, osteopontin mRNA expression was more closely associated with plaque macrophages than VSMCs.[6] We have therefore investigated osteopontin mRNA expression in human VSMCs under different conditions of cell culture, and also in human monocyte-derived macrophages in culture.

Human VSMCs were isolated from the aortic media of heart and liver organ donors (age range 2.5–27 years, studied passage 5–16) by the explant technique, and were cultured in M199 medium supplemented with 20% (v/v) fetal calf serum (FCS). Human monocytes were isolated from 200 ml venous blood by Percoll gradient centrifugation and further selected by adherence to tissue culture plastic after 3 hours incubation in Iscove's medium supplemented with 10% (v/v) autologous serum and insulin (8 μg/ml); the resultant cells were incubated in the same medium for at least 9 days to allow differentiation into macrophages. Osteopontin mRNA levels were assessed by standard Northern analysis of total cytoplasmic RNA using a ^{32}P-labelled human osteopontin cDNA probe.

Human monocyte-derived macrophages (harvested 9–13 days post monocyte isolation) expressed high levels of OP mRNA which were not substantially affected by the inclusion of either human LDL (50 μg/ml) or acetylated human LDL (50 μg/ml) or by culture in 10% (v/v) lipoprotein-deficient serum, for the last 4–6 days prior to harvest. In contrast, osteopontin mRNA expression was not detected in subconfluent or confluent cultures of human VSMCs of any passage maintained in medium containing 20% (v/v) FCS throughout. Osteopontin mRNA was also undetectable following culture for the last 24 hours prior to harvest in medium containing TGFβ (20 ng/ml), angiotensin II (1 μM) of bFGF (10 ng/ml), all of which upregulate osteopontin mRNA expression in cultured rat VSMCs.[5] A low level of osteopontin mRNA expression was detected in one VSMC culture incubated in medium supplemented with activated vitamin D_3 (1 μM) or human LDL (2 mg/ml) for the last 24 hours prior to harvest.

These data highlight an important species difference in osteopontin mRNA expression in VSMCs, and emphasize the need for further experiments using human tissue.

REFERENCES

1. Schwartz, S. M., G. R. Campbell & J. H. Campbell. 1986. Circ. Res. **58:** 427–444.
2. Chamley-Campbell, J. H., G. R. Campbell & R. Ross. 1979. Physiol. Rev. **59:** 1–61.
3. Shanahan, C. M., P. L. Weissberg & J. C. Metcalfe. 1993. Circ. Res. **73:** 193–204.
4. Giachelli, C., N. Bae, D. Lombardi, M. Majesky & S. Schwartz. 1991. Biochem. Biophys. Res. Commun. **177:** 867–873.
5. Giachelli, C. M., N. Bae, M. Almeida, D. T. Denhardt, C. E. Alpers & S. M. Schwartz. 1993. J. Clin. Invest. **92:** 1686–1696.
6. Shanahan, C. M., N. R. Cary, J. C. Metcalfe & P. L. Weissberg. 1994. J. Clin. Invest. **93:** 2393–2402.

Subject Index

Index of Contributors